Developmental Biology

William F. Loomis
University of California, San Diego

Macmillan Publishing Company
New York
Collier Macmillan Publishers
London

Printed in the United States of America

Macmillan Publishing Company
866 Third Avenue, New York, New York 10022

Collier Macmillan Canada, Inc.

Library of Congress Cataloging in Publication Data

Loomis, William F.
Developmental biology.

Includes bibliographies and index.
1. Developmental biology. I. Title.
QH491.L66 1986 574.3 85-13805
ISBN 0-02-371790-4

Printing: 1 2 3 4 5 6 7 8 Year: 6 7 8 9 0 1 2 3 4 5

ISBN 0-02-371790-4

Preface

This presentation is directed at those interested in the mechanisms of development who are familiar with the underlying cellular biochemistry and molecular genetics. I have tried to put the basic processes that are common to all cells in perspective to account for the complex physiological and morphological events that make up embryogenesis. Developmental biology is not defined by clear-cut boundaries but encompasses molecular studies of subcellular compartments, cell shape, and tissue function, as well as analyses of integration of whole organ systems. Some developmental biologists concentrate on single genes while others abstractly conceptualize the processes that generate form and pattern. Between these extremes there are many fruitful approaches. I have attempted to cover the full range of analytical methods since they all illuminate the biological problems with bright but different lights.

This text is organized into three parts preceded by an introductory chapter that reviews classical embryological descriptions. Part I covers the mechanisms available to cells such that they can diverge and take up specialized roles in a multicellular organism. Part II describes some of the embryological processes while posing the mechanistic questions. Part III goes into considerable detail on developmental systems that have proved amenable to advanced techniques. Each chapter is followed by one or more specific examples of the topic in the main section. These are often analyses that have only recently been carried out and are still in the process of being settled. They might be supplemented by reading research articles as they appear in the major journals. A dozen or so related readings, mostly taken from the recent literature, are given for each chapter. Developmental biology is an exciting field in a rapidly expanding phase. However, a coherent view can now be seen.

Study questions are appended to each part. They give some idea of the points that are central to the subject matter covered. At the end of the book I have compiled tables that list subjects discussed in the text as a way of seeing the whole (see Guide to Subjects). It might be worth scanning these from time to time to mark progress. Many of the subjects are brought up several times in different contexts.

The list of those who have helped me to write this text stretches far back. I am indebted to my teachers and advisers starting with my father, W. F. Loomis, who studied hydra. David Bonner taught me to work with genes of *Neurospora*. My adviser in graduate school, Boris Magasanik, showed me the beauty of complex physiological processes in microbial organisms. Maurice Sussman introduced me to *Dictyostelium*. Ed Zwilling showed me chick embryos. While on sabbatical in the laboratory of François Gros, I became familiar with cultured myoblasts. Together these experiences have helped to shape this text. Early portions of the manuscript were reviewed by Peter Bryant, Maarten Chrispeels, Victor Vacquier, and William Wood. Later drafts were reviewed by Bruce Brandhorst, John Gerhart, Joseph Kunkel, Anthony Mahowald, and Ben Murray. Their comments added considerably to the presentation. The cheerful secretarial support of Genette Fuller through drafts and drafts smoothed the way. My thanks to those at Macmillan who worked on the production of this book, especially Gregory Payne, editor, Dora Rizzuto, production supervisor, and Eileen Burke, designer. Throughout, I have benefited from the scientific judgment and encouragement of my wife, Patricia Hasegawa.

W. F. L.

Contents

Figure 1.1. "The she-bear brings forth cubs without shape or feature at all. She has, however, an infinitely clever tongue and with this she licks her shapeless offspring into proper bear shape." Physiologus (anonymous, fifth century).

CHAPTER 1

Introduction and Historical Overview

Only a little over 150 years ago was it realized that an egg is a single cell. Even more recently was it recognized that sperm added their genetic contents to that of the egg. In the middle ages many still thought that bear cubs were "licked into shape" by their mothers. They saw animals give birth to shapeless masses covered in an amniotic sac. Only after diligent licking did the form of the newborn appear (Fig. 1.1). They assumed that morphogenesis was directed by the mother's tongue. We now know it is directed by genes.

It would be nice to describe developmental processes with the precision of physical equations describing ideal gases. However, when one looks at the equations applied to embryos one sees a whole alphabet of ill-defined variables. With some imagination one can chose values that mimic morphological observations, but actually measuring the variables is the meaningful task.

While it is essential to grasp the broad outlines of anatomical changes during embryogenesis, it is often more rewarding to recognize the underlying changes in differentiating cells. By analyzing the molecular differentiations and cytodifferentiations in depth in a few specific cases, our understanding of the processes can be taken toward a mechanistic level. The basic questions are those related to causal links in the chain of events from egg to embryo. When enough of these relationships are clearly grasped, the complexity of the whole process becomes less formidable.

The processes by which groups of cells change and diverge are complex and use many interacting components. When a molecule is implicated in these processes, it can be isolated and described biochemically. Difficulties arise in proving that the specific molecule plays an essential role in the developmental process. Functional assays for a complex process are usually indirect and often one has to settle for only a correlation rather than direct proof. These studies indicate that a given molecule might play a critical function but do not show it beyond a doubt. In certain cases developmental genetics can provide firm evidence for the involvement of a specific gene product. In the best of all cases, a mutation in a gene coding for a known product results in a clear alteration in the sequence of developmental steps. Isolating the normal, wild-type gene and the mutated gene by molecular cloning and showing that a specific change in the nucleotide sequence has caused the mutation puts this approach on an elegant plane. However, the gene product often interacts with other gene products in unforeseen ways and may function in more than one pathway. Therefore, a single mutation in a single gene may have consequences in a multitude of processes. The pleiotropic effects of mutations in developmental genes have often precluded decisive one-to-one assignments. Only when a simple linear process leads to a well-defined and independent event can definitive causal pathways be recognized by this approach. Such pathways usually occur only at the end of the network leading to terminal differentiation. Several of these are presented in Part I.

The steps in dissecting development use every available technique. Enzymes are purified and their catalytic properties studied. Proteins are separated, sized, and studied biophysically. Genes are isolated by cloning and sequenced. RNA molecules are quantitated by hybridization. Mutants that have an altered molecular, physiological, or morphological phenotype are isolated and then characterized with respect to other parameters. Cells are sectioned, cultured, microinjected, and looked at in a variety of microscopes. Tissues are dissected, rotated and put in abnormal places. Some or all of these techniques are brought to bear on problems of development in a specific organism with hopes that insights to a process in one situation will give guidance in the study of related processes seen at other times or in other organisms. Finally, a synthesis of all the available data and concepts has to be distilled from the apparent differences to give a framework for recognizing the outstanding problems.

In Part II embryogenesis is considered from the differentiation of eggs and sperm to the formation of early organs. Homologous processes that occur in a set of organisms including sea urchins, amphibians, birds, and mammals are emphasized. It is important to grasp the relative positions and functions of groups of cells in early embryos that will give rise to discrete tissues.

Terminal differentiations are considered in molecular detail in Part III. A selected set including the formation of limbs and accessory sexual organs as well as the complete development of *Dictyostelium, Drosophila,* and *Caenorhabditis* have been chosen to give an appreciation of the role development plays in the evolutionary fitness of organisms.

A Little History

Human curiosity has always been fascinated by the emergence of tadpoles from frog eggs and plants from little seeds. However, only after the invention of microscopes in the seventeenth century could it be seen that developing embryos and plants passed through a series of stages from the fertilized egg to the emerging form. Detailed observation of varied organisms gradually eroded the common belief that rotten meat produced maggots by spontaneous generation. By the nineteenth century most educated people accepted that all living things come from reproduction of previously existing organisms. Reproduction was a more manageable concept than spontaneous generation but no better understood. How could an apparently simple egg give rise to the exquisite detail of a newborn? For a while, some scientists thought they saw a miniature man in the head of sperm, an homunculus (Fig. 1.2). They thought the homunculus grew into an embryo

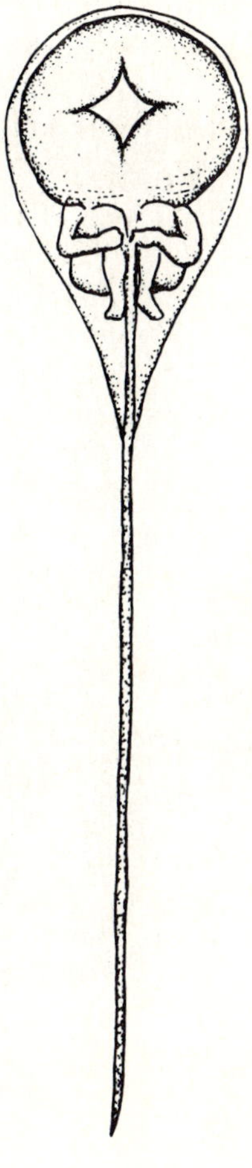

Figure 1.2. Homunculus. In the seventeenth century Niklass Hoertsocker drew a miniature human being thought to be within the head of sperm. It is now realized this was a case of fanciful thinking.

August Weismann 1834–1914

Hans Driesch 1867–1941

Wilhelm Roux 1850–1924

Hans Spemann 1869–1941

Figure 1.3. Experimental embryology was initiated within the last century by Weismann, Roux, Driesch, and Spemann.

within the nutrient-rich environment of the egg. But then the homunculus must carry even smaller copies of man in its own sperm, and on and on to absurdity.

Although this idea was soon discarded, it led to the concept of mosaic development advanced by Wilhelm Roux and August Weismann (Fig. 1.3). Roux carried out experiments on early embryos that suggested that the information to form certain tissues was segregated into one or another cell during the first few cleavages of a zygote. While some of the experiments have since been reinterpreted, in other cases they have held up. Fertilized eggs of some species, such as nematodes (Chapter 16), develop into embryos missing certain tissues when specific cells are removed from the early blastula. However, the Roux–Weismann theory held that the tissues could not be replaced because the remaining cells had lost the genes necessary for them. This theory has now been disproven in all but a few cases. The full complement of genetic information is inherited by all cells but is differentially used in tissues. In those organisms with dramatically mosaic development, use of the genetic potential is stably restricted to a specific subset of genes early in embryogenesis when there are only a few cells but all the genes are there in all the cells. The great majority of species, including humans, follow a regulative style of development in which remaining cells take over the role of cells removed by microsurgery. In these species genetic potential is only restricted late in development, often only during terminal differentiation of cells into a functioning tissue.

The mosaic aspects of development of some tissues have motivated the ongoing search for embryonic localization of determinants. In a few cases, such as the establishment of germ lines that give rise to sperm and eggs, there is evidence for partitioning of components down the specific lineage leading to the definitive tissue. In most cases, the causal conditions that account for cell specialization have not yet been found.

In the last 50 years emphasis has shifted to the study of specific genes. T. H. Morgan (Fig. 1.4) and his collaborators started genetic studies of the fruit fly *Drosophila melanogaster* that have led to the understanding that certain genes direct developmental pathways and others respond to the environment. As will be described in Chapter 15, *Drosophila* genes have been isolated and described in detail, but they have yet to show how the whole process of embryogenesis is integrated. The problem is how to relate a one-dimensional set of instructions (the base sequences in DNA) to the three-dimensional form of the embryo.

Meanwhile, embryologists such as Hans Driesch and Hans Spemann (Fig. 1.3) were carefully observing the consequences of experimental manipulations of embryos. Early embryos were cut, squeezed, ligated, centrifuged, injected, heated, and then allowed to develop as best they could. These studies resulted in an awareness of the interactions of tissues and an appreciation of the role of the cytoplasm in controlling gene expression. Cells of regulative embryos continuously monitor the state of their neighbors and undergo changes depending on the information they receive. Embryogenesis passes through many stages, some with little apparent logic, but arrives at the final form that is adapted to survival. At each step information is

Figure 1.4. T. H. Morgan (1866–1945) as he worked in the Fly lab in 1915.

available for divergence of the cells to specialized functions. The exact nature of this information is a matter of animated debate among theoreticians as well as a major goal of experimentalists. These questions are emphasized in Chapter 14.

Studies in simple cells such as bacteria might appear to be irrelevant to embryology; however, they have elegantly shown the types of molecular interactions that can result in differentiation of a homogeneous population of cells into two or more quite different populations. Studies on bacterial viruses and the control of the *lac* operon in *Escherichia coli* led to models of regulation of gene activity put forth over 25 years ago by François Jacob and Jacques Monod (Fig. 1.5). These models have guided much of the analysis of developmental processes. Nevertheless, the complexity of development of multicellular organisms has made this a lengthy task. Techniques for detailed analysis of minute amounts of material as well as advances in molecular genetics during the last ten years have generated considerable excitement and raised hopes that a clearer view of development will soon appear.

We are still challenged to understand how differentiation controls the sequential patterns of gene expression, which in turn control subsequent differentiation and development. We can show that at various stages of embryogenesis certain cells become irreversibly determined to follow specific developmental pathways, but we cannot yet show how they become determined. These are some of the problems that present studies are attempting to answer.

To use an archeological metaphor, the temple of developmental biol-

Figure 1.5. Jacques Monod and François Jacob sketch out their concept of gene regulation in 1966.

ogy is not yet fully excavated. We know that the structure is there under the sand and we have a collection of stones that appear to be parts of pillars and arches. In a few places we can see how they are connected. For now we have to be content with describing each stone as carefully and completely as possible. We can already see that they fall into groups of similar stones such that the task is simplified. It is difficult to say which stones are the most important. The capstones are held up by the pillars and arches and each plays its role. Perhaps what we would like most is to understand the architectural plans that positioned the components and the scaffolding that put them in place. We cannot yet be certain of the plan in many cases, but there is growing confidence that we see its outline as new stones are uncovered all the time.

Descriptive Vertebrate Embryology

Over the last century, landmarks in embryogenesis have been observed and analyzed by microscopy. Powerful concepts and generalizations have come from these studies that now form the foundations for cellular and molecular approaches. While the early stages in embryogenesis will be discussed in detail in Chapters 7, 8, and 9, the classical descriptions introduce the problems as well as the beauty of developmental pathways.

Genetics has shown that an individual comes into being when a particular sperm fuses with a particular egg. Both sperm and eggs are haploid cells with an assortment of one copy of each gene from the diploid parent. When they fuse, a new diploid organism is made with two copies of each gene. Its genetic potential is completely set and its fate during embryogenesis, and to some extent its life, is determined. Different species have different genes, although many of the genes are surprisingly conserved in their physiological function.

The fertilized egg is an isolated structure and there is no net increase in mass immediately following fertilization. Cellular plasma membrane is synthesized to partition off volumes of the egg into discrete cells. In some species the partitioning is not complete until a fairly large number of partially enclosed cells have been made. During these early cleavages the chromosomes of the newly made diploid nucleus are being rapidly replicated. This is often the most rapid period of DNA synthesis in the life of the organism. As each cell is formed, a nucleus with a full diploid complement of chromosomes is positioned in it. The subunits for synthesis of membrane components are stored in the egg as it forms in the mother. All the enzymes for biosynthesis are also made and stored in the egg. Although new proteins are made, there is no immediate need for most of them during the early stages of embryogenesis. Proteins are initially synthesized on messenger ribonucleic acid (mRNA), which accumulates during the growth and differentiation of oocytes into eggs in the mother and is stored for use following fertilization. These maternal mRNAs can account for greater than 90% of the activated protein synthesis during early cleavage stages. So for the first few divisions, transcription of genes into mRNA plays little or no role in controlling the pattern of protein synthesis.

Blastula Stage

The fertilized egg divides rapidly to generate a large number of cells. In those embryos in which the whole egg divides up into cells (holoblastic division), most of the cell division that will occur during all of embryogenesis occurs during this period of cleavage (Figs. 1.6 and 1.7). Mammalian embryos are an exception in that placental supplies of nutrients permit ex-

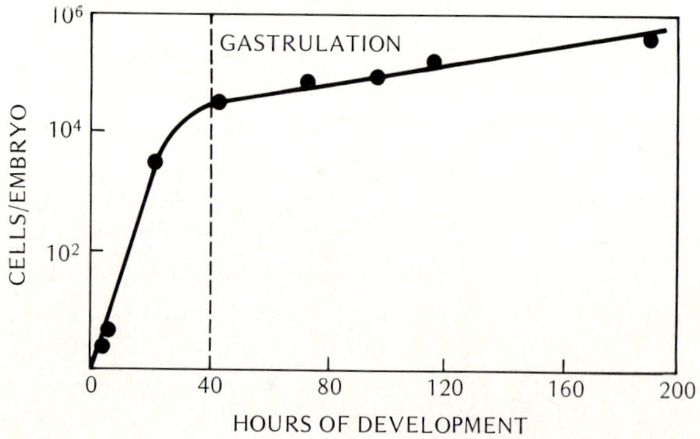

Figure 1.6. Exponential increase in the number of cells during cleavage of frog embryos *(Rana)*. More than half the total number of cell divisions occur during the first day and a half following fertilization. Division slows as gastrulation starts.

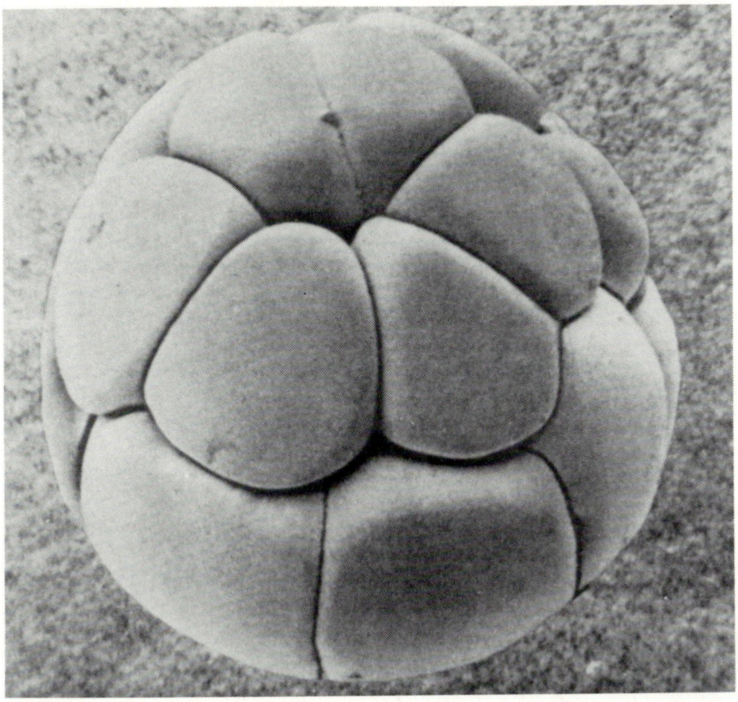

Figure 1.7. A 16-cell blastula of a frog. The smaller cells of the animal hemisphere are uppermost. The embryo is about 2 mm in diameter.

tensive growth late in embryogenesis. By the end of the twelfth round of cell division there are about 4000 cells.

The number of cells increases exponentially as each new cell divides in two. The large egg cell is rapidly subdivided into blastomeres that give rise to different tissues of the embryo, as will be described in Chapter 8. A cavity soon forms near the center of the mass of cells that is called the blastocoel (Fig. 1.8). Transcription of mRNA accelerates gradually as blastula formation continues. By the end of this stage, a considerable number of new gene products have accumulated that play important roles in subsequent development.

Gastrula Stage

The cells that contain more yolk material are larger than others and make up the vegetal hemisphere. At a point that varies between eggs of different organisms, cells invaginate to form an archenteron. This cavity, which has an opening to the outside (the blastopore), will become part of the alimentary canal (Fig. 1.9). The geometry and even the mechanism of gastrulation depends on the amount of yolk initially in the egg. However, as will be described in detail in Chapter 9, in all cases a multilayered form is generated. Thereafter, tissues diverge physiologically as a consequence of expression of cell-type specific genes.

Following gastrulation, the sequence of morphogenetic events differs significantly between embryos of different phyla. For instance, in sea ur-

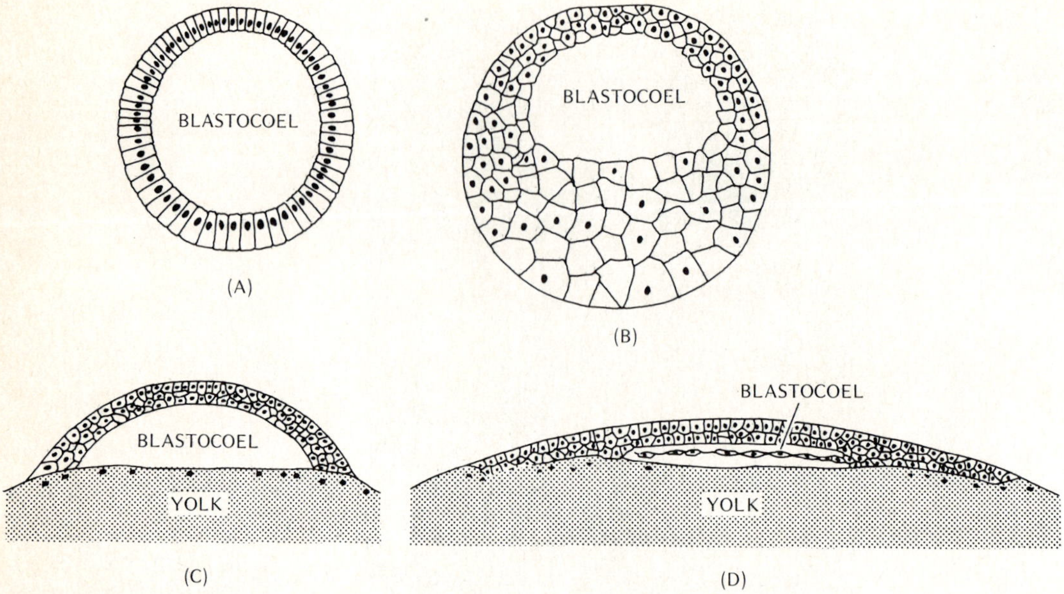

Figure 1.8. The blastocoel in various chordates. Embryos of *(A)* amphioxis and *(B)* amphibians are holoblastic in that the whole egg is divided into smaller cells. In *(C)* teleost fish and *(D)* birds, division is meroblastic in that only a small portion of the fertilized egg is divided into cells. In all cases the blastocoel is invaded by cells during gastrulation.

chins, the embryo develops into a feeding larva without further major cell movements, whereas in vertebrates, an anterior/posterior axis is established by infolding, which generates a dorsal nervous system as well as the sense organs of the head, as will be described in Chapter 10. Detailed sculpturing of the skeleton is different in almost every species, but the basic plan is conserved in amphibians, reptiles, birds, and mammals.

As the archenteron is formed during gastrulation in vertebrates, cells along the dorsal midline condense into a rod-shaped structure referred to as the notochord. Ectodermal cells on the surface that lie over the notochord are induced to differentiate into the neural plate that then rolls up to form the neural tube that will give rise to the central nervous system. Mesenchymal cells on either side of the neural tube give rise to somites that form the vertebrae, ribs, and lateral muscles. The details are presented in Chapter 10.

Once the neural plate is formed, it will give rise to a neural tube even when isolated from the rest of the embryo. Such tissues are said to be determined. As embryogenesis proceeds in vertebrates, different tissues become determined, one after the other. Shortly after they become visible, limb buds become determined to form arms or legs, although the specific pattern of bones is not yet established. Distal structures such as the digits of fingers and toes are formed in response to signals in the limb bud, irrespective of where the bud is placed on the body of the embryo. For this reason the

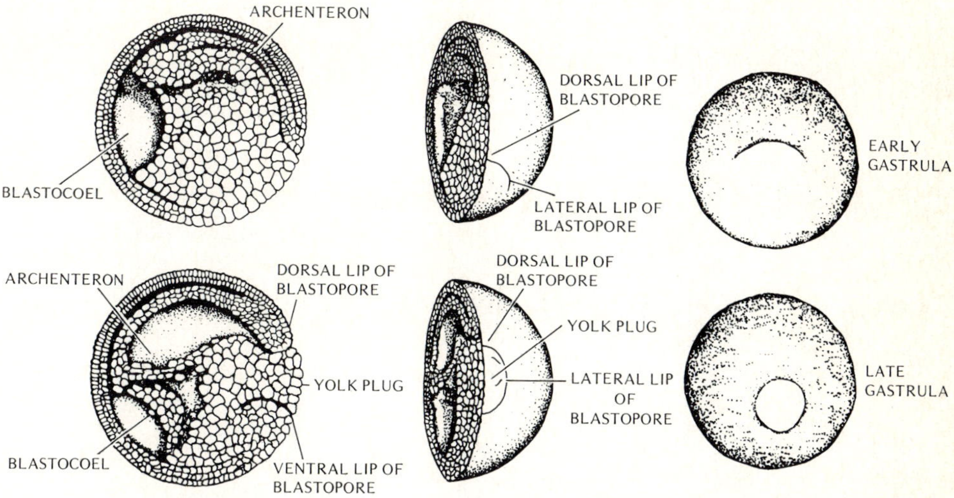

Figure 1.9. Gastrulation in the frog. Sheets of cells move over and through the lips of the blastopore to form the internal structures. Sections cut through the blastopore are drawn showing the stages of invagination as the archenteron expands and partially fills the space previously held by the blastocoel.

limb bud is considered a separate field for pattern formation. The mechanisms by which cells are given positional information and the ways in which they respond to this information are considered in Chapter 14. Before the bones of a limb can be seen, the pattern is irreversibly established. To some extent this is a consequence of our limitations in recognizing the determining processes. Molecular mechanisms have been set that will subsequently result in the massive deposition of cartilage and bone. As techniques are improved to measure smaller and smaller amounts of material, pattern can often be seen at earlier stages.

Fate Map

In any given species the sequence of events during embryogenesis is sufficiently similar from one individual to the next that one can often trace the lineage of a given tissue back to the first few cells in a cleaving egg (Fig. 1.10). These analyses were mostly carried out over the last 50 years by marking specific cells of early embryos with vital dyes and then observing which tissues were stained at later stages. The dyes used do not affect the processes of development but mark the descendants of a specific blastomere. Thus, these lineages tell what happens in normal embryogenesis but do not say what would happen if the specific cell were put in another part of the embryo or removed and allowed to develop on its own. A fate map gives the descendants of a blastomere but does not indicate when their fate is irreversibly determined.

A gastrula has three major cell types: (1) those on the outer surface are called ectodermal; (2) those that invaginated to form the archenteron are

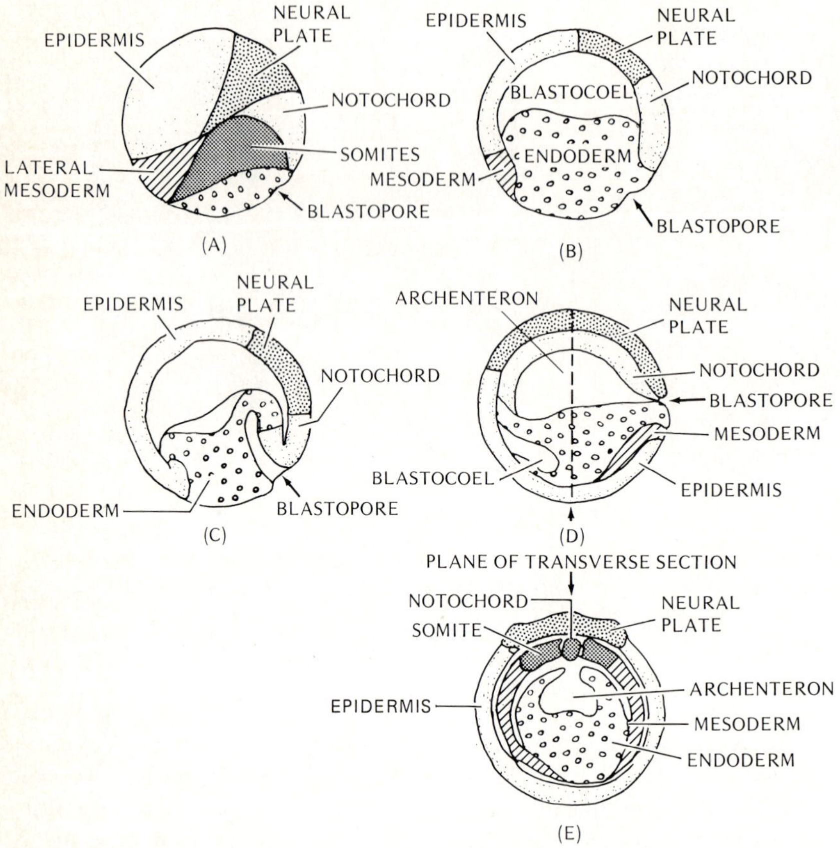

Figure 1.10. Fate maps as gastrulation proceeds in a frog. *(A)* Before invagination has started, the future differentiation of all of the blastula cells can be predicted with certainty. *(B and C)* Cells fated to make up the epidermis surround the yolky endodermal cells of the vegetal hemisphere, which enter through the ventral lip of the blastopore. A deep layer of cells under the surface of the blastopore is fated to form dorsal mesoderm. These cells will give rise to the notochord and somites. *(D and E)* The notochord induces ectodermal cells still on the surface to differentiate into the neural plate.

called endodermal; (3) those between these layers are called mesodermal. The major pathways that these cells follow during subsequent development have been traced for a variety of vertebrate species (Fig. 1.11). Many organs are complex and are constructed of cells derived from more than one of the embryonic germ layers but, in general, internal glands and the gut organs are derived from endodermal tissue, whereas the heart, blood, gonads, muscles, and skeleton are mainly derived from mesodermal tissue. The major organs derived from the ectoderm are epithelial and neural. Neural crest cells are ectodermal in origin but give rise to a variety of internal structures after extended migrations early in embryogenesis. It has been possible to follow

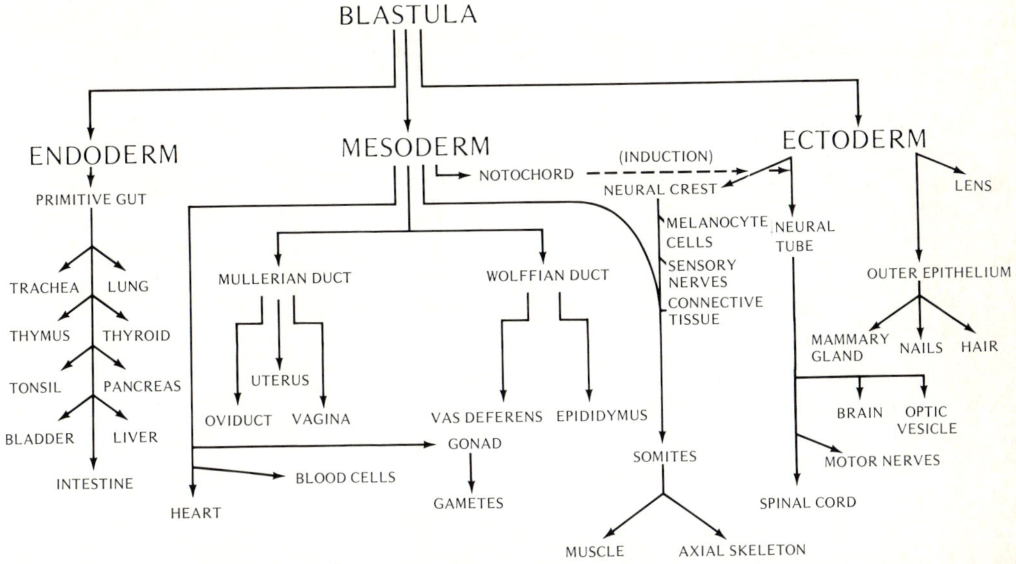

Figure 1.11. Derivation of some vertebrate organs. Following gastrulation three germ layers can be recognized. The inner endodermal cells give rise to the gut and various glands. The initially dispersed mesodermal cells concentrate to form somites, the heart and reproductive organs. The outer continuous layer of ectodermal cells will form the covering of skin as well as specialized external structures such as hair and nails. Neural tissue is also derived from the ectoderm when it forms the neural tube over the notochord. Neural crest cells give rise to a diverse set of tissues.

the pathways of single neural crest cells in avian embryos by substituting small pieces of quail neural plate tissue for chick neural plate tissue. The quail cells can be distinguished from surrounding chick cells by easily recognized differences in nuclear morphology.

As the cells become increasingly specialized to function in one tissue or another, they accumulate proteins adapted to the function of the tissue and take on the size and shape of the cells found in adult tissues. These changes are mediated by selective synthesis and degradation of enzymes and regulatory proteins that control the biosynthetic processes. Differentiation of a specific cell type requires a coordinated change in gene expression. It is not enough to have four out of five enzymes in a specialized biosynthetic pathway; all five must accumulate together. Integration of the genes for complex physiological processes requires that they all respond to a common signal, but it need not be a direct response. Often the initial trigger sets off a sequence of events to which each gene responds in its own way. One of the aims of developmental biology is to recognize the determining components and unravel the interactive relationships giving embryogenesis its precision and adaptability, which leads to the formation of a self-sustaining new individual. Some of the mechanisms of cell differentiation are considered in Part I.

Summary

Classical studies have defined many of the questions of developmental biology: What is the mechanism of differential gene regulation and cellular specialization? How are individual differentiations integrated to form a functioning embryo? What are the determinants that generate the initial differences in embryonic cells and how are they localized? What are the processes that instruct cells as to their position in a developing embryo such that the proper pattern and form is generated? These and related questions are addressed throughout this presentation. Microscopic analyses of fertilized eggs have shown that the haploid male and female pronuclei fuse to form the diploid nucleus of the zygote. A period of rapid DNA replication and nuclear proliferation follows. The zygote divides up into hundreds of cells and expands around a central cavity called the blastocoel. After a dozen or so rounds of cell division, the size of the blastula cells is closer to that of adult cells. Gastrulation generates multiple layers of cells with differences in their recent developmental histories. Differential gene expression at about this stage generates distinct cell types that interact with each other to set in train the pathways leading to definitive tissues. Fate maps can be drawn on the surface of blastulae that indicate which organs will be derived from which blastomeres. In many organisms this fate is not irreversibly established until after stages in embryogenesis. Late in gastrulation some groups of cells will differentiate to their fated functions even if removed or misplaced in the embryo. The molecular nature of this determination is one of the most exciting questions of developmental biology.

While the evolution of complex organisms was first deduced from comparative anatomy of living and extinct species, the similarities in developmental stages between species lends strong support to the ideas eloquently put forward by Charles Darwin. The concepts of random variation and natural selection provide the basis for understanding biology. Comparative embryology has shown that slight variations in cleavage, blastula, and gastrula stages can result in a wide variety of forms. The mechanisms and processes of development in different types of organisms will be compared throughout this presentation.

SPECIFIC EXAMPLES

1. Human Embryogenesis

There is something wonderful in looking at a human egg (Fig. 1.12). We, as humans, know the potential inherent in such a cell. Following the stages of embryogenesis of such an egg to the birth of a human child tells us something directly about ourselves.

Human eggs are large cells but small compared with the eggs of fish, reptiles, or birds. Placental nutrition removes the need for massive amounts

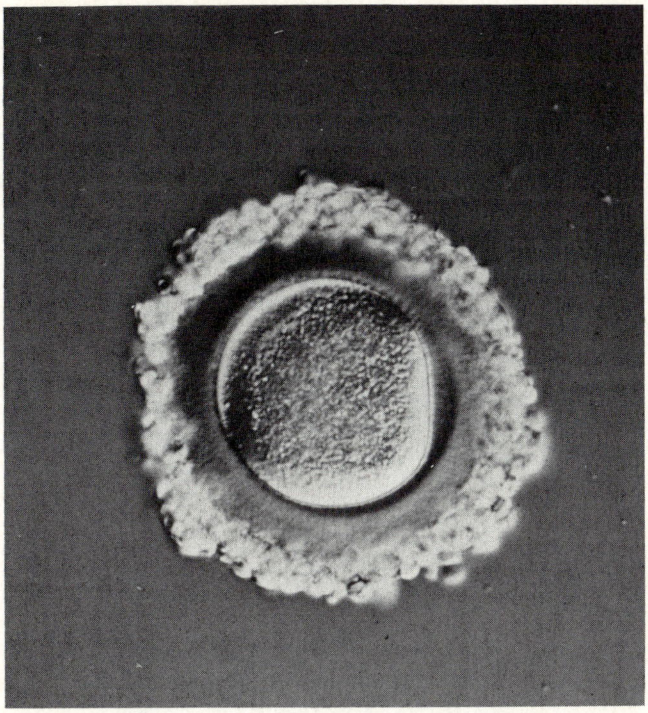

Figure 1.12. A human egg. The large central cell is protected by a protein shell referred to as the zona pellucida. The egg is held in a sphere of much smaller follicle cells.

of yolk. The egg is surrounded by a protective covering, the zona pellucida, and develops within this shell following fertilization (Fig. 1.12). Early divisions divide the egg into a hundred or so smaller cells. At this stage it is referred to as a morula (from the Latin word for mulberry, *morum*). This solid ball of cells expands by actively taking up ions and allowing fluid into the middle and subsequently hatches out of the zona pellucida. The hollow ball has an inner cell mass from which the infant will develop (Fig. 1.13). The outer cells, referred to as the trophectoderm, embed in the uterine wall and give rise to the placenta. These tissues supply nutrients derived from the maternal bloodstream to the inner cell mass through a stalk of cells. The inner cell mass separates to form the embryonic disc (Fig. 1.14).

The embryonic disc forms a bilayer that undergoes gastrulation to generate the tissues of the early embryo. Subsequent stages in mammalian embryogenesis are almost indistinguishable from those of other vertebrates. When the head, eyes, and spinal cord can be seen, it is difficult to distinguish a human embryo from that of a chick, a reptile, or a pig (Fig. 1.15). The head of a human is a bit larger, but the differences are not great.

Development of the Heart

The first organ to start its definitive function in vertebrates is the heart. It develops from mesodermal cells adjacent to the midline. The heart is essential to pump blood through the placental capillary beds and carry oxygen and nutrients to the growing embryo. Structures homologous to gill slits are clearly visible in human embryos at this early stage, but they never func-

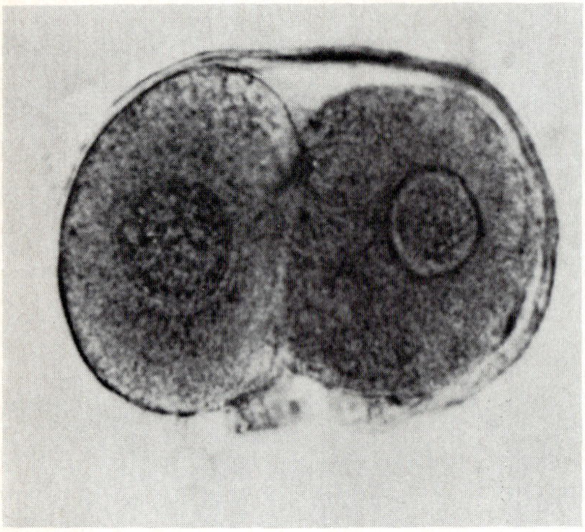

(A)

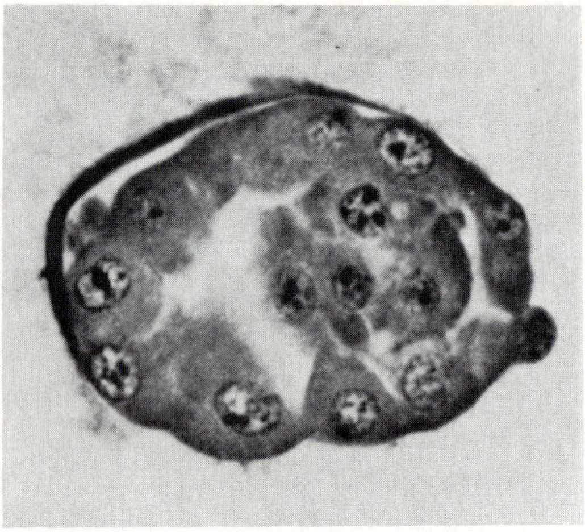

(B)

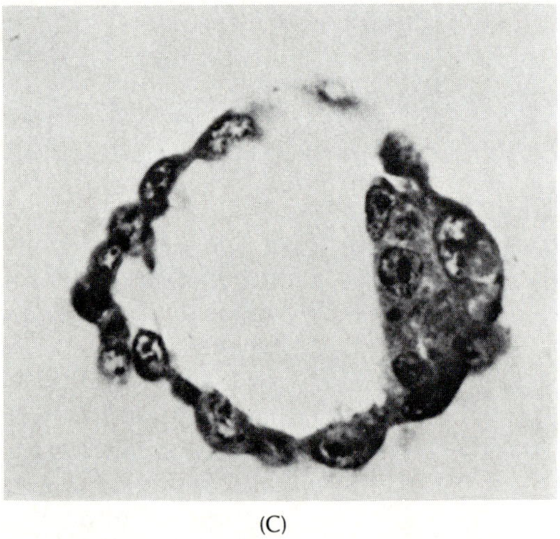

(C)

Figure 1.13. Cleavage of human eggs. *(A)* The two-cell embryo is still within the zona pellucida shell. *(B)* At four to five days of development, the fertilized egg has divided into about 50 cells, each about 20 μm in diameter. *(C)* The blastocoel (also referred to as the blastocyst cavity) enlarges and separates the trophoderm from the inner cell mass. During this process the embryo hatches out of the zona pellucida shell.

tion for exchange of gasses. That occurs only in the placenta. The heart starts to pump very early in development, even before many of the peripheral vessels are ready to receive the stream of blood. The heart grows to fit the size of the embryo and can regenerate portions that are removed. Its development and growth are regulated by the amount of blood flowing through it and the size of the embryo. From an initially straight tube formed within the mesoderm, the heart bends into an S form from which the various chambers will develop (Fig. 1.16).

The function of the heart changes radically at the moment of birth. During fetal life, oxygen is taken up from the placenta and delivered in the bloodstream through the umbilical cords. The lungs serve no function until after birth. Very little blood goes through the lungs in the embryo, and yet

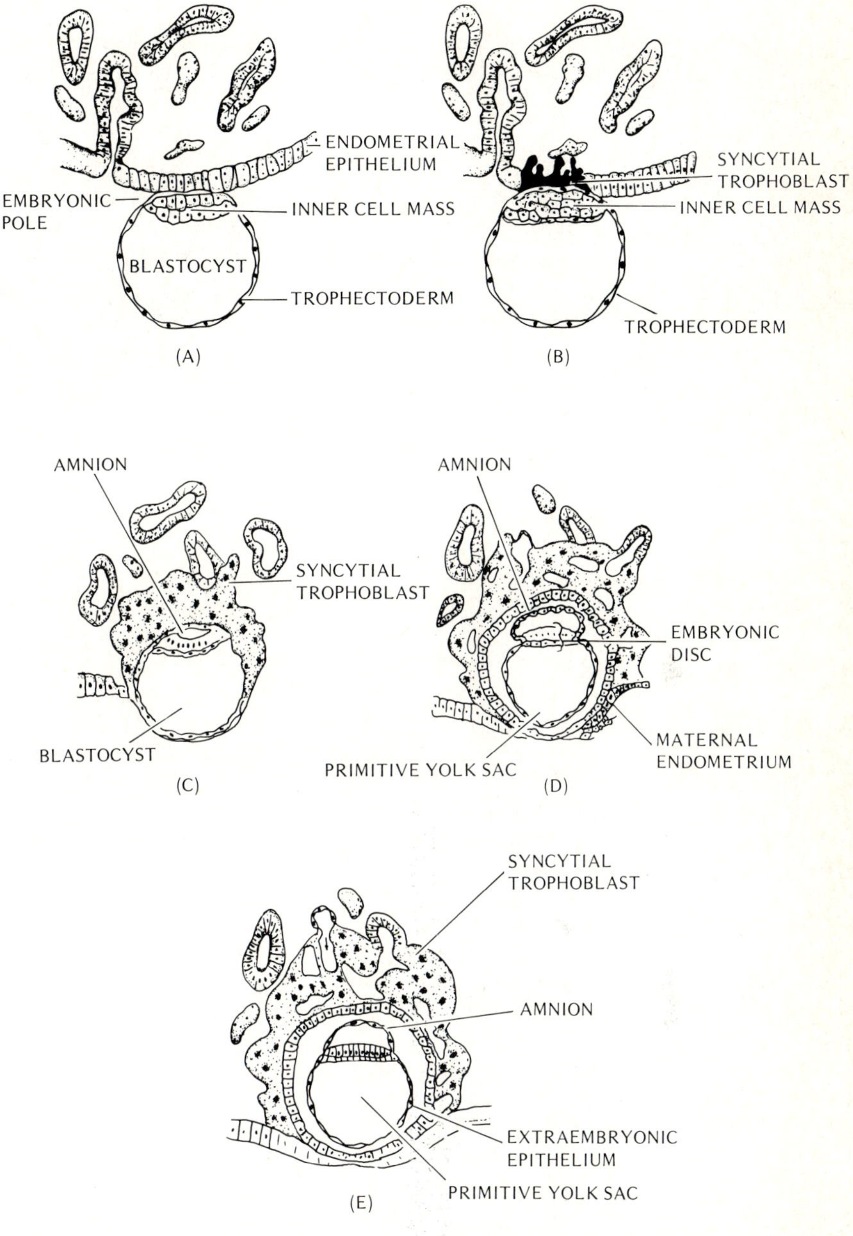

Figure 1.14. Implantation and early development of the human. *(A)* Six days following fertilization the embryo has been swept down the fallopian tube and lodged on the endometrial lining of the uterus. *(B, C, D)* During the next three days the embryo implants in the endometrium. *(C)* The amniotic cavity forms over the embryonic disk at eight days. Above the amnion a syncytial tissue of trophectoderm surrounds small lakes (lacunae) of maternal blood. Below the amnion the embryonic disc consists of two layers of cells, the epiblast and the hypoblast. The cells of the hypoblast spread over the surface of the blastocyst cavity and surround the primitive yolk sac.

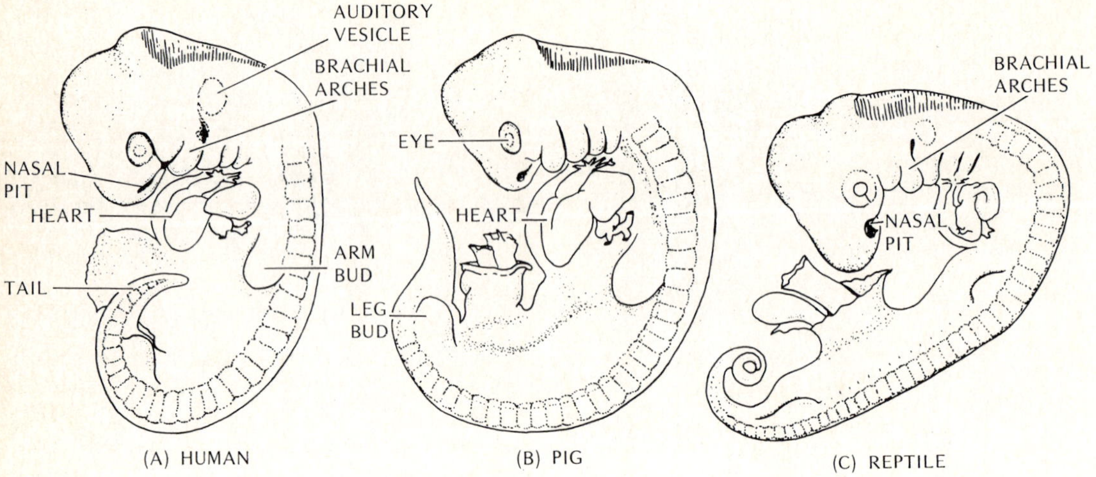

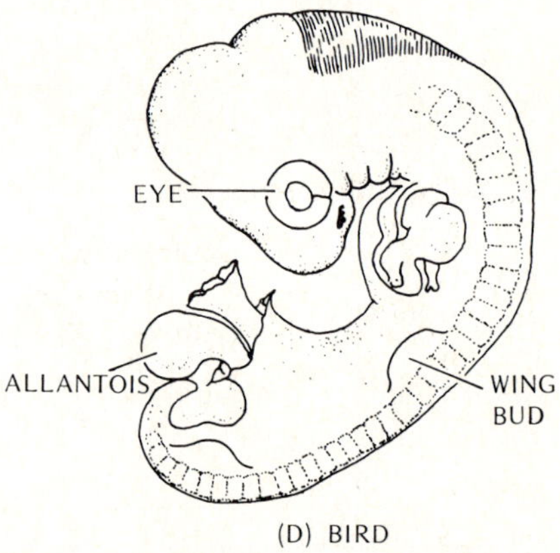

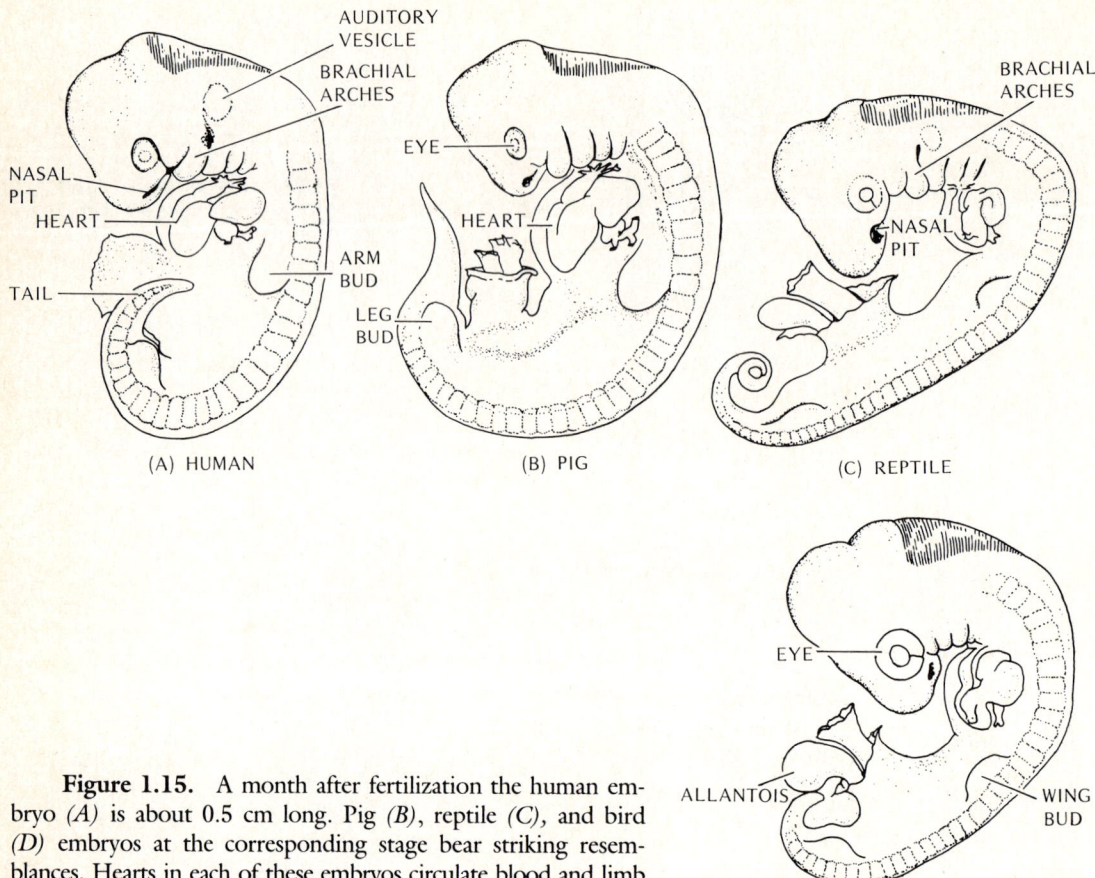

Figure 1.15. A month after fertilization the human embryo *(A)* is about 0.5 cm long. Pig *(B)*, reptile *(C)*, and bird *(D)* embryos at the corresponding stage bear striking resemblances. Hearts in each of these embryos circulate blood and limb buds are developing. The head and eyes are clearly defined.

after birth the right side of the heart must pass the total circulating volume of blood through them. This change occurs immediately at birth. In the fetus the right atrium has an opening to the left atrium, the foramen ovale. Blood from the umbilical cord and other veins enters the right atrium and most passes directly through the foramen ovale to the left atrium. From there it enters the left ventricle and is pumped down the aorta to the body.

As long as the blood pressure in the right atrium is greater than the blood pressure in the left atrium, blood passes through the opening connecting the atria. After birth, when the pulmonary circulation becomes fully established, the pressure in the left atrium becomes higher than that in the right atrium. This forces a flap of tissue over the foramen ovale. The flap eventually fuses with the tissue separating the two atria and for the rest of life the heart works as a paired organ.

There is also a connection between the main pulmonary artery to the dorsal aorta to bypass the lungs during fetal life. At birth there is a violent contraction of this arterial duct, forcing the blood pumped by the right ventricle to flow through the capillary beds of the lungs.

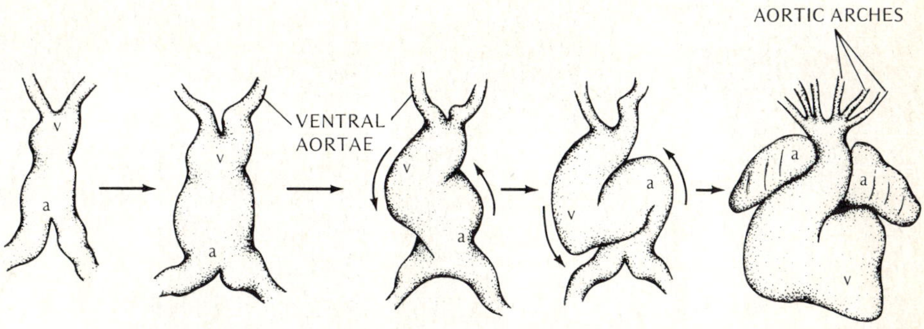

Figure 1.16. During the fourth week of human development, the embryonic heart tubes fuse to form a single heart. The atrium *(a)* is initially below the ventricle *(v)*. Bending brings the atrium to lie above the ventricle. During the fifth week, a septum forms that divides the ventricle in two. The atrium is also divided in two, but the septum has openings connecting the chambers until birth. The first embryonic red blood cells differentiate in the blood islands of the yolk sac and enter the circulation through vitelline veins. Blood is then pumped through the umbilical arteries to the placenta before it returns with nutrients for the embryo.

Eyes form early in embryogenesis from the interaction of bulges of the embryonic brain and overlying ectoderm. The lens and cornea are induced to differentiate. The eyelids form and slowly close the eyes (Fig. 1.17). They open only after birth. The arms and legs also take on easily recognizable shapes early in embryogenesis (Fig. 1.18). Initially, the fingers and toes are webbed but at later stages the cells between them die.

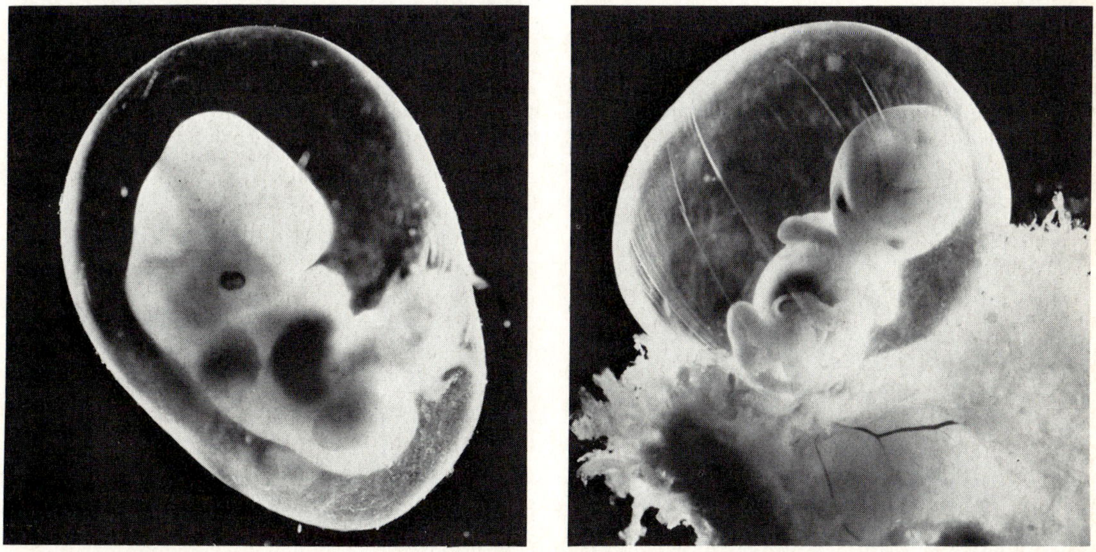

Figure 1.17. A six-week-old embryo is about 2 cm long and weighs about 2 g. Pigmentation of the retina gives the embryo a lifelike appearance.

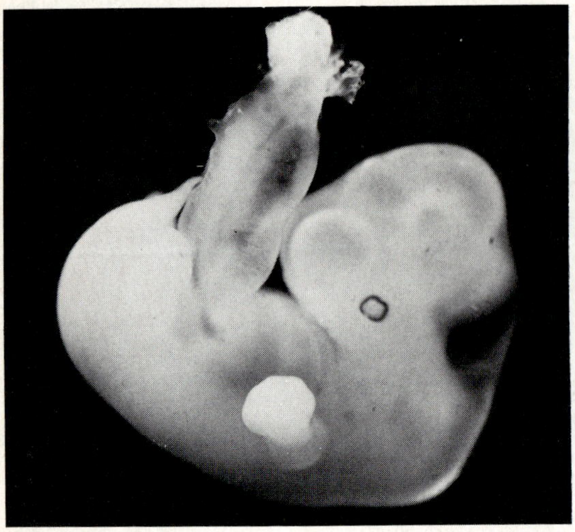

(A)

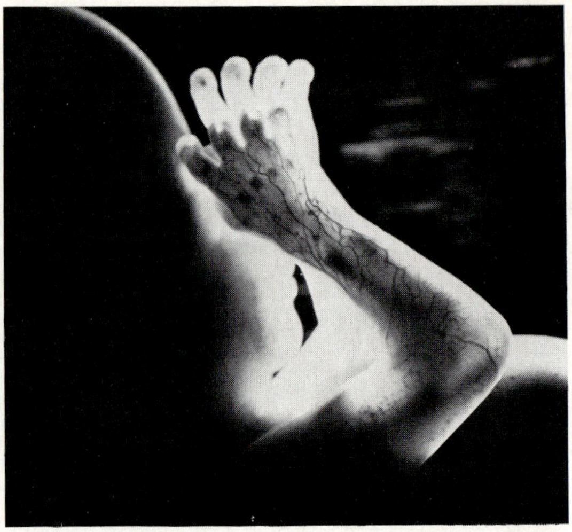

(B)

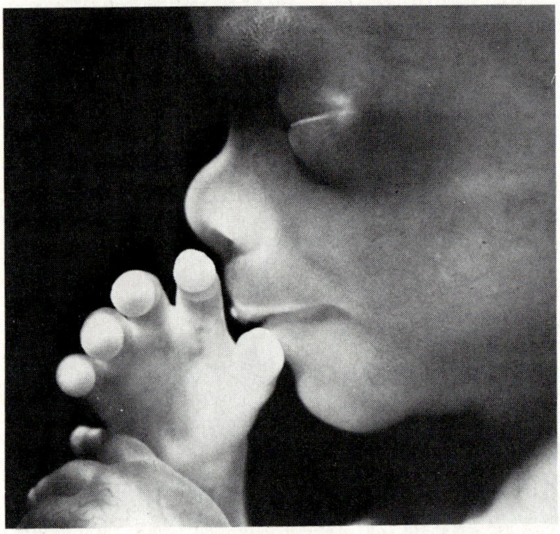

(C)

Figure 1.18. Limbs are apparent as buds with webbed fingers at five weeks *(A)*. The hand is well developed by the eighteenth week *(B)* and the thumb is opposable by the twentieth week *(C)*.

By five months of development the human embryo can move its arms and legs forcefully (Figs. 1.19 and 1.20). This is referred to as quickening of the embryo. During the next four months the embryo grows about four-fold in size. At birth the human infant is much less able to fend for itself than infants of closely related species. It must rely on the care of its parents for many months. Relatively premature birth appears to be necessary to allow delivery of an infant with a head as large as that of humans.

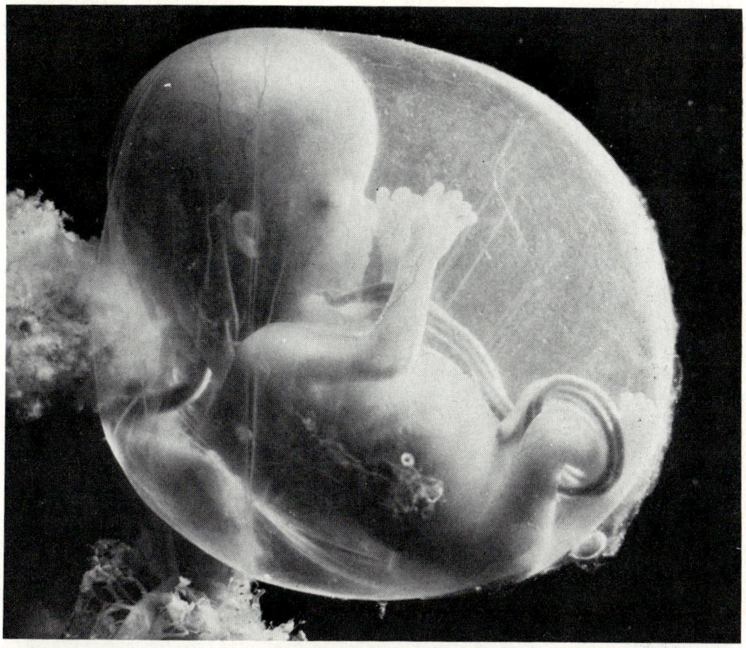

Figure 1.19. A four-month-old embryo is 15 cm long and weighs about 200 g. All the organs have formed and the embryo is growing rapidly within the amniotic sac.

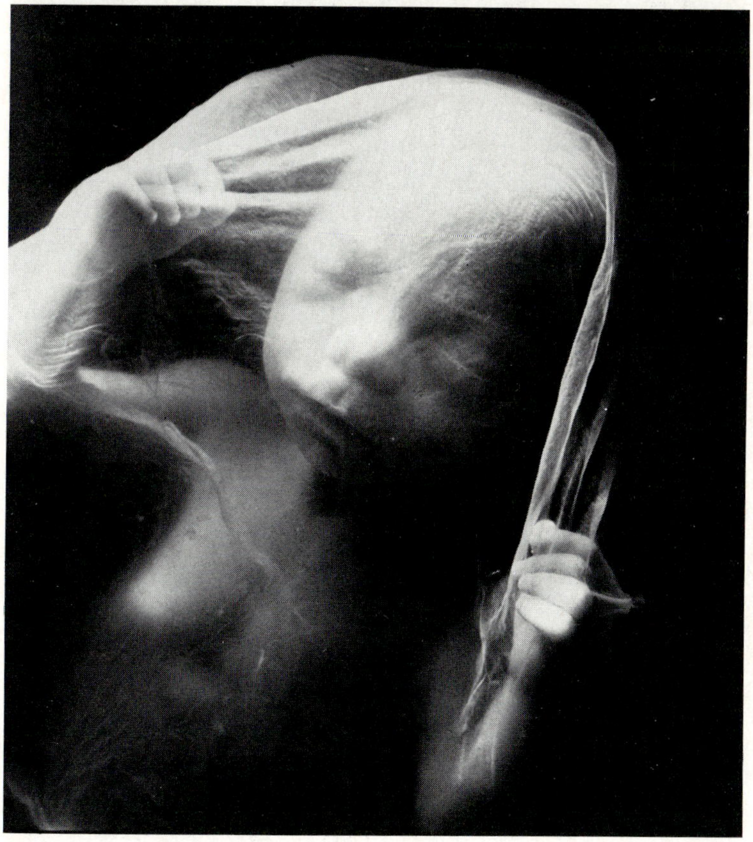

Figure 1.20. A five-month-old embryo is able to move its limbs and swallow liquid. If prematurely delivered, it can survive under intensive care.

2. Plant Embryogenesis

The first multicellular organisms were most likely green. Colonies of algae such as *Volvox* probably evolved soon after the planetary atmosphere contained sufficient oxygen for large eukaryotic cells to arise but before the pO_2 was high enough to sustain compact groups of cells (Fig. 1.21). From simple aquatic beginnings plants colonized the land until, today, they make up the great bulk of the biomass.

Propagation of lower plants is usually simple, involving little more than the growth of germinated spores. In the seed-bearing higher plants of both gymnosperms (pines) and angiosperms (flowering plants) an embryo is formed (Fig. 1.22).

Pollen from male organs is released and sends out a pollen tube when it encounters the female organ, referred to as the ovule. The pollen carries haploid nuclei that enter the female sporangium or egg through an opening referred to as the micropyle. One of the sperm nuclei fuses with the haploid egg nucleus (Figs. 1.23 and 1.24). The resulting diploid zygote then divides several times within the supporting tissue of the seed.

The embryo grows as a solid mass of cells that soon takes on an axial

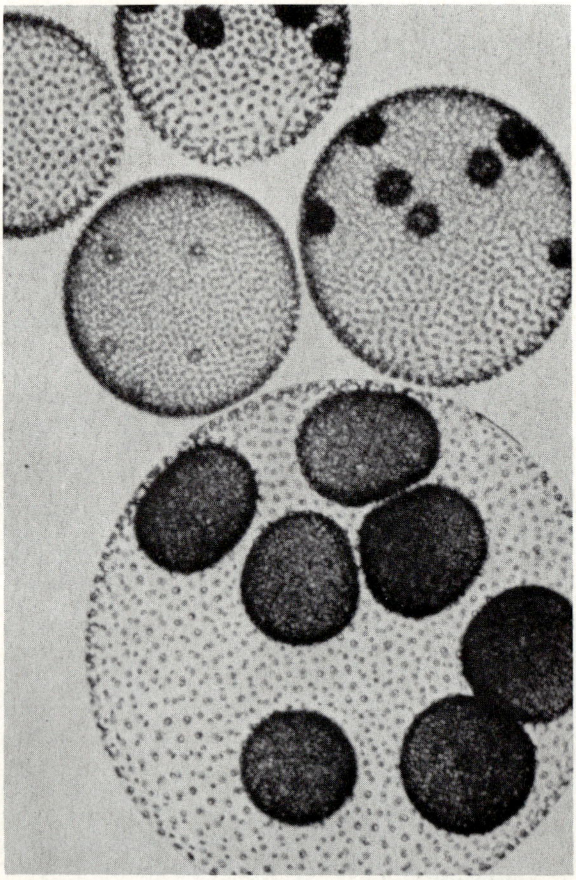

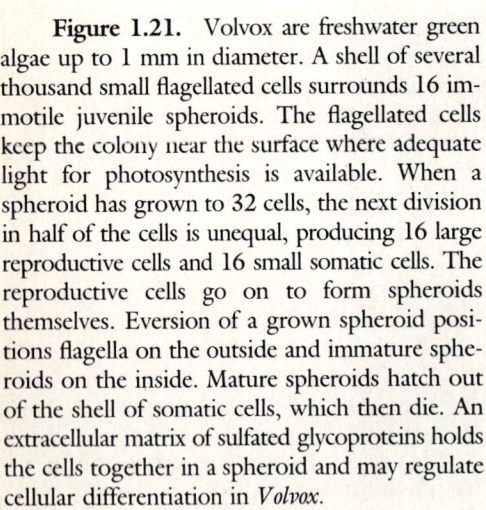

Figure 1.21. Volvox are freshwater green algae up to 1 mm in diameter. A shell of several thousand small flagellated cells surrounds 16 immotile juvenile spheroids. The flagellated cells keep the colony near the surface where adequate light for photosynthesis is available. When a spheroid has grown to 32 cells, the next division in half of the cells is unequal, producing 16 large reproductive cells and 16 small somatic cells. The reproductive cells go on to form spheroids themselves. Eversion of a grown spheroid positions flagella on the outside and immature spheroids on the inside. Mature spheroids hatch out of the shell of somatic cells, which then die. An extracellular matrix of sulfated glycoproteins holds the cells together in a spheroid and may regulate cellular differentiation in *Volvox*.

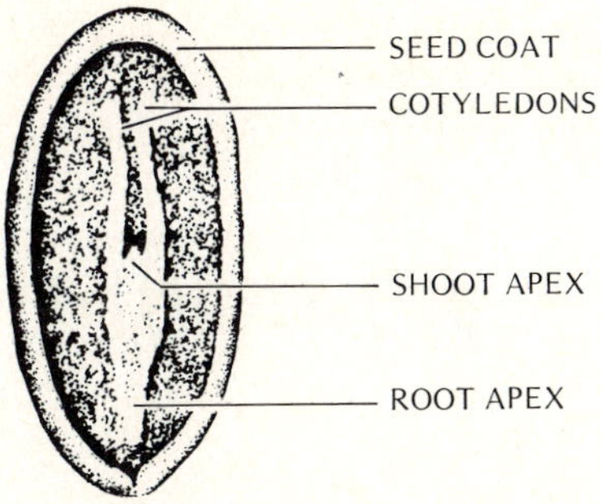

Figure 1.22. Gymnosperm seed. A seed coat protects the embryo. Pines have two or more cotyledons. The shoot and root apices are present in the embryo. Most of the new tissue will grow from these meristematic tissues.

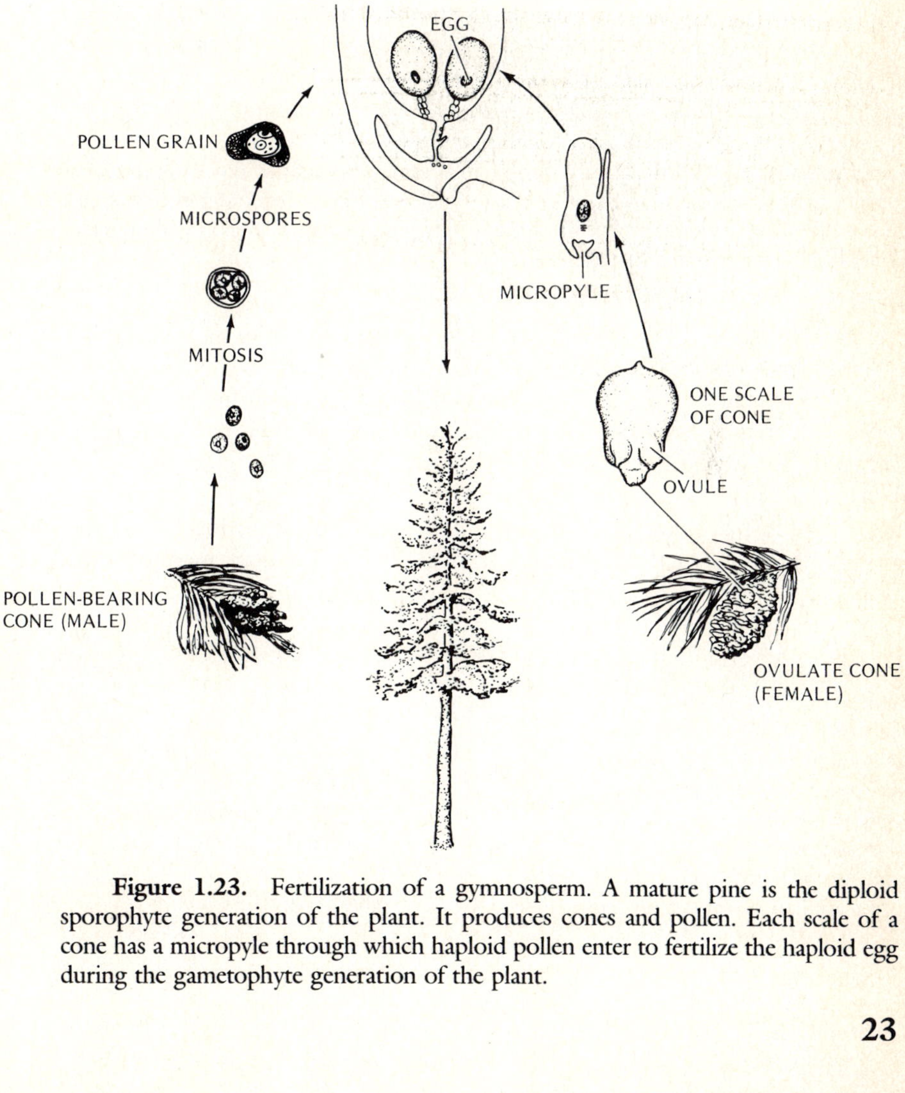

Figure 1.23. Fertilization of a gymnosperm. A mature pine is the diploid sporophyte generation of the plant. It produces cones and pollen. Each scale of a cone has a micropyle through which haploid pollen enter to fertilize the haploid egg during the gametophyte generation of the plant.

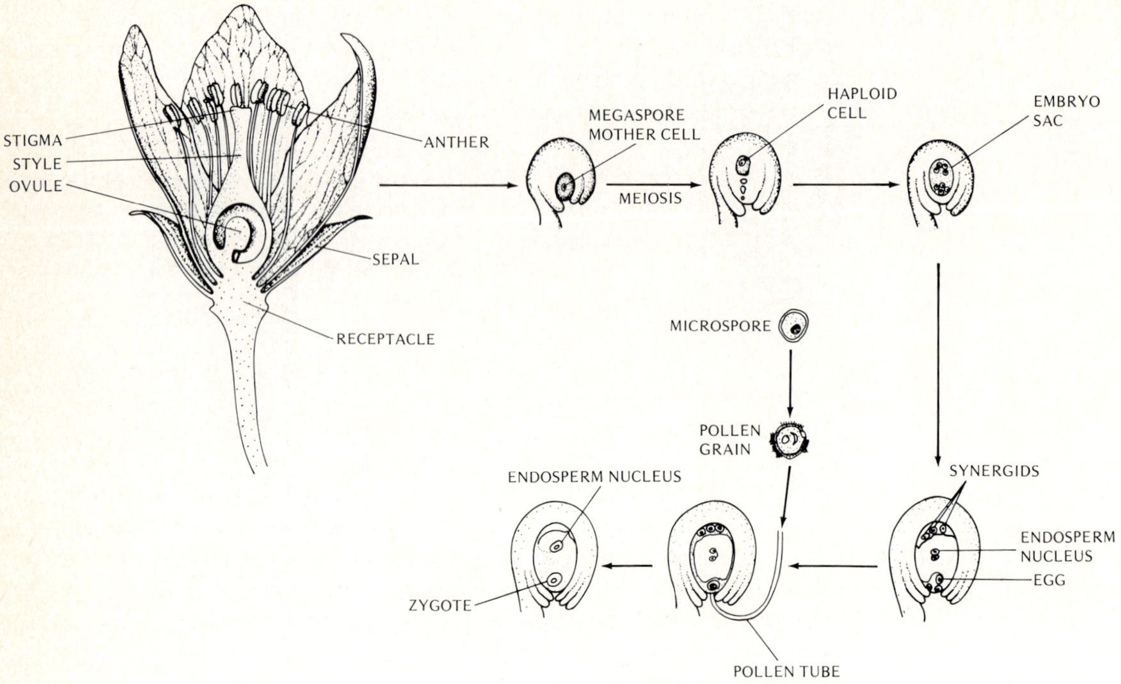

Figure 1.24. Fertilization of an angiosperm. The mature flower is the sporophyte generation of the plant. It produces male and female haploid gametes of the gametophyte generation. Fertilization produces new zygotes that develop within seeds. Pollination is described in detail in Chapter 7.

asymmetry. The end of the embryonic mass nearest the micropyle elongates to form a suspensor. The suspensor attaches to the seed wall and positions the growing embryo. At the opposite end, the cells grow out to form the cotyledons that will nourish the plantlet following germination of the seed. Between the cotyledons is positioned the shoot apex. The root grows out of the tissue just above the suspensor (Fig. 1.25).

In some plant embryos the shoot apex puts out small leaves and the root extends somewhat while still within the seed. However, most of mor-

Figure 1.25. Seed formation in angiosperms. Division of the zygote gives rise to the embryo, which is pushed into the endosperm by elongation of the suspensor. Triploid endosperm cells grow and divide up the endosperm. Seed coats are laid down within the ovary wall. Fruits have enlarged ovary walls.

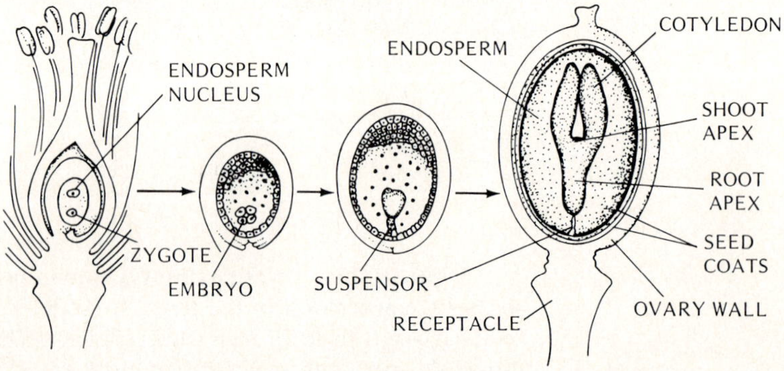

phogenesis occurs following germination. Cell-type differentiation, growth, and expansion of the cells is controlled in seedlings by light and the plant hormones, auxin, gibberellic acid, ethylene, and various cytokinins. Ethylene suppresses leaf growth until the seedling emerges into the air. Auxin is involved in the geotropic and geophobic growth of the roots and the shoots, respectively in seedlings. These are processes of plant physiology rather than of embryogenesis.

Organism Cloning

Plant embryos are much less highly structured than animal embryos, reflecting the simplified tissue arrangements found in somatic tissues of plants. Early embryos of gymnosperms and angiosperms are not much more than masses of cells. Perhaps for this reason, considerable success has been achieved in growing whole plants from single cells of adult plants (Fig. 1.26).

Stem pith has been removed from tobacco plants *(Nicotiana tobacum)* and placed in tissue culture media. The cells grow to form amorphous calluses under these conditions. Single cells can be shaken free from these calluses and grown in isolation. The resulting clumps of cells are each derived from a single cell. They will form plantlets complete with roots and shoots when placed on gelled medium. Complete flowering plants will grow from the plantlets if they are transferred to soil in a flowerpot (Fig. 1.27).

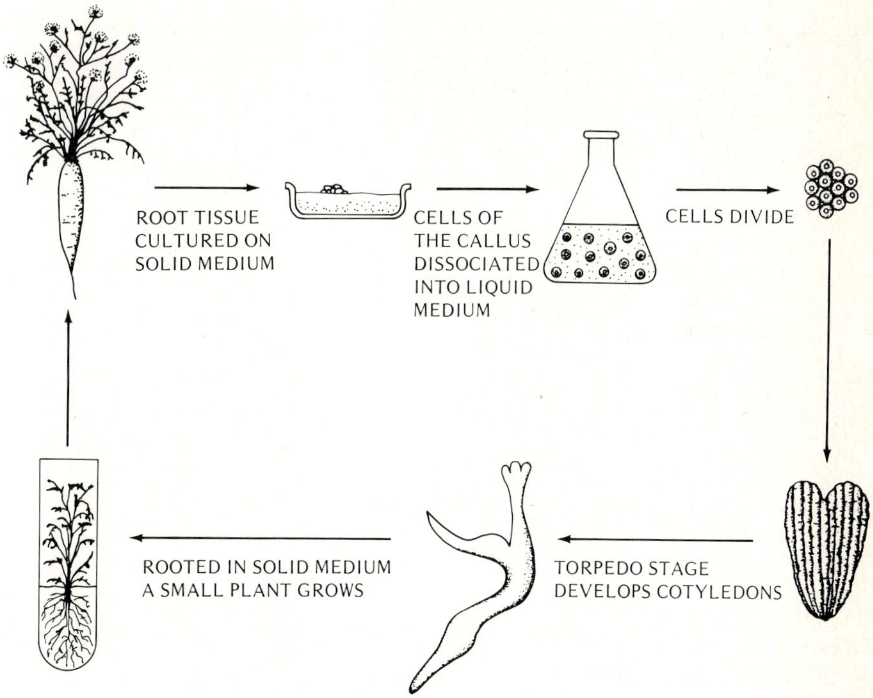

Figure 1.26. Somatic regeneration of a carrot. Phloem cells are specialized to transport nutrients in plants but can be isolated from a mature root and grown to form a callus. Single callus cells grow in suspension to form a new embryo that can be planted. A mature plant complete with roots and shoots will grow out.

Figure 1.27. Tobacco plant grown from a single adult cell. Single cells are first grown in microwells to form a clump (A–C). On a solid support, the clump sends up a shoot (D). When this is transferred to soil, a complete flowering plant will grow (E, F). The proportions of the organs that differentiate from a clump can be altered by varying the pH, the ratio of auxin to cytokinin, or adding certain oligosaccharides derived from cell walls.

The growth of a complete organism from a single differentiated cell dramatically shows that all the genes necessary for the formation of any tissue are present and available in a specific differentiated cell. Only some of the genes are active in any one tissue. A central question to developmental biology is what controls the various sequential patterns of gene expression.

Related Readings

Balinsky, B. (1981). An introduction to embryology (5th edition). W. B. Saunders, Philadelphia.

Brady, T., and Walthall, E. (1985). The effect of suspensor and gibberellic acid on *Phaseolus vulgaris* embryo protein content. Devel Biol. *107,* 531–536.

Browder, L. (1984). Developmental biology (2nd edition). W. B. Saunders, Philadelphia.

Darvill, A., Gollin, D., Chelf, P., and Albersheim, P. (1985). Manipulation of the morphogenetic pathways of tobacco explants by oligosaccharides. Nature *314,* 615–617.

Gilbert, S. (1985). Developmental Biology. Sinauer Assoc., Sunderland, Mass.

Grabowski, C. (1983). Human reproduction and development. W. B. Saunders, Philadelphia.

Green, P. (1980). Organogenesis—a biophysical view. Ann. Rev. Plant Physiol. *31,* 51–82.

Keller, R. (1979). Vital dye marking of the gastrula and neurula of *Xenopus laevis*. Devel. Biol. *51,* 118–137.

Lewin, B. (1985). Genes (2nd edition). John Wiley and Sons, New York.

Mayr, E. (1983). The growth of biological thought. Harvard University Press, Cambridge.

Moore, K. (1977). The developing human (2nd edition). W. B. Saunders, Philadelphia.

Morgan, T. H. (1926). The theory of the gene. Yale Press, New Haven.

Nichols, S., and Laties, G. (1984). Ethylene regulated gene transcription in carrot roots. Plant Mol. Biol. *3,* 393–401.

Ray, P., Steeves, T., and Fultz, S. (1983). Botany. W. B. Saunders, Philadelphia.

Sambrook, J. (ed.) (1985). Molecular Biology of Development. Cold Spring Harbor Symposium Vol. 50.

Spemann, H. (1938). Embryonic development and induction. Yale University Press, New Haven.

Sumper, M. (1979). Control of differentiation in *Volvox carteri:* a model explaining pattern formation during embryogenesis. FEBS Letters *107,* 241–246.

Thimann, K. (1977). Hormone action and the life of the whole plant. University of Massachusetts Press, Amherst, MA.

Wilson, E. B. (1925). The cell in development and heredity. Macmillan Publishing Co., New York.

PART I

The Mechanisms

Development can be thought of as the coordinated sequence of changes in the individual cells. At about the time of gastrulation, embryonic cells differentiate to specialized functions as a consequence of regulated gene expression. Later in development, definitive tissues can be seen in which specific gene products have accumulated to carry out a given role. Components that determine cell-type-specific function are often recognized in adult tissues and then traced back to their appearance during development. The various mechanisms of differential gene expression have been found to have much in common, so that grasping the details in a dozen or so cell types gives insight as to how other differentiated cells probably arise. Likewise, the ways in which specific proteins adapt a given cell type to a unique role in the body are not so diverse that each must be known in detail in embryos of different organisms. For instance, muscle cells differentiate pretty much the same way in worms, flies, chickens, and humans.

This section presents the evidence that all cells in an organism carry the same repertory of genes but that cells of one tissue chose one set while cells of another tissue chose another set. The sets often overlap, but some genes are expressed in only one cell type. These genes have come under the closest scrutiny, such that in many cases the complete nucleotide sequence is known and the function of specific nucleic acid bases clearly demonstrated. The enzymatic and structural specializations that characterize certain terminally differentiated cells are presented so that the goals of embryogenesis can be clearly seen.

CHAPTER 2

Constancy of the Genome

The diversity of cell types is easily recognized in the microscope. For instance, mammalian red blood cells are small and enucleated, muscle cells are multinucleated and striated, whereas cells along the avian oviduct are columnar and adapted for secretion. Biochemical analyses have shown dramatic accumulation of hemoglobin in red blood cells, specific actins and myosins in muscle cells, and ovalbumin in oviducts. These proteins are synthesized only in these specific cell types. Yet it is now known that the genes coding for these specialized proteins are found in all the cells of the embryo. Some mechanism must result in expression of the globin, actin, myosin, and ovalbumin genes in the appropriate cell type and their repression in all other cell types.

Up until about 100 years ago, some embryologists seriously considered that genes for certain processes were selectively positioned only in those precursor cells that would give rise to the tissue type that carried out the specific process. However, shortly after it was recognized that genes are arranged on chromosomes, it was also realized that highly differentiated cells such as those of salivary glands have the same number of chromosomes as a fertilized egg (Fig. 2.1). Therefore, it was thought likely that salivary and other differentiated cells have the same genes as all other cells in the organism. In some cases, including the salivary glands of *Drosophila,* replication of the chromosomes continues after cell division has stopped. This results in highly polytene cells. But it appears that the full genetic complement is present in these cells.

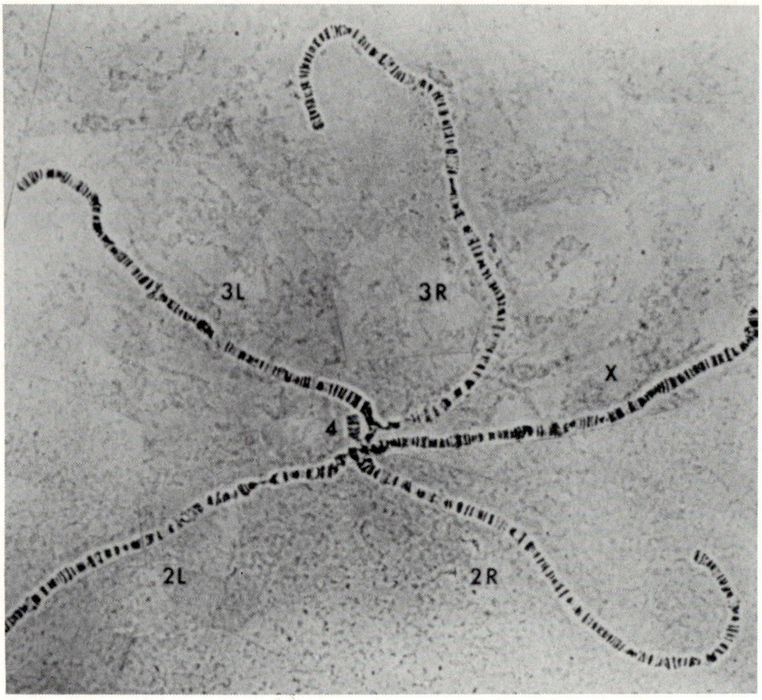

Figure 2.1. Polytene chromosomes of *Drosophila* salivary gland. The centromeres of each of the four chromosomes aggregate in the middle. Both chromosomes 2 and 3 have centrally located centromeres and so have two arms designated L *(left)* and R *(right)*. The X chromosome has its centromere at one end. The small fourth chromosome is located near the middle. Bands can be seen at invariant positions along each chromosome because of variations in the degree of compaction of DNA. Genes can be accurately mapped relative to the banding pattern.

Polytene cells containing multiple copies of the genome are fairly rare, and it has been shown that the great majority of differentiated cells have not only the same number of chromosomes, but the same amount of total nuclear DNA. Further confidence that little or none of the genetic information present in DNA is lost during embryogenesis comes from detailed quantitative reannealing studies. These techniques rely on the ability of a short (~400 base) sequence of DNA to reanneal only with its complementary strand under stringent conditions. Thus, reannealing depends on the potential genetic information present in the exact sequence of the four bases—adenine (A), guanine (G), cytosine (C), and thymine (T). DNA isolated from eggs, sperm, or differentiated tissue will form double-stranded DNA molecules with DNA from other tissues of an individual of the same species when the strands are separated and allowed to come together again randomly. These studies show that greater than 95 percent, and possibly all of the sequence complexity, is conserved throughout embryogenesis. Similar studies with isolated genes, such as those for globin or ovalbumin, have directly shown that they are present in liver, brain, pancreatic, and other cells but are not expressed in those tissues.

There are a few exceptions to the general rule of the constancy of the genome during development, but they are rare and can still be counted on the fingers of one hand. There are cases of precise rearrangement of the genome in a single differentiated cell type; these include the immunoglobin genes in the cells of the specialized immunity system of vertebrates that are described in more detail in the Specific Example of this chapter, and the genes coding for surface antigens of African trypanosomes. There are also cases of specific amplification of genes necessary for the differentiation of oocytes and follicular cells that are discussed in Chapter 6. In all other cases of cell differentiation we are left with the puzzle that all the genes are present but only some are working.

Nuclear Transplantation

Perhaps the most direct test of the completeness of the genome in a differentiated cell is to take out the nucleus and put it into an enucleated egg. If this nucleus can then direct the development of a normal embryo, it clearly carries all the necessary genetic information in a readily available form. Briggs and King initiated these studies in the early 1950s using frog blastula cell nuclei transplanted into frog eggs from which the nucleus had been removed. Over half the transplants developed into feeding tadpoles. Clearly, most nuclei of blastula cells of frogs are totipotent at least up to the tadpole stage. Although if left in place they would have given rise to only a certain subset of tissues, when placed in a fresh egg cell they could give rise to all the normal cell types. These experiments ran into difficulties when nuclei of cells past the gastrula stage were used. However, there are a variety of technical explanations for the lack of success, unrelated to genome potential, such as the slow rate of cell division in differentiated cells compared with that during cleavage. Moreover, nuclei of a rapidly growing renal tumor were found to be capable of directing the development of enucleated eggs to form tadpoles a small percentage of the time.

These experimental tests of the totipotency of nuclei in differentiated cells are so important that extreme efforts have gone into improving them. Gurdon has used eggs of the South African clawed toad, *Xenopus laevis,* for experiments involving the transfer of nuclei from highly specialized gut epithelial cells. In a few cases success was achieved in getting development not only to the tadpole stage but through metamorphosis to adult toads. These results clearly show that at least some tadpole gut epithelial cells can provide the genes for making all the different cell types of an adult. The experimental design is sketched out in Figure 2.2.

Recently, it has been shown that the genome present in mature frog erythrocyte nuclei carries all the genes necessary for embryogenesis and can reactivate these genes at the proper time and in the proper tissues. However, special techniques had to be used to prepare the nuclei of nongrowing, terminally differentiating blood cells such that they could undergo the rapid divisions that occur during the early cleavage stages (Fig. 2.3). Erythrocytes were lysed and their nuclei injected into immature oocytes. Within the next 24 hours the oocyte was induced with progesterone to complete meiosis and mature. At that stage the host egg nucleus was removed and the egg activated. The erythrocyte nuclei replicated and allowed the formation of a blas-

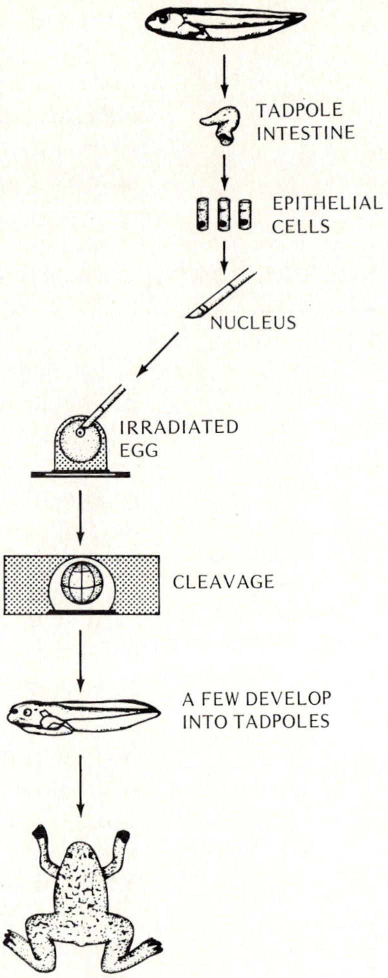

Figure 2.2. Cloning a toad. Rapidly growing cells of tadpole intestine can be used to provide nuclei to eggs in which the resident nucleus has been killed by ultraviolet (UV) irradiation. The injected *Xenopus laevis* egg divides to form a blastula. Nuclei isolated from these blastula cells will give rise, on rare occasions, to normal development of mature frogs when injected into UV-irradiated eggs. To be sure that the nuclei of such toads were from the donor intestinal cells, the donor individuals carried a nuclear marker (1-nucleolus). The cloned toads were found to carry this marker.

tula. Nuclei from single blastoderm cells were then injected into enucleated eggs. Half of the clones were found to develop to the swimming tadpole stage. The most advanced tadpoles swam vigorously and their hearts beat regularly. The eyes fashioned lenses, neural and pigmented retinas, while the gut differentiated into pharynx, stomach, liver, and hindgut. Clearly, the genes necessary for tissue differentiation in a tadpole had been lying quiescent in the erythrocyte genome and only needed the signals present in maturing eggs and embryos to be reactivated.

The erythrocytes used in this experiment had accumulated large amounts of hemoglobin, yet their genomes were fully capable of making all the other proteins of an embryo. During their terminal differentiation, erythrocytes had preferentially expressed their globin genes to the exclusion of almost all other genes. The mechanisms that result in the choice of which genes are expressed and which repressed underlie much of development.

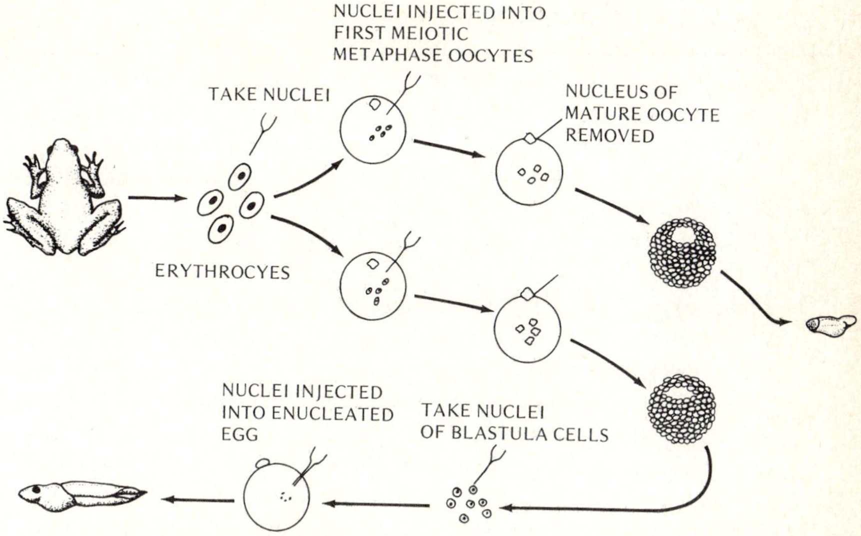

Figure 2.3. Tadpoles from erythrocyte nuclei. Oocytes of *Rana pipiens* were injected with blood cell nuclei and allowed to mature for a day. The resident egg nucleus was then removed. Development then proceeded to the prehatched tadpole stage *(upper)*. Nuclei from some of the nuclear transplant blastulae were injected into enucleated eggs. Half of the clones developed into swimming tadpoles.

Differential Gene Expression

The techniques of molecular biology have been used during the last ten years to determine the steps in macromolecular synthesis at which the expression of genes present in all cells is restricted to a few cell types characterized by accumulation of specific gene products. To a large extent this approach has depended on the ability to incorporate fairly small fragments of the genome into vectors able to replicate in bacteria, such as plasmids and bacterial viruses. By recognizing a sequence not much larger than a gene carried by a cloned vector, the DNA which codes for a developmentally regulated product can be isolated in sufficient amounts for biochemical use. And the purified sequence is free of all other sequences found in the embryo. This gene or gene fragment can be used to hybridize to complementary sequences found in the mRNA of that gene. These techniques have shown that the accumulation of specific proteins is most often regulated by controlling the transcription of mRNA from the specific gene. However, this is only one of the steps in the biosynthesis of a protein that is potentially subject to specific regulation. A short list is given in the following section of the steps known to be regulated for one gene or another.

Steps in Gene Regulation

1. **Initiation of transcription.** The association of RNA polymerase with the start of a given gene is the most commonly regulated step. Sequences near or within genes bind specific proteins, which then increase or decrease the probability that RNA polymerase will start to catalyze the polymerization of an mRNA copy of the gene.

2. **Processing of the mRNA.** The primary transcript of most eukaryotic genes must be enzymatically altered before it can function in protein synthesis to direct the translation of a specific protein. Although many of these modifications occur to almost all mRNA molecules, such as addition of a modified nucleotide base at the start of the molecule (5′ capping) and the addition of several hundred adenine residues to the other end (3′ poly-A), others such as removal of sequences within the transcript (intron excision and splicing) have been found to be developmentally regulated. The mRNA molecules that result from altered splicing patterns direct the synthesis of distinctly different proteins.

3. **Stability of mRNA.** Different gene products have significantly different half-lives, resulting in dramatic differences in the levels to which they accumulate. Two genes may be transcribed at the same rate but their mRNAs will accumulate to very different levels if one is broken down twice as rapidly as the other. These differences in stability are mostly determined by the nucleotide sequences built into the specific mRNA molecules. However, under several conditions it has been found that the half-life of a given mRNA depends on the differentiated state of the cell.

4. **Transport to the ribosomes.** mRNA molecules are synthesized in the nucleus but must be transported to the cytoplasm before they can direct protein synthesis on ribosomes. Although the evidence is somewhat murky at present, it appears that the transport of some specific mRNA sequences is more efficient at particular stages of development than at others. Such a selective process at the nuclear pores could clearly have significant effects on the rate at which given proteins are synthesized.

5. **Translational efficiency.** The probability of an mRNA associating with ribosomes and directing the translation of a specific protein depends to a large extent on the nucleotide sequence of the mRNA molecules. Thus, an efficiently translated mRNA will direct the synthesis of many more copies of its protein than will a less efficiently translated mRNA even if the abundance of each mRNA is identical. However, the sequence of the mRNA is built into the DNA sequence of the gene. Changes in the efficiency of translation of given

mRNA molecules that depend on the state of differentiation of the cell have been reported, but the mechanisms for such regulation are far from clear.

6. **Protein stability.** Finally, it should be remembered that proteins do not usually last as long as the life of the cell. Many proteins are innately unstable and accumulate to a lesser extent than more stable ones. There are cases in which the stability of specific proteins differ markedly in different cell types. Moreover, many proteins are secreted into vesicles or out of the cell altogether at rates dependent on the differentiation stage. The routing of proteins is often mediated by post-translational modifications such as the covalent attachment of carbohydrate side chains or the polypeptide ubiquitin to specific proteins. These processes clearly affect the level to which different proteins will be found to accumulate in different cell types. In some cases we are able to follow a sequence of particular molecular interactions that account for differential gene expression whereas in others we may know only the broad outlines of the mechanisms for several years to come.

Summary

In all but a few organisms the full diploid complement is present in cells of every tissue. Transplantation of nuclei from highly differentiated cells to oocytes or eggs has shown that in some cases, at least, no essential genetic information has been lost during embryonic development. Nuclei of some adult cells can direct all of normal embryogenesis if properly conditioned. Divergence of cell types is the consequence of differential gene expression rather than preferential gene retention. Gene expression is most commonly regulated by controlling the rate of transcriptional initiation. However, the expression of some genes is regulated post-transcriptionally and others post-translationally.

SPECIFIC EXAMPLE

Rearrangement of Immunoglobin Genes

It is estimated that mammals generate several million different antibodies that protect them from infections. Rather than carry several million genes, each coding for a different antibody, a combinatorial system has evolved by which the millions of antibodies are coded for by only a hundred or so genes. The mix and match process involves the subunits of antibodies as well as the DNA sequences coding for different portions of the subunits themselves. This is one of the rare exceptions in which the primary sequence of

DNA is altered during differentiation. The techniques developed to show rearrangement of immunoglobin genes have been used to show that other genes are not rearranged during differentiation in vertebrates.

Antibodies are composed of four polypeptides, two identical subunits of 56,000 daltons (56 kd) and two identical subunits of 26,000 daltons (26 kd) (Fig. 2.4). The larger subunits are referred to as heavy chains and the smaller subunits are referred to as light chains.

There are thousands of different heavy chains and thousands of different light chains. They can randomly associate with each other and thus generate the product of their diversity ($10^3 \times 10^3 = 10^6$ combinations). The specificity of antibody binding is dependent on the amino acid sequence in the N-terminal portion of both the immunoglobin light chains and the immunoglobin heavy chains. Thus, the products of a few thousand genes can generate a few million different antibodies by a combinatorial process.

There are tumors in mice derived from antibody-producing cells. These plasmacytoma cells can be grown in the laboratory and are found to produce copious amounts of a single antibody. Although a mouse produces millions of different antibodies, a plasmacytoma cell line produces only a single one. This and other data indicate that each differentiated lymphocyte in the mouse produces only a unique immunoglobin heavy chain and a unique immunoglobin light chain.

Amino acid sequencing of a large number of unique light and heavy chains produced by different plasmacytoma cell lines showed that most of

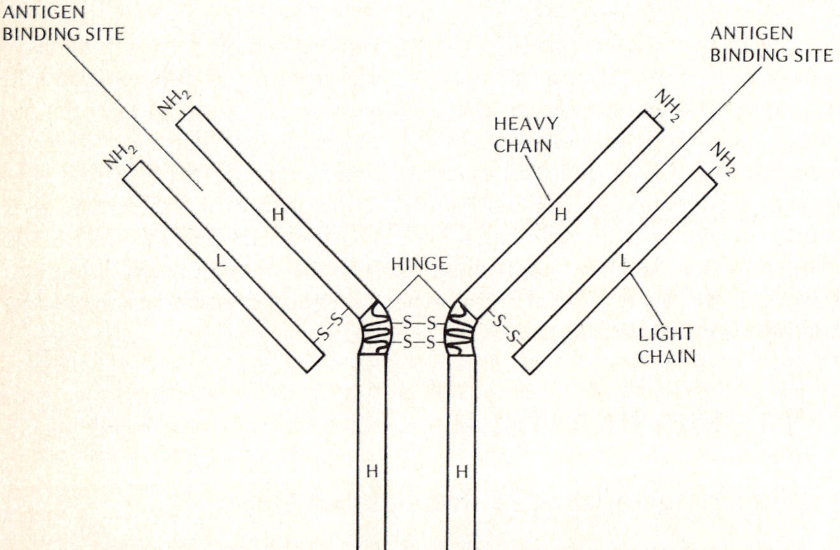

Figure 2.4. Antibody molecules are composed of two identical heavy chains *(H)* and two identical light chains *(L)* held together by disulfide bonds. Antigens are held in sites formed by the amino terminal (NH_2) regions of both heavy and light chains.

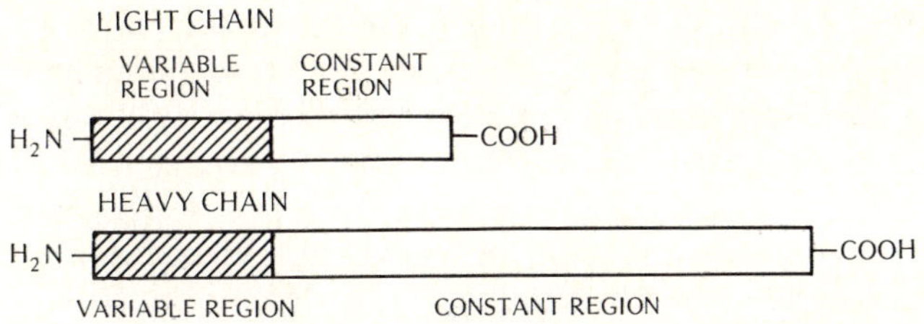

Figure 2.5. Constant and variable regions of immunoglobins. Almost all the differences between various immunoglobins are found in the first 100 amino acids at the amino-terminus. Light chains have a constant region of 115 ± 5 amino acids, whereas heavy chains have a constant region three or four times as long depending on the class of antibody.

the differences between different chains were clustered in the N-terminal portions of both heavy and light chains (Fig. 2.5).

The surprising conservation of the carboxy-terminal portions of different immunoglobin chains raised the possibility that the variable and constant regions might be coded for by different segments of DNA and then somehow joined together by a covalent peptide linkage. There were three likely possibilities: (1) the joining of separate polypeptides, (2) the joining of separate mRNA molecules, and (3) the joining of separate DNA sequences. When these possibilities were first being considered, there was no evidence for such processes ever occurring. But it was worth looking for.

Translation of purified plasmacytoma mRNA *in vitro* using the protein synthetic components of reticulocyte cells showed that complete heavy and light chains were made. There were no separate variable or constant polypeptides. These results ruled out the joining of finished polypeptides.

The next step involved purifying the mRNA for a single light chain from a specific plasmacytoma, MOPC321. This cell line produces a light chain of the κ class, as defined by the amino acid sequence of the constant region. Proteins are synthesized from their N-terminus to their C-terminus by translating their mRNA from the 5′ end of the RNA toward the 3′ end. Thus, the 5′ half of an mRNA molecule codes for the N-terminal half of a protein and the 3′ half codes for the C-terminal half. The κ light chain mRNA was cleaved in the middle and the 3′ half isolated. Then the whole mRNA as well as the separated 3′ half were radioactively labeled so that hybridization to the DNA sequences that code for them could be measured.

Two chains of nucleic acid will bind to each other if their sequences are complementary. The adenine (A) bases will bind to thymine (T) bases and the guanine (G) bases will bind to cytosine (C) bases. Since both DNA and RNA are linear arrays of only four bases, a strand of RNA will form a hybrid with a complementary strand of DNA. Likewise, two complementary DNA strands will hybridize. But the sequences must be exactly complementary; for example, AGGCTTA will hybridize with TCCGAAT but not with TGGAATT or any other sequence. About 30 bases must be complementary to form a stable hybrid molecule. Because any of 4 bases can occur at any position, a run of 30 bases will occur by chance only once (at most) in the whole genome. The conditions for recognizing stable hybrids can be adjusted to be very stringent or to be less stringent and allow mismatched hybrids to be seen. Usually high stringency is used so that only the specific DNA sequence that originally coded for the mRNA can be recognized. This technique was used to recognize the DNA that codes for the κ light chain.

The DNA in chromosomes is millions of bases long. The question was not whether the variable and constant regions were coded for on the same chromosome but whether they were coded for by two genes or a single one. Luckily, there is a way to cut a long stretch of DNA into gene-sized pieces.

Restriction Enzymes and Southern Blotting

Bacteria are always susceptible to infection by viruses and stray pieces of DNA from other organisms. The bacteria protect themselves with enzymes that immediately break down foreign DNA. These restriction endonucleases recognize four or six base sequences and cleave the DNA within these sites. Each species of bacteria has an enzyme that recognizes a particular sequence, and the restriction enzymes differ in the sites they recognize. There are more than a hundred different restriction enzymes available now that cleave at different sites. Those enzymes that recognize four bases, such as MboI (GATC), cleave on the average every 256 bases ($4^4 = 256$). Those that recognize six bases, such as BamH1 (GGATCC) cleave on the average every 4096 bases ($4^6 = 4096$). Of course, these sites are not completely randomly distributed along the DNA and, therefore, DNA fragments of many different sizes are produced when a long strand of DNA is cleaved by restriction enzymes. Since most genes are a few thousand bases long, the restriction fragments are of the right size to separate genes.

DNA fragments that range in size from 1000 bases to 20,000 bases [1 kilobase (kb) to 20 kb] can be separated from each other by electrophoresis in 0.6 percent agarose gels. The polygalactose chains of agarose slow up the migration of long DNA fragments more than they slow up shorter chains, and so after a few hours of electrophoresis the fragments are separated as the smaller ones have migrated further than the larger ones. The distance between specific restriction sites along the DNA chain determines the size of the fragment in which a given sequence lies.

Once the DNA is cleaved by a restriction enzyme and the fragments separated by size, the DNA can be transferred to nitrocellulose, where it will be immobilized. The sheet of nitrocellulose with bound DNA fragments on it is then used in hybridization experiments to recognize specific sequences. This technique is named Southern blotting after its developer, Edward Southern.

DNA was isolated from whole embryos and treated with the restriction enzyme BamH1. The fragments were separated and a Southern blot was prepared. When the blot was hyridized to the intact mRNA for κ light chain, two fragments lit up: one of 6.5 kb and one of 9.2 kb. However, when the 3′ half that codes for the constant region was used to hybridize to the Southern blot, only the 9.2-kb fragment lit up (Fig. 2.6). The DNA sequence coding for the constant region seemed to be separated from the DNA sequence coding for the variable region by a BamH1 site.

When this same experiment was repeated using DNA of the plasma-

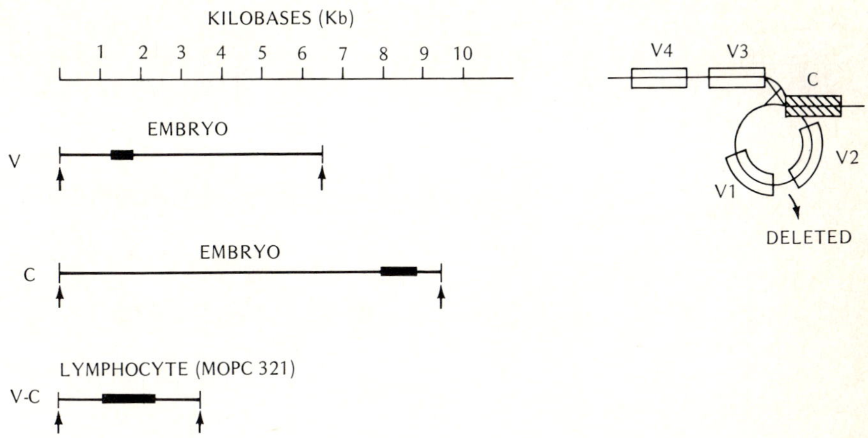

Figure 2.6. Rearrangement of immunoglobin genes. In mouse embryos the sequence coding for the variable region (V) of the κ light chain is found on a 6.5 kb BamH1-cleaved DNA fragment while the sequence coding for the constant region (C) is found on a 9.2 kb BamH1 fragment. In cells that have differentiated so as to produce antibody (MOPC 321), both sequences are found together on a 3.5 kb BamH1 fragment, showing that there had been rearrangement of the genome during this differentiation. There is evidence that a large (>100 kb) piece of DNA separating the V and C regions in uncommitted lymphocytes is deleted by a specific recombination event during differentiation that brings the V and C regions together. A cell has several variable regions to chose from.

cytoma producing κ light chains, very different results were found. Both intact κ chain mRNA and its 3′ half hybridized to a single 3.5 kb fragment (Fig. 2.6). The BamH1 site between the constant and variable coding sequences had been lost in the DNA of these differentiated cells. And the flanking BamH1 sites were closer by 5.7 kb. Somehow the DNA was rearranged in this region of the genome to bring the sequences coding for the variable and constant regions together. The most likely mechanism for rearrangement involves site-specific recombination of the variable *(V)* region and constant region *(C)* to form the *V–C* joint. The DNA that separated the two regions in the genome appears to be deleted in the differentiated cell.

Similar studies on more than 100 other mRNAs isolated from differentiated cells have turned up only a few other cases of regulated gene rearrangement. One interesting case involves the genes coding for surface proteins in African trypanosomes. These parasites elude the immune system of their hosts by a mechanism similar to that used in the immune system itself. In mammals, gene rearrangement is known only to affect the immunoglobin genes and the lymphocyte receptor that elicits an immune response. The evolution of large, long-lived organisms required the formation of a versatile defense mechanism. Rather than evolve thousands of genes for different immunoglobins, mechanisms arose to take a few genes and combine them in various ways to produce the millions of different antibodies.

There is more than one variable gene in the genome that can be joined to the constant κ gene. In the mouse genome there are over a dozen V_k

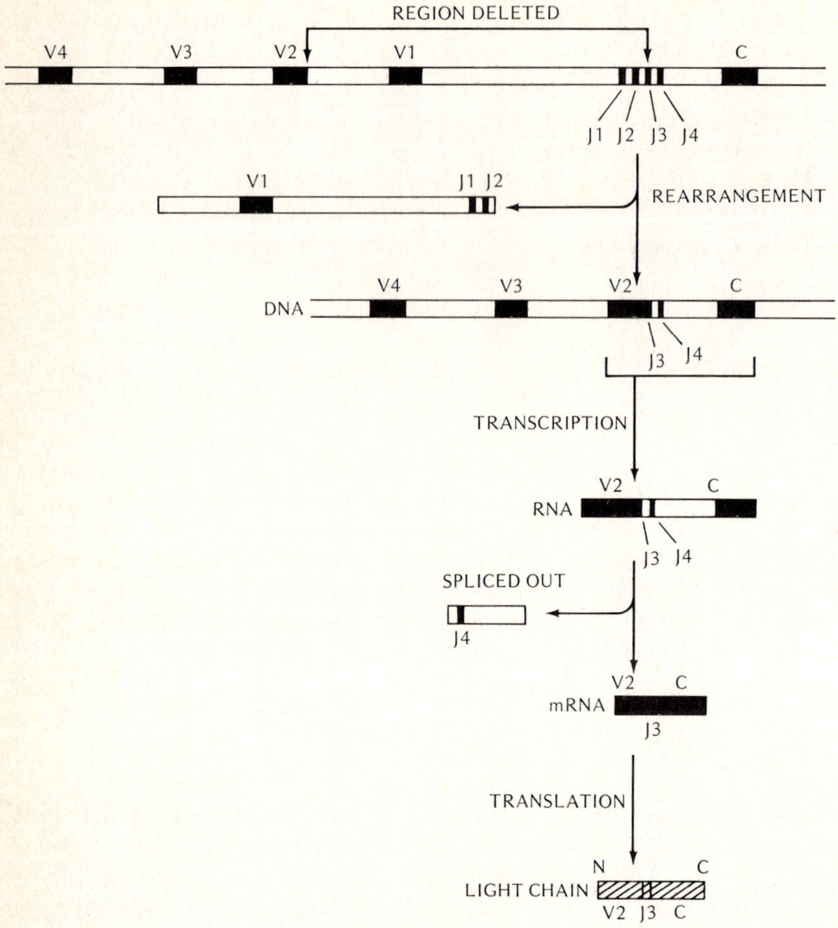

Figure 2.7. Steps in making a specific immunoglobin. Recombination fuses a chosen V region (V2 in this case) with a chosen joint regoin (J3 in this case). There are four J regions close to the constant region of mouse κ chain (C). The intervening DNA is discarded during B-lymphocyte differentiation. Transcription starts 5′ to the V region and reads through the joint regions to the end of the C region. Intervening sequences containing extra joint regions are cleaved from the transcript and the ends of the RNA ligated. Processed mRNA is then translated into a specific κ light chain.

genes and a single C_k gene on chromosome 16. During differentiation the V_k genes are rearranged to come closer to the constant gene. They actually recombine with one or another of four small (13 amino acid) *J* regions that are just upstream (going toward the 5′ end) of the sequence coding for the constant region. A chosen *V* gene might recombine with J_3, deleting J_1 and J_2. Transcription starts upstream of the *V* region, proceeds through J_3 and J_4, and on down past the *C* region (Fig. 2.7). The RNA sequences separating the end of J_3 and the beginning of *C* are subsequently cut out, and the ends of the RNA are spliced to give the final *V–J–C* sequence of the mRNA for κ light chain. Thus, both the joining of RNA molecules and the joining of separate DNA sequences are involved in producing an immunoglobin mRNA.

This combinatorial approach to diversity can choose any of a dozen or so *V* genes, and any of four *J* genes to attach to the C_k gene. Somatic mutations in the *V* genes and alternate recombinational sites in the *J* genes produce still more diversity. The same processes of recombination of diverse sequences rearrange the genes for the variable and constant regions of the heavy chains and other classes of light chains. This is a highly evolved and efficient set of steps to use available genomic complexity to produce a vast array of different proteins. But in vertebrates it seems to occur only during differentiation of the immune cells. In the rest of the cells the genome is constant.

Related Readings

Alberts, B., Bray, D., Lewis, J., Raff, M., Roberts, K., and Watson, J. (1983). Molecular biology of the cell. Garland, New York.

Born, W., Yague, J., Palmer, E., Kappler, J., and Marrack, P. (1985). Rearrangement of T-cell receptor β-chain genes during T-cell development. Proc. Nat'l Acad. Sci. *82,* 2925–2929.

Briggs, R., and King, T. (1952). Transplantation of living nuclei from blastula cells into enucleated frogs' eggs. Proc. Natl. Acad. Sci. *38,* 455–463.

Britten, R., and Kohne, D. (1968). Related sequences in DNA. Science *161,* 529–540.

DiBernardino, M., Hoffner, N., and Etkin, L. (1984). Activation of dormant genes in specialized cells. Science *224,* 946–952.

Foster, J., Stafford, J., and Queen, C. (1985). An immunoglobin promotor displays cell-type specificity independently of the enhancer. Nature *315,* 423–425.

Gurdon, J. (1962). The developmental capacity of nuclei taken from intestinal epithelium cells of feeding tadpoles. J. Embryol. Exp. Morph. *10,* 622–640.

Hozumi, N., and Tonegawa, S. (1976). Evidence for somatic rearrangement of immunoglobulin genes coding for variable and constant regions. Proc. Natl. Acad. Sci. *73,* 3628–3632.

Lewis, S., Gifford, A., and Baltimore, D. (1985). DNA elements are asymmetrically joined during the site-specific recombination of kappa immunoglobin genes. Science *228,* 677–685.

Manser, T., Huang, S., and Gefter, M. (1984). Influence of clonal selection on the expression of immunoglobulin variable region genes. Science *226,* 1283–1288.

McClintock, B. (1984). The significance of responses of the genome to challenge. Science *226,* 792–801.

McConaughty, B., and McCarthy, B. (1970). Related base sequences in the DNA of simple and complex organisms. Biochem. Genet. *4,* 425–446.

Myler, A., Allison, J., Agabian, N., and Stuart, K. (1984). Antigenic variation in African trypanosomes by gene replacement or activation of alternate telomeres. Cell *39,* 203–211.

Perlmutter, R., Kearney, J., Chang, S., and Hood, L. (1985). Developmentally controlled expression of immunoglobin V_H genes. Science *227,* 1597–1601.

Reynaud, C., Anquez, V., Dahan, A., and Weill, J. C. (1985). A single rearrangement event generates most of the chicken immunoglobin light chain diversity. Cell *40,* 283–291.

Ritchie, K., Brinster, R., and Storb, U. (1984). Allelic exclusion and control of endogenous immunoglobin gene rearrangement in k transgenic mice. Nature *312,* 517–520.

Snodgrass, R., Kisielow, P., Kiefer, M., Steinmetz, M., and von Boehmer, H. (1985). Ontogeny of the T-cell antigen receptor within the thymus. Nature *313,* 592–595.

Van der Ploeg, J., Cornelissen, A., Michels, P., and Borst, P. (1984). Chromosome rearrangements in *Trypanosoma brucei.* Cell *39,* 213–221.

CHAPTER 3

Regulation of Transcription

Most genes are contiguous sequences of 500 to 10,000 base pairs usually interspersed in a large amount of flanking DNA. Transcription of a gene must start at the beginning and terminate at the end of the gene. There are specialized signals that indicate these "start" and "stop" signs. About 30 bases before the start of most genes there is a sequence of 4 or more bases in the DNA related to TATA (T = thymine; A = adenine), which appears to signal RNA polymerase to initiate polymerization of a complementary strand of RNA from the DNA sequence just downstream. RNA is polymerized by forming a phosphodiester bond from the ribonucleotide triphosphate to the 3′ position on the growing transcript and thus grows in the 5′ to 3′ direction.

Promotors and Operators

In bacteria the promotor directs initiation of transcription about ten bases downstream. RNA polymerase binds to the promotor sequence and takes up an active configuration. Mutations that change the base sequence within the promotor sequence reduce polymerase binding as well as transcription of the contiguous gene. But promotors just signal the start of a gene, not whether it should be transcribed under one condition and not under others. Control signals have been found to be specific sequences of a

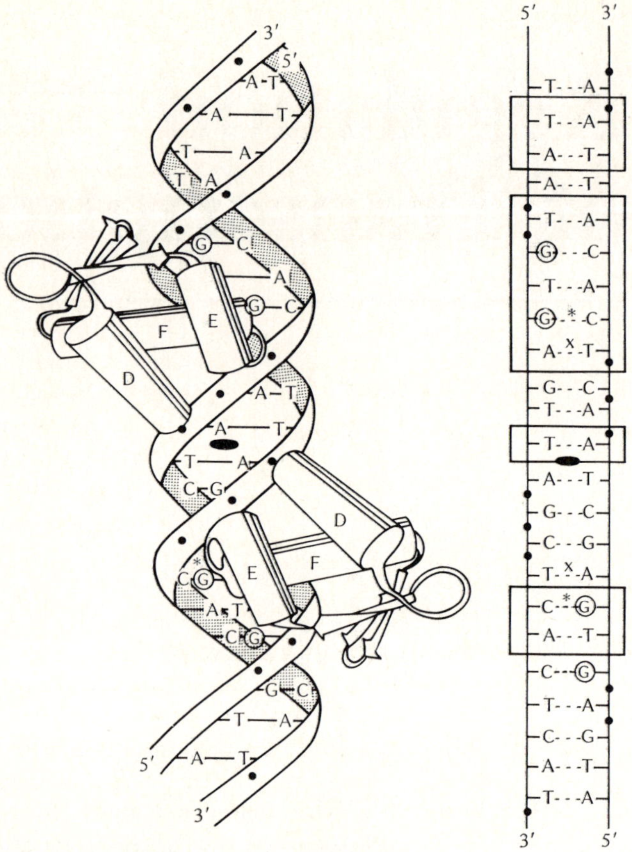

Figure 3.1. Regulator–operator interaction. Two molecules of catabolite-activating protein (CAP) bind to a stretch of 24 base pairs (bp) in the *lac* operon of *E. coli*. The primary sequence of the DNA and a representation of this sequence in the B-form of double-helix are presented. CAP has three α-helical regions (D, E, F) connected by random coils of protein. It associates with DNA in the major groove and protects the circled Gs from chemical reactions. Mutations at the starred Gs result in low CAP affinity. Ethylation of the phosphates marked with dots prevents CAP binding. Boxed sequences are >75 percent conserved in eight different CAP sites. The sequence is an imperfect palindrome centered on the boxed T . . . A and so can symmetrically bind two molecules of CAP. Association of cAMP with CAP greatly increases its affinity to this sequence. Transcription of adjacent genes is dependent on CAP binding.

dozen or so bases located within several hundred bases of the point of transcriptional initiation. These short sequences, referred to as operators, control transcription on the DNA molecule they are part of but do not affect other pieces of DNA. Genetically this is referred to as *cis*-acting. About 25 years ago it was recognized that there are gene products, proteins, whose only function is to control the expression of other genes. These proteins—repressors and activators—bind to *cis*-acting operators and determine the frequency of initiation of transcription. Since regulatory proteins can diffuse within a cell, they can control operators found on any DNA molecule. Genetically this is referred to as *trans*-acting. In some cases the operator region is very close to the promotor and the repressor/activator protein interacts directly with the RNA polymerase bound to the promotor, but in other cases the operator region is fairly far removed (up to several hundred bases) and no direct molecular interaction appears to occur. The signal may be passed down the DNA molecule by alterations in its structure.

Genetic and molecular manipulation of operators and promotors has shown they will regulate whatever gene they are next to and thus function independently of the gene they control. Selective pressures have kept specific pairs together, since the regulation of a gene is as important as the function of the gene product.

For instance, just upstream of the transcriptional start site of several bacterial genes that code for sugar-metabolizing enzymes there is a site at which a specific protein must bind for transcription to occur at the maximal rate. The protein referred to as catabolite gene activator protein (CAP) acts as a positive regulator of the *lac* and *gal* operons. However, it functions only in association with the small molecule, cyclic adenosine monophosphate (cAMP). The concentration of cAMP in the cell, which is high when the cells are carbon-source starved and low when they are metabolizing a readily available sugar such as glucose, thereby controls the rate of expression of *lac* and *gal* genes by interacting with CAP. This protein of 22,000 daltons (22 kd) complexed with cAMP functions as a dimer to recognize a stretch of about 24 base pairs of DNA in the normal B-form (right-handed) (Fig. 3.1). When the complex associates with DNA, RNA polymerase binds to a site next to it and can transcribe an adjacent gene starting at the 5′ end of the mRNA. If the sequence that binds CAP is artificially put in front of other genes, they then become dependent on CAP and cAMP for expression, which may be detrimental to the survival of the cell.

Steroid Hormone Induction

While many genes have been clearly shown to be regulated during embryogenesis, in only a few cases have the regulatory components been recognized in eukaryotic cells. One of the best understood processes concerns steroid hormone induction of specific genes.

There are a variety of steroid hormones that function in vertebrates, and many of these regulate developmental processes such as the proliferation of cells in the chick oviduct as well as stimulation of the initiation of transcription of the ovalbumin gene. Hormones, by definition, are circulating molecules and so are presented to all cells. The steroid hormones enter the cells since they are sufficiently lipid soluble to cross the cell membrane. However, specific steroid hormones affect only certain target cells and leave most cells unaffected. To a large extent this can be explained by the presence or absence of unique steroid-binding proteins in target cells. The steroid hormones do not appear to affect gene expression on their own, but only after associating with proteins that have the ability to bind them tightly. The target cells contain these steroid-binding proteins and thus are able to respond to this hormonal induction.

When a steroid hormone enters a target cell, it is tightly bound by the receptor protein. The protein-steroid complex binds to many different sites along the genome, only some of which affect transcriptional initiation. Within a few hundred bases of the point of initiation of transcription there are regions to which the steroid-binding protein associates. Each region contains multiple binding sites about 12 bases long. The sequence 5′ ACAANN TGTTCT 3′ where N is any of the four bases or a slight variant of it occurs in these binding sites. Binding to these sites rapidly stimulates transcription at the normal initiation site or any other promotor region put nearby by molecular engineering. This clearly shows that regulation is independent of the genetic information controlled and that only selective pressures keep them linked. Various sites in the operator region have been altered by molecular biological techniques to determine whether they are necessary for regulation. It turns out that one binding site usually plays the dominant role. Changes in its sequence abolish steroid induction of contiguous genes. Similar techniques have shown that the orientation of the operator does not matter for regulation. That is, the particular sequence works just as well when pointing at the gene it regulates as when pointing away. This by itself tells us something about the mechanism by which association of the steroid-binding protein to the DNA regulates RNA polymerase 300 bases away. It appears that the association changes the environment of the DNA over a relatively long distance. For these reasons such orientation-independent, long range *cis*-acting regions are often referred to as enhancers.

DNA is not present as a naked double-stranded molecule within cells. It is always associated with DNA-binding proteins. In eukaryotes the bulk of the protein bound to DNA consists of multiple copies of the five major histones, H1, H2a, H2b, H3, and H4 (Fig. 3.2). The DNA is wrapped around nucleosomes, which are made up of two copies each of H2a, H2b, H3, and H4. Nucleosomes associate in higher-order helices to form solenoids (Fig. 3.3). This compacts DNA about 50-fold relative to the fully extended form. In this form it is referred to as chromatin. Histones are basic proteins that have been highly conserved since the first appearance of nucleated cells and function to organize the structure of DNA strands. In the vicinity of certain genes the nucleosomes appear to preferentially associate with specific short sequences of DNA. Since the amount of DNA that can wrap around a nucleosome is limited to about 145 base pairs, the positioning of one nucleosome determines the arrangement of nearby nucleosomes. This phasing of nucleosomes results in some sequences being in nucleo-

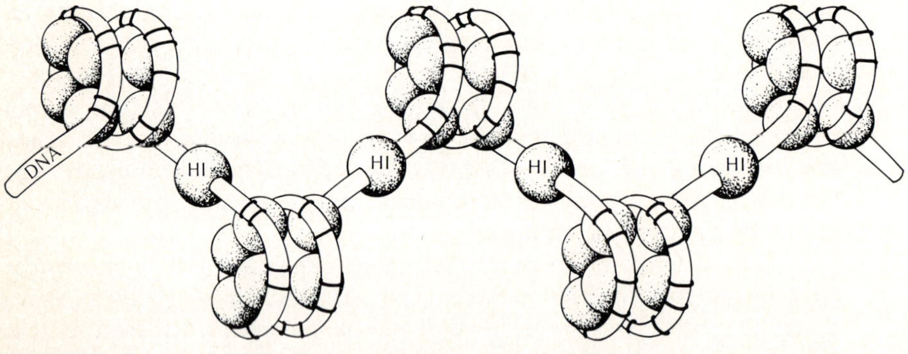

Figure 3.2. Nucleosome model. DNA wraps twice around cores of histone aggregates. Histone H1 is more loosely associated with the DNA linking the nucleosomes. Nucleases preferentially cut DNA in the linker regions between nucleosomes.

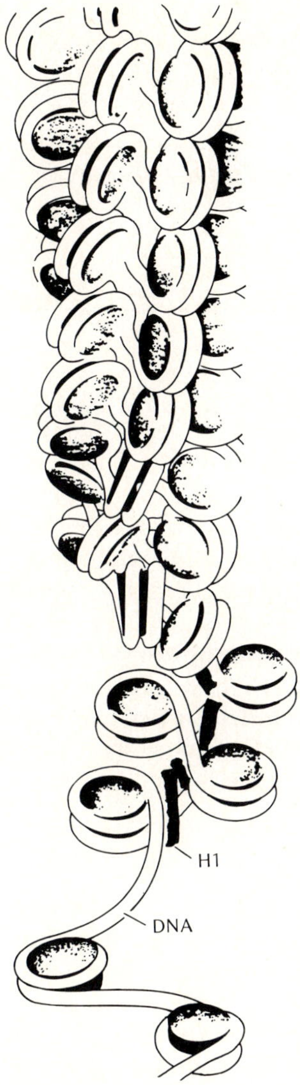

Figure 3.3. Solenoid of nucleosomes: a model of possible arrangements of nucleosomes as found in chromatin. About six nucleosomes form a turn in a helical array at high salt concentration *(top)*. At lower salt concentration *(bottom),* histone H1 dissociates from a central polymer and the nucleosomes zigzag.

somes and other sequences being in the regions between nucleosomes where they may be more available to transcription factors. Moreover, association of DNA-binding proteins, such as steroid receptors, can modify the chromatin structure in the region. The evidence for this is derived from measurements of the susceptibility of specific regions of DNA to attack by enzymes such as DNase. This enzyme will rapidly hydrolyze DNA in solution when it is free of DNA-binding proteins but acts much more slowly on DNA that is in association with nucleosomes. When chromatin of the operator region that binds steroid-binding protein is treated with DNase in the absence of added steroid, the DNA is relatively protected from hydrolytic attack. But when steroid is added, the configuration of the DNA is changed so that it is rapidly attacked by DNase. This indirect analysis of the detailed

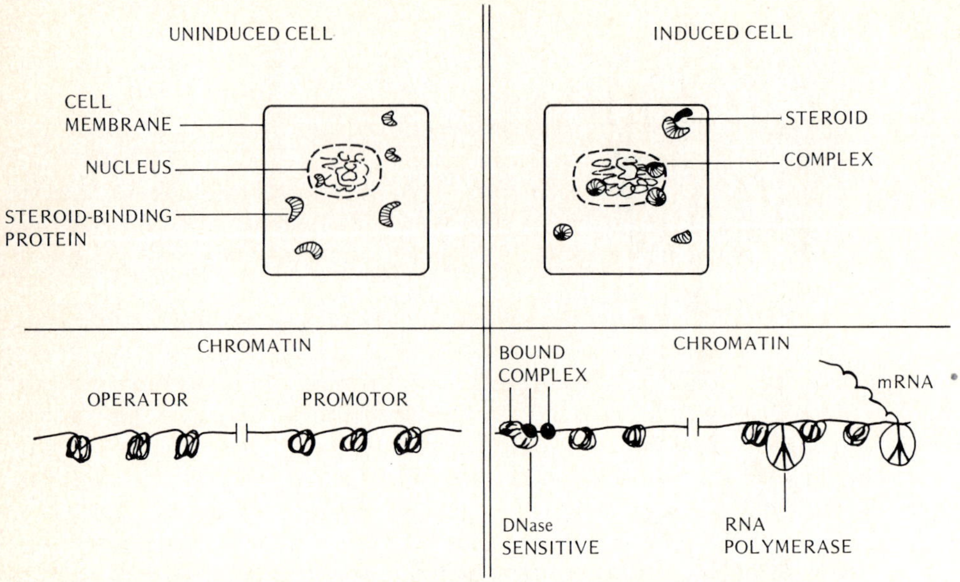

Figure 3.4. Schematic view of steroid induction. Uninduced cells carrying steroid-binding proteins do not transcribe certain genes. Steroids enter the cell and bind to their protein receptors. Steroid-binding protein complexed with a steroid (●) associates with a specific region of DNA and enhances transcription of contiguous genes by RNA polymerase. Binding of inducer complexes is cooperative such that once the first complex is bound, others bind more readily. This results in a sharp threshold for induction.

form of the DNA shows that steroid induction changes the association of that region of DNA with histones. Such a change in chromatin structure may be essential for the stimulation of transcription at contiguous promotors.

This process of steroid regulation of transcriptional initiation can be thought of schematically as shown in Figure 3.4. Although not all the details are well understood or fully proven, it is highly likely that the broad outlines actually indicate the processes that go on in the cell.

The configuration of DNA is altered not only around genes that are steroid induced but also around genes that are regulated during the process of development. For instance, the erythropoeitic cells, which can differentiate to give rise to red blood cells, show significantly greater DNase susceptibility of the DNA near globin genes than do cells of other tissues not destined to transcribe the globin genes. It can be inferred that specific DNA-binding proteins are perturbing the DNA structure near the start of the globin genes in erythropoeitic cells but not in other cells—say those of liver or muscle tissue. Several other cases of altered DNase sensitivity near genes following their induction have been recognized, and the components involved in this change are now being analyzed.

Control of 5*S* RNA Genes

Different polymerases are involved in transcription of the genes that code only for RNA such as those for ribosomal RNAs or the small transfer RNA, which function in the translation of proteins. One such small RNA, called 5*S* for its sedimentation characteristics, is an essential component of ribosomes. The gene that codes for 5*S* RNA is transcribed by RNA polymerase III.

During oogenesis in frogs, excess 5*S* RNA accumulates so that there will be enough to form a large number of new ribosomes. The 5*S* RNA is not kept as a naked molecule in the cell but associates with a specific protein that binds it with high affinity. This protein has been purified from these ribonuclear protein complexes. The protein binds not only to 5*S* RNA but also to the DNA of the gene that codes for 5*S* RNA. When mixed with purified DNA containing the 5*S* gene and highly purified RNA polymerase III, it stimulates transcription dramatically. This protein, now referred to as Transcription Factor IIIA, was found to bind right in the middle of the gene. However, it is a small gene of only about 120 bases and it is quite possible that Factor IIIA can interact directly with RNA polymerase III positioned at the start of the gene (Fig. 3.5). Factor IIIA is a 40 kd protein that binds to a 50-base-pair internal control region of 5*S* ribosomal genes.

Although only a few genes serviced by RNA polymerase III have been studied in detail, it appears that they all contain regulatory regions (operators) within themselves.

In the case of regulation of the 5*S* gene, we can see an added loop to the regulatory pathway, namely, a feedback control on the transcription of

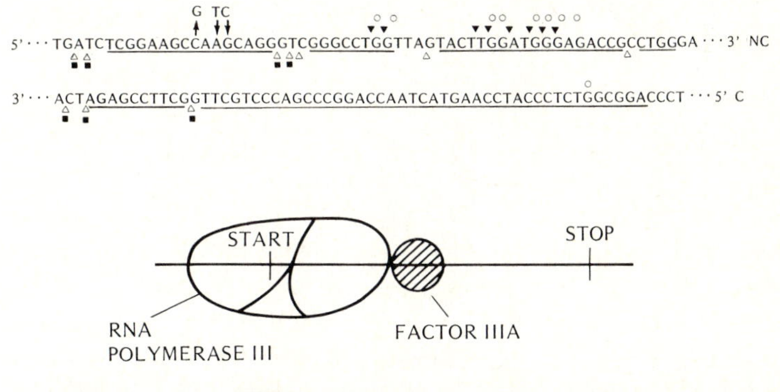

Figure 3.5. TFIIIA activates transcription of 5*S* RNA genes. A 40-kd transcription factor binds to each *Xenopus* 5*S* RNA gene within the transcribed region. Essential contact points include 8 bases (▼) as well as 8 phosphates (○) between bases on the non-coding strand in this region. Oocyte 5*S* RNA genes bind TFIIIA more weakly and differ at three bases from somatic 5*S* RNA genes. Several sites in this region become hypersensitive to DNAase-1 when TFIIIA is bound (■).

the gene by the amount of 5*S* RNA that has accumulated. Factor IIIA has a higher affinity for 5*S* RNA than for the DNA that codes for it. As long as there is excess Factor IIIA in the cell, it can stimulate transcription. But as soon as sufficient 5*S* RNA is made that all of the Factor IIIA is bound to it, transcription of the gene stops. Therefore, the level to which the gene is transcribed is determined by the amount of its RNA product and the level of expression of the gene coding for Factor IIIA. It turns out that Factor IIIA is synthesized to a much greater extent in differentiating oocytes than in any other cell type. It accumulates to a level of 3×10^9 molecules in mature eggs. This accounts for the high rate of transcription of the 5*S* gene during oogenesis and the accumulation of 5*S* RNA. Division of a fertilized egg into hundreds of thousands of cells reduces the concentration of TFIIIA to 10^4 molecules per cell. At this concentration it stimulates transcription only from those 5*S* genes with a high affinity to TFIIIA (somatic genes), as will be further discussed in Chapter 6.

Yeast GAL Genes

There is genetic evidence in microbial organisms that the mechanism of specific gene regulation can be quite complicated. One of the best understood of these genetic studies involves the regulation of the genes that code for galactose metabolizing enzymes in yeast. These genes are transcribed only when the yeast cells are growing in the presence of galactose. Three separate enzymes coded for by GAL1, GAL7, and GAL10 genes are necessary before galactose can be used as an energy source. Thus, there is a selective advantage to inducing them all coordinately. One alone would not do any good. It turns out that there is another gene, GAL4, which makes a protein necessary for transcription of each of the galactose enzyme genes, GAL1, GAL7, GAL10. Mutations in GAL4 result in no expression of the other GAL genes even in the presence of galactose. But galactose does not interact directly with the GAL4 protein. What galactose does is dissociate the product of yet another gene, GAL80, from a complex with the GAL4 protein. The complex, GAL4.GAL80, does not activate GAL1, GAL7, and GAL10 but, when dissociated, the GAL4 protein does stimulate transcription of these genes (Fig. 3.6). The GAL4 protein binds to sequences closely related to CGGTCAACAGTTGTCCGAGCGCT about 300 bases upstream of the transcriptional start sites of GAL1, GAL7, and GAL10. The contiguous genes can then be expressed. Thus, GAL4 protein is a required positive regulator whereas GAL80 protein is a purely negative regulator.

What this system in yeast points out is that a change in the physiology of a cell that is adaptive to a new environmental state often requires the coordinated expression of several genes. There are several ways in which this can be accomplished, including the mechanism just discussed for the GAL genes of sharing the requirement for the same activator. For instance, each gene might have to bind a distinct activator, but each of the activators might be sensitive to the same effector, such as cAMP, or other small molecules in

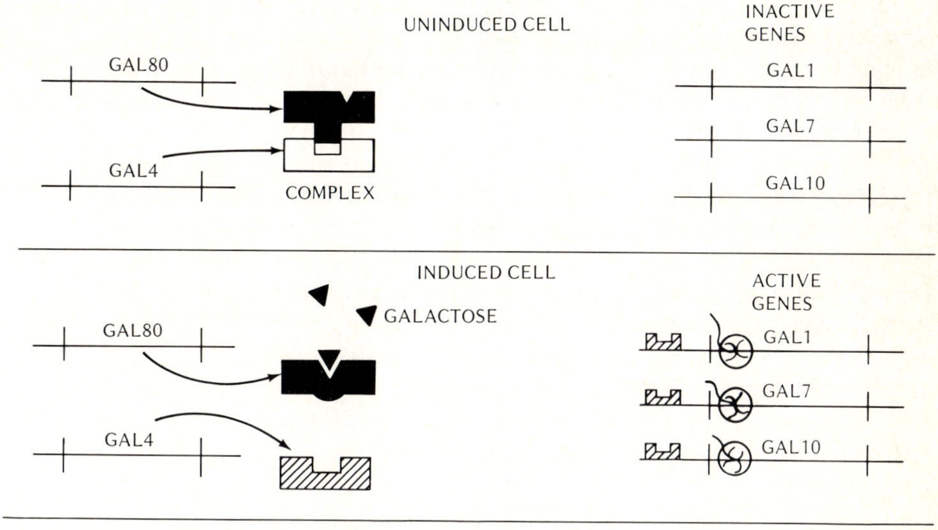

Figure 3.6. Positive regulation of yeast GAL genes. The GAL80 gene product associates with the GAL4 gene product and keeps it from activating transcription of the structural genes, GAL1, GAL7, and GAL10, for galactose metabolism. Galactose enters from the medium and dissociates the GAL80 product. GAL4 protein then binds to each of the structural genes and coordinately stimulates transcription. The GAL 1 and GAL 10 structural genes are closely linked in the yeast *Saccharomyces cerevisiae* but are transcribed off opposite strands.

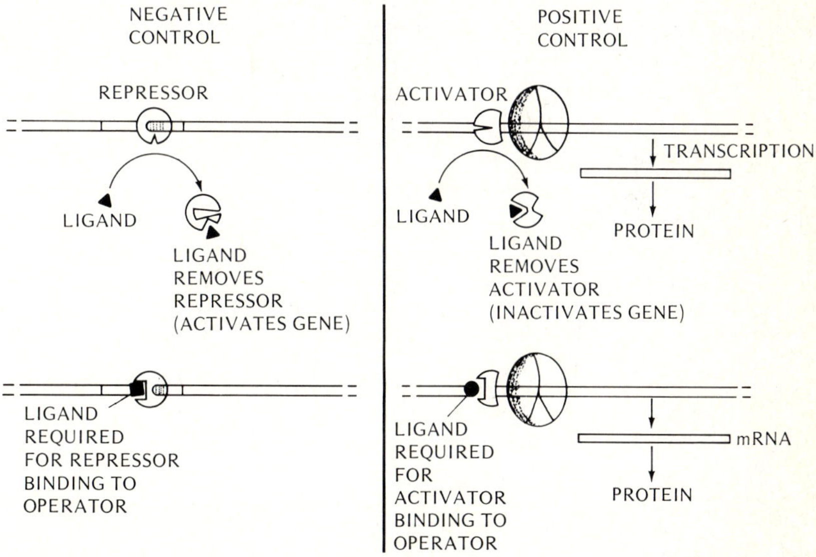

Figure 3.7. Positive and negative control. Activators and repressors bind to small molecule effectors. The complex may have either a greater or lesser affinity for a specific sequence of DNA than in the absence of its effector. Deletion of a gene coding for an activator always results in lack of expression in a positive control system.

the cell. Transcriptional control is positively regulated for some genes and negatively regulated for other genes (Fig. 3.7). When a positive regulator is bound to the DNA, it acts in *cis* to allow transcription of nearby sequences. When a negative regulator is bound to DNA, contiguous genes are repressed.

The integration of a selectively advantageous pattern of gene expression at a given stage of differentiation might require some genes to be regulated by more than one regulatory protein. In the *lac* operon of the bacterium *E. coli* we have a clear case of just that: expression of the genes necessary for metabolism of lactose requires not only lactose to remove a repressor molecule from the operator but also the presence of cAMP to bind an activating protein to a separate operator region (Fig. 3.8). cAMP in these cells indicates the degree of starvation that the cells are experiencing. Likewise, there are regulatory proteins known that affect the point at which transcription stops. Under some conditions transcription is terminated before a spe-

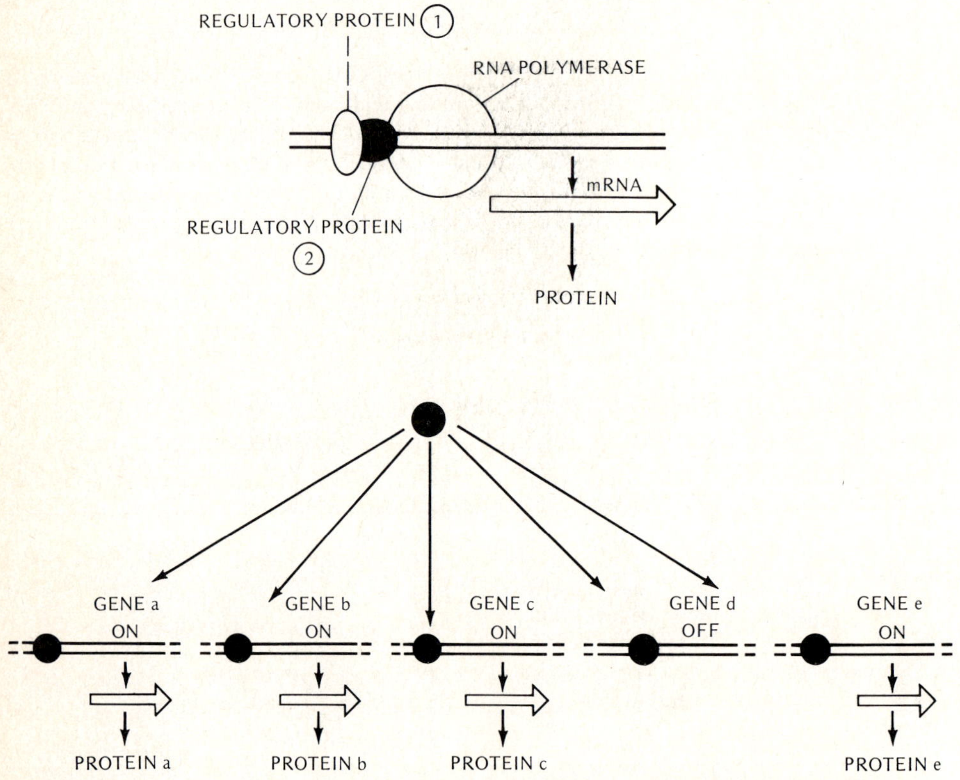

Figure 3.8. Double control. Some genes are regulated by two independent systems. Separate proteins bind to different regions of the DNA and affect transcription of contiguous genes. Cooperation between regulatory protein ① and regulatory protein ② often occurs such that binding the second protein is facilitated. When the protein-nucleic acid association is relatively stable, cooperativity can result in a determined state that is hereditary in the cells. In some cases the same regulatory protein binds to multiple genes, stimulates transcription of some, and represses others.

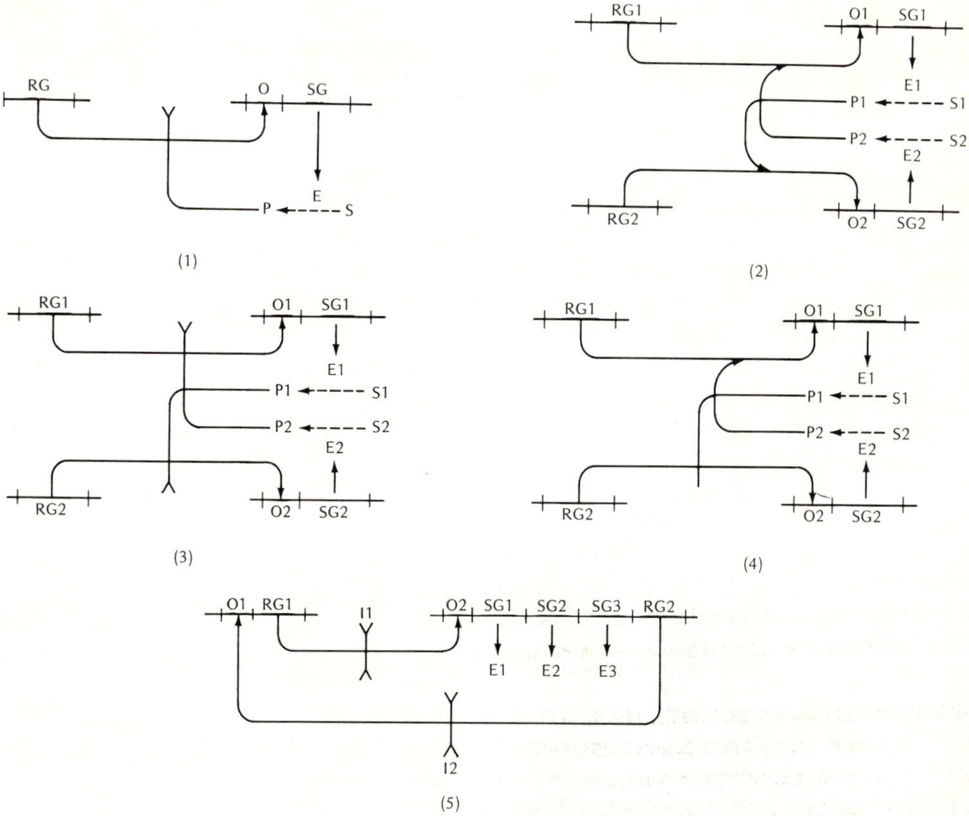

Figure 3.9. Monod–Jacob models. Regulatory genes *(RG)* make *trans*-acting products that bind to operator *(o)* sequences to control structural genes *(SG)*. Five models of negative control circuits are presented.

In model 1, the structural gene makes an enzyme *(E)* that catalyzes the synthesis of its own inducer. The *lac* operon of *E. coli* is an example of this (β-galactosidase catalyzes the synthesis of allolactose from lactose; allolactose binds the *i* gene repressor and allows transcription of the β-galactosidase gene). This circuit amplifies a weak induction signal.

In model 2, two structural genes, each with its own regulatory gene, are considered. Enzyme E1 catalyzes synthesis of a corepressor of *SG*2. Likewise enzyme E2 catalyzes synthesis of a corepressor of *SG*1. Whichever gene is expressed first will keep the other one permanently off.

Model 3 is similar to model 2, except that the products of the reactions catalyzed by each of the enzymes induce synthesis of the other. In this case, each gene will be on no matter which was expressed first.

In model 4, the product of the reaction catalyzed by E1 induces *SG*2 while the product of the reaction catalyzed by E2 is a corepressor of *SG*1. This will result in expression of *SG*1 followed by expression of *SG*2 and repression of *SG*1. This flip-flop circuit will oscillate between expression of one gene and the other with a periodicity determined by the stability of the enzymes and the products of the reactions they catalyze.

Model 5 brings in higher-order complexity in which two regulatory genes (*RG*1 and *RG*2) regulate expression of each other. Each is assumed to be rendered nonfunctional when inducer I1 or I2 is present. If I1 is added, the structural genes as well as *RG*2 are expressed. These genes need not be contiguous as long as each is controlled by the same operator sequence (O2). Expression of *RG*2 represses transcription of *RG*1 and ensures continued expression even in the absence of inducer I1. This is a stable physiological state with the genes *SG*1, *SG*2, and *SG*3 permanently on. However, if inducer I2 is added in the absence of inducer I1, *RG*1 would be expressed and repress the structural genes. The *RG*2 repressor could regulate a barrage of other genes not part of the model and integrate genetic differentiation.

cific gene is read whereas under other conditions transcription continues on through the gene. The decision to stop or not is determined by the concentration of specialized proteins that affect the RNA polymerase molecule. Combining several regulatory systems exponentially increases the number of possible conditions to which genes can respond.

Clearly, various circuits can be conceived which will result in different patterns of expression of given genes. Cross-repression will result in either one gene or the other being on, but never both; cross-induction will result in both being on no matter which one was induced first. Many such circuits were described 25 years ago by Monod and Jacob (Fig. 3.9). The challenge is to see which ones have actually been selected for in various embryonic tissues.

Similar models can be drawn invoking positive regulatory mechanisms. Almost any pattern of gene expression can be generated by variations on these simple circuits, which involve only reversible associations of regulatory proteins and short stretches of DNA in the operators.

Summary

Proteins with specific affinity to a short stretch of bases in DNA control the rate of transcription of contiguous genes. In eukaryotes the DNA is compacted into chromatin of several different forms. There is evidence that specific protein–nucleic acid interactions are necessary to open up regions of chromosomes such that individual genes in that region can be transcribed when given the proper signals. Coordinate expression of genes can be achieved by a common activator protein controlling the transcription of a barrage of genes. Control can be mediated by either positive or negative regulators. Hierarchies of regulatory processes can lead to an enormous diversity in patterns of gene expression.

SPECIFIC EXAMPLES

1. Regulation of Globin Genes

Patients with anemia have been the source of a large amount of genetic and molecular information concerning the genes that code for the proteins of hemoglobin. In adults most hemoglobin is made of two copies each of two related but distinct proteins, α- and β-globin, each associated with a heme molecule that carries oxygen in red blood cells (Fig. 3.10). Coordinate expression of the globin genes is essential for efficient hemoglobin synthesis. Moreover, during development there is a temporal sequence of expression of different genes coding for globin chains that adapts the oxygen-carrying capabilities of hemoglobin to the embryonic and newborn conditions.

Globin mRNA is very abundant in terminally differentiating erythro-

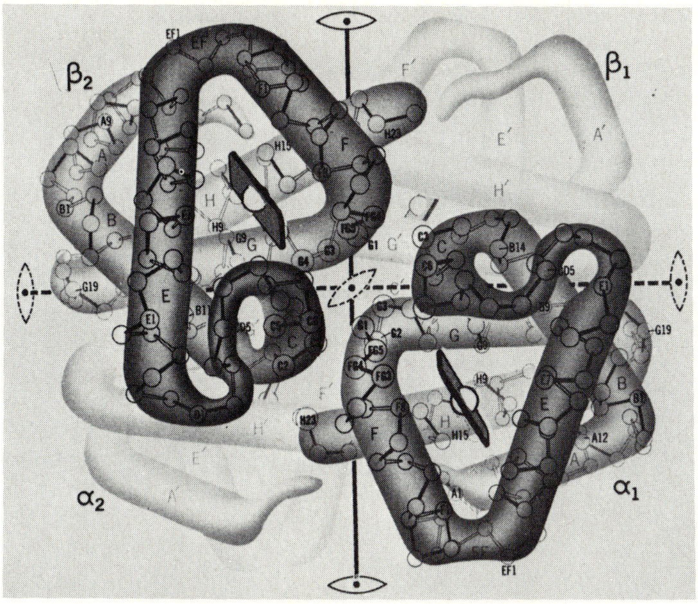

Figure 3.10. Three-dimensional structure of hemoglobin. Two alpha (α) and two beta (β) chains associate in a unique way to make hemoglobin. Each chain carries a heme group to which oxygen binds in the lungs where the pO_2 is high. The oxygen dissociates in the capillaries, where the pO_2 is lower. The protein molecule "breathes" slightly when oxygen binds to one heme group, which enhances binding to the other heme groups.

poietic cells and so is easier to purify than most other mRNA molecules. This has allowed the determination of the complete nucleotide sequence (Fig. 3.11).

DNA Sequencing

There are two convenient methods for sequencing nucleic acids: (1) cleavage of specific bases by chemical reactions and (2) premature termination of a DNA copy by incorporation of dideoxynucleotides. In both techniques one end of the polynucleotide is radioactively labeled to mark it. The fragments that carry the terminal nucleotide are all labeled but are of different lengths, depending on where the chemical cleavage occurred or the copy reaction stopped. The different length fragments can be separated on 12 percent polyacrylamide gels by electrophoresis. The separation is so good that a fragment 100 bases long can be easily seen to migrate ahead of a fragment 101 bases long. If the 101st base is an adenine (A), then cleavage of adenines generates a band at the position of 100 bases (Fig. 3.12). Likewise, incorporation of a dideoxythymidine at position 100 will generate a band at the position of 100 bases. The technique allows the sequence of several hundred bases to be read off a gel because the chemical cleavages or dideoxy incorporations are adjusted to occur infrequently and randomly along the polynucleotide, thereby generating a ladder of fragments ranging from one to several hundred bases (Fig. 3.12).

The sequence of rabbit β-globin mRNA was determined by the dideoxy method using labeled poly-dT as a primer for reverse transcriptase. The poly-dT hybridized to the poly-A tail at the 3′ end of purified mRNA and was then elongated into a cDNA copy by the action of RNA-dependent DNA polymerase. The cDNA was made in the presence of four normal deoxynucleotide triphosphates mixed with a small amount of one of the dideoxynucleotide triphosphates. At random points in the synthesis of a cDNA

mGpp	pAC	ACU	UGC	UUU	UGA	CAC	AAC	UGU	GUU	UAC	UUG	CAA	UCC	CCC	AAA	ACA	GAC	AGA	AUG
GUG	CAU	CUG	UCC	AGU	GAG	GAG	AAG	UCU	GCG	GUC	ACU	GCC	CUG	UGG	GGC	AAG	GUG	AAU	GUG
Val	His	Leu	Ser	Ser	Glu	Glu	Lys	Ser	Ala	Val	Thr	Ala	Leu	Trp	Gly	Lys	Val	Asn	Val
GAA	GAA	GUU	GGU	GGU	GAG	GCC	CUG	GGC	AGG	CUG	CUG	GUU	GUC	UAC	CCA	UGG	ACC	CAG	AGG
Glu	Glu	Val	Gly	Gly	Glu	Ala	Leu	Gly	Arg	Leu	Leu	Val	Val	Tyr	Pro	Trp	Thr	Gln	Arg
UUC	UUC	GAG	UCC	UUU	GGG	GAC	CUG	UCC	UCU	GCA	AAU	GCU	GUU	AUG	AAC	AAU	CCU	AAG	GUG
Phe	Phe	Glu	Ser	Phe	Gly	Asp	Leu	Ser	Ser	Ala	Asn	Ala	Val	Met	Asn	Asn	Pro	Lys	Val
AAG	GCU	CAU	GGC	AAG	AAG	GUG	CUG	GCU	GCC	UUC	AGU	GAG	GGU	CUG	AGU	CAC	CUG	GAC	AAC
Lys	Ala	His	Gly	Lys	Lys	Val	Leu	Ala	Ala	Phe	Ser	Glu	Gly	Leu	Ser	His	Leu	Asp	Asn
CUC	AAA	GGC	ACC	UUU	GCU	AAG	CUG	AGU	GAA	CUG	CAC	UGU	GAC	AAG	CUG	CAC	GUG	GAU	CCU
Leu	Lys	Gly	Thr	Phe	Ala	Lys	Leu	Ser	Glu	Leu	His	Cys	Asp	Lys	Leu	His	Val	Asp	Pro
GAG	AAC	UUC	AGG	CUC	CUG	GGC	AAC	GUG	CUG	GUU	AUU	GUG	CUG	UCU	CAU	CAU	UUU	GGC	AAA
Glu	Asn	Phe	Arg	Leu	Leu	Gly	Asn	Val	Leu	Val	Ile	Val	Leu	Ser	His	His	Phe	Gly	Lys
GAA	UUC	ACU	CCU	CAG	GUG	CAG	GCU	GCC	UAU	CAG	AAG	GUG	GUG	GCU	GGU	GUG	GCC	AAU	GCC
Glu	Phe	Thr	Pro	Gln	Val	Gln	Ala	Ala	Tyr	Gln	Lys	Val	Val	Ala	Gly	Val	Ala	Asp	Ala
CUG	GCU	GAC	AAA	UAC	CAC	UGA	GAU	CUU	UUU	CCC	UCU	GCC	AAA	AAU	UAU	GGG	GAC	AUC	AUG
Leu	Ala	His	Lys	Tyr	His														
AAG	CCC	CUU	GAG	CAU	CUG	ACU	UCU	GGC	UAA	UAA	AGG	AAA	UUU	AUU	UUC	AUU	GC-polyA		

Figure 3.11. Primary sequence of rabbit β-globin. The mRNA is capped at the 5′ end with a m^7Gpp group and tailed at the 3′ end with about 150 adenines (poly-A). Translation starts at the first AUG triplet codon. The methionine coded by this AUG is subsequently cleaved off. Translation stops at the first termination codon (UGA). Addition of poly-A occurs just downstream of the signal AAUAAA in this and many other mRNAs.

copy from the mRNA, a dideoxynucleotide was incorporated thereby stopping the elongation. Gel separation of the fragments gave the lengths of the fragments that ended in the nucleotide present as a dideoxy derivative. The reaction was repeated four times, each time with a different dideoxynucleotide, to give the complete sequence of the 3′ end of globin mRNA. By using primers complementary to sequences of β-globin mRNA nearer to the 5′ end, the full sequence of 589 bases in rabbit β-globin was determined.

The first base transcribed in rabbit β-globin is A. The molecule is subsequently enzymatically modified by the addition of a cap, m^7Gppp, and methylation of the first adenosine. The first 53 bases are untranslated. The next 441 bases code for the 147 amino acids of β-globin. There are 95 bases in an untranslated sequence at the 3′ end. A homopolymer of about 150 adenines is added enzymatically to the 3′ end following transcription to give the finished β-globin mRNA. This was the first mRNA fully sequenced. Hundreds of other mRNA molecules have since been sequenced.

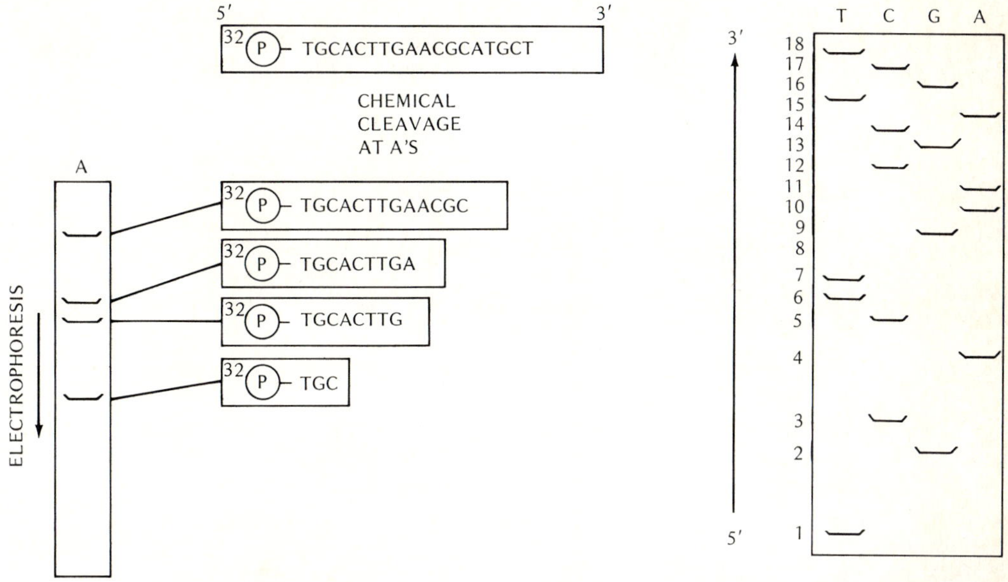

Figure 3.12. DNA sequencing. A DNA segment is labeled at the 5′ end with radioactive phosphate ($^{32}PO_4$) by first removing the resident phosphate by treatment with phosphatase and replacing it with labeled phosphate in a reaction catalyzed by polynucleotide kinase using γ-labeled ATP. The segment is then cleaved with a restriction enzyme and the radioactive fragments separated by size on a gel. There will be only two fragments each with its 5′ end uniquely labeled. The fragments are then partially cleaved by chemical reactions that hydrolyze the DNA after one of the four bases. Since the chemical reaction is not allowed to go to completion, cleavage occurs randomly at the base chosen. Oligonucleotides of different length are generated that can be separated by electrophoresis in acrylamide gels. Only radioactive fragments carrying the 5′ end are seen by autoradiography of the gel.

When the fragments are separately treated with chemicals that cleave specifically after T, C, G, or A and the reaction products run on the same gel, the sequence of the DNA fragment can be read starting at the first base at the bottom. Several hundred bases can be read from a single gel.

The same techniques can be used to sequence cloned fragments of chromosomal DNA. When this was carried out for the β-globin gene, it was found that the gene was considerably longer than the 589 bases of β-globin mRNA. The primary RNA transcript included two intervening sequences (introns) that are spliced out before the mRNA is translated (Fig. 3.13).

The human β-globin gene is initially transcribed into a pre-mRNA of about 1600 bases. The first 50 bases at the 5′ end of the molecule are not translated. These are followed by 90 bases that are expressed into the first 30 amino acids of β-globin (exon 1). The next 130 bases of the primary transcript are spliced out and are referred to as IVS1 (intervening sequence 1) or intron 1. The second exon is 201 bases long and is followed by 850 bases of IVS2 that are spliced out to form the finished mRNA. The third

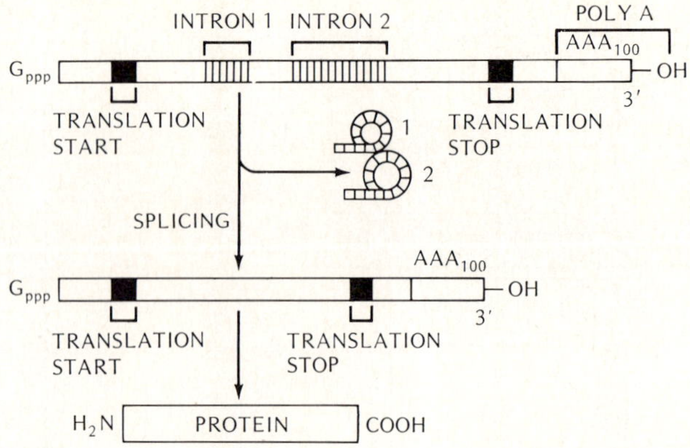

Figure 3.13. Processing of globin transcript. Two intervening sequences (intron 1, intron 2) are spliced out of the primary transcript to generate globin mRNA. At the 5′ end of the introns a G-G bond is cleaved and the first G of the intron is transferred to an A within the intron. It forms a lariat structure by a 2′5′ A–G linkage. At the 3′ end of the intron the linkage from G is cleaved and the two exon RNA sequences are ligated. Processed mRNA associates with ribosomes and is translated starting at the start codon (AUG). The triplet code is translated into globin protein. An mRNA molecule can be translated many times and is therefore catalytic.

exon is followed by 132 untranslated bases before the post-transcriptionally added poly-A tail at the 3′ end.

The first intron occurs between codons for amino acids 30 and 31 and the second intron occurs between amino acids 104 and 105. The splicing must be exact to the base or the reading frame will be offset or codons lost. The splice joints in each case consist of 5′ exon *GG/GU* . . . intron . . . *AG/* . . . exon 3′. The *GU* . . . *AG* pairs occur at the ends of all introns found in eukaryotic transcripts. Several anemic patients with β^{o}-thalassemia fail to produce β-globin. They have been shown to have mutations that affect the nucleotides at the splice junctions of β-globin. One of these mutations is a G→A change at the first base of IVS1. This base change completely blocks splicing out of the first intron, and no β-globin can be synthesized. A G→C change at the fifth base of IVS1 reduces splicing out of the first intron but does not abolish it. Another mutation was found that abolished normal splicing of IVS2. The *AG* dinucleotide at the 3′ end of the intron was changed to *GG* and could no longer function as an acceptor site in splicing.

Accurate splicing requires recognition of both the 5′ and 3′ ends of the introns. There are short RNA molecules, referred to as U1 and U2, that are present in all eukaryotes in small nuclear ribonuclear protein particles (snRNP) involved in recognizing the joints. The RNA in these particles hybridizes to short sequences found in the introns and determines where cleavage of pre-mRNA occurs. The 3′ signal includes a sequence surrounding an adenine base to which the guanine at the 5′ end of the intron is linked (2′5′) following cleavage of the 5′ splice point to form a lariat intermediate (Fig.

3.13). Both IVS1 and IVS2 of human α and β-globin genes form the lariat 18 to 36 bases upstream of the 3′ splice point at sequences of CT*X*ATC where X is G, A, or C. Such lariat attachment sequences occur within 55 bases of the 3′ splice point in all introns of vertebrate genes analyzed so far. This sequence can be recognized by a pentanucleotide sequence of U2 RNA in snRNP.

About half of all of the mutations analyzed so far that block β-globin synthesis have been found to affect the splicing pattern. Some mutations alter the normal splice joints whereas others create new splice sites in inappropriate places. In several cases a single base change generates a donor or an acceptor site. The mRNA processed in this way does not code for functional β-globin.

Thirty-two bases upstream of the transcriptional start site of β-globin is a sequence referred to as the "TATA box." Similar sequences are found at this position preceding almost all eukaryotic genes. They affect the frequency of transcriptional initiation and the choice of base at which polymerization starts. Anemic patients of several different racial backgrounds have been found to carry mutations in this control element (Table 3.1). The mutations lower the rate of transcription severalfold but do not abolish it.

These mutations clearly show that the TATA box plays a significant role in the transcription of the β-globin gene in humans. The sequence ATA is conserved at −30 in all globin genes of humans, mice, goats, and rabbits and appears to be essential for normal expression of these genes.

Further upstream, 87 bases 5′ to the transcriptional initiation site, is another sequence conserved in all globin genes: 5′CCAAT3′. This has been called the "cat box" with some humor. A C→G mutation at −87 of the β-globin gene has been found in a patient with β-thalassemia that results in a tenfold decrease in the rate of transcription of the β-globin gene. However, the residual transcription that does occur starts at the normal 5′ base. This mutation points out the essential role the "cat box" plays in β-globin transcription. Recent evidence points at a sequence within the translated portion of the human β-globin gene that is subject to positive regulation in erythroid differentiation.

Clustered within 55 kb on human chromosome 11 are seven sequences of DNA that are homologous to β-globin. Each of these sequences cross-

Table 3.1 TATA Box* Mutations of Human β-Globin

	−30
Normal wild type	5′ CATAAAA 3′
American Black mutation	5′ CAT*G*AAA 3′
Chinese mutation	5′ CATA*G*AA 3′
Kurdish Jew mutation	5′ CATA*C*AA 3′

*The sequence ATAA serves as the TATA box for transcription of human β-globin.

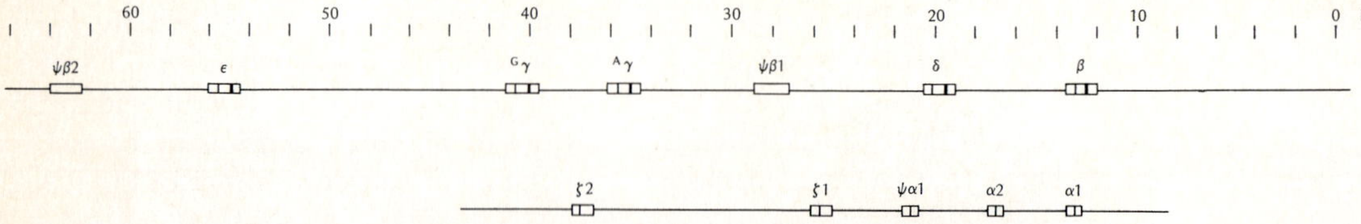

Figure 3.14. Human β-like and α-like genes. Within 60 kb there are seven β-like sequences. Two are pseudogenes that make no globin (ψB2, ψB1). Embryonic globin is coded for by ϵ and two γ genes. Adults express the δ and β genes. Unlinked to this cluster, there are five α-like sequences within 40 kb. One is a pseudogene (ψα1). There are two embryonic genes (δ1 and δ2) as well as two adult genes (α1 and α2). Each of these globin genes has two introns and three exons, indicating that they evolved from a similar progenitor gene.

hybridizes with authentic β-globin, and they are referred to as β-like genes (Fig. 3.14). Two of them are nonfunctional genes that do not code for proteins. They are duplications of β-globin genes that have sustained mutations leading to loss of function and are referred to as pseudogenes. One of the other β-like genes, δ, is expressed at a low rate in adults. δ-globin is 95 percent homologous to β-globin and functions in a minor adult homoglobin. There are two genes, $^{G}\gamma$ and $^{A}\gamma$, that differ from each other by only a single amino acid and function during most of embryogenesis to form fetal hemoglobin. Finally, there is a β-like gene, ϵ, that is 84 percent homologous to β-globin. This gene is expressed only during the first few weeks of human embryogenesis.

Each of these β-like genes has two intervening sequences, a short IV1 between codon 30 and 31 and a longer IV2 between codons 104 and 105. This arrangement is so strikingly conserved that it is clear that all the β-like genes have evolved from a common ancestor. The conservation of nucleo-

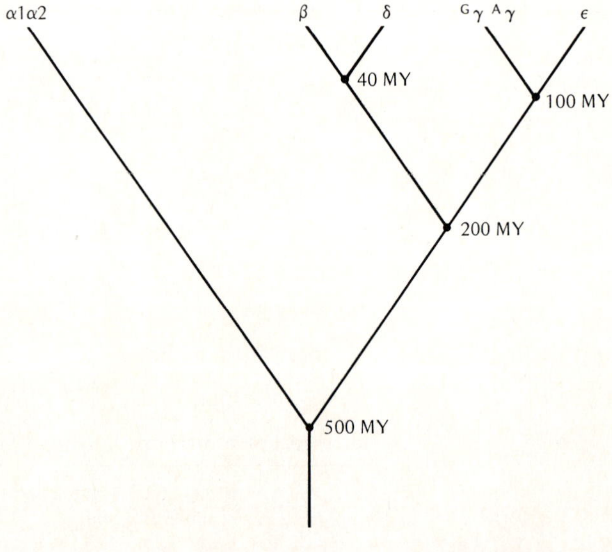

Figure 3.15. Evolutionary tree of human hemoglobin genes. Distances along the branches are additive and proportional to evolutionary time. About 500 million years ago, α-like and β-like genes diverged from a progenitor that had duplicated. β and δ genes probably diverged from each other only 40 million years ago as judged by the number of nucleotide differences and similarities. The β-gene is expressed at a higher rate than the δ-gene. The progenitor gene for both β- and δ-globin diverged from the progenitor of the embryonic globin genes, ϵ and γ, about 200 million years ago. The two γ genes diverged very recently since they differ by only a single amino acid.

tide sequence makes this conclusion even stronger. During the last 200 million years a common β-like sequence has duplicated and diverged to generate five different genes that are expressed at different stages of development in humans (Fig. 3.15).

On a separate human chromosome (number 16), five related genes are clustered within 30 kb of each other (Fig. 3.14). These code for the α-like globins of hemoglobin. The arrangement of two intervening sequences as well as the nucleotide sequences of the α-like genes makes it clear that they have evolved from an ancestral gene common to both α- and β-globins. There are two α genes that function during both fetal and adult life. They are almost identical, but both are necessary to produce enough α-globin for normal function. An individual unable to express one of these genes, α2, due to mutations in the gene is severely anemic (α-thalassemia). The two other α-like genes, ζ1 and ζ2, both differ somewhat from the α genes and function during the first few weeks of human development to make embryonic hemoglobin (Fig. 3.16). The hemoglobin present in early embryos differs in its association with oxygen from that present in the fetal phase. Likewise, adult hemoglobin is adapted to associate with oxygen at the higher partial pressure found in air rather than in the maternal circulation of the placenta. Changes in the six types of globin genes (ϵ, ζ, α, β, γ, δ) fine tunes this important physiological function.

Molecular analysis of mutations that occur in humans has pinpointed the specific base sequences needed for correct splicing as well as transcriptional initiation. These studies provide direct evidence for the function of both the "cat box" and the "TATA" box for expression of the β-globin gene. Although it is not possible to do genetic crosses with humans carrying interesting mutations, molecular pedigree studies in affected families have allowed the mutations to be traced. The frequency of mutations affecting globin expression is high in several races because of the selective advantage of

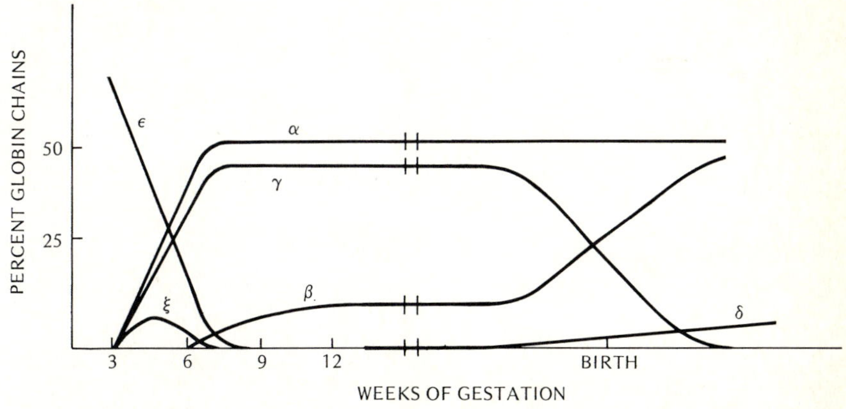

Figure 3.16. Expression of human hemoglobin genes. In the first few weeks of development ϵ and ζ genes are transcribed. They are then replaced by γ and α gene products. Both are expressed until birth, when γ gene transcripts disappear and are replaced by β and δ transcripts. The α gene continues to be expressed from the first month of development throughout life.

individuals heterozygous for the mutations when infected with malaria. In ancient times the shores of the Mediterranean Sea were endemically infested with malaria, and thalassemic mutations were selected in the populations that lived on its shores. Although malaria is no longer a serious problem around the Mediterranean, the mutations persist and lead patients carrying them to come into clinics for help. Since so many people are diagnosed in detail, even to the point of having their genes sequenced, a wealth of molecular genetic studies are available for the elucidation of the regulatory mechanisms of globin genes in humans.

2. Coordinate Regulation of the *qa* Genes of *Neurospora*

The filamentous fungus, *Neurospora crassa*, has served gentics well (Fig. 3.17). Rare mutants can be recovered from up to a billion conidia by selecting for growth in simple media. The mutations can then be mapped by sexual crosses with high precision. The generalization "one gene, one enzyme" was derived from studies with *Neurospora*.

This orange bread mold grows rapidly on a wide variety of carbon sources. One of these is quinic acid. It takes three separate enzymes to con-

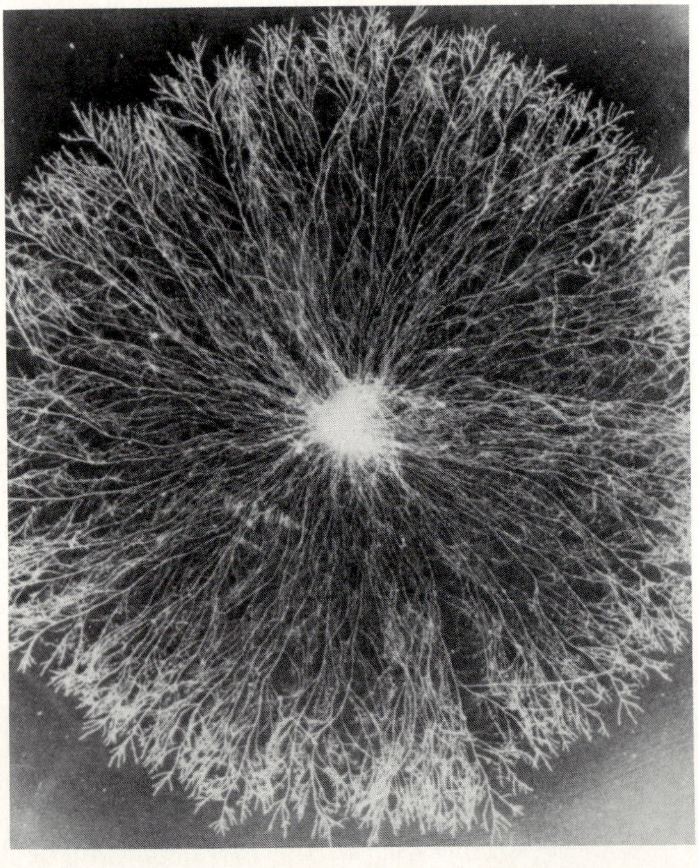

Figure 3.17. *Neurospora crassa* grows on simple, defined media by putting out hyphae many centimeters in length. Haploid nuclei divide mitotically in the branched filaments that have a tough chitinous cell wall.

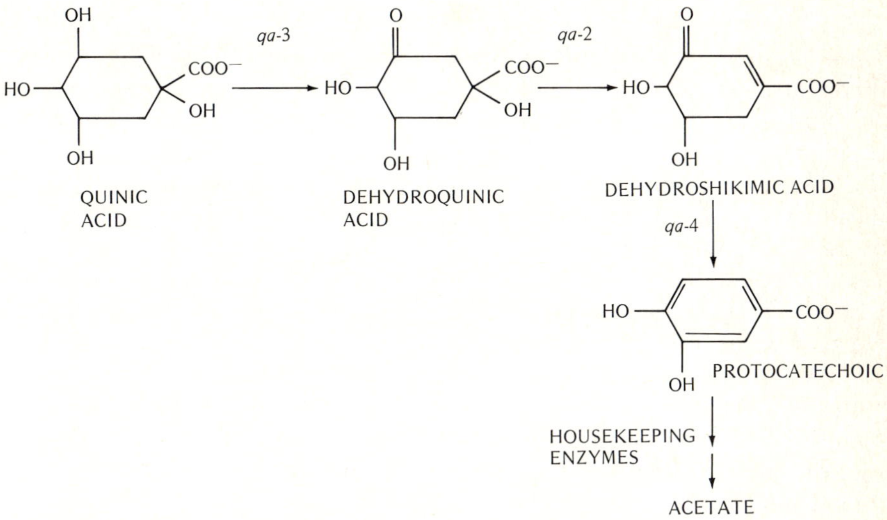

Figure 3.18. Metabolism of quinic acid. The cyclic carboxylic acid, quinic acid, is converted to protocatechoic acid in reactions catalyzed by three enzymes coded for by the genes, *qa*-3, *qa*-2, and *qa*-4. Protacatechoic acid is further metabolized by housekeeping enzymes present in *Neurospora* at all times.

vert quinic acid to a metabolite that can be further metabolized by the housekeeping enzymes of the cell (Fig. 3.18).

Quinic dehydrogenase converts quinic acid to dehydroquinate; dehydroquinase catalyzes the formation of dehydroshikimic, and dehydroshikimic dehydrase forms protocatechoic. This last compound is converted to acetate by enzymes necessary for growth on many different carbon sources. But each of the first three enzymes is needed only when the cells are growing on quinic acid. All three are necessary and the lack of any one of them precludes growth on that carbon source. Of course, they can grow on other carbon sources such as succinate.

Cells growing on succinate make very little of the quinic acid metabolizing enzymes. Soon after the cells are suspended in medium containing quinic acid, the specific activity of all three enzymes increases about 1000-fold. They are coordinately induced by quinic acid. A series of genetic and molecular biological experiments has delineated the mechanism by which these enzymes are regulated.

The structural genes for the enzymes of quinic acid metabolism were recognized by analyzing mutant strains that, unlike wild-type strains, are not able to grow on quinic acid. About a hundred independent mutations were genetically mapped to Linkage Group VII. They were found to occur in four loci *qa*-1, *qa*-2, *qa*-3, and *qa*-4. Mutations in a given locus, say *qa*-3, are genetically complemented by mutations in other loci but not by other *qa*-3 mutations. Each locus makes a separate protein: *qa*-3 mutants lack detectable quinate dehydrogenase, *qa*-2 mutants lack detectable dehydroquinase, and *qa*-4 mutants lack detectable dehydroshikimate dehydrase. These results explain why strains carrying a mutation in *qa*-2, *qa*-3, or *qa*-4 do not grow on quinic acid: these are the structural genes for the enzymes.

Mutations in *qa*-1 also result in lack of growth on quinic acid. Strains carrying *qa*-1 mutations make none of the three enzymes. Yet these mutations map at a separate locus from the structural genes. Revertants of these mutations can be selected by spreading 10^7 conidia on a plate containing quinic acid as the sole carbon source. A few cells grow and form colonies. These revertants were found to once again make all three of the enzymes of quinic acid metabolism. Many of these revertants synthesized the enzymes not only when presented with quinic acid but also when growing on succinate in the absence of quinic acid. The regulation of *qa*-2, *qa*-3, and *qa*-4 genes was altered in the revertants of *qa*-1. The expression of these genes is constitutive in strains carrying the *qa*-1^c mutations. The effects of *qa*-1^c on the regulation of the other genes were found to be dominant in heterozygous strains, carrying both wild-type *qa*-1^+ and *qa*-1^c. These results have been interpreted as indicating that *qa*-1 makes a protein that recognizes the presence of quinic acid in the environment and actively stimulates expression of *qa*-2, *qa*-3, and *qa*-4 genes. The *qa*-1^- mutants fail to make a functional positive regulator, and the *qa* genes cannot be expressed. The *qa*-1^c mutants make a positive regulator that no longer depends on the presence of quinic acid in the environment. Confirmation of this interpretation has required isolation of the genes by cloning them in bacteria.

Gene Cloning

The first step in cloning a gene is often the hardest. How do you select for the right piece of DNA when the genome carries enough DNA for 20,000 genes or more? The stratagies used for different genes depend on the characteristics of the gene and the cell. In the case of the *qa* genes, it was possible to select for the *qa*-2 gene function in the bacterium *Escherichia coli. E. coli* uses the enzyme dehydroquinate hydrolase in the biosynthesis of aromatic amino acids. Strains of *E. coli* had been isolated that fail to grow unless given tryptophan, phenylalanine, and tyrosine. Some of these carry mutations in a locus, *aro*D, that is the structural gene for dehydroquinate hydrolase. This enzyme catalyzes the same reaction that the *qa*-2 gene product does, namely, the conversion of dehydroquinate to dehydroshikimate. It was thought that if the *Neurospora qa*-2 gene could be inserted into a bacterial plasmid carried by an *aro*D mutant of *E. coli,* the transformed bacteria would grow in the absence of aromatic amino acids.

DNA of *Neurospora* was cut with the restriction enzyme HindIII (Fig. 3.19). This enzyme recognizes a specific sequences of six bases (AAGCTT) present, on the average, once every 4096 bases and cleaves the DNA after the first A. The restriction fragments from *Neurospora* were enzymatically ligated to the DNA of the bacterial plasmid pBR322. This plasmid can infect bacteria and grow stably within the cell (Fig. 3.19). A population of *aro*D bacteria was transformed with the mixture of plasmids, each carrying different fragments of the *Neurospora* DNA. The population was spread on plates lacking aromatic amino acids. One cell grew and gave rise to a population that was shown to carry a 4 kb fragment of *Neurospora* DNA in the plasmid.

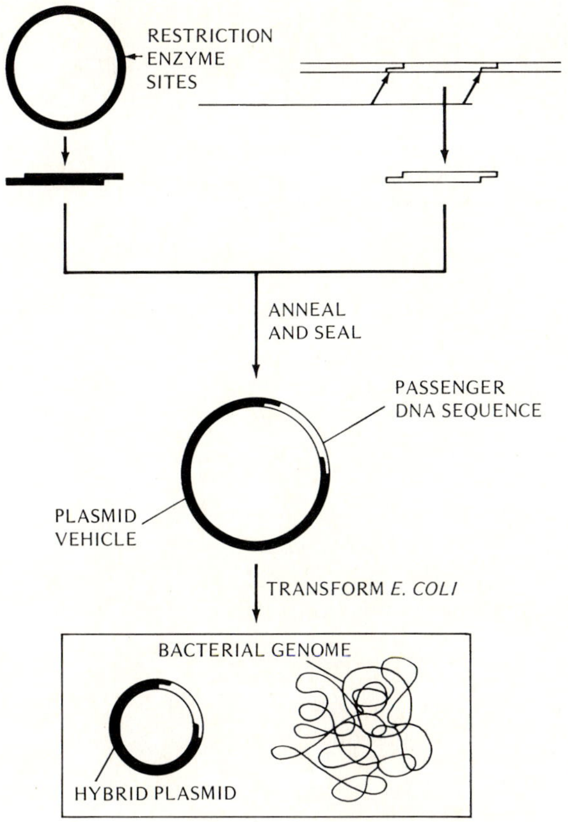

Figure 3.19. Cloning of DNA fragments. Bacteria often harbor small circular plasmids that replicate autonomously in the cell. Plasmid DNA can be cut with a restriction enzyme and annealed to fragments of DNA from another source such as *Neurospora*. The DNA pieces are then ligated. Recombinant DNA can be reinserted into a bacterial host where it will replicate as an autonomous plasmid. Restriction enzymes cut asymmetrically, leaving a few single-strand bases that will hybridize to complementary bases of the other strand. Bacterial viruses can also be used as cloning vectors.

The presence of this fragment was shown to be essential for the growth of the bacteria in the absence of aromatic amino acids and to direct the synthesis of the *Neurospora* enzyme, dehydroquinase, in the bacterial cells. The 4 kb fragment was replicated in the bacterial strain. It was now separate from all other *Neurospora* DNA.

The cloned *qa*-2 sequence was then used to recognize large pieces of *Neurospora* DNA that contained the *qa*-2 gene. The cloned *qa*-2 sequence will hybridize only to the authentic *qa*-2 gene and will tag it and adjacent genes. *Neurospora* DNA was partially digested with the restriction enzyme EcoRI, and the large fragments were cloned on a bacterial plasmid. The large fragments were screened by hybridization to the radioactively labeled cloned *qa*-2 sequence previously isolated. A fragment of 36.5 kb was found that hybridized with the smaller 4 kb sequence of *qa*-2. The large cloned fragment was used to transform *Neurospora* cells that carried mutations in *qa*-1, *qa*-2, *qa*-3, and *qa*-4. The 36.5 kb fragment carried each of these genes and permitted growth on quinic acid. Further analysis of this cloned fragment positioned the genes (Fig. 3.20). They all occur within an 18 kb stretch of DNA.

The sequences around the points at which translation starts for the *qa*-2 and *qa*-3 genes were determined chemically using the technique of

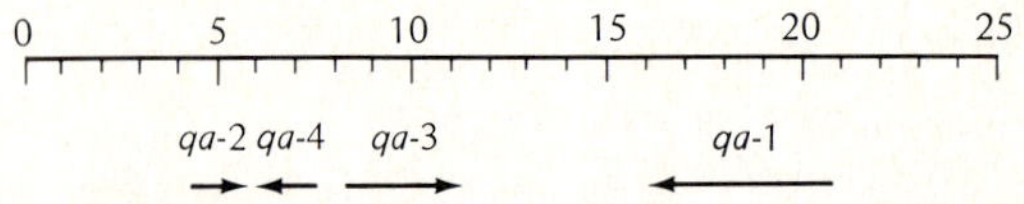

Figure 3.20. The *qa* genes of *Neurospora* are clustered within a 17-kb stretch of DNA but are independently transcribed; *qa*-1 is a positive regulator of the structural genes and may have remained closely linked, so that recombination during meiosis does not often separate it from the genes it controls.

Maxam and Gilbert. When the sequences of these two genes are compared, many conserved bases are found in the 100 nucleotides upstream of the translation initiation sites (Fig. 3.21). These sequences might be recognized by the *qa*-1 gene product for positive regulation of transcription, but further molecular genetic studies are necessary to be sure.

One way to find out how the *qa*-1 protein normally acts to facilitate transcription of the *qa* genes is to isolate mutations, rendering the *qa* genes independent of the presence of wild-type *qa*-1. The enzyme made by *qa*-2 is able to replace the enzyme necessary for aromatic amino acid biosynthesis in *Neurospora* as well as in *E. coli*. In *Neurospora* the *arom*-9 gene normally makes the biosynthetic enzyme. Mutants carrying deletions in *qa*-1 as well as *arom*-9 mutation do not grow in the absence of aromatic amino acids. Prototrophic mutants were selected from a mutagenized population of such *Neurospora* cells.

At a frequency of about 10^{-6}, revertants were recovered that expressed the *qa*-2 gene even in the absence of functional *qa*-1 gene. Five mutants were studied in some detail (Fig. 3.22). None made the full amount of dehydroquinase. The best one made 45 percent of the amount of enzyme seen in wild-type cells. It was found to have moved a new piece of DNA up to a point 378 bases before the transcriptional initiation site. But transcription started at the normal place. Somehow the new sequence enhanced transcription of *qa*-2 and freed it of dependence on *qa*-1 protein.

One of the *qa*-1 independent strains was found to have a C/G to A/T transversion 200 bases upstream of the transcriptional start site. Although this strain makes only 6 percent as much dehydroquinase as wild-type cells, it does so in the absence of *qa*-1 protein.

Among the other revertants were found two duplications of sequences about 100 bases upstream of the transcriptional start site, and a replacement of 9 bases by 7 other ones in the same area. It is far from clear how these

```
                                              –50                                        –20          –10          +1
QA2 5'   ATAGTGCGGCGGCATCTTTCGGACGCATTCCCTGTTGCGCCCATCTCCCACAAGCCCATCGCACCCAACCAGAGGTACCAAACACAATGGCGTCCCCCCGTCACA
QA3 5' ATTTTCATTCCGGTCTTCTGTCGAATCTTGATTTTCGAGTGACTCTGACTTCTCATAGCCACATACACCACACAATCAAGCATATATCACCATGTCGACAGCAACCACCA
```

Figure 3.21. Homologies of *qa*-2 and *qa*-3. Although the structural genes *qa*-2 and *qa*-3 are separated by several thousand bases, considerable conservation of the sequence can be seen in the portion preceding the translated region. The first codon (ATG) is number +1. The stars indicate conserved bases.

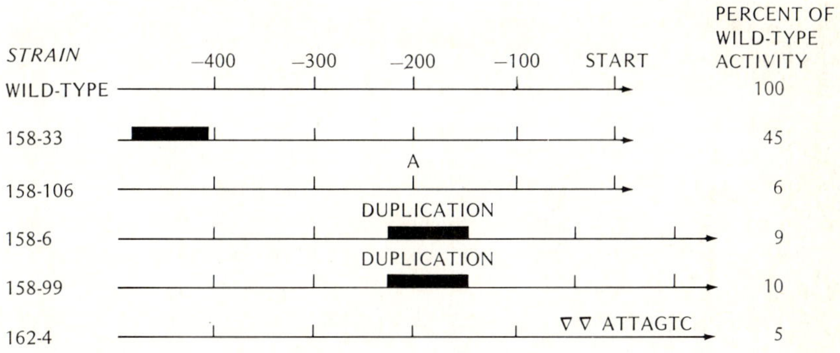

Figure 3.22. Mutations rendering *qa*-2 independent of *qa*-1. Five *qa*-1 independent isolates were analyzed. Unlike the wild-type, they contained quinic dehydrogenase, although the *qa*-1 gene was deleted. However, even when induced they made less than the maximum level of enzyme measured in induced wild-type cells. One *qa*-1 independent strain (158-33) was linked to a new sequence about 400 bases upstream of the transcriptional start site *(solid box)*. Two other strains (158-6, 158-99) carried duplications of about 100 bases *(solid box)*. The other two strains had fewer base changes.

mutations render *qa*-2 independent of *qa*-1, but they might form new RNA polymerase entry sites.

These studies of quinic acid metabolism in the bread mold *Neurospora* are beginning to delineate the control sequences that drive the genes and have recognized a positive control gene, *qa*-1, that integrates the coordinate expression of these genes when quinic acid is added to the environment. They shed light on the development of *Neurospora* only indirectly by giving us confidence that *cis*-acting mutations do affect the transcription of genes and that this is controlled by a dedicated regulatory protein, the *qa*-1 gene product. Cells of *Neurospora* differentiate to form conidia and the female sexual organ, the protoperithecium. These are complicated processes that are not explained just by knowing how a single set of genes is regulated. Nevertheless, it is hoped that when the genes controlling differentiation of *Neurospora* and other organisms are recognized, they will be found to be regulated in a manner not too different from that functioning for the *qa* genes.

Related Readings

Antonarakis, S., Orkin, S., Cheng, T., Scott, A., Sexton, J., Trusko, S., Charache, S., and Kazazian, H. (1984). β-Thalassemia in American blacks: novel mutations in the "TATA" box and an acceptor splice site. Proc. Natl. Acad. Sci. *81,* 1154–1158.

Brown, D. (1984). The role of stable complexes that repress and activate eukaryotic genes. Cell *37,* 359–365.

Chada, K., Magram, J., Raphael, K., Radice G., Lacy, E., and Constantini,

F. (1985). Specific expression of a foreign β-globin gene in erythroid cells of transgenic mice. Nature *314,* 377–380.

Charnay, P., Treisman, R., Mellon, P., Chao, M., Axel, R., and Maniatis, T. (1984). Differences in human α- and β-globin gene expression in mouse erythroleukemic cells: the role of intragenic sequences. Cell *38,* 251–263.

Church, G., Ephrussi, A., Gilbert W., and Tonegawa, S. (1985). Cell-type-specific contacts to immunoglobin enhancers in nuclei. Nature *313,* 798–801.

Efstratiadis, A., Posakon, J., Maniatis, T., Lawn, R., O'Connell, C., Spritz, R., DeRiel, J., Forget, B., Weissman, S., Slightom, J., Blechl, A., Smithies, O., Baralle, G., Shoulders, C., and Proudfoot, N. (1980). Cell *21,* 653–668.

Giniger, E., Varnum, S., and Ptashne, M. (1985). Specific DNA binding of GAL4, a positive regulatory protein of yeast. Cell *40,* 767–774.

Jost, J., Seldran, M., and Geiser, M. (1984). Preferential binding of estrogen-receptor complex to a region containing the estrogen-dependent hypomethylation site preceding the chicken vitellogenin II gene. Proc. Natl. Acad. Sci. *81,* 429–433.

Miesfeld, R., Okret, S., Wikstrom, A., Wrange, O., Gustafson, J., Yamamoto, K. (1985). Characterization of a steroid hormone receptor gene and mRNA in wild-type and mutant cells. Nature *312,* 779–781.

Monod, J., and Jacob F. (1961). Telenomic mechanisms in cellular metabolism, growth, and differentiation. Cold Spring Harbor Symp. Quant. Biol. *26,* 389–401.

Moore, D., Marks, A., Buckley, D., Kapler, G., Payvar, F., and Goodman, H. (1985). The first intron of the human growth hormone gene contains a binding site for glucocorticoid receptor. Proc. Natl. Acad. Sci. *82,* 699–702.

Peschle, C., Mavillo, F., Care, A., Migliaccio, G., Migliaccio, A. R., Salvo, G., Samoggia, P., Petti, S., Guerriero, R., Maninucci, M., Lazzaro, D., Russo, G., and Mastroberardino, G. (1985). Hemoglobin switching in human embryos: asynchrony of $\zeta \rightarrow \alpha$ and $\epsilon \rightarrow \gamma$ globin switches in primitive and definitive erythropoietic lineage. Nature *313,* 235–238.

Renkowitz, R., Shutz, G., Von der Ahe, D., and Beato, M. (1984). Sequences in the promotor region of the chicken lysozyme gene required for steroid regulation and receptor binding. Cell *37,* 503–510.

Richmond, T., Finch, J., Rushton, B., Rhodes, D., and Klug, A. (1984). Structure of the nucleosome core particle at 7Å resolution. Nature *311,* 532–537.

Scheidert, C., and Beato, M. (1984). Contacts between hormone receptor and DNA double helix within a glucocorticoid regulatory element of mouse mammary tumor virus. Proc. Natl. Acad. Sci. *81,* 3029–3033.

Swift, G., Hammer, R., MacDonald, R., and Brinster, R. (1984). Tissue-specific expression of the rat pancreatic elastase I gene in transgenic mice. Cell *38,* 639–646.

Tyler, B., Geever, R., Case, M., and Giles, N. (1984). *Cis*-acting and *trans*-

acting regulatory mutations define two types of promoters controlled by the *qa-1F* gene of *Neurospora*. Cell *336,* 493–502.

West, R., Yocum, R., and Ptashne, M. (1984). *Saccharomyces cerevisiae* GAL1-GAL10 divergent promotor region: location and function of the upstream activating sequence UAS_G. Mol. Cell Biol. *4,* 2467–2478.

Zaret, K., and Yamamoto, K. (1984). Reversible and persistent changes in chromatin structure accompany activation of a glucocorticoid-dependent enhancer element. Cell *38,* 29–38.

CHAPTER 4

Physiological Differentiation

The function of a tissue requires more than the coordinated expression of a set of genes. It often involves dramatic changes in the structure of cells and of their membranes, as well as regulation of their growth. A large number of specialized processes are known in different cell types. We will go into some detail on a few (muscles, lens of the eye, liver cells) to highlight the range of problems faced during embryogenesis. Later, when we consider chosen systems, we will cover more of the integrative processes that adapt cells to the functioning of specific tissues.

Muscle Cells

All metazoans must be able to move parts or all of their bodies to feed, reproduce, and avoid predators. In fact, all growing eukaryotic cells have a requirement for movement, at least of internal organelles such as chromosomes and vesicles. Movement in almost every case involves the function of the contractile proteins, actin and myosin, in conjunction with several ancillary proteins such as α-actinin, tropomyosin, and calmodulin as well as the attachment of actomyosin filaments to membrane structures. It would not do much good to have a contractile filament if it was not attached and pulling on something. The coordination of these proteins specialized for movement starts by regulation of differential gene expression, mostly at the level of initiation of transcription, but extends to complex interactive relationships that determine the state of polymerization and association of the various components.

In the skeletal muscles we see the most highly specialized differentiation of muscle cells. Striated muscles consist of very large syncytial cells containing hundreds of nuclei that were formed by fusion of individual myoblasts. Up and down these myotubes run filaments that can be seen easily in the electron microscope. These filaments consist of polymers of actin and myosin. These proteins are very similar to the actins and myosin found in nonmuscle cells but have certain specialized differences. The muscle actin and muscle myosin light chains are the products of specialized genes expressed preferentially only in differentiating muscle cells. Thus, activation of these specialized genes is one of the events essential to muscle differentiation. Protein synthesis committed to these proteins increases from negligible levels up to about 20 percent of total protein synthesis as growing myoblasts differentiate and fuse to form myotubes.

There are a variety of other biochemical changes that occur at the same time as the accumulation of the contractile proteins that adapt the metabolism of the cell to the function of being a muscle. Besides the normal energy stores found in all cells, muscle cells accumulate a special reserve in the form of creatine phosphate. When demand is placed on the pool of adenosine triphosphate (ATP) for the work of contraction, creatine phosphate replenishes the ATP by donating its high-energy phosphate. The creatine phosphate, in turn, is synthesized by a reaction catalyzed by creatine phosphokinase (CPK). Not too surprisingly, the specific activity of CPK increases considerably during differentiation of myoblasts. Thus, the cells have specialized not only to be able to contract forcibly but also to provide the enzymatic machinery that can support this work.

When muscle cell differentiation is looked at in the microscope, the most dramatic event is the transformation of apparently nondescript growing cells into large multinucleated myotubes. The process goes in stages: first, the myoblasts elongate and align with one another; next, they fuse to form a syncytium; and then the nuclei congregate to one side of the myotube as the characteristic striations arise from polymerization of actomyosin filaments and their attachment to the Z bands, which run across the myotubes (Fig. 4.1). At about that point one can see spontaneous twitching of the myotubes even when differentiation has occurred in the isolation of a glass or plastic dish. The muscle is fully functional.

Culture conditions that permit differentiation of myoblasts, isolated from avian and mammalian embryos, to fuse into myotubes have been worked out. Primary myoblasts require the buffered salts with essential amino acids and sugars that all vertebrate cells require when cultured *in vitro* in dishes. Together with proteins and growth factors found in serum, such a medium will keep the cells intact and healthy and will support a moderate amount of growth. Fusion of myoblasts further requires that the cells be deposited on a thin layer of collagen and that calcium be present in the medium. When the myoblasts are at a sufficient density on the plate, they align, fuse, and accumulate functional actin and myosin (Fig. 4.2). These results clearly in-

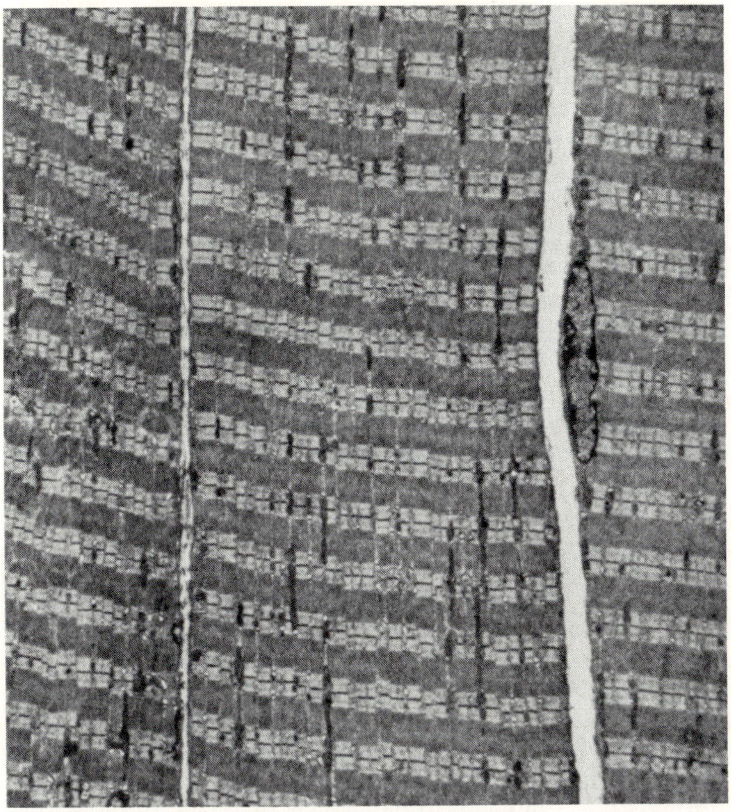

Figure 4.1. Striated muscle. The large syncytial cells are traversed by Z-lines at which actin microfilaments are anchored by several proteins. Microfilaments extend about halfway to the next Z-line and are surrounded by myosin filaments. Relative movement of the actomyosin draws the Z-line together and gives rise to muscle contraction. The muscle is serviced by the sarcoplasmic reticulum where calcium is sequestered.

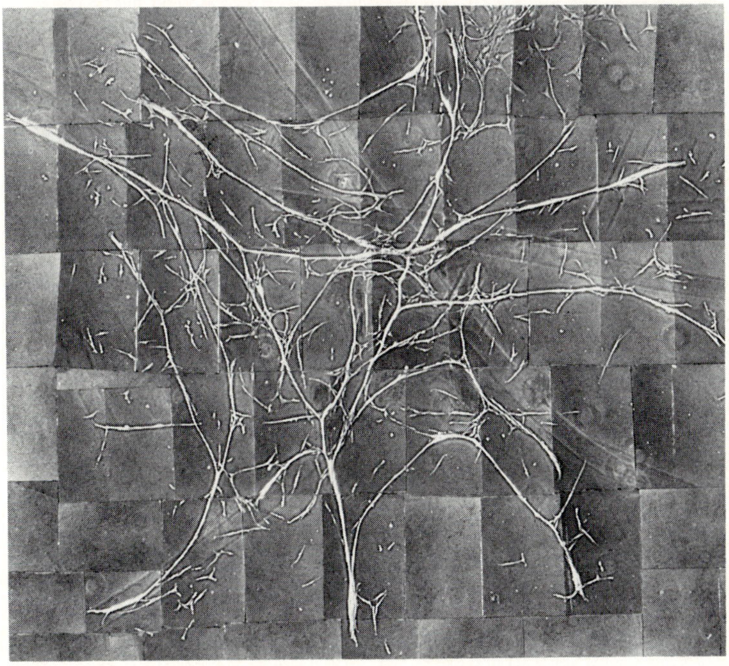

Figure 4.2. Myotubes in culture. Myoblasts can be cultured in dishes until they become confluent. They then fuse to form syncytial myotubes that actively contract.

dicate that myoblasts have the full capacity to undergo terminal differentiation in the absence of other cells or intercellular signals and only require the above-mentioned components to do so. Primary myoblasts do not appear very different at the morphological level from a large number of other cells in the young embryos. They would be hard to distinguish from fibroblasts, many of which will differentiate into connective tissue and specialize in collagen production rather than contractile proteins. Myoblasts can be separated from fibroblasts in minced embryonic tissue because they adhere less rapidly to plastic surfaces. The fibroblasts stick rapidly allowing the myoblasts which are still only loosely attached to be collected. This difference in adhesive properties is only one of the many more subtle differences between myoblasts and fibroblasts. When the two cell types are separately cultured under identical conditions, only the myoblasts differentiate into muscle cells. Something fundamental has determined that myoblasts have the potential for muscle differentiation while fibroblasts have not been so determined. The determined state of myoblasts is stable through many cell divisions, as shown by the ability of permanent cell lines derived from myoblasts to grow for years in the laboratory as long as the cell density is kept low. Yet, they will differentiate into myotubes when placed at high cell density under the appropriate conditions. Cell lines derived from other kinds of embryonic tissue will not differentiate into muscle cells under these or any conditions yet known. We will return several times to this intriguing question of determination and its causes.

Knowledge of the conditions necessary for muscle differentiation and the availability of cloned myoblast cell lines has permitted experiments designed to see whether each of the various aspects of myoblast differentiation are all controlled by the same stimuli or some processes are regulated independently of others.

By keeping primary myoblasts sufficiently far away from each other that they could not fuse, it was found that accumulation of muscle actin, myosin, and CPK occurred normally even though the cells had not fused into myotubes. Likewise, temperature-sensitive mutants of a permanent myoblast cell line were found to accumulate myosin and CPK even at the nonpermissive temperature at which fusion did not occur. Clearly biochemical specialization typical of muscle cells is independent of the process of cell fusion. Differentiation of myoblasts under a variety of conditions has always been found to result in coordinate expression of muscle actin, myosin, and CPK. It may be that these genes share many regulatory molecules such that conditions which activate one activate all.

Lens Cells

Vertebrate eyes use a specialized group of cells to focus light onto the retina. These lens cells are highly differentiated in that they predominantly consist of a small family of related proteins, the crystallins (Fig. 4.3). These proteins are found nowhere else in the organism. They make up about 80

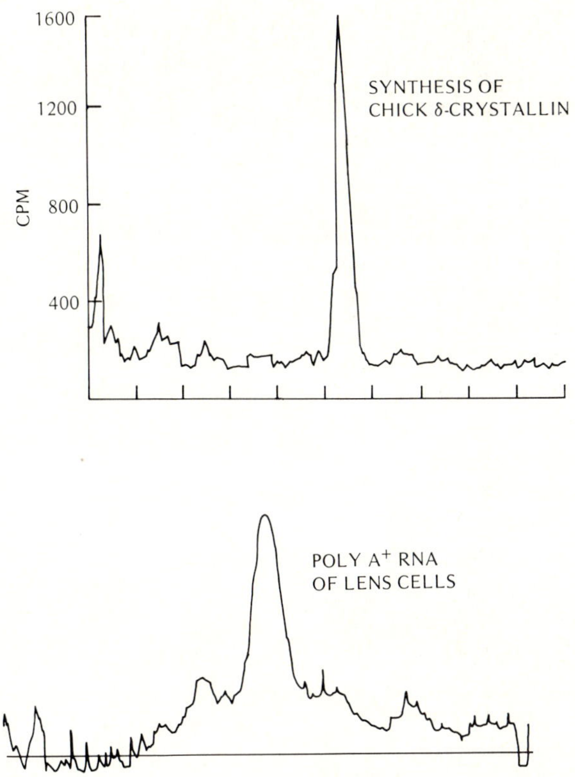

Figure 4.3. Lens fiber crystallin. Total proteins were isolated from chick lens fibers and electrophoretically separated. A protein stain showed exclusively δ-crystallin while analysis of newly made proteins by counting those that had incorporated ^{3}H-valine also showed predominantly δ-crystallin being synthesized. Poly-A$^+$ RNA of lens fiber showed only a single major band. This mRNA codes for δ-crystallin.

percent of the total protein in lens cells and become associated in a semicrystaline manner with each other, in which form they uniformly refract light that passes efficiently through them. The lens cells have further specialized differentiations, which result in their taking up a highly flattened and elongated shape so that they align with each other in layers many cells deep. The cells also reduce the amount of components that might absorb or scatter light, such as mitochondria or small vesicles. They are completely specialized to one function—to pass light into the eye.

During embryogenesis of chickens *(Gallus domesticus)*, δ-crystallins constitute 70 percent of the soluble protein in the lens. There are two tandemly arranged δ-crystallin genes in the chicken genome. One codes for a δ-crystallin of 48 kd and the other for a 45 kd δ-crystallin. The 48 kd protein is synthesized on an mRNA that is 1572 bases long, not counting the poly-A tail that is 200 to 300 bases long. Translation starts at position 87 and ends at position 1427, leaving a 142 bp 3′ untranslated region. The amino acid sequence predicts that more than half of the protein is in the form of α-helices. Twenty-eight bases upstream of the transcriptional starting point of the gene there is a TATA box, and 68 bases upstream there is a CAAT box that may be part of the promoter.

Mice have a single αA-crystallin gene but produce two αA-crystallins that differ by a 23-amino-acid-sequence insert. These crystallins are synthesized from mRNA molecules transcribed from the αA-crystallin gene that

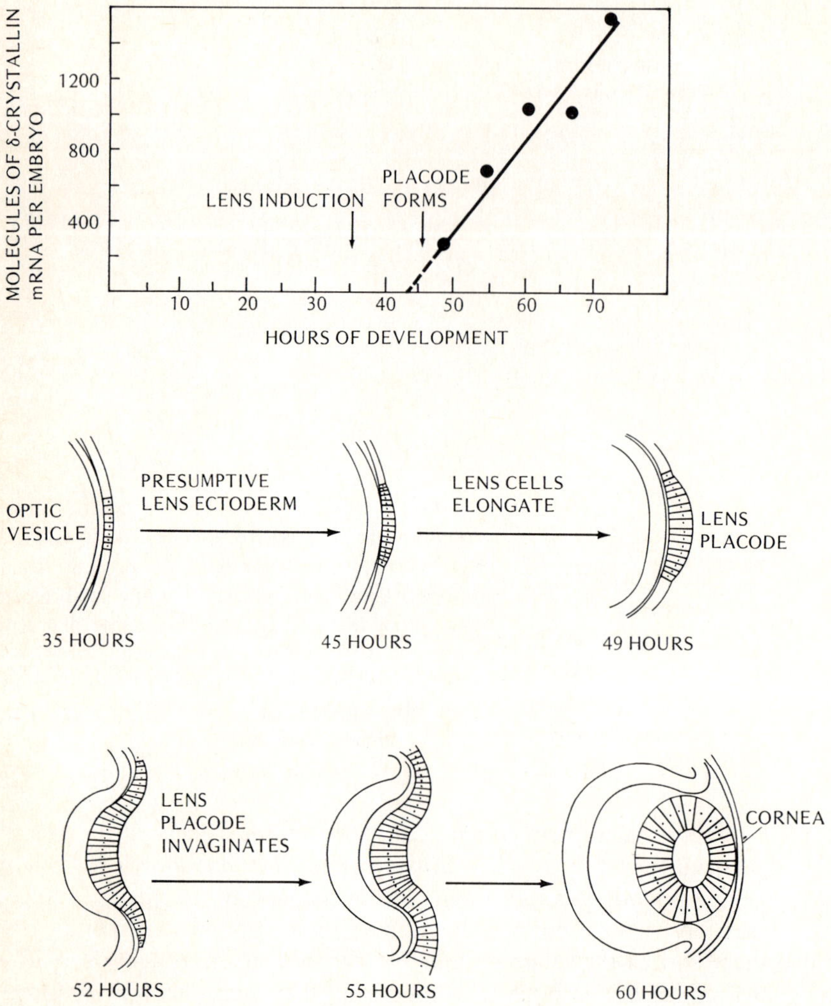

Figure 4.4. Differentiation of the lens. At 35 hours of development in the chick, the optic vesicle is closely opposed to overlying ectoderm but no morphological differentiation of the lens can be seen. The δ-crystallin genes are not expressed at this stage. At 45 hours of development, cells of the lens placode begin to elongate and the δ-crystallin genes are activated. For the next 24 hours δ-crystallin mRNA accumulates until it makes up the bulk of the total mRNA and is present at 1500 copies/cell. During this period the lens placode invaginates and sinks below the ectodermal layer that will form the cornea. Lens specific transcription is controlled by a *cis*-acting DNA sequence present within 364 bases upstream of the transcription start site. The abundance of δ-crystallin mRNA continues to increase until one month after hatching. By then each lens contains 10^{11} molecules of this mRNA.

have slightly different patterns of splicing. The mRNA coding for the smaller αA-crystallin has an intron of 1376 bp excised between the codons for amino acids 63 and 64. The mRNA coding for the larger αA-crystallin is spliced from codon 63 to a 69 bp sequence 266 bp into the intron that codes for 23 amino acids. The remainder of the intron is then spliced out. This alternate splicing pattern appears to account for the production of two polypeptides from a single gene. Whether or not these different αA-crystallins serve selectively advantageous functions in rodent lens is not known for sure, but these results point out the genetic subtleties that can be utilized in cellular differentiation.

The crystallin genes are present in all cells, such as those of liver, thymus, or skin, but are repressed except in the differentiating lens cells. During embryogenesis the neural tube in the head bulges out to form the optic cup. When this neural tissue comes close to overlying ectoderm, a poorly understood induction mechanism induces changes in the ectodermal cells that result in activation of the crystallin genes, among other things. It has been directly shown that the rate of synthesis of crystallin mRNA is stimulated, most likely because of an increase in the rate of initiation of transcription of these genes (Fig. 4.4). This results in a rapid rate of synthesis of the crystallins. Once made, these proteins are stable. So they continue to accumulate for days and then remain and function for years. One of the major challenges of developmental biology is to understand and define the processes of induction by which juxtaposition of neural tissue of the optic bulge with ectodermal cells can direct these cells to lens differentiation.

Liver Cells

The liver is a major changing house of the body. The hepatic vein brings in nutrients from the brush border cells of the intestine. Liver cells efficiently absorb the nutrients and metabolize them into the components then in demand by the body. Since liver cells are exposed to a rapidly varying nutritional environment dependent on the recent diet, they must adapt their enzymatic machinery often. Fasting results in rapid mobilization of glycogen stored in liver cells (Fig. 4.5).

Rats fed a high-protein diet excrete much more urea than rats fed a low-protein diet. Urea production requires five enzymes of the urea cycle. All five enzymes have been found to increase in specific activity when rats are shifted from a diet of 15 percent casein to one of 60 percent casein. Induction of one of these enzymes, arginase, has been studied in detail. It accumulates in the liver cells of rats fed a high-casein diet as a result of an increased rate of synthesis as well as an increase in stability of the enzyme. Thus, liver cells adapt to the change in diet by two independent molecular mechanisms, which both lead to an increase in the relative concentration of arginase.

Likewise, rats given high doses of the amino acid tyrosine rapidly adapt

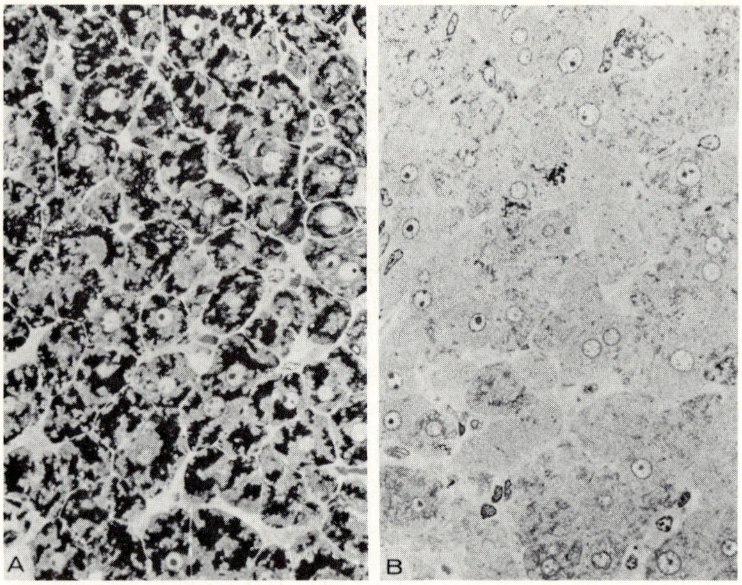

Figure 4.5. Changes in glycogen content in liver cells of fasting rats. *(A)* The liver of an animal that fasted for 2 hours contains 8 percent glycogen. *(B)* The liver of an animal that fasted for 21 hours contains 0.9 percent glycogen.

by increasing the specific activity of the enzyme that metabolizes tyrosine, tyrosine amino transferase (TAT). The induction of TAT in liver cells is not directly mediated by tyrosine but is controlled by raised levels of steroid hormone in the animal. Increased tyrosine levels stimulate adrenal cells to secrete glucocorticoid hormones into the bloodstream, and these induce the TAT genes in liver cells but not in other cells of the body. Liver cells contain steroid-binding proteins, which mediate the induction of TAT and other genes. The steroid-binding protein associates with the steroid hormone and stimulates transcription of the TAT gene by binding at a site adjacent to the gene. The mRNA that codes for TAT is present at about a tenfold greater level in steroid-induced liver cells than in unstimulated cells. Both the rate of transcription of the TAT gene into mRNA and the stability (half-life) of the mRNA are increased in steroid-induced cells.

The liver is also the site for detoxification of the bloodstream. One class of toxic compounds to which organisms are exposed is the heavy metals such as zinc and cadmium. In response to increased levels of these metals, liver cells synthesize a protein, metallothionein, which tightly binds the heavy metal and sequesters it for the remainder of the life of the organism. The protein accumulates as a consequence of increased transcription of metallothionein genes.

There are several genes in the human genome that code for metallothionein (MT). One of them, MT-I, has been cloned and studied in detail. Transcription of MT-I is regulated not only by heavy metals in the environment but also by steroid hormones (Fig. 4.6).

There is a sequence of about 20 bases present 250 bases upstream of the transcriptional start site of MT-I at which steroid hormone–receptor

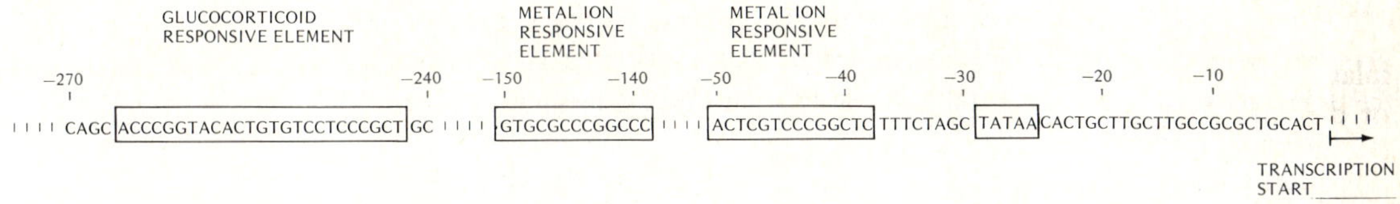

Figure 4.6. Human metallothionein promotor. The hMT-I is a gene that is regulated by both glucocorticoids and heavy metals. Transcription starts at base number 1. A TATA box occurs 25 bases upstream. Sites necessary for the metal ion response (MRE) are boxed. A 20-base sequence further upstream (GRE) is protected when the glucocorticoid receptor is bound.

complex binds and stimulates transcription. The nucleotide sequence in this *cis*-acting control element is similar to the consensus sequence recognized in other steroid hormone induction sites. Closer to the MT-I gene, at −150 to −138 and again at −50 to −38, are two sequences involved in metal ion responsiveness. The nucleotide sequences in these control elements are almost identical and resemble a consensus sequence of metal ion induction sites found in other metallothionein genes. The identity of the heavy metal receptor that binds to these control elements is not yet known.

Heavy metals in the bloodstream induce the MT gene in liver cells but not in many other cells. What is the basis for this variation between cells carrying the same genome? There is some indication that the DNA sequences near the MT gene are covalently modified in cells in such a way as to allow or preclude induction. The evidence points to methylation of the DNA in the upstream flanking regions of the MT gene being decisive.

In metazoans the cytosine residues are sometimes methylated at CG pairs. When such a sequence is replicated, the newly made duplex will be methylated on the CG of the parental strand but not on the daughter strand GC. A maintenance methylase will rapidly methylate the new GC cytosine, giving fully methylated DNA. However, a sequence containing unmethylated CG will not be methylated *de novo* with anything near the same frequency. Thus, the pattern of DNA methylation will be a relatively stable hereditary property of the cell.

Metallothionein genes that are inducible have been found to have unmethylated sequences in the proximal 5′ flanking regions, whereas similar genes in cells that fail to response to heavy metal induction have been shown to have these sequences methylated. If nonresponsive cells are treated with compounds that interfere with DNA methylation, they then become responsive, and this property is inherited through many successive generations. Their MT-associated DNA is altered in methylation and they respond much as liver cells do.

The interpretation of these results is that during embryogenesis liver precursor cells were instructed to alter their methylation pattern near the MT and other genes so that these genes would be responsive to exogenous signals. Because liver cells grow through many generations during embryogenesis and, in fact, throughout the whole lifetime of the organism, this property must be a stable differentiation of liver cells, that is, they are determined in such a way that they are able to induce genes that other cells are unable to express.

Methylation of control regions of genes may be one mechanism of cellular determination but it is clearly not the only one. There are other cases of determination in organisms in which methylation has not been implicated. Perhaps other stable and transmittable differences in the DNA account for those cases, but they have yet to be analyzed at the molecular level.

Metabolic Integration

Many of the specialized functions carried out by different tissues do not require accumulation of specific gene products to extremely high concentrations but only slight changes in the expression of a few dozen genes. The great majority of the pattern of gene expression in cells of the dermis is almost the same as that in liver. Likewise, in early embryogenesis one cannot easily point to the critical differences that distinguish primary mesenchymal cells that lead the invasion at gastrulation. These cells seem to be expressing most of the same genes that are active in cells on the surface of the blastula. There are some differences in gene expression between these early cell types but they are hard to recognize because they constitute a minor proportion of the total activity.

Nevertheless, changes in a few proteins can have dramatic consequences for the function of a cell. Many of the cytoskeletal components of cells (microtubules, intermediate filaments, microfilaments) are in an unstable equilibrium with their subunits. If the components, such as tubulin, increase in a cell, a major extension of the microtubular array may occur. This can result in the elongation of the cell.

Association of proteins with each other, activation of one enzyme by another, degradation of an enzyme by a specific protease, and modification of a protein by a protein kinase are all nonlinear processes by which a small change in the proportion of a specific gene product may be amplified to have major consequences.

At yet another level, the pattern of metabolic flow within a cell may be radically altered by fairly minor changes in one or a few of the enzymes that catalyze specific interconversions. The flux of small molecules along a given pathway of intermediary metabolism or biosynthetic metabolism is controlled not only by the relative catalytic activity of the respective enzymes but also by the concentration of the intermediates that serve as substrates. The cellular concentration of almost all compounds of metabolism is less than that necessary for saturation of the enzymes that use them as substrates. To put it another way, the substrates are at a concentration less than the Km of the enzymes. Therefore, from classical enzymology we know that doubling the substrate concentration will almost double the rate of the reaction without any need for a change in the enzyme concentration. Likewise, reducing a substrate by drawing it off along another pathway will decrease the flux along the original pathway. One can think of small molecule metabolism as the flow of water through a maze of interconnecting canals—raising a dike in one will alter the flow not only in that canal but in many others. In this way relatively small changes in only a few pathways can have far-reaching effects on the overall pattern of distribution.

On the other hand, the interdependence of pathways and components of organelles dampens the consequences of differences on the final behavior. The end result is that several quite different configurations can equally well serve the same purpose. Of course, there are configurations that do not work, such as the complete lack of an essential component or a radical departure from the path as the result of a great excess of a controlling component. But between these extremes there are many solutions to such biological problems as optimum growth in a complex environment or the integrated function of tissues in a complex organism. Redundancy is built into responsive biological systems that use many separate units to achieve a goal. Therefore, individual cells that all carry out the same role may do so in significantly different ways in the same tissue of a given species. Selective pressure is applied to the end result and not to how it is achieved. Only when we accurately quantitate single components on a cell-by-cell basis do we see the individual variations. We get an average when the whole tissue is analyzed.

Specialization of cellular physiology involves dramatic changes in some cases and minor changes in others, and it is often fine-tuned by specific feedback regulation by both macromolecules and small molecules of metabolism onto the biosynthetic processes of both. The details of these subtle regulatory processes differ from cell type to cell type as well as during the sequence of embryonic stages. Some of the molecular changes underlying the development of *Dictyostelium* are considered in this light in Chapter 11.

Summary

Cells in tissues with specialized functions are often characterized by dramatic accumulation of mRNA from a few genes. These transcripts direct a high rate of synthesis of the specialized proteins of the cells. Muscle cells accumulate far more actin, myosin, creatine phosphokinase, and a dozen other proteins than cells of other tissues. Conditions necessary for a high rate of expression of these genes have been studied *in vitro* in wild-type and mutant myoblasts. Lens cells accumulate a small family of proteins referred to as crystallins. Some of the crystallin genes generate more than one mRNA species by differential splicing. Liver cells can be induced by steroid hormones to accumulate several proteins, including tyrosine amino transferase (TAT) and metallothionein. Steroids associate with specific receptor proteins and increase both the rate of transcription of these genes and the stability of the mRNAs made. The metallothionein genes are also induced by heavy metals. Methylation of certain sites near the metallothionein genes occurs in cells unable to respond to either steroids or heavy metals and may determine their lack of expression in inappropriate tissues.

SPECIFIC EXAMPLES

1. Actin and Myosin Gene Expression in *Drosophila*

There are six different actin genes in the fruit fly, *Drosophila melanogaster*. Two are expressed throughout embryogenesis and give rise to the actins needed in all cells. Two are expressed predominately in late embryos and in the grub larvae and most likely make larval skeletal muscle actin. Two others are expressed only in pupae and give rise to the actins required for flight muscles in the adult fly. The portions of these genes that are translated into actin are about 85 percent homologous to each other and are similar to actin genes found in other organisms. In fact, the first actin sequence was isolated from the eukaryotic amoeba, *Dictyostelium discoideum*, and then subsequently used to recognize actin genes in *Drosophila* by cross-hybridization. Actin is the most abundantly synthesized protein in early development of *Dictyostelium* and thus can be conveniently isolated from these cells.

Actin-specific sequences can be isolated by cloning cDNA to actin mRNA. The first step in this procedure is to isolate mRNA from *Dictyostelium* cells a few hours after the initiation of development. By passing a preparation of RNA over a column of bound poly-T (oligo-dT cellulose) the mRNA can be purified. Almost all mRNA molecules in eukaryotes have a run of about 100 adenine bases (poly-A) added to the 3′ end following transcription. Ribosomal RNA molecules (28*S* and 18*S*) make up about 96 percent of the total RNA in a cell but conveniently do not have poly-A at their 3′ ends. Nor do tRNAs. The poly-A^+ mRNA is bound to the column while the bulk of RNA flows through. The mRNA can then be eluted by reducing the ionic strength of the eluant.

Purified poly-A^+ RNA can be copied into DNA by addition of reverse transcriptase that is isolated from retroviruses. The cDNA copy is tailed with single strands of oligo-dC at each end with the enzyme terminal transferase. A bacterial plasmid, pBR322, is cleaved with the restriction enzyme Pst, which cuts at the single CTGCAG site. The linear plasmid DNA is tailed with single strands of oligo-dG at each end to generate complementary sequences that will bind to the dC-tailed cDNA. The cDNA and plasmid molecules are allowed to hybridize to each other and covalently linked by the enzyme DNA ligase.

The recombinant DNA is then used to transform bacteria by selecting for tetracycline resistance. The plasmid carries a gene *(Tet^r)* conferring resistance to tetracycline. Those bacteria that take up the plasmid become drug resistant and each one carries a single clone. Several hundred independent

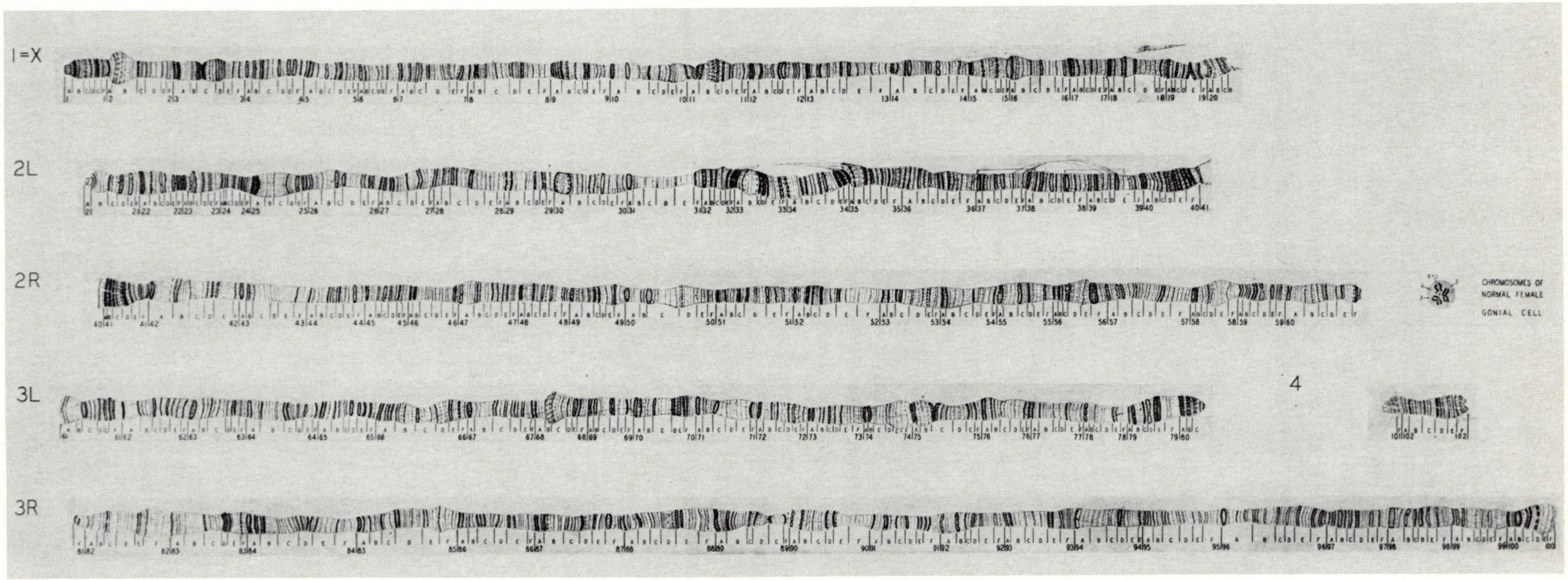

Figure 4.7. Cytogenetic map of *Drosophila melanogaster*. Chromosome 1 (the X) is cataloged into 20 sections, each of which is further subdivided into 6 divisions lettered A through F based on the sequence of bands. Chromosome 2 has 40 sections (21 to 60); chromosome 3 also has 40 sections (61 to 100); the small chromosome 4 has only two sections (101 to 102). Homologous polytene chromosomes are intimately paired in salivary gland cells. Genetic mapping by recombinational frequency has generated a parallel map.

transformed bacterial colonies are then screened for the presence of *Dictyostelium* sequences by hybridization to radioactively labeled *Dictyostelium* DNA. The screening can be carried out right on the plate on which the bacteria grow by lysing the cells and denaturing the DNA.

Since actin mRNA makes up 3 percent of all the mRNA in *Dictyostelium* cells, only a few dozen clones had to be screened. Several cDNA plasmids were found that would hybridize to actin mRNA as judged by their ability to direct actin synthesis. A rabbit reticulocyte lysate provided all the components for protein synthesis (ribosomes, tRNAs, etc.) and synthesized actin protein when hybrid-selected mRNA was added. The actin was recognized by its size and isoelectric point. Sequencing of the actin cDNA confirmed that there were several hundred bases that coded for a protein almost identical to actin from other organisms in amino acid sequence. The *Dictyostelium* actin sequence could be used with confidence to screen for actin genes in other organisms by hybridization.

Drosophila DNA was partially digested by the restriction enzyme EcoRI and the fragments were attached to a λ phage vector (Charon 4). These recombinant phage were grown in bacteria and the DNA in each plaque was screened directly on the plate with labeled *Dictyostelium* actin probe. DNA of one recombinant phage was found to hybridize well with the probe. It carried a 17.5-kb insert of *Drosophila* DNA in the middle of which was 3.3 kb of an actin gene.

The actin mRNAs of *Drosophila* range in size from 1.6 to 2.2 kb. An intervening sequence of 1.65 kb is spliced out of the original transcript to give the final mRNA. The particular actin sequence cloned comes from a gene at salivary chromosome band 5*C* on the X chromosome. The portion of the cloned gene coding for actin protein is 85 percent homologous to five other actin genes and hybridizes to them. They were mapped by *in situ* hybridization to salivary bands 5*C*, 42*A*, 57*A*, 79*B*, 87*E*, and 88*F* (Fig. 4.7). The actin genes at these loci are designated by these band names. Transcribed sequences to the 3′ end of the translated portion are different for each mRNA and allow expression of each gene to be distinguished by hybridization of cloned 3′ fragments (Fig. 4.8).

Actin-5*C* mRNA is present throughout development and accumulates as three different-sized molecules of 2.2, 1.95, and 1.7 kb. Actin 42*A* is also expressed throughout development as a 1.7 kb mRNA. Both of these genes are also expressed in nonmuscle cell lines grown in tissue culture. It was concluded from these results that these genes supply the nonmuscle actin required in all cells.

mRNA from actin 57*A* and 87*E* are most abundant during late embryogenesis, decrease during larval growth, and reaccumulate during late pupal development (Fig. 4.8). These genes are expressed when the larval musculature is differentiating and then being restructured, and they probably give rise to skeletal muscle actin.

mRNAs from actin 79*B* and 88*F* appear only during pupal development (Fig. 4.8). These genes are most active in the leg and thorax where the flight muscles are localized (Fig. 4.9). They are not expressed in the head, abdomen, or ovary and thus seem to be specialized to flight muscles. Mu-

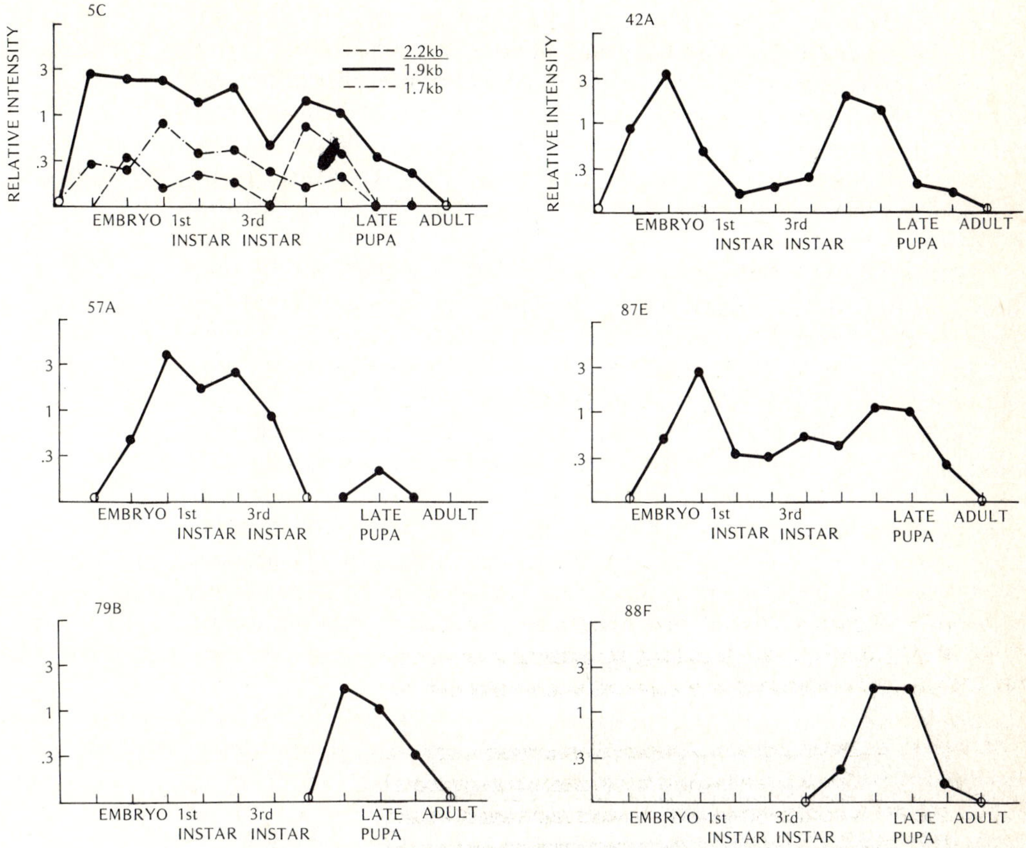

Figure 4.8. Relative accumulation of various actin mRNAs during *Drosophila* development. The amount of hybridization of the different cloned probes to total poly-A$^+$ RNA isolated at different stages is plotted on a logarithmic scale. The actin 5C probe bound to mRNA of three different sizes.

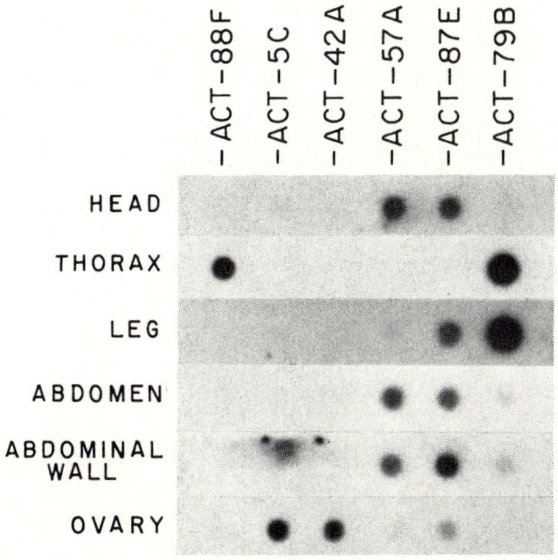

Figure 4.9. Organ specificity of various actin mRNAs in *Drosophila*. RNA extracted from isolated organs was copied into radioactively labeled cDNA, which was then hybridized to dots of DNA from each clone immobilized on nitrocellulose. The amount of radioactivity bound to each dot blot indicates the relative prevalence of the homologous mRNA.

tations in the actin gene at 88*F* result in flightless flies with aberrant myofibrils in the thoracic flight muscles, which supports this conjecture.

Each of these genes codes for similar if not identical 42,000 dalton (42 kd) actin. The differences in expression suggest either that the actins they code for are specialized in a subtle way for the different functions they carry out or, alternatively, that all code for functionally identical actins but are members of different teams of genes brought on line when a given tissue differentiates.

In contrast to the six actin genes, there is only a single myosin heavy-chain gene in *Drosophila*. However, the myosin gene gives rise to three different mRNA molecules (7.2, 8.0 and 8.6 kb) that differ primarily in their patterns of splicing at the 3′ end. One of these mRNA molecules is found only when the principal flight muscles of the adult are made.

During embryonic development of *Drosophila,* myosin is synthesized at a maximal rate at 15 hours. Myosin is an exceptionally large protein (200 kd) and requires an mRNA of at least 7 kb. To isolate cloned myosin sequences, mRNA greater than 7 kb was purified from 15-hour embryos. Labeled copies of this mRNA were used to screen a partial EcoRI genomic DNA library cloned in phage λ Charon 4. The gene turns out to be very large (19 kb) and contains at least nine introns that are spliced out to give

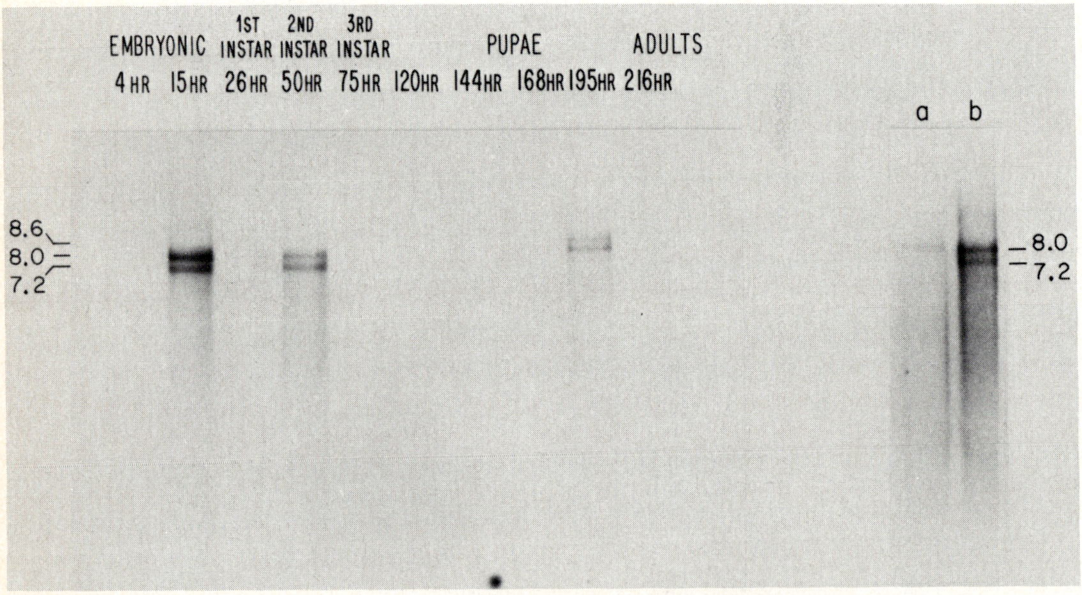

Figure 4.10. Developmental Northern of myosin heavy chain mRNA. At various times after egg deposition, RNA was isolated from *Drosophila* embryos, larvae, and pupae, size separated, and hybridized on nitrocellulose to a 6-kb Eco RI fragment of λDmMHC-1 that was radioactively labeled by nick-translation. The MHC-1 fragment recognized two mRNA species in the range of 7 to 8 kb in a 15-hour embryo, second instar larvae, and late pupae but found no homologous RNA at other stages.

the final mRNA molecules. Only a single sequence for myosin heavy chain occurs in the *Drosophila* genome and is present at salivary band 36*B*, yet differential splicing can generate three different myosin mRNAs.

Northern Blotting

The myosin gene is expressed at three distinct periods in *Drosophila* development: (1) at 15 hours of embryogenesis, (2) in second instar larvae, and (3) late in pupation (Fig. 4.10).

In this experiment RNA was isolated at various stages of development and separated by size on a 1 percent agarose gel by electrophoresis. After transfer of the RNA to nitrocellulose, it was hybridized with labeled DNA of the cloned myosin gene. Only myosin mRNA hybridizes, and the amount of radioactivity bound to the different-sized mRNA gives an indication of their abundance. This is referred to as a Northern analysis to distinguish it from the Southern analysis in which DNA fragments of various sizes are analyzed. The different *Drosophila* myosin mRNAs code for different myosin proteins since splicing differences at the 3′ end result in different carboxy-terminal amino acid sequences.

In rat and human genomes there are 13 nonmuscle myosins. These are flanked by repetitive sequences. Although repetitive sequences have often been proposed as common regulatory sequences, the repetitive sequences flanking the myosin genes of man and rats differ significantly. Yet early development of these two mammals is almost identical. The lack of conservation of the repetitive sequences indicates that such sequences probably do not regulate myosin gene expression.

2. Differentiation in the Pancreas

When food leaves the stomach, it is attacked by a barrage of enzymes secreted into the small intestine by the pancreas. The major pancreatic enzymes are amylase, lipase, chymotrypsin, trypsin, carboxypeptidase, and nucleases. These digestive enzymes are synthesized at a high rate in pancreatic cells and secreted into ducts that lead to the top of the small intestine. The pancreas is the second largest gland in mammalian organisms.

At 9.5 days of gestation in the mouse, the pancreatic diverticulum can be seen as a small protuberance from the layer of epithelial cells lining the gut. Surrounding mesodermal cells condense around the diverticulum and proliferate. Over the next few days the epithelial sheet grows surrounded by mesodermal cells and branches (Fig. 4.11). By 15 days of gestation acinar structures with collecting ducts are evident (Fig. 4.12). Endocrine B cells separate as islets and start to produce insulin and other hormones.

The exocrine cells that line the ducts accumulate massive amounts of rough endoplasmic reticulum (RER) (Fig. 4.13). These internal membranes are studded with functioning ribosomes and are indicative of cells involved in rapid secretion of proteins. The enzymes destined for secretion are concentrated as inactive forms (zymogens) in large dense bodies referred to as zymogen granules (Fig. 4.13). These granules concentrate on the side of the cell facing the lumen of the ducts. Pancreatic cells in adults have the same polarized appearance as those in the pancreas of 20-day embryos.

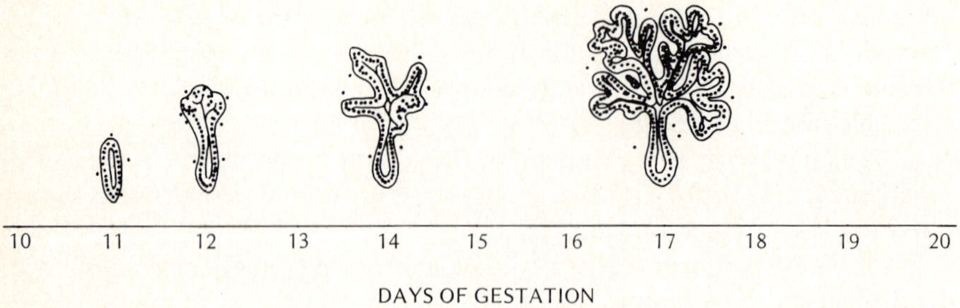

Figure 4.11. Morphogenesis of pancreatic gland. During the second and third week of gestation of the mouse embryo, the original diverticulum grows and takes on a characteristic branching pattern.

Conditions have been found that permit differentiation of pancreatic rudiments *in vitro.* Both biochemical and cytological changes follow the same program as is seen *in vivo* in the embryo. Isolated pancreatic rudiments allow various analyses and experimental treatments to be carried out with precision. For instance, this technique has shown that differentiation of pancreatic epithelial cells requires the presence of mesenchymal tissue. Mesenchyme taken from pancreatic, salivary, or other rudiments will support differentiation of pancreatic epithelial cells. A protein fraction (MF) has been isolated from mesenchymal tissue that interacts with the surface of pancreatic cells to permit both biochemical and morphological differentiation. The exact nature of this interaction has not yet been elucidated.

Mouse pancreatic tissue accumulates large amounts of amylase between 14 and 20 days of embryonic development. This has been shown to be the result of accumulation of mRNA directing a high rate of synthesis of the enzyme. RNA was extracted from pancreatic cells at various times and used

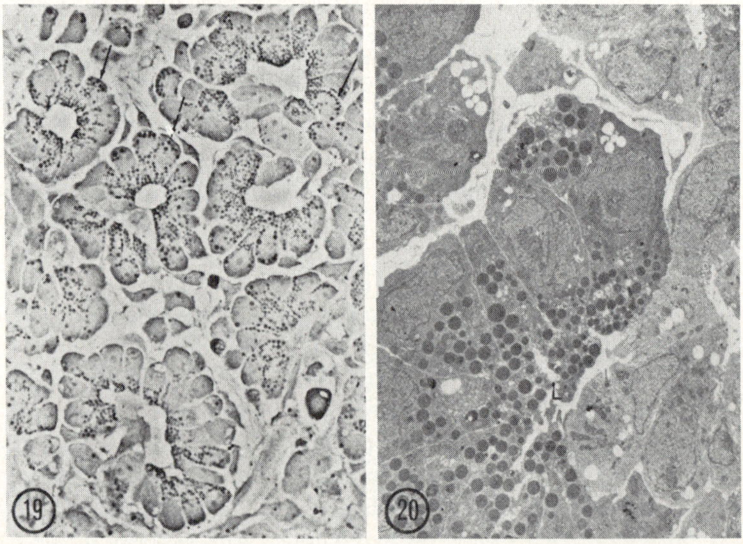

Figure 4.12. Differentiation of explanted pancreas. Thirteen-day mouse embryonic pancreas was explanted and cultured for an additional 6 days (19 days total). Large acinar lumens formed. Cytoplasm of the cells filled with zymogen granules. A day later in culture (20 days total) the acini resemble those in adult pancreas. The zymogen granules have become dense.

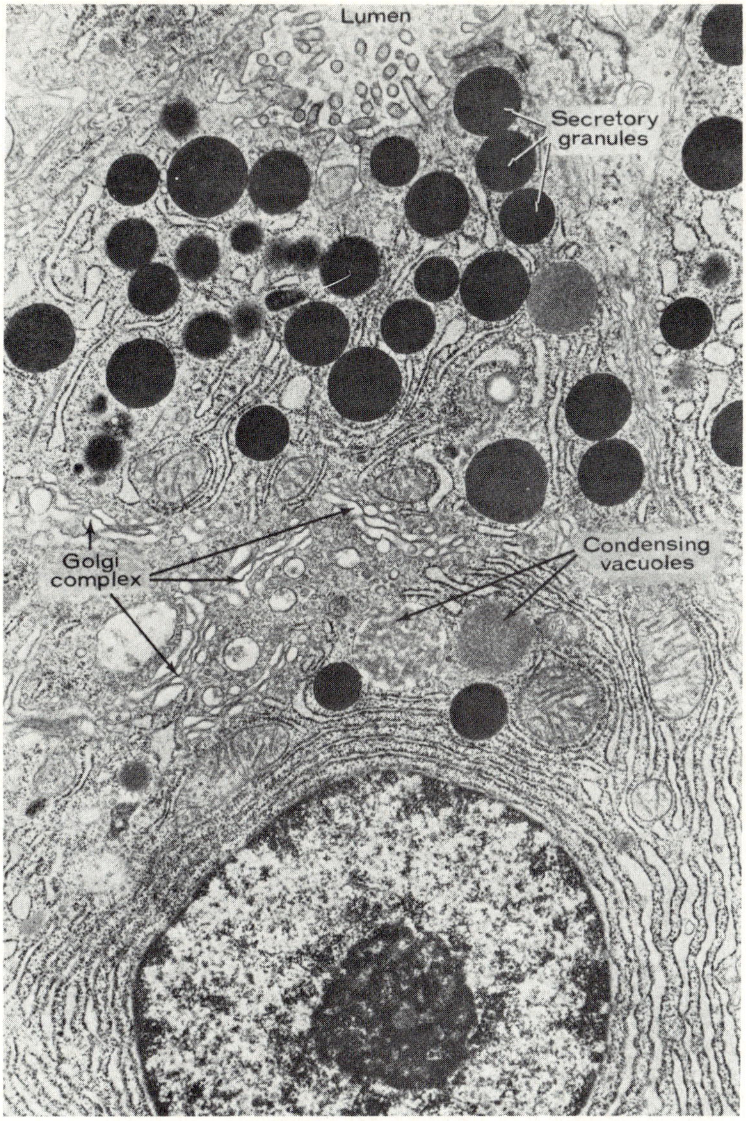

Figure 4.13. Pancreatic acinar cell. The layers of granular endoplasmic reticulum (ER) are closely packed in parallel arrays. Pancreatic enzymes are synthesized on membrane-bound ribosomes and translocated into the lumen of the ER. From there they are moved to the Golgi apparatus for post-translational modification before being stored in secretory granules. The lumen is at the top.

to direct cell-free protein synthesis in an *in vitro* translation system prepared from rabbit reticulocytes. There was a 100-fold increase in amylase mRNA activity between 14 and 20 days of development.

Two-Dimensional Display of Proteins

The proteins synthesized by embryonic cells of the pancreas can be recognized when they are separated on gels in two dimensions (2D) (Fig. 4.14). Cells are labeled for six hours with ^{3}H-leucine and then their proteins extracted with detergent. The solublized proteins are put on the top of a small tube containing ampholytes dissolved in an acrylamide gel. Ampholytes are small peptides of various isoelectric points (pK_I) that diffuse during electrophoresis until they reach their pK_I. They establish a pH gradient. Proteins

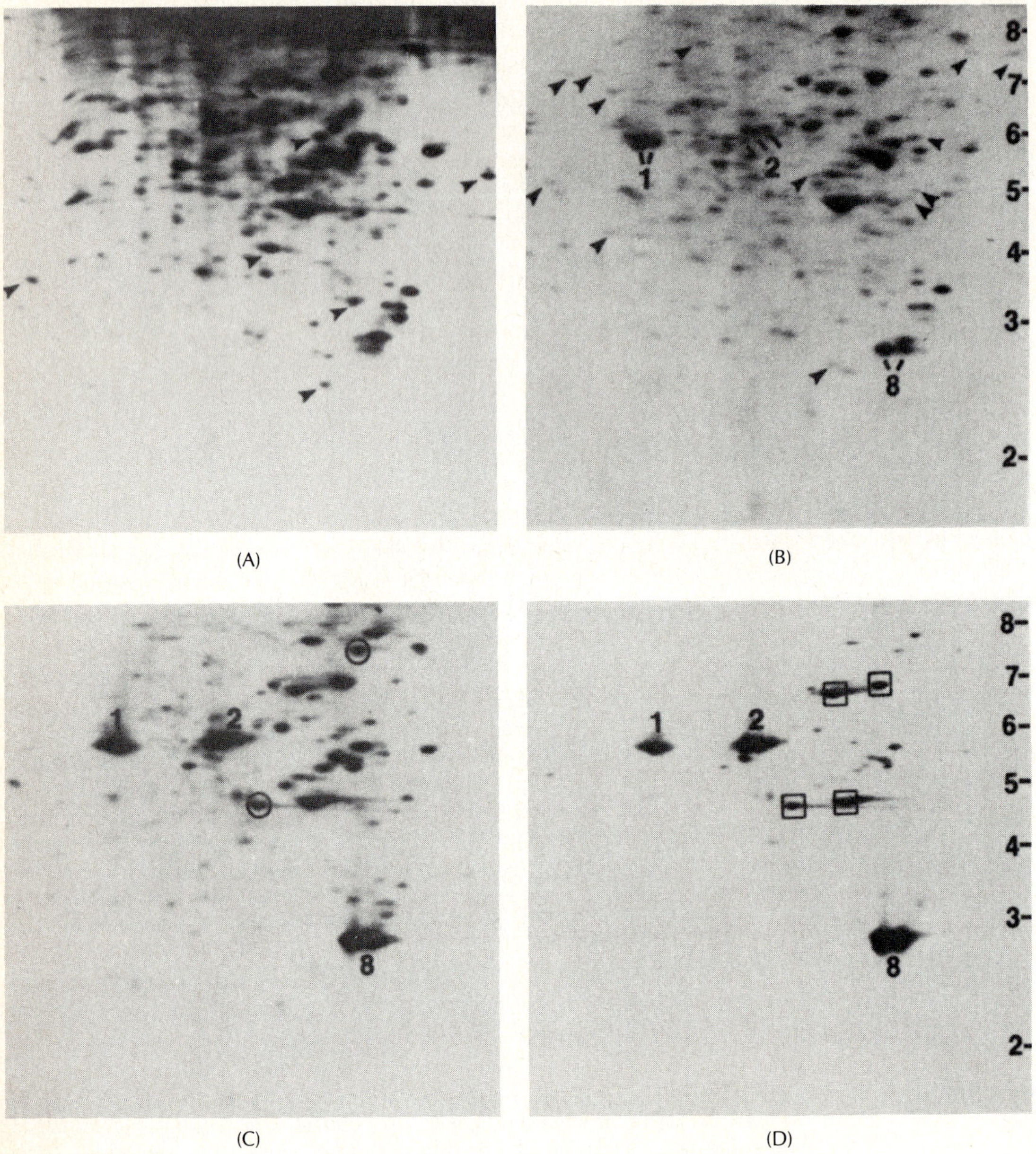

Figure 4.14. Synthesis of pancreatic proteins. Fourteen-day embryonic pancreas was explantated and cultured: *(A)* Zero days, *(B)* two days, *(C)* four days, *(D)* six days, before being labeled with ^{3}H-leucine for six hours. Newly synthesized proteins were separated by isoelectric point in the horizontal dimension and by size in the vertical dimension. Autoradiography of the gels showed the rapidly synthesized proteins. Boxed spots and proteins 1, 2, 8 come to predominate. Protein 8 is chymotrypsinogen.

enter the gel more slowly and migrate electrophoretically until they reach the region of the gel where the pH results in the net negative charges balancing the net positive charges. Individual proteins are concentrated in a few hours at the gel position of their intrinsic pK_I. The tube gel is then placed on a slab gel that does not contain ampholytes but has the detergent SDS dissolved in the polyacrylamide. Proteins migrate in SDS gels under an electric field at rates determined by their size. Smaller proteins, such as those of 20 kd, migrate rapidly through the pores of the gel while larger proteins, such as those of 100 kd, migrate much more slowly and remain near the top of the gel. By separating proteins on the basis of these quite distinct properties, pK_I and size, up to 1000 different proteins can be distinguished as falling at discrete spots on a 2D gel.

Autoradiography of 2D gels on x-ray film gives a picture of the relative rate of synthesis of different proteins during the labeling period. Cells of a 14-day pancreas synthesize at least 200 proteins at a high rate (Fig. 4.14). No detectable synthesis of amylase or chymotrypsinogen can be seen. The rate of synthesis of six of these prevalent proteins decreases dramatically in the next 2 days of development. Amylase and chymotrypsin synthesis can first be seen at 16 days. These pancreatic enzymes are synthesized at ever-increasing rates during the following 4 days until they account for at least a quarter of the amount of all the proteins being made (Fig. 4.14). Two-dimensional gel analysis of newly synthesized proteins gives a global quantitative view of the pattern of protein synthesis in a differentiating tissue.

Although chymotrypsinogen synthesis is turned on at 16 days of gestation, trypsinogen synthesis does not occur until after 20 days of gestation. This indicates a clear temporal sequence in the biochemical differentiation of pancreatic cells. Chymotrypsinogen and trypsinogen are inactive forms of the pancreatic enzymes chymotrypsin and trypsin that are stored in zymogen granules. Thus, these proteases do not attack each other while being stored in a concentrated form in the pancreatic cells. When liberated into the intestine, they are activated by the removal of an inhibitory portion of the protein and can assist in breaking down ingested proteins.

Sequence analysis of five genes coding for proteases expressed at high levels in pancreatic cells has shown that each is preceded by a TATA box about 30 bases 5′ to the transcriptional initiation site and, more interestingly, each of these genes has a sequence closely related to TCAGGGCACCTGTCCTTTTCC a few hundred bases upstream. This sequence is found at bases −217 to −197 of the rat chymotrypsin *B* gene. A cloned gene carrying 274 bases that precede the chymotrypsin gene is transcribed at a high rate in a rat exocrine pancreas tumor cell line (AR4-2J) that produces chymotrypsin from its endogenous genes. However, when the cloned gene is cut down so that it carries only 192 bases preceding the chymotrypsin gene, it is expressed tenfold less frequently. These results directly indicate that there is a *cis*-acting sequence essential for expression of the chymotrypsin *B* gene between bases 192 and 274 upstream of the transcriptional initiation site. The region includes the 21-base consensus sequence found in all pancreatic protease genes.

The transcriptional control region of the chymotrypsin gene works well in chymotrypsin-producing cells (AR4-2J) but not in fibroblasts (CHO) nor in hamster pancreatic endocrine cells (HIT) that produce insulin but not chymotrypsin. Thus, the control region functions in a cell-type specific manner. When a 302-base sequence 5′ to the rat insulin gene was tested in these cell lines, it was found to efficiently direct transcription in insulin-producing β-cells (HIT) but not in the chymotrypsin-producing cells (AR4-2J) nor in the fibroblast cell line (CHO). Thus, the insulin gene also carries a *cis*-acting cell-type specific region that is found upstream of the transcriptional initiation site.

Neither the orientation nor the precise positioning of these control regions relative to the transcription units are crucial for cell-type specific regulation. The insulin sequence (−250 to −160) can even be moved to the 3′ end of a gene and still direct cell-type specific transcriptional regulation. These properties differ from those of bacterial operators that must be 5′ to the genes they control but are similar to those of viral enhancer sequences that affect transcription over extended regions of many kilobases in either direction. Thus, these pancreatic genes may be regulated by enhancers that respond to cell-type specific positive regulatory factors.

Related Readings

Bernstein, S., Mogami, K., Donady, J., and Emerson, C. (1983). *Drosophila* muscle myosin heavy chain encoded by a single gene in a cluster of muscle mutations. Nature *302,* 393–397.

Falkenthal, S., Parker, V., and Davidson, N. (1985). Developmental variations in the splicing pattern of transcripts from the *Drosophila* gene encoding myosin alkali light chain result in different carboxyl-terminal amino acid sequences. Proc. Natl. Acad. Sci. *82,* 449–453.

Hejmancik, F., Beebe, D., Ostrer, H., and Piatigorsky, J. (1985). δ- and β-crystallin mRNA levels in the embryonic and posthatched chicken lens: temporal and spatial changes during development. Devel. Biol. *109,* 72–81.

Karin, M., Haslinger, A., Holtgreve, H., Richards, R., Krauter, P., Westphal, H., and Beato, M. (1984). Characterization of DNA sequences through which cadmium and glucocorticoid hormones induce human metallothionein-II_A gene. Nature *308,* 513–519.

Karlick, C., Mahaffey, J., Coutu, M., and Fyrberg, E. (1984). Organization of contractile protein genes within the 88F subdivision of the *D. melanogaster* third chromosome. Cell *37,* 469–481.

King, C., and Piatigorsky, J. (1983). Alternate RNA splicing of the murine αA-crystallin gene: protein-coding information within an intron. Cell *32,* 707–712.

Nickerson, J., and Piatigorsky, J. (1984). Sequence of a complete chicken δ-crystallin cDNA. Proc. Natl. Acad. Sci. *81,* 2611–2615.

Piatigorsky, J. (1984). Lens crystallins and their gene families. Cell *38,* 620–621.

Pinset, C., and Whalen, R. (1985). Induction of myogenic differentiation in serum-free medium does not require DNA synthesis. Devel. Biol. *108,* 284–289.

Richards, R., Heguy, A., and Karin, M. (1984). Structural and functional analysis of the human metallothionein-I_A gene: differential induction by metal ions and glucocorticoids. Cell *37,* 263–272.

Schimke, R., Brown, M., and Smallman, E. (1963). Turnover of rat liver arginase. Annu. N.Y. Acad. Sci. *102,* 587–601.

Shani, M., Zevi-Sonkin, D., Saxel, O., Carmen, Y., Katcoff, D., Nudel, U., and Yaffe, D. (1981). The correlation between synthesis of skeletal muscle actin, myosin heavy chain, and myosin light chain and the accumulation of corresponding mRNA sequences during myogenesis. Devel. Biol. *86,* 483–492.

Stuart, G., Searle, P., Chen, H., Brinster, R., and Palmiter, R. (1984). A 12-base pair DNA motif that is repeated several times in metallothionein gene promotors confers metal regulation to a heterologous gene. Proc. Natl. Acad. Sci. *81,* 7318–7322.

Tomkins, G., Thompson, E., Hayashi, S., Gelehrter, T., Grauner, D., and Peterkofsky, B. (1966). Tyrosine transaminase induction in mammalian cells in tissue culture. Cold Spring Harbor Symp. Quant. Biol. *31,* 349–360.

Van Nest, G., Raman, R., and Rutter, W. (1983). Effects of dexamethasone and 5-bromo-deoxyuridine on protein synthesis and secretion during *in vitro* pancreatic development. Devel. Biol. *98,* 295–303.

Walker, M., Edlund, T., Boulet, A., and Rutter, W. (1983). Cell-specific expression controlled by the 5′ flanking region of insulin and chymotrypsin genes. Nature *306,* 557–561.

CHAPTER 5

Cellular Differentiation

Cells come in many sizes and shapes as well as many kinds. The mechanisms that give rise to these differences as well as the consequences of these differences make up the problems of cell differentiation. We will discuss differentiations of epithelial, nerve, and red blood cells to emphasize how specializations in embryogenesis can lead to quite different functions.

Epithelial Cells

Epithelial cells are those that line surfaces. Many of them are highly flattened so as to cover as much surface as possible and are referred to as squamous cells. Others, such as the brush border cells of the intestine, are columnar and terminate in the gut lumen with highly distended villi (Fig. 5.1).

The villi provide the maximum cellular surface for absorption of nutrients from the intestine. Each cell presents several hundredfold more surface to the nutrients passing by and therefore can take up proteins, lipids, carbohydrates, and ions hundreds of times more efficiently than if only a flat surface were presented. Some of the morphological differences between brush border cells and squamous cells result from differences in the cytoskeleton. Microfilaments run the length of each villus and keep that portion of the cell distended. The cytoskeleton that structures each microvillus is a com-

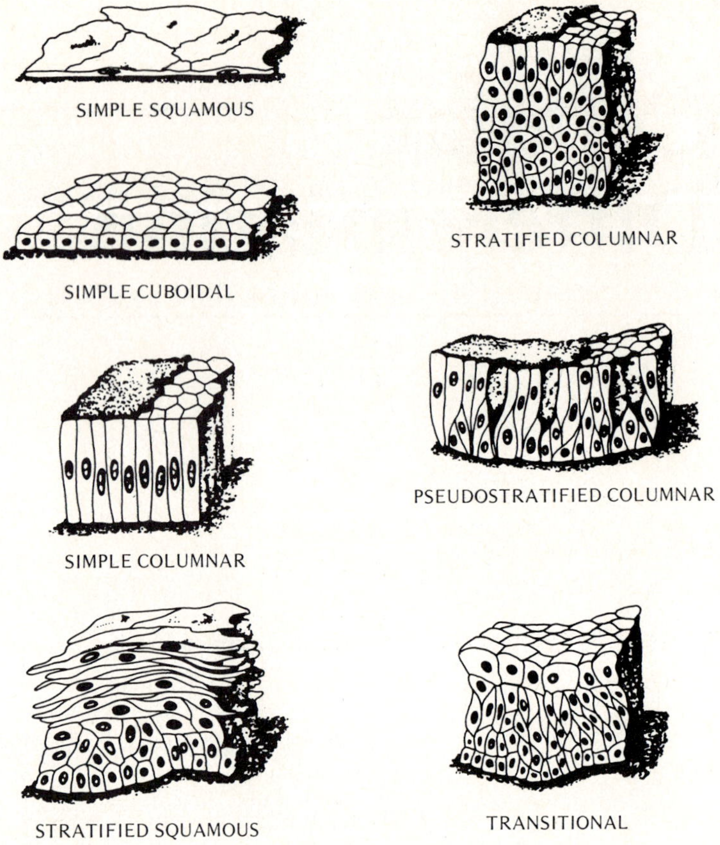

Figure 5.1. Epithelial cell shapes. Squamous cells are highly flattened and are stratified in some tissues such as skin. Other epithelial cells are columnar such as those of the intestinal lining. There are intermediate, transitional shapes in some epithelial tissues.

plex scaffolding extending from a cap at the end to the underlying intermediate filaments, microfilaments, and microtubules (Fig. 5.2). A variety of structural proteins are assembled in specific arrangements to hold the microfilaments in place. The microfilaments themselves are long polymers of actin molecules that have self-associated. The proteins fimbrin and tropomyosin bundle the filaments together. α-Actinin and vinculin appear to anchor them to the cellular cytoskeleton. Association with myosin permits movement of these cellular structures. As cells of the intestine are replaced, these complex structures are formed specifically on the side facing the lumen of the gut. There is reason to consider that when an intestinal epithelial cell divides at least some of the cortical structures are inherited intact and direct the formation of replicas as the daughter cells grow.

Although we cannot point to specific genes whose activities account for the gross difference between squamous and brush border cells, we are confident they are responsible. Some of them will probably include genes for actin-binding proteins, which anchor the microfilaments to the tips of the villi and provide a foundation for the microfilaments just under the villi. Specialized structures can be seen in electron micrographs at these positions but their composition and molecular structure are still unknown.

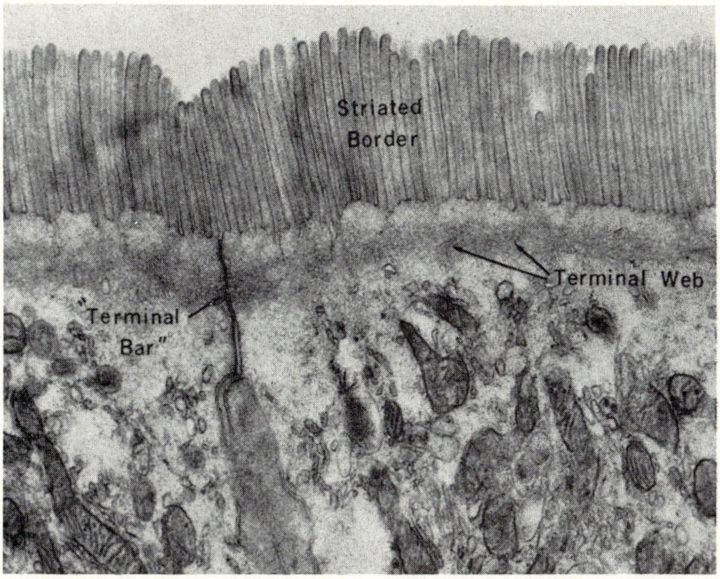

Figure 5.2. Microvilli of brush border cells from intestinal epithelium form a striated border. Each process is 80 to 90 nm in diameter and up to 1 μm in length. Microfilaments run the length of each villus and end in a terminal web just below the surface of the cell. Close aposition of adjacent cells forms a terminal bar.

The membrane of brush border cells is also highly differentiated: nutrients are rapidly taken up at the tip and sides of villi and then transported across the cell and donated to the capillary bed, which lies just under the brush border. Once in the bloodstream, the nutrients are carried to the liver by the hepatic portal circulation. The brush border cells are highly adapted to pinocytosis of the nutrient fluid as well as the direct absorption of small molecules. Permeases and molecular-transport-proteins appear to be particularly concentrated in the membrane of these cells. Thus, the protein content of the membrane has been adapted to the specialized function of these cells. Even the lipid content of the membrane has been specialized in the villi to allow optimal uptake.

The intestine also contains many things that are best kept out of the body, such as viruses and bacteria. The brush border cells are welded together to form an impermeable sheet. The welds consist of certain structures visable in an electron micrograph, which form rings around each cell and are embedded in the plasma membranes of both adjacent cells. Closest to the intestinal lumen, just below the region of microvilli, is the zonula occludens or tight junctions, followed by desmosomes consisting of a dozen or so specific proteins to which intermediate filaments attach. Further down the sides of the cells there are gap junctions connecting adjacent cells (Fig. 5.3). Spanning the two membranes at a gap junction is a protein pipe called a connexon. It is made of six interlocking protein molecules surrounding a central cavity about 10 to 20Å wide. Small molecules can diffuse between cells through these gap junctions. These specialized membrane proteins block the passage of all molecules in the extracellular space. The only way a molecule reaches the bloodstream is to enter the brush border cells and then be secreted to the capillaries.

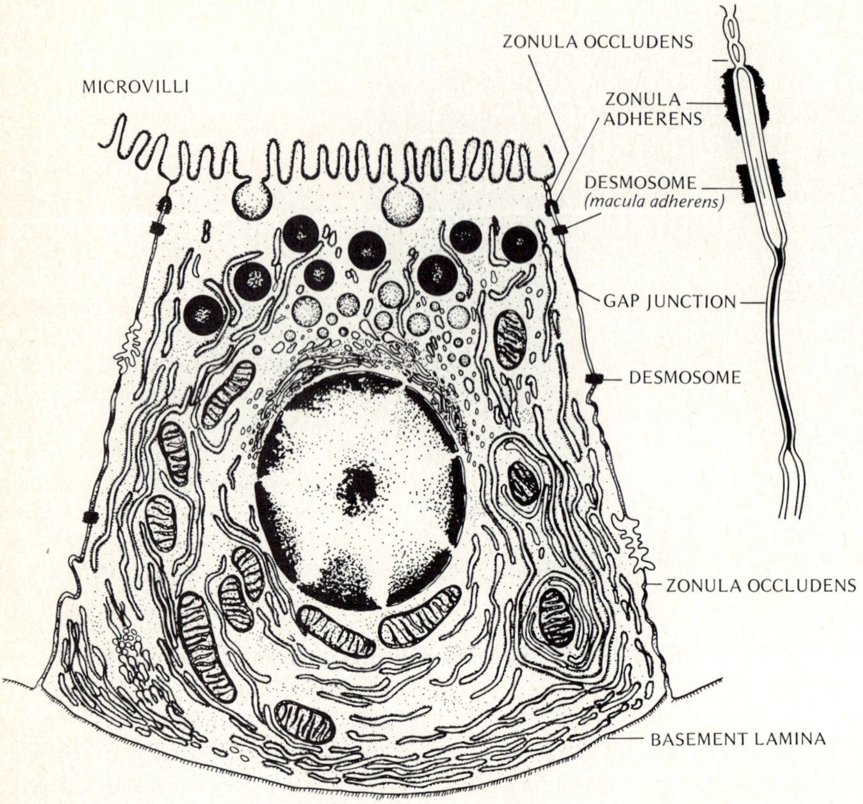

Figure 5.3. Gut epithelial cell. The intestine is lined with cells that lie over a basement lamina and extend microvilli into the lumen of the intestine. When the apical border is magnified, several specialized structures can be seen attaching the cells to one another. The zonula occludens is a band at which adjacent membranes are tightly opposed. Desmosomes are found further below, followed by a band of gap junctions that provide direct communication of small molecules between cells.

Adhesion

One mechanism of adhesion of brush border cells can be easily seen in electron micrographs: the gap junctions (Fig. 5.4).

The membranes of two cells connected by gap junctions are held together by the interaction of the proteins of a connexon. Six molecules of 54 kd associate in hexagonal arrays with a central pore (Fig. 5.5). These proteins span the lipid membrane of the cell. When connexons on two cells are juxtaposed, they associate and open their pores to allow passage of metabolites and ions. The pores of a gap junction are too small for proteins or nucleic acids to diffuse between the cells, but small molecules freely diffuse and metabolically couple the cells. Gap junctions not only allow rapid ionic communication between adjacent cells but also hold the cells together strongly. Densely packed gap junctions show up on electron micrographs as electron

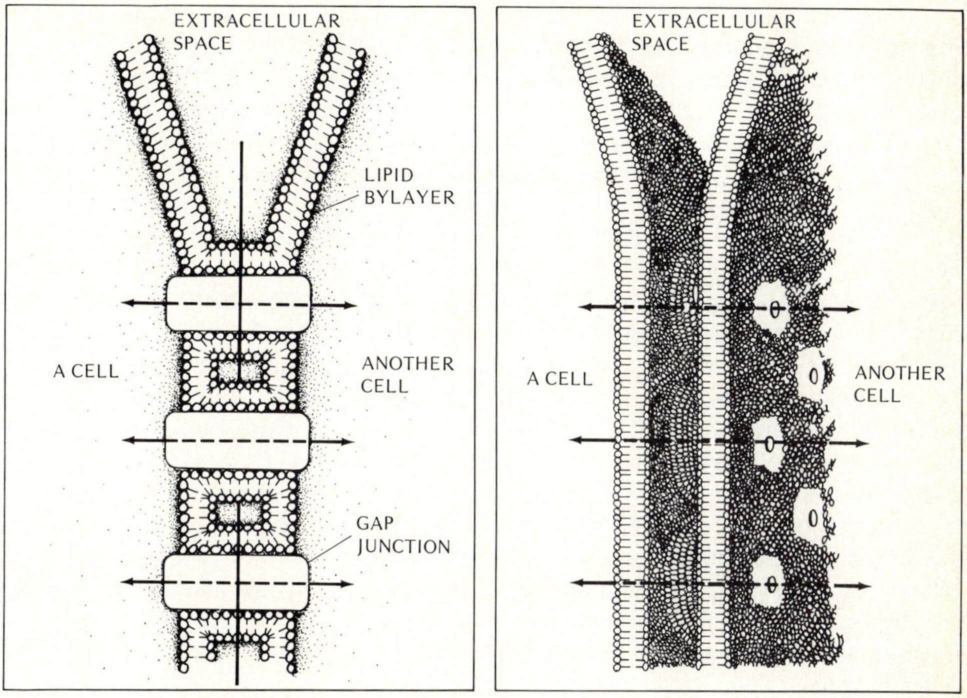

Figure 5.4. Two views of gap junctions. Intercellular channels allow movement of small molecules from one cell to the next with no leakage to the extracellular space.

opaque material spanning a 20-A gap between cells. The cells are essentially riveted together. Cells of *Xenopus* and mouse embryos are connected by gap junctions during the blastula stage. These junctions provide strength to the tissue roofing over the blastocoel and allow intercellular interaction to be mediated by diffusion of small molecules.

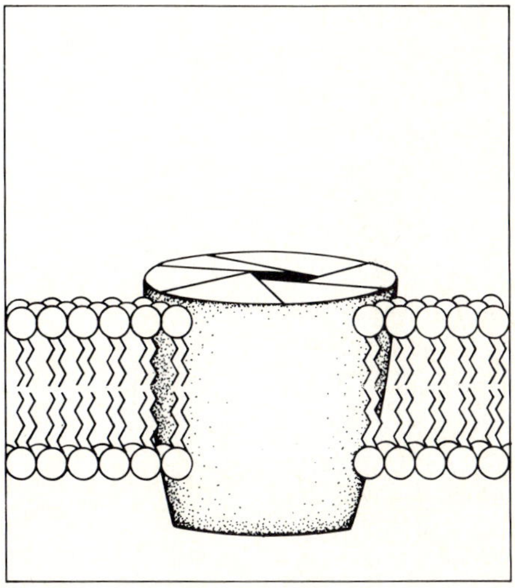

Figure 5.5. Gap junction pore. A connexon is formed from six subunits. When two such units on adjacent cells come together, the external faces of the connexons bind to each other and the central channel enlarges. This model is derived from detailed x-ray crystalographic analysis and is accurate to 2Å. The lipid bilayer is shown with the nonpolar chains as wavy lines and the polar heads on the surface. The membrane is 85Å across.

Many cell types adhere to each other by more subtle molecular mechanisms, which cannot be seen but only inferred. By immunizing rabbits or mice with specific cells or membranes, antibodies have been found that will bind to the surface of cells and block their adhesion to other cells of the same type. Some degree of cell-type specificity is found among the cells that bind each other. Thus, nerve cells bind most strongly to nerve cells and liver cells bind most strongly to liver cells. Antibodies to nerve cells block their self-adhesion but have little effect on liver cells. Likewise, antibodies that bind to the surface of liver cells block cell–cell adhesion specifically among liver cells without affecting adhesion among nerve cells. In each case, unique membrane glycoproteins appear to be the target for the adhesion-blocking antibodies. These molecules are cell-type specific with one kind being found on nerve cells and another on liver cells. As development of an embryo proceeds, the carbohydrate modification on these molecules changes and this may play a role in how these cells associate with each other and related cells. Although much still remains to be determined concerning the molecular mechanism of cellular adhesion, it is presently thought that specific proteins are inserted into the membrane and modified by addition of one or another carbohydrate side chains. These molecules bind to each other, and since they are anchored in the cell membrane, they hold the cells together. This gives integrity to a tissue made up of similar cells and can help segregate cells of different types.

Nerve Cells

Cells in the neural tube and on the neural crest, which form during early embryogenesis, are not obviously specialized at that stage for the role they will play—to be the nerve cells of the organism. They are not especially large or elongated until their terminal differentiation starts at later stages. Many nerve cells become highly elongated by extending a long process, which will become the axon. Some of these are up to a meter in length. Shorter projections become the dendrites. The projections pass chemical and electrical signals to each other and tell the head what the foot is doing. The processes are extended and maintained by polymerization of tubulin into microtubules. All cells have microtubules, but nerve cells have many more arranged in long filaments that hold the cells out in an extended form. There are multiple genes coding for β-tubulin, one of the subunits of microtubules. At least two of them are preferentially expressed in neural cells of the rat and may code for tubulin especially adapted to function in neurite extension. There is much more that determines the shape of specific nerve cells. During development they are often seen rapidly extending processes, then retracting them, then putting them out again, sometimes on another path. They seem to follow landmarks on the structures they pass through, and the final shape is stabilized by interaction with cells along the way.

With such long processes, nerve cells must be able to transport materials back and forth over long distances within themselves. There is rapid axonal as well as retrograde transport of both small molecules and macromolecules in nerve cells that appears to require specialized mechanisms of motive force within the cells.

Signals are passed from one end of a nerve cell to the other very rapidly. These signals are referred to as electrical since the membrane potential changes as the stimulus passes like a wave traveling down the axon. In molecular detail, the concentration of Na^+ and K^+ is rapidly changing due to the presence of potential-sensitive Na^+ ports and specialized pumps that return the internal Na^+ concentration to its unstimulated (low) level. Although many cell types contain a few of these membrane proteins, nerve cells have differentiated to have them at a high density.

When the electrical signal reaches the end of an axon, the stimulus must be passed on to the next cell. In many cases a small molecule is released into the space between it and the next cell. There is a considerable number of different neurotransmitters, as these small molecules are called. One of them is acetylcholine. It is synthesized in the nerve cell and accumulated in small specialized organelles called synaptic vesicles. These are found clustered at the endplate (Fig. 5.6). Here is another biochemical and cytological differ-

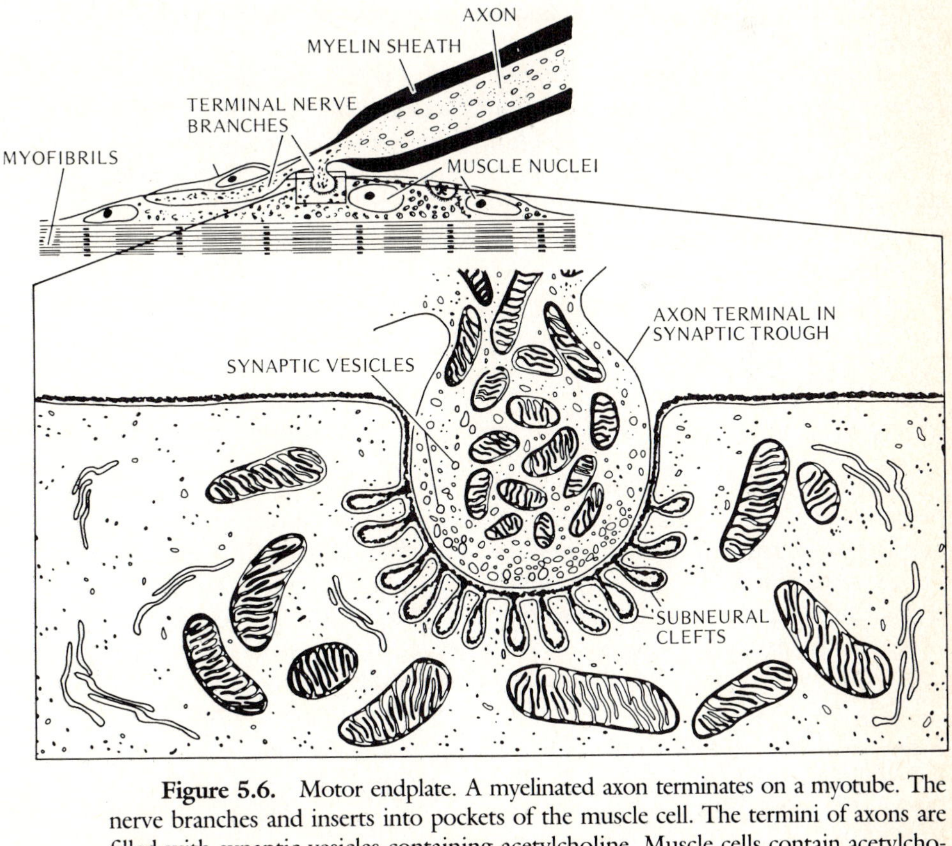

Figure 5.6. Motor endplate. A myelinated axon terminates on a myotube. The nerve branches and inserts into pockets of the muscle cell. The termini of axons are filled with synaptic vesicles containing acetylcholine. Muscle cells contain acetylcholine receptors concentrated on the surface lining subneural clefts.

entiation of nerve cells: (1) alteration of the enzymatic make-up to be able to synthesize relatively large amounts of acetylcholine; (2) construction of a special membrane system to package the neurotransmitters. When an electrical signal reaches the synaptic cleft, acetylcholine is liberated into the extracellular space. There, the neurotransmitter diffuses across the short distance to the surface of the adjacent cell. Both nerve and muscle cells have unique acetylchoine receptors on their surfaces, usually clustered under the endplate. The receptor consists of a complex of five different proteins that together form a cation channel across the membrane. When acetylcholine is bound by one of the subunits, the channel opens and allows Na^+ to rush in. The resulting electrical signal is then passed on down the nerve or muscle cell.

The expression of the genes responsible for the subunits of acetylcholine receptor is regulated so that it is high in nerve and muscle cells but low in other cells. Therefore, only these cells are responsive to stimuli of acetylcholine. Some types of nerve cells do not use acetylcholine as a neurotransmitter but use other small molecules such as norepinephrin. These cells have the enzymatic machinery for production of norepinephrin and secrete it onto cells with specialized norepinephrin receptors. In this way, mutually responsive systems can differentiate by controlling the expression of only a small number of genes. The modulation and routing of the gene products to different parts of cells can then adapt them to very specific functions.

Embryonic neural cells can differentiate to become noradrenergic or cholinergic, depending on what hormones and inducing proteins they are exposed to. A specific protein of 45 kd has been found that induces cells to be cholinergic. The production of the 45 kd protein is regulated in turn by such hormones as insulin, epidermal growth factor (EGF), and corticosteroids. Thus, a cascade of induction processes can finely regulate neural differentiation.

Blood Cells

Another type of highly specialized cell is known to us every time we prick our fingers: red blood cells. These small cells are red because they contain large quantities of the red protein, hemoglobin. During their differentiation greater than 80 percent of total protein synthesis goes into making α- and β-globin, the proteins of hemoglobin. The globin mRNAs are some of the best understood of all because they are relatively easy to purify from erythropoietic tissues; very few other genes are being expressed.

The red blood cells of mammals are about as terminally differentiated as can be. They not only have broken down most of their proteins so that the amino acids can be used for globin synthesis but have also extruded their nuclei, precluding any further cell division (Fig. 5.7). They are bags of hemoglobin playing the absolutely essential job of conveying oxygen to internal tissues and removing carbon dioxide. But after their functional life of

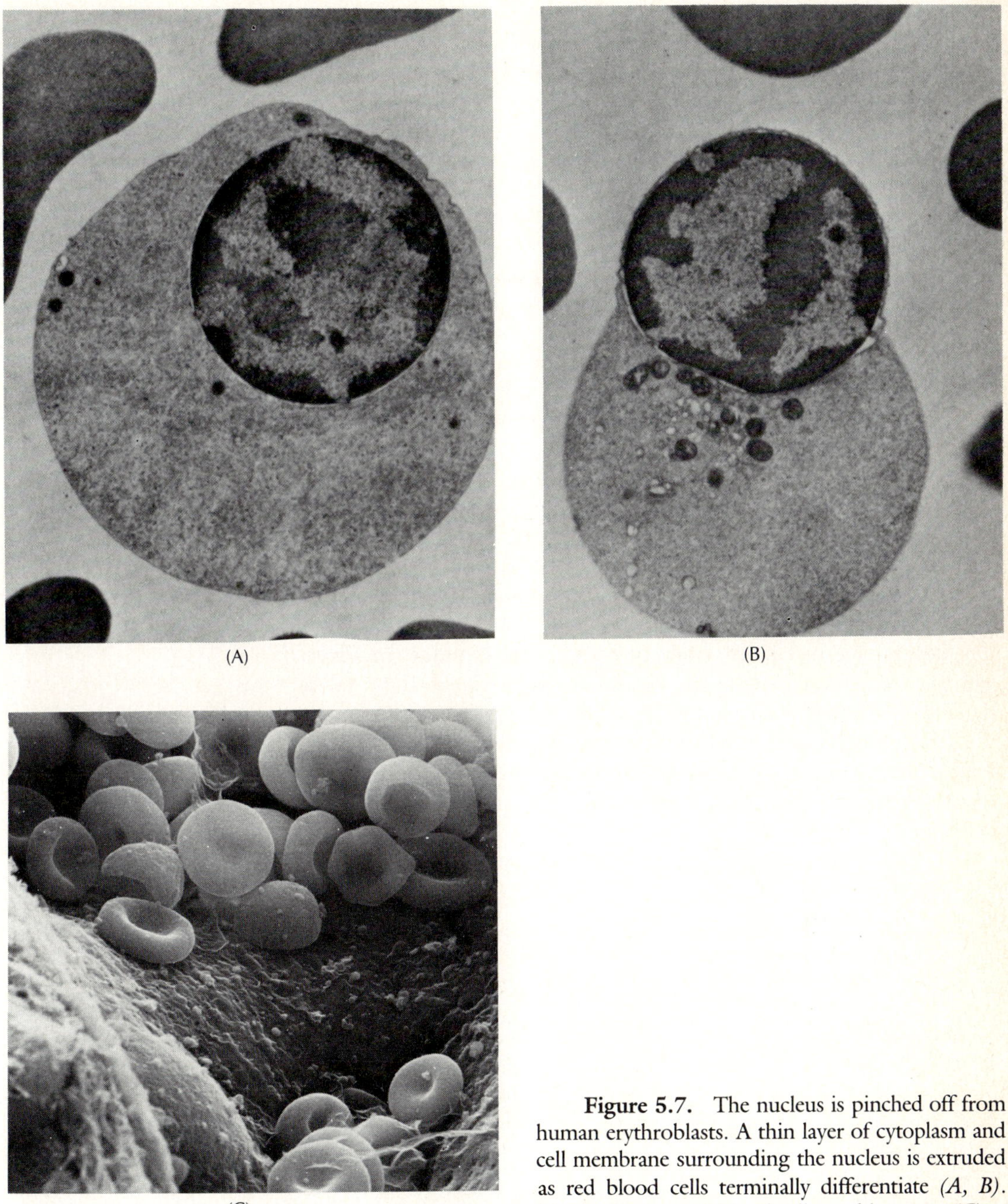

(A) (B) (C)

Figure 5.7. The nucleus is pinched off from human erythroblasts. A thin layer of cytoplasm and cell membrane surrounding the nucleus is extruded as red blood cells terminally differentiate *(A, B)*. Mature human red blood cells are biconcave *(C)*.

a few months, they are broken down in the liver and the components are redistributed throughout the body for other purposes.

These very characteristics raise some interesting questions. How is the timing of hemoglobin accumulation linked to extrusion of the nucleus? What regulates growth of the cells that give rise to red blood cells? What signals the degratory process and concomitant specialization that will preclude further growth?

So far almost nothing is known about extrusion of the nucleus. It happens in mammals but not in birds, reptiles, or amphibians. Under severe anemic conditions in mammals, oxygen transport is carried out by circulating erythrocytes that still contain their nuclei and much of their biosynthetic and metabolic machinery. Ridding themselves of their nuclei may just allow red blood cells to shrink to a smaller size and thereby prolong their functional lives as they course through tiny capillaries.

Immature erythrocytes have specific cell surface sites at which they attach to one of the large proteins of extracellular matrices, fibronectin. Their affinity to the matrix may keep them in the bone marrow until maturity. When they are prematurely released in anemic animals, they circulate to the spleen, where the fibronectin-coated membrane is modified by macrophages. The phospholipid and cholesterol content of erythrocyte membrane is also reduced as they mature in the spleen. The affinity of immature erythrocytes to fibronectin is quite specific. Although they adhere strongly to surfaces coated with this large glycoprotein, they do not attach to laminin or Type I or Type IV collagen, other extracellular matrix proteins. In mammals, fibronectin is a major component of the interstitial matrix. It appears that adhesion to the extracellular matrix plays an important role in localizing differentiating erythrocytes.

Because red blood cells must be replaced throughout the life of the organism, there must be precursor cells that not only give rise to these terminally differentiated cells but also replenish their own stock. These progenitor cells are referred to as erythropoietic stem cells. A variety of other tissues require continuous replenishment, such as sperm and skin cells, and each of these has it own stem cells. So the problem of regulating stem cell proliferation and differentiation is a general one.

In most adult mammals the erythropoietic stem cells reside in the mar-

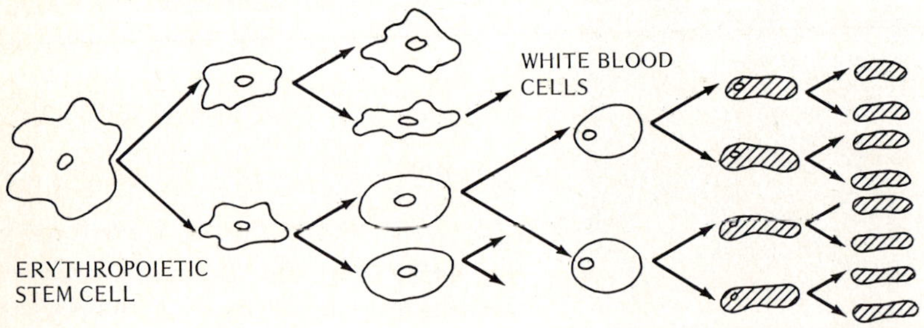

Figure 5.8. Stem cell growth and differentiation. Erythropoietic stem cells are represented as irregularly shaped nucleated cells. At the first division one daughter remains a stem cell while the other will differentiate into blood cells including red blood cells. At this stage no morphological differences distinguish these cells. After the next division the differentiating cells can be recognized. They continue to divide while accumulating hemoglobin *(shading)*. After the last division mammalian red blood cells extrude their nuclei. Erythropoeitic stem cells also give rise to white blood cells.

row of long bones. There they grow and divide but about half of the daughter cells become committed to erythrocyte differentiation. If more than half of the daughter cells went down the pathway to red blood cells, after a while, there would be no more stem cells. Likewise, if much fewer than half the daughter cells differentiated into blood cells, the population of stem cells would continue to increase. The process of stem cell proliferation and differentiation is outlined in Figure 5.8.

An erythrocyte precursor initially does not look very different from the stem cell that gave rise to it but it is fundamentally different. It is destined to divide only about four more times, and during those generations it will accumulate massive amounts of hemoglobin. Even before the first of those divisions the erythrocyte precursor begins to synthesize globin mRNA and accumulates ribosomes and other components necessary for a high rate of protein synthesis. During the next two cell generations both α- and β-globin mRNA accumulate to levels hundreds of times higher than are found in any other cell type. Likewise, some of the enzymes that are rate limiting for heme biosynthesis accumulate because for each globin molecule made a heme molecule must also be made. This is a clear case in which dramatic specialization of physiology proceeds even while cell division continues.

The rate of red blood cell production adapts to the prolonged need of the organism. A sea-level creature taken to the Andes or Tibet is exposed to only about a quarter of the partial pressure of oxygen (pO_2) it is normally used to, and it will be distinctly out of breath. Within a week or so the organism adapts by producing more red blood cells to facilitate oxygen transport to the tissues. At first sight one might expect that the stem cells would increase the frequency at which they commit daughter cells to erythropoietic differentiation, but this would deplete the stem cell population. What actually happens is that tissue anaerobiosis results in the release of a 34 kd glycoprotein, erythropoietin, into the circulatory system, and this hormone results in a decreased frequency of commitment of cells to the erythropoietic pathway. The stem cell population expands. Then the frequency returns to the previous, unstimulated level and a greater number of cells differentiate into red blood cells. This is now a new steady state.

During erythropoietic differentiation the various specializations must be kept in balance: heme production must equal globin production, and α-globin production must equal β-globin production. In the case of the globins it is known that the mRNA for α-globin accumulates to a somewhat higher level than the mRNA for β-globin. However, the two globin proteins are made at almost exactly the same rate because the β-globin mRNA is translated more efficiently than the α-globin mRNA.

Transcription of the α- and β-globin genes is hundreds of times greater in erythropoietic cells than in any other cells. How is their transcription coordinated? The genes are far apart and each is transcribed into a separate mRNA. They both have the "TATA" start signal, as do most other genes. Further upstream no long runs of similar bases are seen except at about 80 bases upstream of the transcriptional initiation site there is a sequence [CCAAT] in all globin genes (the "cat box"). However, "cat boxes" are seen at about that position preceding many genes that are not coordinately expressed along with the globin genes. Thus, the sequence CCAAT does not uniquely indicate globin genes. There is evidence derived from the expression of cloned α- and β-globin genes transformed into normal cells as well as into erythropoietic precursor cells (MEL cells) that indicates that the chromatin carrying these genes is altered during the differentiation of stem cells so as to derepress them. α-Globin genes are then expressed at a high rate. The β-globin genes, on the other hand, appear to require a second signal that occurs only in cells actively synthesizing globin proteins. The β-globin genes, but not the α-globin genes, have *cis*-acting sequences within their transcribed portions that require the function of a cell-type specific positive regulatory factor. Thus, the regulation of α- and β-globin genes differ fundamentally. This might be a consequence of the major difference in developmental expression of these genes: α-globin is synthesized throughout most of embryological development as well as in adult cells, while β-globin is synthesized maximally only in adult cells.

At the same time that transcription of the globin genes is activated during differentiation of erythropoietic cells, transcription of genes coding for a major membrane protein (band 3) is activated. This 95 kd protein facilitates chloride/bicarbonate cotransport such that bicarbonate produced from absorbed CO_2 in red blood cells is exchanged with serum chloride in the capillary beds of peripheral tissues. In the lungs the bicarbonate is taken up from the serum, converted to CO_2 by carbonic anhydrase of red blood cells, and exhaled. The ability to exchange bicarbonate ions for chloride ions allows the full volume of the serum to be utilized for transport of CO_2 out of the body. Band 3 makes up 20 to 50 percent of the total amount of integral membrane proteins of red blood cells. There are about a million band 3 molecules per red blood cell. This is at least a thousandfold higher concentration than is found in any other cell type. While massive accumulation of hemoglobin allows efficient transport of oxygen from the lungs to the tissues, the dramatic accumulation of band 3 in red blood cell membranes allows efficient transport of CO_2 as bicarbonate ions from the tissues to the lungs.

There are many more aspects to the differentiation of blood cells and these studies could (and have) filled up books focused on the development of blood cells. The few points we have touched on here give some idea of the complexity of the problems faced in differentiation of even a single cell type and illustrate mechanisms that may be used in the differentiation of many other cell types.

Summary

The function of many cell types is intimately dependent on their structure. Epithelial cells form an impermeable sheet by nature of belts of tight junctions and desmosomes that hold them to each other. They integrate their metabolic activity by exchange of small molecules through gap junctions that connect the cytoplasm of one cell to that of the next. Nerve cells maintain their highly elongated processes with scaffolds of microtubules. They localize receptors for neurotransmitters to regions of synapse. Red blood cells shrink to a size and shape compatible with the small capillaries they flow through. However, even the best adapted soon break and must be replaced. Stem cell lines continuously produce cells that differentiate to specialized blood cells. The probability that one of the two daughter cells of a division will differentiate is controlled by a variety of feedback systems. Once committed to differentiate, an erythrocyte will activate α- and β-globin genes and divide only a few more times.

SPECIFIC EXAMPLES

1. Cortical Structures in Ciliates

One of the largest groups of unicellular organisms is the ciliates. These eukaryotes range up to a millimeter in diameter and are equipped with rows of cilia that direct bacteria and small algae to an oral apparatus where they are engulfed (Fig. 5.9). Most organisms of this size and complexity are made of many cells, but the ciliates are unicellular.

Coordination of the beats of many cilia so as to direct food particles to a mouth requires rapid signaling along the length of the whole organism. In a multicellular organism, this would entail the evolution of efficient neural connections between the cells. Probably before this occurred a mechanism evolved to support the increased size found in ciliate cells. The genetic ma-

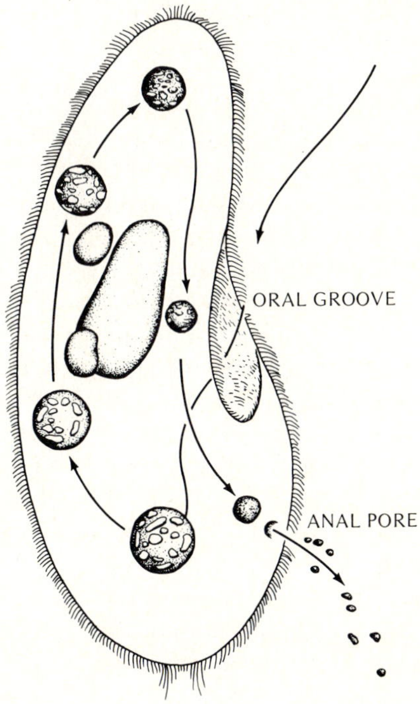

Figure 5.9. Paramecium. Bacteria are swept into the oral groove by cilia and engulfed in a food vacuole. The vacuole circulates through the large cell, growing gradually smaller as the bacteria within it are digested and absorbed. Debris is gotten rid of through the anal pore.

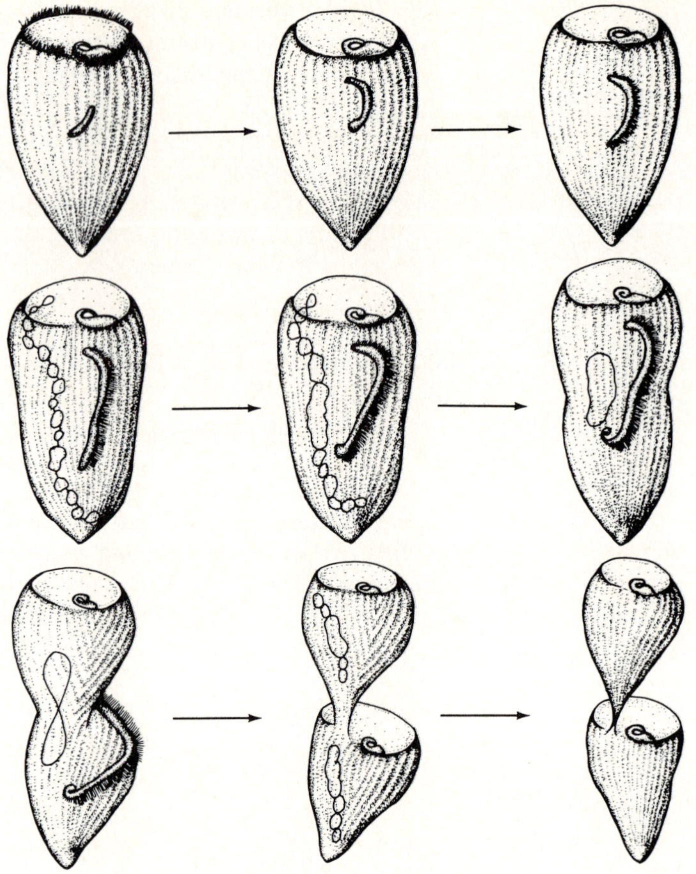

Figure 5.10. *Stentor.* This large complex cell duplicates its structures exactly and divides. The primordium elongates and puts out a row of cilia; the macronucleus with its multiple copies of genes contracts to the middle; both the macronucleus and the cell constrict; the primordium organizes into a new oral apparatus and the macronucleus expands into each half; finally, the daughter cells separate.

terial in the nucleus is amplified many hundreds of times and kept in a macronucleus. In this way the transcriptional needs of a cell hundreds of times larger are met by extra copies of each gene. A cell of this size faces further problems in the processes of cell division (Fig. 5.10). How can a complex cell divide in two and yet endow both daughter cells with functional copies of the necessary structures? The answer was found in the ability of cortical structures to direct the assembly of similar structures without the need for detailed genetic instructions. Membranes carry hereditary information just as genes do because membranes are not formed *de novo* but always as extensions of other membranes. The ability of a particular array of membrane components to replicate itself is particularly clearly seen in the ciliates. There are thousands, perhaps millions, of different species of ciliates each with its own shape and size. The details vary between different species but have been most intensively studied in the exceptionally large cells of *Stentor* and the relatively simple *Tetrahymena* and *Paramecia*.

A decapitated *Stentor* heals and soon forms a new primordium arrayed in relationship to the remaining rows of cilia. The primordium then migrates to the top and forms a new anterior end complete with a whorl of concentric stripes and ciliary rows (Fig. 5.11). This movement of the pri-

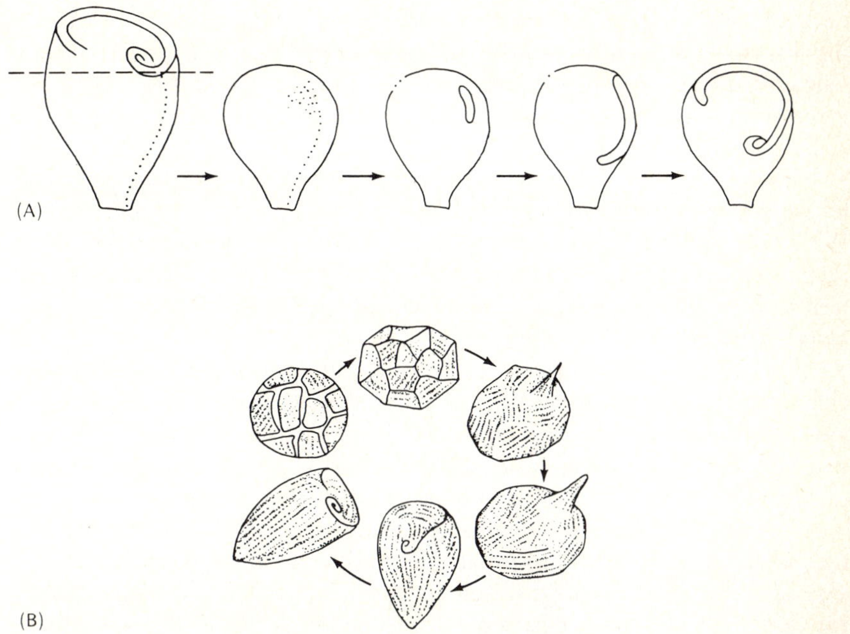

Figure 5.11. Regeneration in *Stentor*. *(A)* When the head is excised, a new primordium appears and soon reforms the oral apparatus. *(B)* When both the head and the tail are excised and the cortex cut with a fine glass needle, the patches reorient and a new primordium regenerates a typical head structure.

mordium and rearrangement of ciliary rows is similar to the processes seen during cell division. The cortex carrying the ciliary rows can also be cut into many small disoriented pieces. Within a few hours the ciliary rows reorient relative to each other to establish lines running from one end of the cell to the other. A new primordium is formed and migrates to the anterior, and within a day the cell is feeding again. The small plates of cortex seem to find their equilibrium state when lined up with others and the whole pattern responds by reorganizing a primordium that will form the oral apparatus. It is a most resilient cell.

Grafting a piece of cortex from one cell of *Stentor* to a different posi-

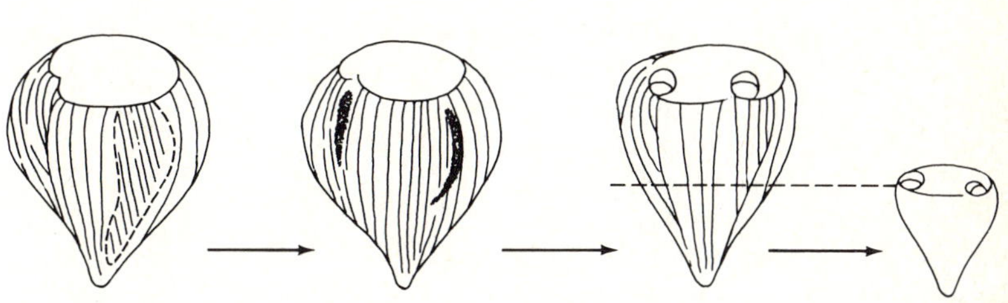

Figure 5.12. Double *Stentor*. Grafting a cortical segment from the fine-line zone into the wide-stripe region of a decapitated *Stentor* results in double oral regeneration, one from the host and one from the graft primordium. Decapitation of this double *Stentor* is followed by double regeneration.

tion on another cell disrupts the regular spacing of ciliary rows and leads to the formation of a cell with two oral apparatuses (Fig. 5.12). Such cells divide normally, producing two deranged daughter cells at each division. The genes have not changed, only the cortical structures. These propagate in a stable hereditary manner.

The genetics of *Tetrahymena* are amenable to laboratory breeding. Different strains exhibit recognizably different cortical patterns of ciliary rows. When two different cells exchange their nuclei, they do not change their stripes. Nor do their progeny for hundreds of generations. The cortical pattern seems to be completely independent of the nuclear information. It is perpetuated by the cortex alone.

The cortex of *Tetrahymena* is about 2 μm deep and contains rows of cilia, each cilium with its own basal body or kinetosome (Fig. 5.13). The kinetosomes and attached cilia are only formed in a row of preexisting kinetosomes. Thus, at division the pattern of rows is replicated. Exactly how this happens is not yet known, but clearly structure determines structure.

Yet cortical structures are also genetically controlled. A two-faced mutant of *Tetrahymena* has been isolated and called *janus*. Wild-type cells have a single ventral oral apparatus and contractile vacuole pores situated near the posterior end of two ciliary rows situated about 90° to the right of the oral apparatus. Homozygous *janus* cells *(jan/jan)* have two oral apparatuses on opposite sides of the cell in mirror-image orientation. On one side of the cell, to the right of one oral apparatus and the left of the other, contractile vacuole pores appear near four ciliary rows in a broad arc. *Janus* cells are

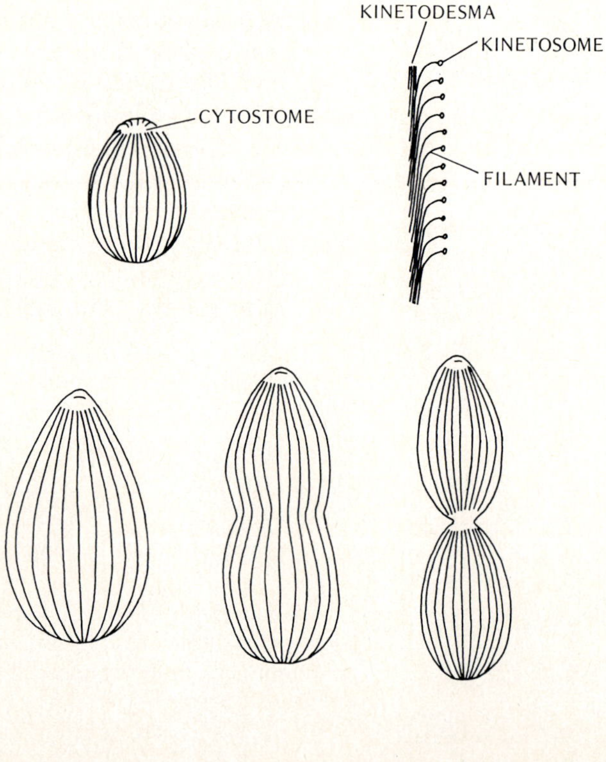

Figure 5.13. Paramecium. A terminal cytostome shows the polarity of this ciliate. Kineties run the length of the cell. Each kinety consists of rows of basal bodies (kinetosomes) from which filaments extend. When these cells divide, kineties elongate by multiplication of kinetosomes in the middle of the cell. Constriction and formation of a new cytostome produces a new daughter cell.

regarded as having their dorsal side transformed into a reversed ventral one. The mutant *jan* allele appears to modify a reference dorsal border. Analysis of symmetry-reversals that result from altered gene action and those that result from geometrical anomalies in wild-type cells promise to shed light on the mechanisms of cellular pattern determination.

2. Cytoskeletal Structures

A cell is more like a gel than a solution. The concentration of protein alone is such that convection is severely limited. Holding it all together is a network of polymerized proteins arrayed in microscopically visible filaments and tubules. The cytoskeleton is not structured like the bony skeleton we are familiar with under the skin of large vertebrates, but it serves some of the same functions at the much smaller scale of the single cell. It controls aspects of the shape of the cell and can transmit force exerted in one place to a distant response in another part of the cell. It can surround and protect an organelle much as the rib cage protects our viscera. Unlike a skeleton of bones, the cytoskeleton can rapidly change its form and even its composition. It adapts in minutes to changes in cell physiology. From a developmental biology point of view, we want to know how it changes during differentiation. We know the names and sizes of many of the components (Table 5.1), but so far we only have a hazy idea of how they interact to give the varying shapes and functions of differentiated cells.

Table 5.1 Some Cytoskeletal Components

Structure	Size, nm	Components	Subunit Size, kd	Properties
Microfilaments	7	actin	42	long filaments
		fimbrin		bundles actin filaments
		profilin	16	restricts polymerization of actin
		villin		Ca^{++} dependent
		gelsolin		fragments actin filaments
		tropomyosin	35	binds actin filaments
		myosin	210	moves actin filaments
		α-actinin	95	cross-links actin filaments
		filamin	250	
Intermediate filaments	8–10	vimentin	55	gives tensile strength
		keratin	40–65	tonofilaments of epithelial cells
		neurofilaments	70, 140, 210	strengthen nerve axons
Microtubules	25	α, β-tubulin	50	hollow tubes
		dynein	400	bends microtubules
		(MAP)	200–220	microtubule-associated protein

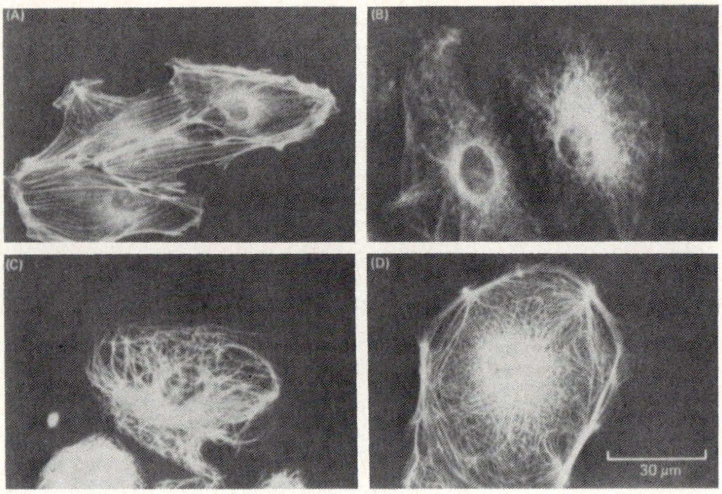

Figure 5.14. Cytoskeletal structure. Antibodies to *(A)* actin, *(B)* vimentin, *(C)* tubulin, and *(D)* keratin were added to permeabilized kangaroo epithelial cells and visualized by fluorescent labels. Actin is arrayed in microfilaments, vimentin in intermediate filaments, tubulin in microtubules, keratin in intermediate filaments. Each type of structure has its own characteristic pattern.

The cytoskeleton is a complex network of protein filaments that connect different parts of cell membranes to each other (Fig. 5.14). The major filaments are of three sizes: 7 nm microfilaments, 8 to 10 nm intermediate filaments, and 25 nm microtubules. Each is formed by polymerization of a basic subunit: actin for microfilaments; vimentin for intermediate filaments; tubulin for microtubules. Antibodies have been raised that are monospecific for each separate protein. By coupling the antibodies to a fluorescent dye, one can visualize the pattern of each cytoskeletal structure in a cell (Fig. 5.14).

Cells, such as kangaroo rat epithelial cells, can be made permeable to antibodies, incubated in the presence of specific antibodies coupled to a fluorescent dye, washed, and then observed in a fluorescent light microscope. The filamentous array of each cytoskeletal component gives a characteristic pattern. Microfilaments stretch from one end of a cell to the other and are often referred to as stress fibers. Intermediate filaments radiate from points near the periphery and the nucleus. Microtubules crisscross the cell from an organizing center near the Golgi apparatus. None of these filaments enters the nucleus. This technique of immunohistochemistry allows one to directly visualize the cytoskeleton in a single cell at a specific time.

In a living cell the microfilaments and microtubules are changing all the time. The rate of polymerization and depolymerization is a function of the subunit concentration in the cell. In most cases the concentration of actin and tubulin is slightly above that resulting in net polymerization. However, specific proteins regulate the growth and dissolution of filaments. Profilin binds to actin found beneath the acrosome in sperm heads and keeps it from polymerizing until the sperm nears an egg. Villin and gelsolin fragment microfilaments when triggered by Ca^{++} ions. Actin filaments are bundled together by fimbrin, while α-actinin and filamin form flexible links between microfilaments. Tropomyosin is a rod-shaped protein that binds along the length of actin filaments. Myosin actually moves microfilaments relative to each other using the energy derived from hydrolysis of ATP.

The function of microfilaments can be most easily recognized by altering them with specific compounds. The cytochalasins bind to one end of actin filaments and block the addition of subunits. This paralyzes the cell; there is no further movement, cytokinesis, or phagocytosis. Folding of epithelial sheets is sensitive to cytochalasin. Clearly, microfilaments play essential roles in cell movement and cell division. This conclusion is further supported by the localization of microfilaments in stress fibers and in a contractile ring separating two daughter cells at division. Likewise, arrayed microfilaments are seen in pseudopods engulfing material by phagocytosis. Sheets of epithelial cells often have an array of microfilaments just beneath the inner surface before involution. It is not yet known how this subcellular localization is achieved, but the involvement of microfilaments in subsequent processes is clear.

Phalloidin is a poison that inhibits depolymerization of actin filaments. Interestingly, it blocks cell movement. It can be concluded that the dynamic state of microfilaments is essential for cell migration.

The orientation of microfilaments is controlled in part by interaction with α-actinin and filamin (Fig. 5.15). The association is nonsymmetric and can establish polarity in the cytoskeleton. Upon division of a cell several properties, such as the pattern of movement, are partially inherited by both daughter cells. This may be imprinted by the structure of the cytoskeleton.

Intermediate filaments are much more stable than microfilaments or microtubules. Once vimentin polymerizes it deploymerizes very slowly. Thus, unlike the other components of the cytoskeleton, intermediate filaments are not in a dynamic equilibrium. The subunits are specialized to different cell types. For instance, neurofilaments are found exclusively in neural cells. Although the intermediate filaments in different cell types appear similar, they may carry out specialized functions.

Antibodies have been raised to vimentin that bind to intermediate filaments and cause them to collapse onto the nuclear membrane. Surprisingly, the cells continue to grow, divide, and move. It can be concluded that the pattern of intermediate filaments is not essential for the characteristics of cells observed in the laboratory.

Microtubules are ubiquitous but are especially prominent in cilia and flagella in which they are arranged down the length. Dynein arms stretch between doublets of polymerized tubulin and can cause bending by movement of adjacent microtubules. A variety of microtubule-associated proteins (MAP) have been described that may control the positioning and polymerization of microtubules.

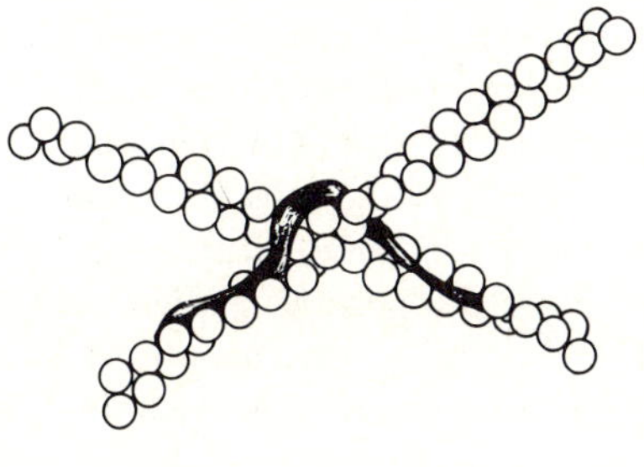

Figure 5.15. Filamin links microfilaments. A three-dimensional asymmetric network of actin filaments with the viscosity of a gel is formed by flexible links of filamin.

Colchicine is a drug that binds to tubulin subunits and blocks polymerization. Mitosis is inhibited by colchicine. This is not surprising since microtubules can be seen to connect each chromosome with the centriole at mitosis. The movement of chromosomes to the poles is not inhibited by cytochalasin but is inhibited by colchicine. This result indicates that chromosomal movement is not mediated by microfilaments but depends on microtubules.

Taxol is a drug that shifts the equilibrium of tubulin and microtubules toward polymerization. It also results in inhibition of mitosis. This suggests that the continuous cycle of polymerization and depolymerization of microtubules is essential for mitosis and may play several other roles.

Although is it clear that the insides of a cell are highly structured, we know little of how the structures are adapted to changing roles. The components of a cell are seldom free to diffuse about as in an aqueous dilute solution but move about, over, and through each other. A better understanding of developmental processes at the cellular level will undoubtedly come from biochemical and biophysical studies of the components of the cytoskeleton and their interactions.

3. Extracellular Matrices

Cells of most tissues are embedded in an extracellular matrix composed of various amounts of hyaluronic acid, proteoglycan, and structural glycoproteins like collagen, fibronectin, laminin, chondronectin, and elastin. These components are produced by the cells themselves and secreted. The cells continue to interact with these molecules and those produced by other cells. Thus, there is a structural and functional continuity from the interior of the cell through the cell membrane to the extracellular matrix (ECM).

The ECM is often defined as the structurally stable material that lies under epithelia and surrounds connective tissue cells because it can be easily visualized there in electron micrographs. However, when methods were improved for the biochemical analysis of extracellular materials, the macromolecular components of the ECM were found associated with all sorts of tissues. Antibodies that specifically bind to a given molecule such as fibronectin or one of the classes of collagen were particularly helpful in recognizing ECM around cells. Thick layers of ECM are often referred to as basement laminae. They function to give strength, resiliency, and cohesiveness to tissues. A principal component of basement laminae is collagen.

Collagen is actually a heterogeneous class of proteins with common chemical and physiological properties. In mammals there are 10 to 20 genes coding for collagenous proteins. The Type Iα2 collagen gene of chickens is 40 kb in length and includes about 50 exons separated by intervening sequences that are spliced out in the process of generating mRNA molecules of about 6 kb. These are some of the largest RNAs in the cells. They code for preprocollagen of about 180 kd. These polypeptides are secreted and, like most secreted proteins, have a hydrophobic signal peptide at the N-terminus that results in their polysomes being bound to the rough endoplasmic reticulum (RER) membrane. The protein is secreted into the lumen

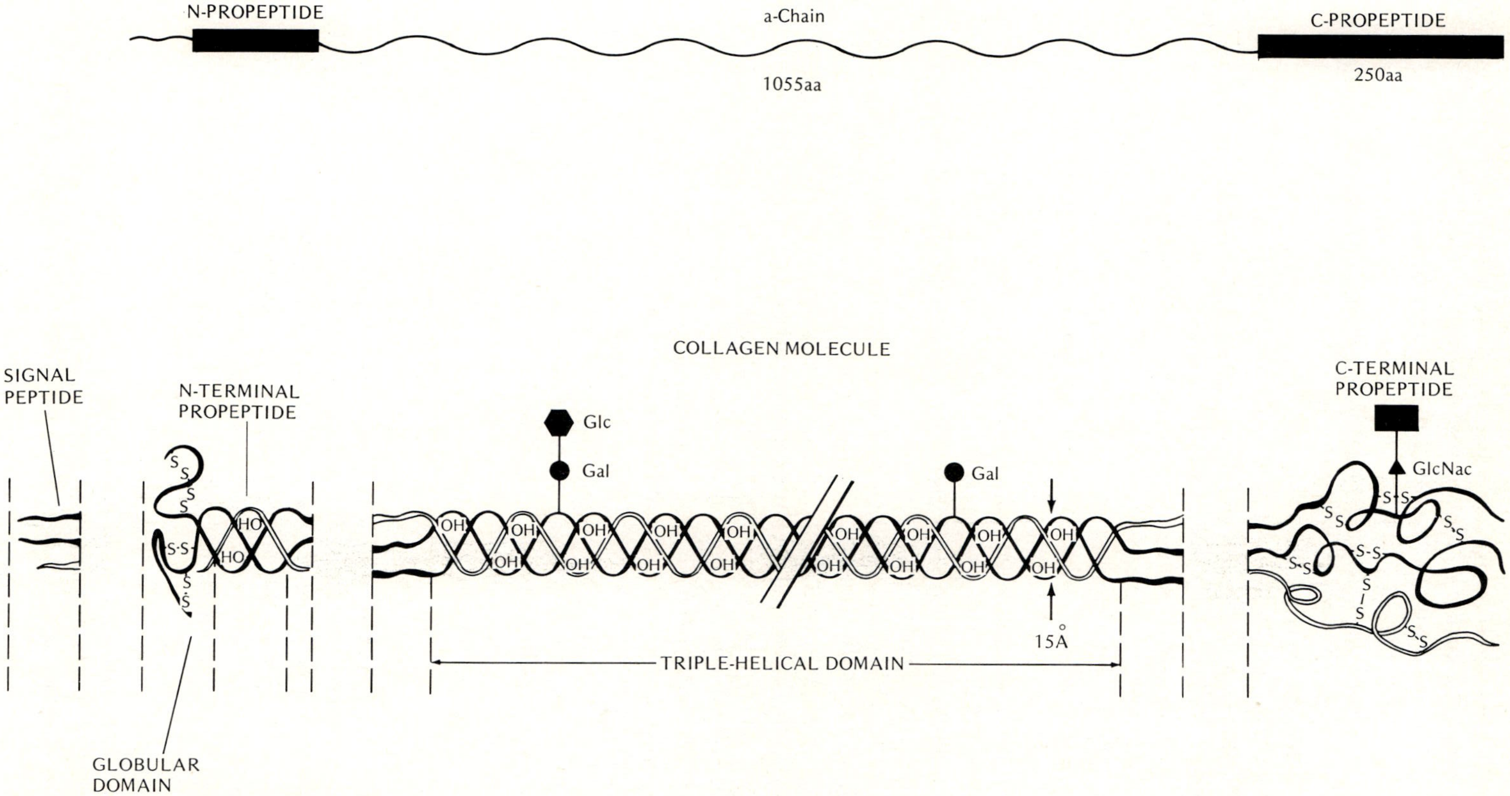

Figure 5.16. Structure and processing of Type I collagen. The collagen mRNA is translated into a polypeptide of about 1055 amino acids. The hydrophobic signal peptide is cleaved off the N-terminal even as the remainder of the protein is being synthesized. Procollagen is modified in the rough endoplasmic reticulum (RER) by conversion of proline to hydroxyproline and further modified in the Golgi apparatus. The N-terminal and C-terminal propeptides are cleaved by specific proteases before the proteins associate in a triple helix to form collagen.

of the RER during translation and the 15 to 25 amino acid signal peptide is cleaved off. A globular domain of about 100 amino acids is left at the N-terminus of procollagen. This portion is subsequently removed by a specific enzyme, procollagen N-protease, to generate collagen. If this portion of the protein is not removed due to a genetic defect that results in a lack of procollagen N-protease activity, the basement lamina of skin contains strikingly abnormal collagen fibrils. Sheep and cattle with this genetic abnormality have extremely fragile skin that rips easily. Lambs and calves with the abnormality die from infection of the large wounds generated by the least trauma.

Normally both the N-terminal peptide and a C-terminal peptide are cleaved off procollagen. The resulting collagen proteins associate with each other in a triple helix (Fig. 5.16). Within the helical region, almost every third amino acid is glycine (structure: Gly-X-Y) and X and Y are frequently proline. About half of the prolines are enzymatically converted to hydroxyproline in a reaction requiring iron and ascorbic acid (vitamin C) as cofactors. Some of the lysines are also converted to hydroxylysine and a dissacharide is added to the hydroxylysine in some types of collagen in some tissues. Oligosaccharide side chains are added to asparagine residues from the dolichol intermediate to generate a diversity of highly modified proteins. Within the triple-helical region, the three polypeptides are covalently linked by disulfide bonds to generate the strong 15-Å diameter collagen fibrils seen in electron micrographs. The fibrils associate with each other to form larger collagen fibers.

Collagen makes up more than half of the total proteins in some adult organisms since it is the major protein component of tendons, bones, and cartilage. Control of its synthesis, secretion, and self-association has much to do with the final form of the animal. There are five major types of collagen distributed in different tissues. The regulation of the genes coding for these different extracellular proteins is clearly cell-type specific.

In the chick embryo, collagen is first detected at gastrulation. Fibers of collagen appear around the notochord and are produced by it. Mesenchymal cells move over the basement laminae. The ECM has also been implicated in the interactions of epithelial and mesenchymal tissues. Various cell types such as myoblasts have a requirement for attachment to collagen on surfaces before they will proliferate. In many cases this attachment is mediated by another molecule of the ECM, fibronectin.

Plasma fibronectin is a soluble protein consisting of two identical chains of 215 kd that are cross-linked by a disulfide bond. Cellular fibronectin is a very similar but not identical molecule present on the cell surface and in the ECM. Both fibronectins have three functional domains. They are coded for by different mRNAs derived from the same gene by alternative splicing events within the coding region. The domains include:

I. A globular portion that can be cross-linked to collagen.
II. Another globular portion that binds collagen noncovalently.
III. An elongated region that binds to cells and carries cysteine residues that can form covalent S-S bonds to either cells or ECM (Fig. 5.17).

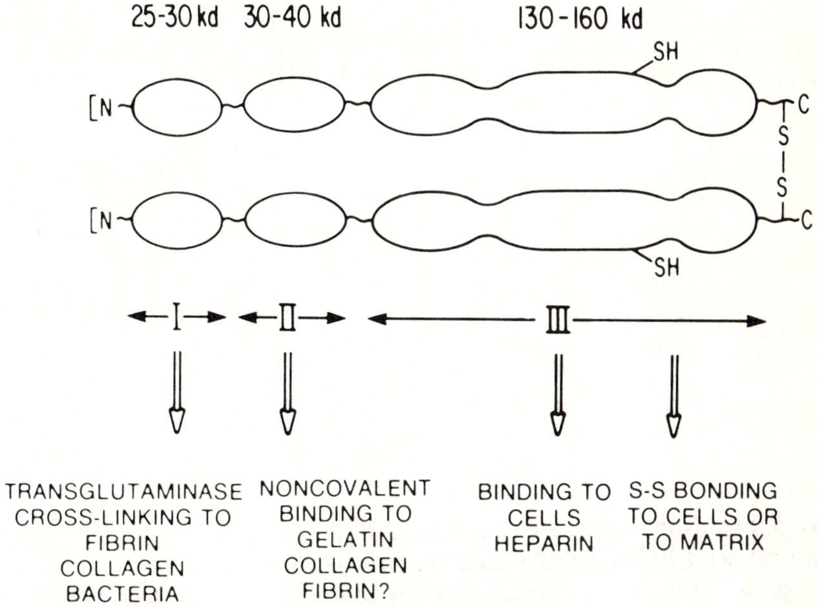

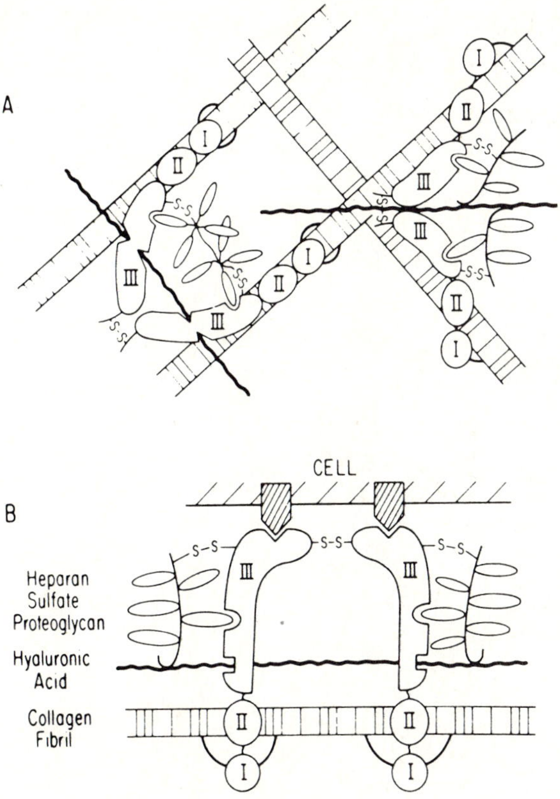

Figure 5.17. Structure and function of fibronectin. The size of domains I, II, and III differ slightly in fibronectins of different species but can be separated by proteolytic enzyme cleavage. Domain I has transglutaminase activity and binds to collagen. Domain II associates noncovalently to collagen. The elongated domain III binds both heparin and cells. Sulfhydryl bonds can form to cells or matrix near the C-terminus. Fibronectin organizes collagen *(thick bands)*, heparin *(wavy lines)*, and glycosaminoglycans [GAGs] *(lobed chains)* (A). Domain III may hold cells to the extracellular matrix (B).

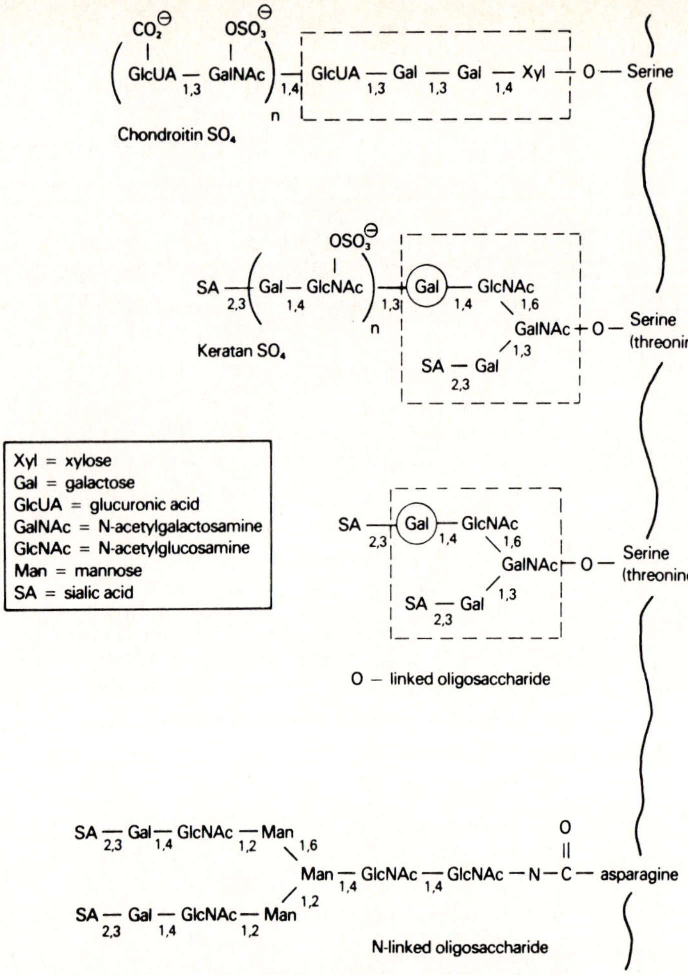

Figure 5.18. Structure of cartilage proteoglycans. The core protein of about 250 kd *(wavy line)* is decorated with long repeating dissacharides of chondroitin sulfate and keratin sulfate each characteristically linked to serine residues through specific sugar moieties. Complex carbohydrates are N-linked to asparagine moieties. An electron micrograph shows the bushy molecules (M) of proteoglycan associated with a central strand of hyaluronic acid (HA). GAGs associate with HA at one end, where most of the modifications are N- and O-linked oligosaccharides. Further along the core protein most of the modifications are keratin sulfate (KS). At the far end most of the modifications are chondroitin sulfate (CS).

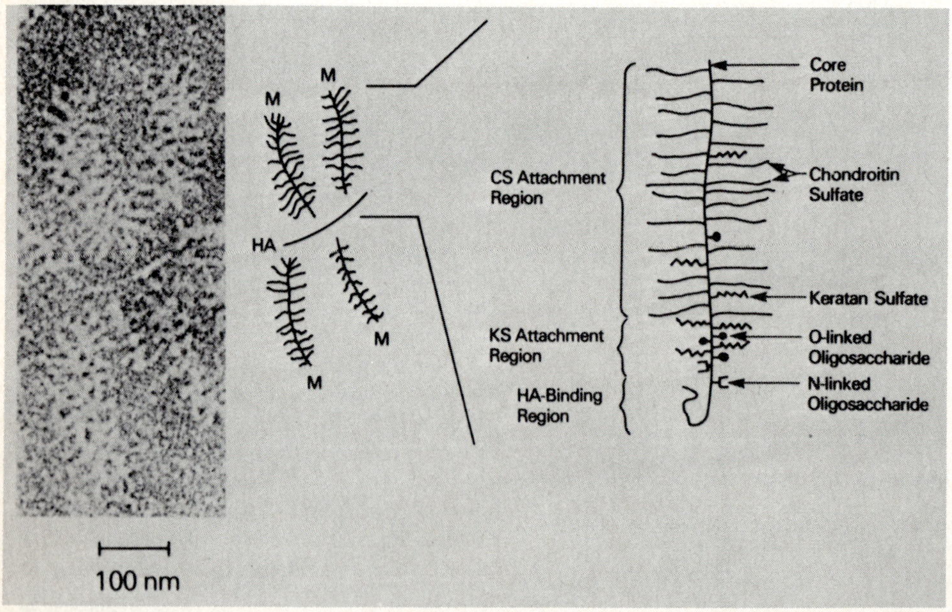

The molecule looks superbly engineered to cross-link collagen fibers and attach cells to them.

Fibronectin also has specific affinity to the polysaccharide, hyaluronic acid $(GlcNacGlcUa)_n$. This acidic carbohydrate is found in the ECM of many tissues and binds to the elongated region (III) of fibronectin. This same region also binds and can be cross-linked to proteoglycans (Fig. 5.17).

Proteoglycans are major components of many extracellular matrices and contain a core protein to which glycosaminoglycans (GAGs) are covalently attached. Cartilage contains a large amount of a complex proteoglycan. The core protein is a molecule of 250 kd to which about 40 long carbohydrate side chains (GAGs) are post-translationally attached (Fig. 5.18). There is a central region of 49 amino acids where serine alternates with glycine and chondroitin sulfate is attached.

Extended proteoglycans are about 0.3-μm long and associate with hyaluronic acid chains up to 1 μm in length. Hydrogen bonding between the carbohydrate moieties stabilizes intermolecular association of the polysaccharides. Such complexes could easily span the distance between cells in most tissues (Fig. 5.19). The proteoglycans are kept extended by charge repulsion generated by the sulfates on chondroitin sulfate and keratin sulfate side chains so that they resemble bottle brushes.

Mutations affecting proteoglycan synthesis have been found in mice. One such mutation, cartilage-matrix deficiency (cmd), results in cartilage containing less than 5 percent of the normal amount of proteoglycan. Homozygous mutant embryos (cmd/cmd) have foreshortened and malformed limbs and die at about the time of birth. Another mutation results in decreased sulfation of chondroitin in the cartilage and liver. This genetic lesion results in small, deformed (brachymorphic) mice. Clearly, the proteoglycans play essential roles in shaping the embryo.

During morphogenesis of epithelial organs such as lungs and kidneys,

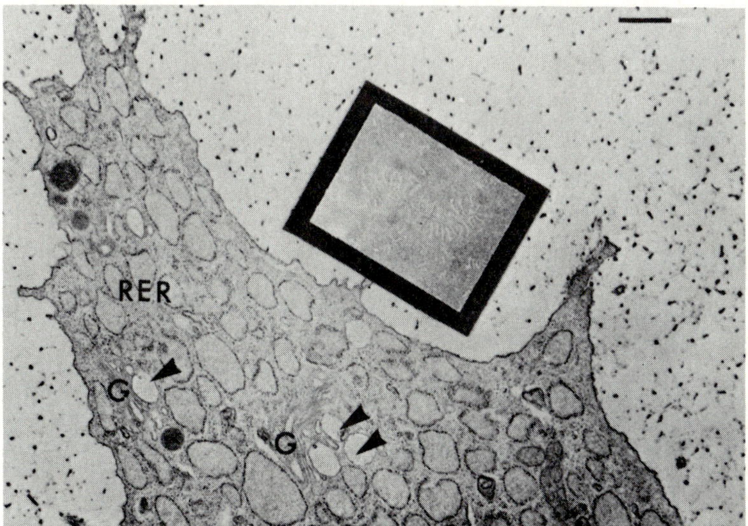

Figure 5.19. Hyaluronic acid organizes GAG complexes up to 3 μm in length *(insert)*. The relative size of this matrix component to that of a chondrocyte can be seen in the electron micrograph made at the same magnification. Golgi region (G), rough endoplasmic reticulum (RER), and secretory vacuoles *(arrows)* are marked. Bar = 0.6 μm.

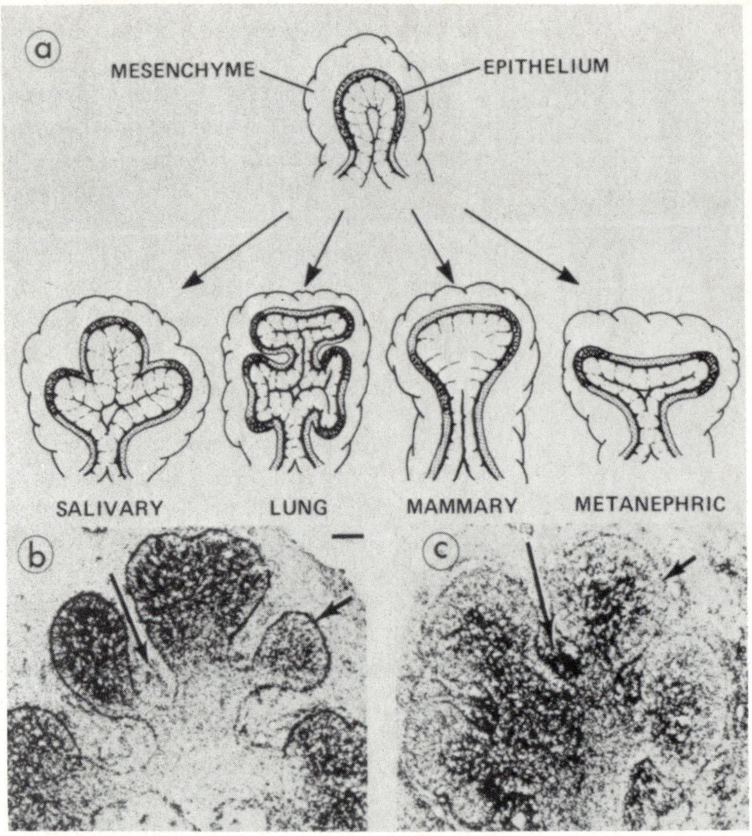

Figure 5.20. Synthesis and accumulation of GAG in several epithelial organs. *(A)* Regions of highest synthesis *(dark shading)* were recognized by labeling with ^{3}H-glucosamine. Regions of accumulation *(light stippling)* were recognized by staining with Alcian blue. *(B)* Autoradiogram of 13-day-old mouse embryo submandibular glands labeled for 2 hours. *(C)* Autoradiogram of the same tissue labeled for 2 hours and chased for 6 hours with unlabeled glucosamine. The label initially appears at the ends of the lobes *(short arrows)* and then accumulates in the clefts *(long arrows)*.

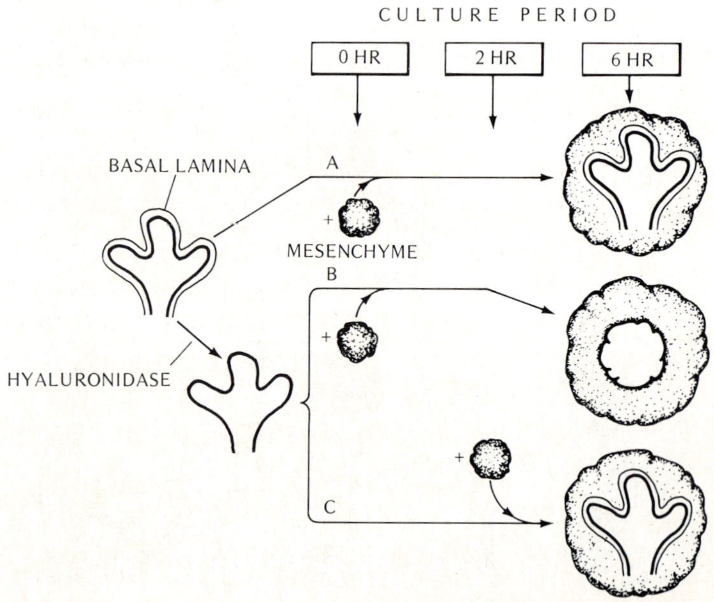

Figure 5.21 Morphogenetic effects of basal lamina. Thirteen-day mouse embryo submandibular epithelium was freed of mesenchymal tissue. *(A)* When mesenchyme was added back, morphogenesis proceeded normally. *(B)* When the basal lamina was removed by hyaluronidase and mesenchyme was added back, the epithelium lost its branching. *(C)* When basal lamina was regenerated by the epithelium for two hours before mesenchyme was added back, branching persisted.

epithelial layers surrounded by mesenchymal tissue take up characteristic lobular shapes (Fig. 5.20). Glycosaminoglycan synthesis is highest on the ends of the lobules but accumulates in the clefts as the tissue grows and expands. If the mesenchyme is stripped away and the basal lamina removed from the surface of the epithelial tissue by treatment with hyaluronidase before adding back mesenchyme, the lobes are lost. However, if the hyaluronidase-treated epithelium is allowed to regenerate a basal lamina and then recombined with mesenchyme, branching continues (Fig. 5.21). These results indicate that basal lamina GAG is essential for lobular morphogenesis and that mesenchyme inhibits its resynthesis. The mesenchyme appears to control turnover of the basal lamina, which in turn contributes to the changing shape of the epithelium.

ECM at the interface of cells and tissues controls shape, differentiation, relative movement, and a variety of physiological properties. Elasticity of structures such as blood vessels and lungs is ensured by incorporating rubberlike fibers made of the 80 kd protein elastin into the extracellular matrix. Deposition of the X-shaped protein laminin in the ECM leads to adhesion of epithelial cells to Type IV collagen. Chondrocytes attach preferentially to Type II collagen by secreting chondronectin (Fig. 5.22). In these and other ways, tissues are shaped to their characteristic functions. Throughout development and, indeed, throughout adult life, the extracellular matrices are

Figure 5.22. Condrocytes in their matrix. Cells are embedded in an extensive extracellular matrix containing Type II collagen. Chondronectin holds the cells to the collagen fibers (SEM).

formed, broken down, and remolded. During metamorphosis of tadpoles into frogs, collagenases are mobilized to break down the collagen fibers of tail skin as the tail is resorbed. Another example is the replacement of Type I collagen by Type II collagen in the development of long bones. Healing of a wound also requires remodeling of the basal lamina. These mechanisms are used in subtle ways throughout the processes of embryogenesis.

Related Readings

Arnott, S., Mitra, A., and Raghunathan, S. (1983). Hyaluronic acid double helix. J. Mol. Biol. *169,* 861–872.

Beisson, J., and Sonneborn, T. (1965). Cytoplasmic inheritance of the organization of the cell cortex in *Paramecium aurelia.* Proc. Natl. Acad. Sci. *53,* 275–282.

Bond, J., Robinson, G., and Farmer, S. (1984). Differential expression of two neural cell-specific β-tubulin mRNAs during rat brain development. Mol. and Cell Biol. *4,* 1313–1319.

Daursky, C., Richa, J., Solter, D., Knudsen, K., and Buck, C. (1983). Identification and purification of a cell surface glycoprotein mediating intercellular adhesion in embryonic and adult tissue. Cell *34,* 455–466.

Frankel, J. (1982). Global patterning in single cells. J. Theor. Biol. *99,* 119–134.

Frankel, J., Nelsen, E., Bakowska, J., and Jenkins, L. (1984). Mutational analysis of patterning of oral structures in *Tetrahymena.* A graded basis for the individuality of intercellular arrays. J. Embry. Exp. Morph. *82,* 67–95.

Hatta, K., Okada, T., and Takeichi, M. (1985). A monoclonal antibody disrupting calcium-dependent cell-cell adhesion of brain tissues: possible role of its target antigen in animal pattern formation. Proc. Natl. Acad. Sci. *82,* 2789–2793.

Hay, E. (1981). Cell biology of extracellular matrix. Plenum Press, New York.

Kappas, A., Song, C., Levere, R., Sachson, R., and Granick, S. (1968). The induction of δ-aminolevulinic acid synthetase *in vivo* in chick embryo liver by natural steroids. Proc. Natl. Acad. Sci. *61,* 509–513.

Lee-Huang, S. (1984). Cloning and expression of human erythropoietin cDNA in *Escherichia coli.* Proc. Natl. Acad. Sci. *81,* 2708–2712.

McMahan, U., and Slater, C. (1984). The influence of basal lamina on the accumulation of acetylcholine receptors at synaptic sites in regenerating muscle. J. Cell Biol. *98,* 1453–1473.

Mooseker, M. (1983). Actin binding proteins of the brush border. Cell *35,* 11–13.

Moss, B., and Ingram, V. (1968). Hemoglobin synthesis during amphibian metamorphosis. J. Mol. Biol. *32,* 493–504.

Murray, B., Hemperly, J., Gallin, W., MacGregor, J., Edelman, G., and Cunningham, B. (1984). Isolation of cDNA clones for the chick neural cell adhesion molecule (N-CAM). Proc. Natl. Acad. Sci. *81,* 5584–5588.

Patel, V., Ciechanover, A., Platt, O., and Lodish, H. (1985). Mammalian reticulocytes lose adhesion to fibronectin during maturation to erythrocytes. Proc. Natl. Acad. Sci. *82,* 440–444.

Piez, K., and Reddi, A. (1984). Extracellular matrix biochemistry. Elsevier Publishing, New York.

Pytela, R., Pierschbacher, M., and Ruoslahti, E. (1985). Identification and isolation of a 140 kd cell surface glycoprotein with properties expected of a fibronectin receptor. Cell *40,* 191–198.

Rutishauser, U. (1984). Developmental biology of a neural cell adhesion molecule. Nature *310,* 549–554.

Schwarzbauer, J., Paul, J., and Hynes, R. (1985). On the origin of species of fibronectin. Proc. Natl Acad. Sci. *82,* 1424–1428.

Trelstad, R. (1984). The role of extracellular matrix in development. Alan R. Liss Publishing, New York.

Yamada, K., and Kennedy, D. (1984). Dualistic nature of adhesive protein function. J. Cell Biol. *99,* 29–36.

Study Questions
Part I

1. What is the evidence that some differentiated cells carry all the genetic information necessary to make all the other types of cells in an organism?
2. Give three separate mechanisms which will result in the increase in the amount of a specific protein in a cell.
3. What controls 5*S* RNA synthesis in frog oocytes?
4. Give two specific cases of cell differentiation in response to experimental alteration of the environment.
5. Give two proteins in eukaryotic organisms necessary for expression of other genes, that is, positive regulatory gene products.
6. Describe in detail how steroids stimulate the rate of synthesis of ovalbumin in chick oviducts.
7. Give an experiment which indicates that the pattern of DNA methylation is inherited in a fairly stable manner.
8. How can a cell specialize to respond to acetylcholine?
9. Why is it important that regulation of transcription of α-globin and β-globin genes be coordinate?
10. Give two examples of cell types that coordinately synthesize several proteins during terminal differentiation and give the proteins involved.
11. In males of certain mammals, sperm production increases tenfold every year during a brief period. Give a plausible mechanism that could account for this observation.
12. Rats injected with tyrosine synthesize tyrosine amino transferase (TAT) ten times faster in their liver cells than control rats. However, there is no effect of tyrosine injection on TAT synthesis of rats whose adrenals have been removed. What can you conclude?
13. Give two examples of cell types that can grow and yet are determined.
14. Describe the experimental approach that can be used in the analysis of transcriptional control of actin and myosin genes during myoblast fusion.
15. What is the mechanism by which red blood cells accumulate to a great extent in an individual after moving to a higher place (hypobaric conditions)?
16. Give three examples of dramatic differentiation of the plasma membrane of cells.
17. Define:
    a. maternal mRNA
    b. mesoderm
    c. morula
    d. trophectoderm
    e. suspensor

f. pollen tube
g. endosperm
h. intron (intervening sequence)
i. polytene chromosome
j. vimentin
k. actin
l. colchicine

18. How is the endoderm formed in:
    a. amphioxis embryos?
    b. frog embryos?
    c. bird embryos?
19. What are the steps in growing a carrot plant from cells of a carrot root?
20. Explain why a unique gene is found on only a small number of fragments of different sizes when a genome is cleaved by a restriction enzyme.
21. Give three steps in processing a primary RNA transcript to produce functioning mRNA.
22. What is the evidence that the "TATA box" affects transcription of a human β-globin gene?
23. What similarities indicate that all of the globin genes evolved from duplicates of an ancestral gene?
24. How are the three enzymes necessary for quinic acid metabolism coordinately regulated in *Neurospora?*
25. How can a single gene code for more than one protein?
26. How can you distinguish between mRNA molecules transcribed from different actin genes?
27. Give three enzymes that accumulate dramatically in differentiating pancreas cells.
28. What is the evidence that the chymotrypsin *B* gene in mice is regulated by a cell-type specific enhancer?
29. Describe an experiment that can demonstrate that cortical structures of ciliates propagate in a stable hereditary manner independently of the genes of the cell.
30. What is the evidence that the structural pattern of intermediate filaments is not crucial to the growth of cells in culture?
31. How does fibronectin attach extracellular collagen fibers to cells?
32. Give four post-translational modifications of collagen.
33. What does the phenotype of the *janus* mutant of *Tetrahymena* indicate concerning the organization of the cortex in this ciliate?
34. In what ways might the extracellular matrix mediate epithelial–mesenchymal interactions?

PART II

The Processes

The study of embryology follows the processes by which a fertilized egg gives rise to the differentiated cells of an embryo. Slowly, form and symmetry become manifest. An embryo is not a jumble of cells but a regular arrangement at every stage. The potential to generate an individual is present at each step in morphogenesis. Gradually the outlines of the organism can be recognized. The shapes and structures of the cells change radically during embryogenesis and give insight into how tissues are generated. The pathways leading to each final form can often be traced by judicious microscopy coupled with experimental manipulation. Anatomical descriptions give a three-dimensional picture, but development involves a fourth dimension: time. The structures change in a defined sequence that is essential for overall development.

The initial stages in embryogenesis are chiefly determined by the characteristics of the sperm and eggs of the species. These characteristics are established during differentiation of the germ cells in the gonads. Only after several divisions of the fertilized egg does differential gene expression start to play a role in divergence of cell types. The results from a variety of experiments make it clear that transcription of the embryo's genes is essential for continuation along the developmental pathway that is appropriate for that species. However, there do not appear to be specific "gastrulation genes" or "neurulation genes." These processes occur as the result of subtle alterations in the mix of genes that are expressed. New recipes are made by varying the amounts of the ingredients. Later in embryogenesis, when highly specialized cells are organized into tissues, specific genes begin to play major roles in differentiation. These will be further explored in Part III.

There is evidence that the steps in embryogenesis are controlled by preceding events. Therefore, it makes sense to start with the differentiation of the germ cells—sperm and eggs. The process of fertilization has received a lot of attention since germ cells are readily available in many species and the questions are clearly formulated. Moreover, the answers are of medical and breeding interest. Embryogenesis starts in earnest when the zygote cleaves. The rate of increase in complexity picks up at gastrulation. By the time the definitive organs appear, it is often difficult to keep in mind all the processes that are going on simultaneously. Therefore, we take a few cases in some detail. A truly comprehensive study of organ morphogenesis is best done with the embryos in front of you such as in a laboratory course. In Part III the development of sexual organs and limbs will be considered.

CHAPTER 6

Sperm and Eggs

Sperm Structure

Sperm cells are highly differentiated to transport a haploid nucleus to an egg and initiate development of the embryo. Sperm come in many different shapes but most are quite small cells (Fig. 6.1).

In vertebrates the structures of sperm are all variations of a general form consisting of a bullet or hook-shaped head propelled by a long tail (Fig. 6.2).

At the very front of the sperm head there is an acrosomal vesicle containing enzymes that function during fertilization to allow penetration of the sperm through the extracellular layers surrounding the egg. A prominent Golgi apparatus synthesizes the acrosomal proteins and packages them within a membrane vesicle. As sperm differentiation proceeds, the acrosomal vesicle enlarges and finally dissociates from the Golgi apparatus. Just below the acrosomal vesicle in sperm of most marine invertebrates and many vertebrates there is a bed of unpolymerized actin that is triggered to explosively polymerize as the sperm nears the egg surface. Actin polymerization extends the acrosome so that it presents a large surface for fusion with the egg membrane. The accumulation of a store of actin is one of the clear molecular differentiations in sperm.

The nucleus lies beneath the acrosome and its underlying bed of actin. Sperm nuclei are highly compact to allow streamlining. One of the mecha-

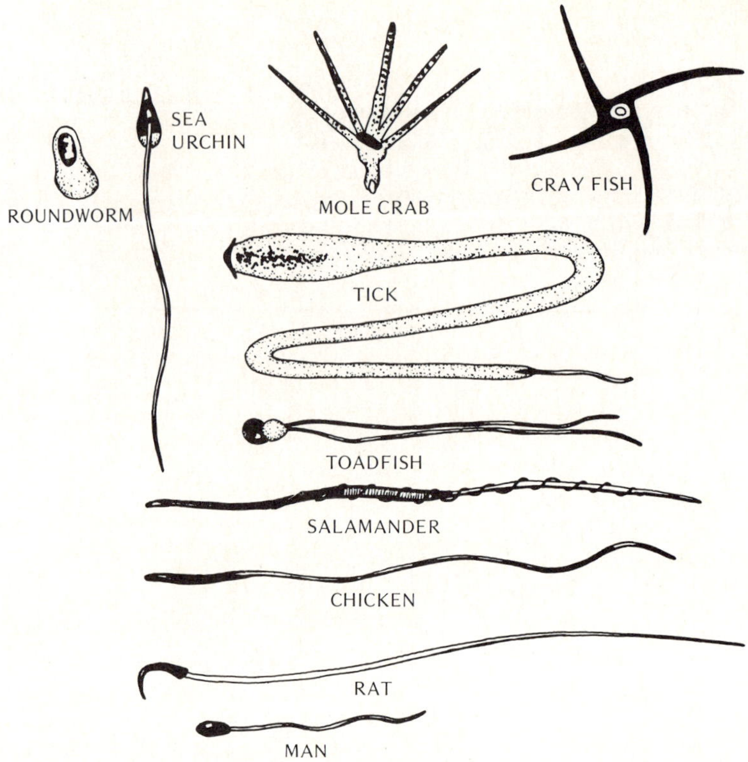

Figure 6.1. Sperm. Many forms are used to inseminate various species. The structures are adapted to internal fertilization or binding to free eggs depending on the reproductive strategy. Clearly, the relative size of sperm is unrelated to the size of the adult.

nisms that give rise to the extreme condensation of the DNA in the nucleus of vertebrate sperm is the replacement of histones by other small, basic proteins called protamines. While it has not yet been shown that DNA is wound in a tighter coil around protamines than the usual nucleosome configuration, this seems quite likely. During the maturation of spermatocytes into sperm, mRNA molecules coding for protamines have been found to accumulate in trout. There appears to be a considerable lag following synthesis of these mRNAs before they are translated into protamines, perhaps because of post-transcriptional regulation of splicing. Once the protamines are made, they enter the nucleus and replace the histones. This is a fascinating case of controls first affecting transcription of specific genes and then affecting the processing of the transcripts to give rise to functional protamine mRNA.

Centrioles are positioned posterior to the nuclei. The sperm centriole enters the egg during fertilization along with the nucleus and together they organize the contractile system of the egg so as to direct fusion of the sperm and egg nuclei and establish the first cleavage plane of the fertilized egg.

Extending from the centriole to the tip of the sperm tail is an array of microtubules, which forms a flagellum. There are nine doublets in a ring surrounding two central microtubules (Fig. 6.3). The dynein arms connecting the microtubular doublets are responsible for the thrashing movement of the tail, which propels the whole sperm. The subunits of the microtu-

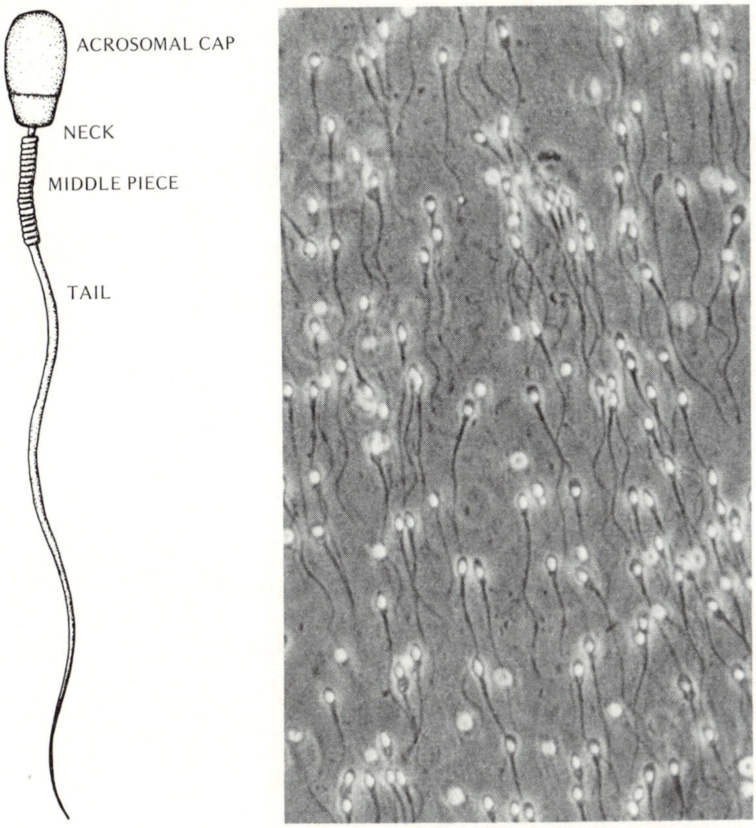

Figure 6.2. Human sperm. The nucleus is found in the head surrounded by an acrosomal cap. The tail is attached at the neck. In the middle piece, mitochondria surround the flagellum. Many sperm are ejaculated simultaneously.

bules in sperm tails are the familiar tubulin proteins. However, at least in the fly *Drosophila melanogaster,* the tubulin that is made during spermiogenesis is made from cell-type specific tubulin genes. This was discovered by analysis of a certain class of male sterile mutant flies. These flies were found to be normal in all ways except that their sperm lack motility. Molecular analysis showed that they carried alterations in a structural gene for tubulin. Other tubulin genes were still functional and supplied the tubulin needed for microtubules in other cells of the organism, but these were not used in differentiating sperm. The plausible interpretation of these observations is that the gene mutated in the male sterile flies makes a tubulin specifically selected for the specialized function of sperm tails. Although similar in many physical parameters to the tubulin found in somatic cells, sperm tubulin may have subtle differences that make it better able to propel a sperm. Alternatively, the sperm tubulin gene may not be significantly different from other tubulin genes but the regulatory sequences that control the time and extent of expression of the sperm tubulin gene may be part of the genetic program that integrates sperm differentiation. The other tubulin genes may be silent in differentiating sperm because they are not activated under those conditions.

Movement of any sort burns up a considerable amount of energy. The power source for sperm motility is provided by mitochondria closely ap-

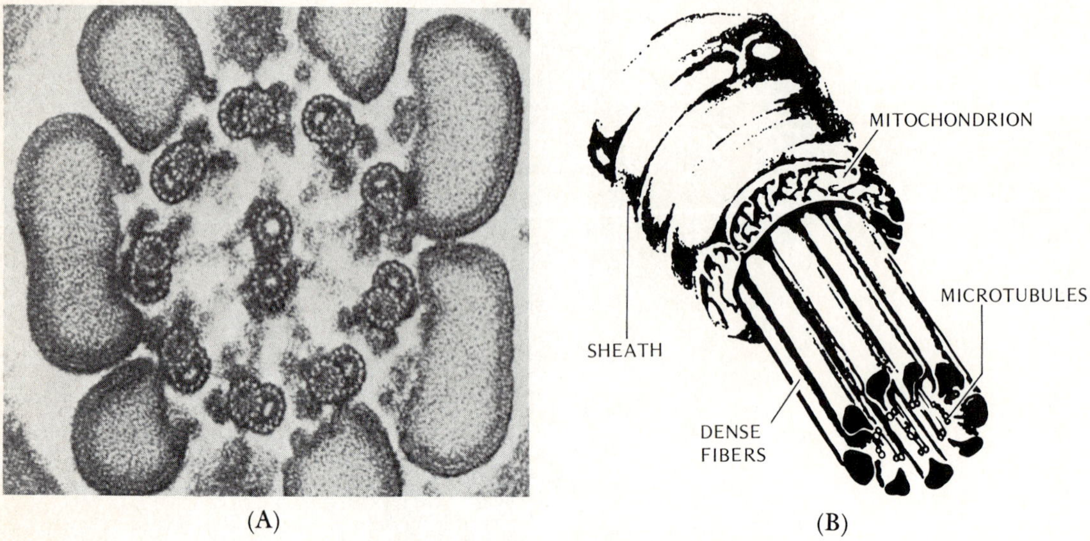

Figure 6.3. Mammalian sperm middle piece structure. *(A)* The ciliary pattern of nine pairs of microtubules surrounding a central pair is seen in an electron micrograph. Dynein arms connect the microtubules and generate movement. *(B)* Outer dense fibers surround the axoneme. These are surrounded in turn by mitochondria.

posed to the sperm tail and often spirally wrapped around it. The mitochondria are found just below the sperm head and so make the proximal portion of the tail broader. This region is referred to as the middle piece. In many organisms neither the tail nor the middle piece enter the egg during fertilization. All of the mitochondrial genes of the embryo are provided by the egg. Mitochondria carry a small genome of about 30 genes, most of which are responsible for components of the mitochondria themselves. Since the mitochondrial genome is provided by the egg, mitochondrial genes are inherited in a strictly maternal manner. This has some interesting consequences for the population genetics of the mitochondrial genome.

Sperm Stem Cells

Sperm cells are produced throughout adult life just as red blood cells are. In both cases specific stem cells both proliferate and give rise to terminally differentiated cells. The stem cells of the sperm line are found in the testes surrounded by somatic cells, which do not differentiate into sperm. Sertoli cells are wrapped around differentiating sperm cells within the seminiferous tubules (Fig. 6.4). In the spaces around the tubules are found Leydig cells.

The growth and differentiation of the stem cells is controlled in part by

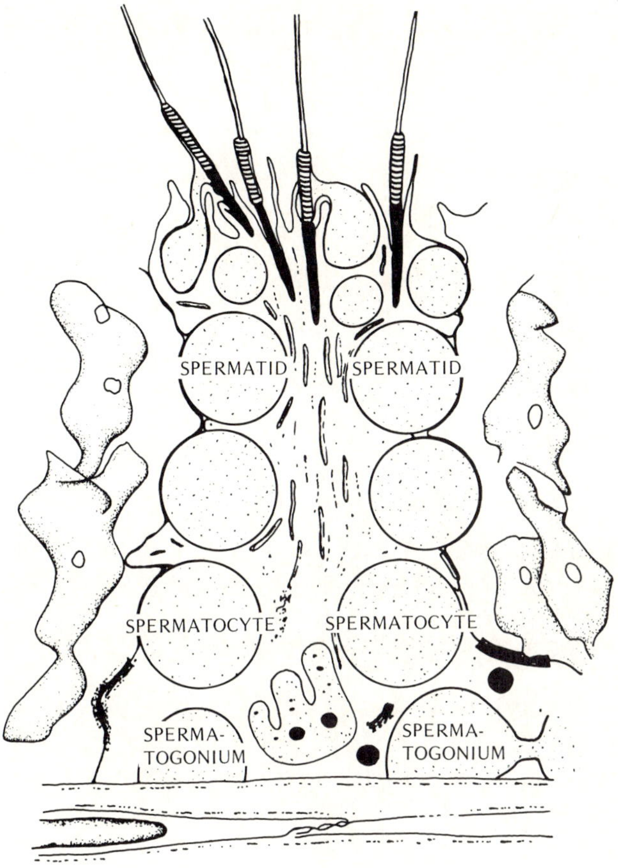

Figure 6.4. Spermatogonial differentiation. Flattened against the basal lamina of a seminiferous tubule are spermatogonia. A Sertoli cell sits over them. Spermatocytes given off by the spermatogonia are surrounded by and almost within the Sertoli cell. Cells in progressively later stages of differentiation are found at progressively more apical positions. Mature spermatozoa are released into the lumen of the tubule from the tip of the Sertoli cell.

the steroid hormone, testosterone, made by the Leydig cells. A testosterone-binding protein is produced by the Sertoli cells which appears to mediate this control.

When a daughter cell of the stem cell line becomes committed to sperm differentiation it will undergo only about five or six mitotic divisions to produce a nest of spermatogonia cells. While the nuclei divide and separate normally, cell division in spermatogonia is incomplete, so that cytoplasmic bridges are left connecting the cells. The nuclear divisions in this syncytial mass are synchronous as a result of the free diffusion of controlling molecules among the connected nuclei. At each division the amount of cytoplasm surrounding each nucleus is less and less as they prepare to form the small sperm.

These mitotic divisions are followed by two meiotic divisions, which reduce the nuclei to haploid genomes of spermatocytes (Fig. 6.5). At this stage one can first see the flagellum extending from the cell. As the sperm head is formed, it separates from the bulk of the cytoplasm. When differentiation of sperm is complete, they pinch off from the residual bodies and are released free into the lumen of the seminiferous tubules.

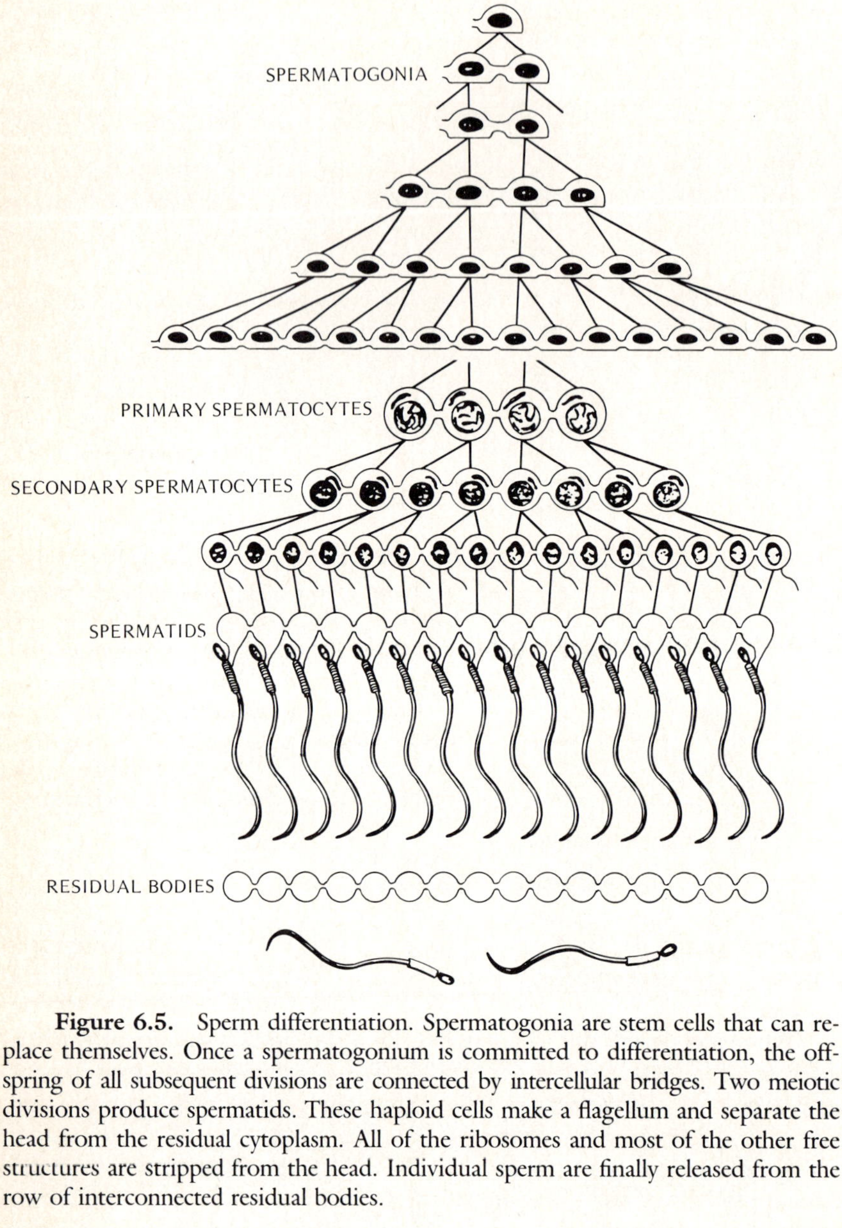

Figure 6.5. Sperm differentiation. Spermatogonia are stem cells that can replace themselves. Once a spermatogonium is committed to differentiation, the offspring of all subsequent divisions are connected by intercellular bridges. Two meiotic divisions produce spermatids. These haploid cells make a flagellum and separate the head from the residual cytoplasm. All of the ribosomes and most of the other free structures are stripped from the head. Individual sperm are finally released from the row of interconnected residual bodies.

Oogenesis

Eggs are large cells. They have to contain all of the material necessary to make a feeding offspring. Of course, some are bigger than others but they are all highly differentiated with respect to size. The fact that mammalian eggs are fairly small is simply the consequence of early feeding from the maternal blood supply through the placenta. Non-mammals must carry sufficient reserves to support the formation of a thousand or more differentiated cells. The reserves are stored as yolk, which, in the case of birds, can make up the bulk of the cell.

Differentiating eggs, referred to as oocytes, have evolved several ways in which to accumulate massive amounts of cytoplasm. In some occytes, materials are picked up directly from the bloodstream. For instance, in chickens vitellogenin is synthesized in the liver of laying hens and travels in the bloodstream to the ovaries. Oocytes take up vitellogenin and cleave it into phosvitin and lipovitellin, the major components of yolk platelets. The vitellogenin gene is active in the liver cells of laying hens but inactive in roosters. However, if the liver cells of male chickens are treated with the steroid hormone estrogen, transcription of the vitellogenin gene is initiated at a high rate. There are several DNA sites upstream of active vitellogenin genes that are sensitive to DNase in estrogen-treated cells that are relatively resistant to the entry of the probe enzyme in uninduced cells. These DNase hypersensitive sites may be generated as a consequence of binding of estrogen-binding regulatory proteins to the upstream (5′) regions of the gene.

In other cases differentiating oocytes are surrounded by follicle cells that transport proteins from the bloodstream to the oocyte as well as donate material they synthesize themselves (Fig. 6.6). However, organelles and complexes such as ribosomes are too large to cross the cell membrane and must all be made within the oocyte.

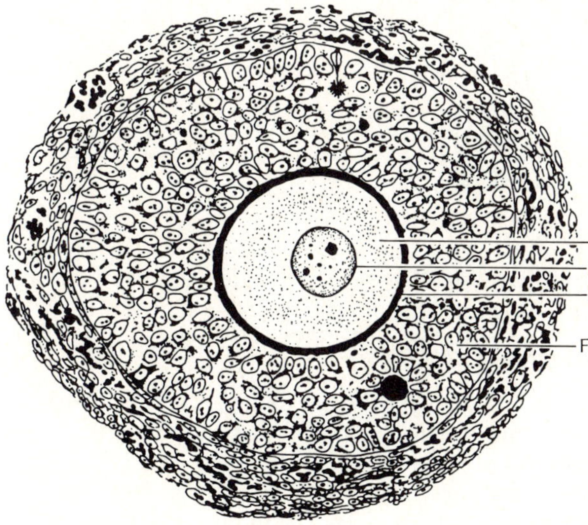

Figure 6.6. Growing human follicle. The large cell in the middle is the egg or ovum. It is almost full size but will grow another 20 percent before being released. The zona pellucida is well formed. Follicular epithelial cells surround the egg in the ovary (×375).

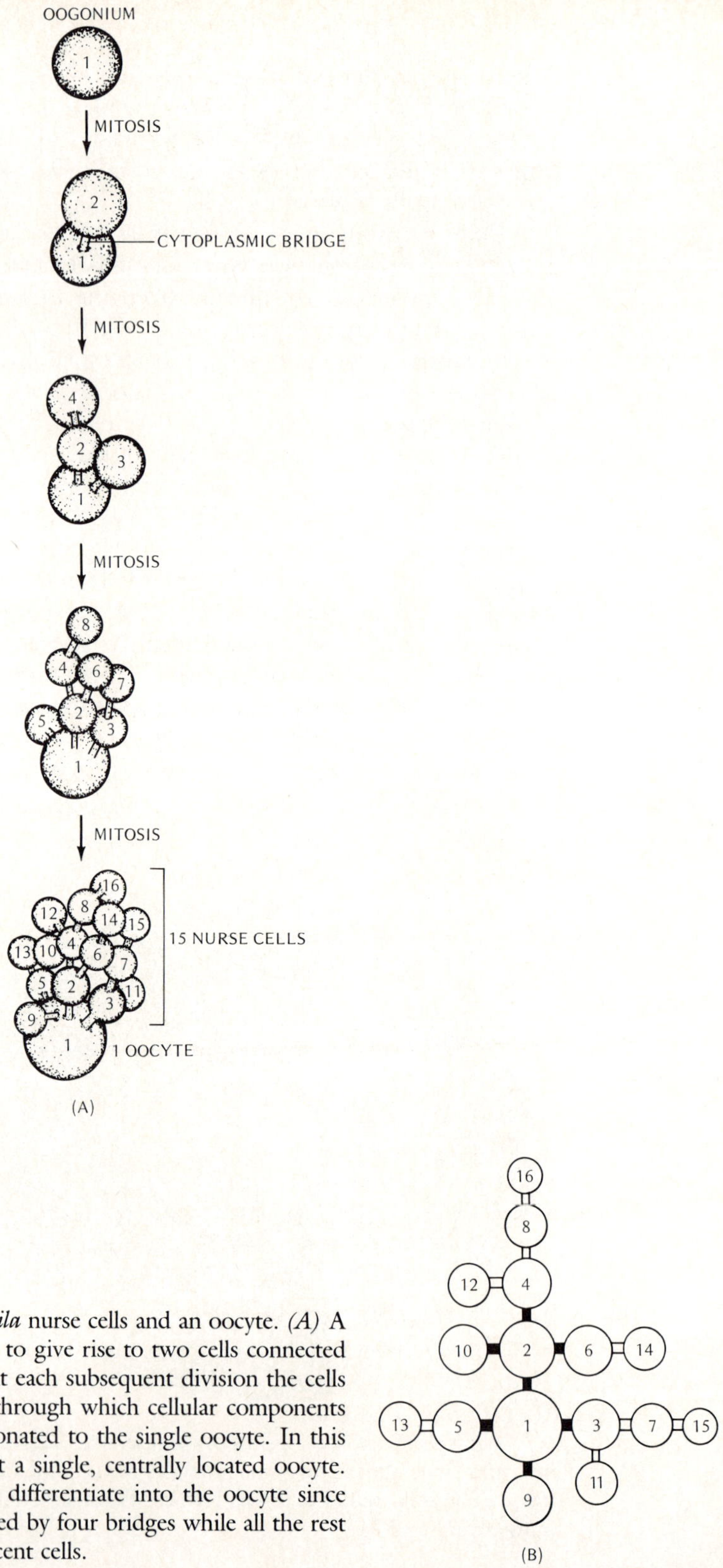

Figure 6.7. *Drosophila* nurse cells and an oocyte. *(A)* A primary oogonium divides to give rise to two cells connected by a cytoplasmic bridge. At each subsequent division the cells are connected by bridges through which cellular components including ribosomes are donated to the single oocyte. In this way 15 nurse cells support a single, centrally located oocyte. *(B)* Either cell 1 or 2 can differentiate into the oocyte since these two cells are connected by four bridges while all the rest have fewer bridges to adjacent cells.

In some organisms, such as many insects, the nuclei of primary germ cells divide several times but the cells remain connected by cytoplasmic bridges. The nucleus connected by the most bridges then differentiates into an oocyte while the other nuclei function as nurse cells. In this way a single nucleus can have the benefit of a dozen or so supporting nuclei. Transcription in the nurse cell nuclei provides the ribosomal RNAs as well as mRNAs, which are carried through the bridges to the swelling cytoplasm of the oocyte. The bridges connect nurse cells to the oocyte in an asymmetric way such that certain portions of the egg may have received more substance than other portions. This asymmetry within the egg is used in some cases to establish the axial polarity of the developing embryo (Fig. 6.7).

Biosynthetic Patterns in Oocytes

While many proteins are supplied to oocytes from somatic tissue, most differentiating oocytes are also actively engaged in producing the components needed during the first few hours of embryogenesis. Hybridization analyses have shown that the amount of potential genetic information that is expressed and present as RNA molecules is higher in oocytes than in any other differentiating cell. Most of this complexity is present in RNA molecules that are at a low abundance; that is, there are only a few molecules of that specific mRNA per cell. It is not yet clear whether this extensive expression of the genome plays an essential role in either the egg or the embryo.

Oocytes also contain many different RNA molecules at moderate abundance—around 1000 molecules per cell. These RNAs appear to be functional mRNA molecules, which associate with ribosomes and direct the translation of many different proteins. There are a score or so of high-abundance mRNA species that code for such proteins as histones, tubulins, and actin. They are present at several thousand copies per cell. These structural proteins are in great demand immediately following fertilization, as the DNA replicates exponentially and new cells are made every hour or so. Since these mRNAs were made in the differentiating oocyte while still in the mother, they are referred to as maternal mRNA. They direct the great majority of protein synthesis in an embryo during the cleavage stages. These maternal mRNAs are translated both before and after fertilization, although they are translated more rapidly after fertilization. When the patterns of abundant proteins synthesized before and after fertilization are inspected following separation in two dimensions they are barely distinguishable (Fig. 6.8).

Until recently it was thought that the specialized pathways that various blastomeres follow might be the consequence of localization of specific maternal mRNA species in one cell or another during cleavage. However, it now appears that each cell produced by the first few divisions shares equally from the pool of maternal mRNA, at least those that are moderately or very abundant. We have to look at processes that occur following fertilization to spot the steps that lead cells off onto specialized pathways.

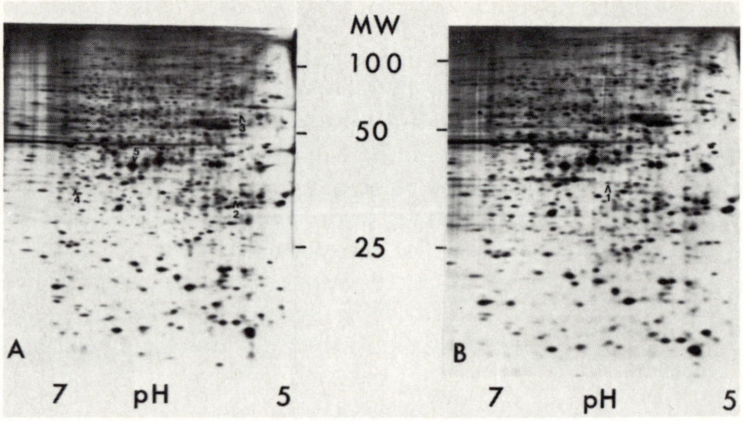

Figure 6.8. Proteins synthesized by unfertilized eggs *(A)* and 16-cell embryos *(B)* of sea urchins. Extracts of ^{35}S-methionine-labeled cells were separated in two dimensions and autoradiographed. Only a single protein (spot 1) was found to be made in embryos but not in eggs. Several hundred other proteins were abundantly synthesized both in unfertilized eggs and in early embryos.

As mentioned previously, ribosomes are not transported across oocyte membranes and must be made during the growth of the oocyte. Eggs are not only big, but their cytoplasm is well loaded with ribosomes to permit high rates of protein synthesis during early embryogenesis. Ribosomes are about half protein and half ribosomal RNA (rRNA). Synthesis of ribosomal proteins does not seem to present much of a problem during oogenesis because they are made catalytically from their respective mRNAs. However, the rRNA of the large and small ribosomal subunits, 28*S*, 5*S*, and 18*S* RNAs, are the direct products of the transcription of their genes. Although most species carry several hundred copies of the gene that codes for 28*S* and 18*S* RNA, in many species this is insufficient for the rate of rRNA synthesis needed during oogenesis. The rate of transcription is limited by the number of genes that can be used as templates. In amphibians and other organisms, a mechanism has evolved to amplify the number of genes for rRNA specifically during differentiation of oocytes. A specialized process of DNA replication singles out the portion of the genome that carries the rRNA genes and directs rolling circle replication (Fig. 6.9). This is a process by which a circular DNA is used as a template for synthesis of a long copy containing many replicas of the DNA in a tandem array. It is a process used in bacteria but it has only been found to occur in eukaryotes during amplification of rRNA genes. Rolling circle replication generates several hundred thousand copies of the rRNA genes in frogs. These copies are then transcribed to give the very high rate of rRNA synthesis observed in frog oocytes.

The other ribosomal RNA, 5*S*, is transcribed off a separate set of genes from the one for 28*S* and 18*S* RNA. Evolution of frogs has used a different approach to solve the need for a rate of 5*S* RNA synthesis equal to that of 28*S* and 18*S* RNA during oogenesis. Although only a few hundred genes for 5*S* RNA are transcribed in adult somatic cells, another gene family is used specifically for 5*S* RNA transcription in occytes, and there are 20,000 copies of this oocyte 5*S* gene family.

As discussed in Chapter 3, transcription of 5*S* genes is catalyzed by RNA polymerase III in conjunction with TFIIIA. Part of the mechanism that ac-

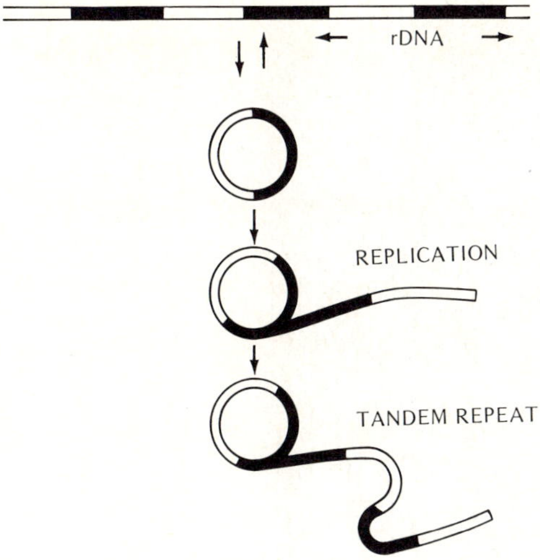

Figure 6.9. Amplification of rDNA. The genes for the large ribosomal RNAs are tandemly arranged in *Xenopus* oocytes. During differentiation one unit gives rise to a circular copy that is replicated by a rolling circle mechanism, generating a concatemer of repeating units. Reinsertion of the copy could rectify the sequences of the genomic copies.

counts for the extremely rapid synthesis of 5*S* RNA in oocytes is TFIIIA, which accumulates to high levels, specifically in oocytes. The family of 20,000 5*S* genes transcribed during oogenesis requires a higher concentration of TFIIIA for activation *in vitro* than the smaller family of 5*S* genes transcribed in adult cells. In oocytes, TFIIIA is at a sufficiently high concentration that both families of 5*S* genes are active, and since the oocyte family is so large 5*S* RNA is synthesized at a very high rate mostly from the oocyte genes. All transcription stops when eggs of the amphibian, *Xenopus laevis,* mature and only starts up again following the mid-blastula transition when there are about 4000 cells in the embryo. TFIIIA is not synthesized during this period of embryogenesis; therefore, its concentration falls from about 3×10^9 molecules/cell to 10^6 molecules/cell. Competition for TFIIIA between somatic 5*S* genes and oocyte 5*S* RNA favors the former, which have a higher affinity to TFIIIA. As a result, the oocyte 5*S* genes become increasingly inactivated until they are completely silent in adult somatic cells where the concentration has fallen to 10^4 molecules/cell of TFIIIA. Only the somatic 5*S* genes are transcribed in adult cells, with the exception of differentiating oocytes. This molecular example points out how regulation of expression of a gene for a regulatory protein (TFIIIA) can account for dramatic switching of transcription patterns on whole families of genes that differ chiefly in their affinity to the factor.

Maternal Determination

The structure of an egg determines several aspects of the subsequent course of embryogenesis. The most dramatic asymmetry of eggs concerns the localization of yolk. Some eggs, such as those of birds, contain so much yolk that the non-yolky cytoplasm makes up only a small cap in these very large cells. The egg pronucleus is positioned in the cytoplasmic portion and

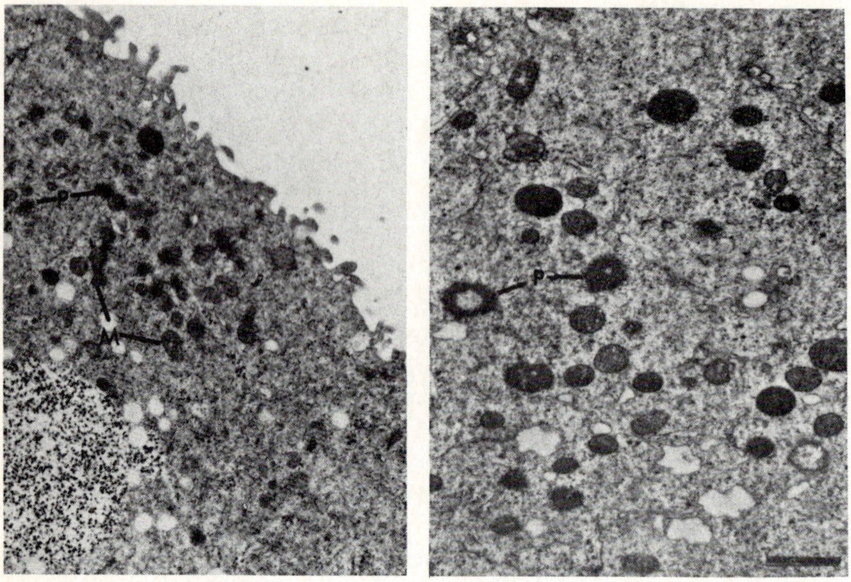

Figure 6.10. Polar granules. At the posterior tip of *Drosophila* embryos 30 minutes after fertilization, a concentration of polar granules (P) can be seen within a few microns of the surface. These granules are attached to mitochondria (M) until the nuclei migrate to the surface.

all of the embryonic cells are formed there following fertilization. In smaller less yolky eggs, such as those of sea urchins, no clear demarcation of the yolk can be seen. The yolk is spread throughout the egg. However, because of the difference in density between yolk and the other cytoplasmic components, one half of the egg contains more yolk than the other. Following fertilization in sea urchin eggs, cleavage separates the yolky half from the less yolky half. Cells in the former soon take on different properties and are referred to as vegetal cells whereas those in the less yolky half are referred to as animal cells. The blastopore forms at the vegetal pole and establishes the axis of the embryo. Although many processes are involved in establishing the embryonic axis, yolk distribution plays a very significant part.

Eggs of some organisms have components other than yolk positioned in an asymmetric manner, and these may direct the differentiation pathway of cells which end up including them. One clearly defined case has been recognized in *Drosophila* eggs. Since these cells differentiate in association with asymmetrically connected nurse cells, they have several characteristics that distinguish the anterior from the posterior end. When *Drosophila* eggs are analyzed in the electron microscope, they can be seen to have a concentration of distinctive granules (polar granules) localized at the posterior pole (Fig. 6.10). When this portion of the *Drosophila* egg becomes compartmentalized in cells of the blastula, it directs the subsequent division and differentiation of the pole cells. They ultimately form the adult germ cells—eggs or sperm. Microtransplantation of pole plasm containing polar granules has shown that the information to direct cells toward germ cell differentiation is completely contained in this region and does not depend on any interactions with other portions of the egg or the developing embryo.

The determination of germ cells seen in *Drosophila* eggs appears to be a rare case. It is possible that such determination is unique to germ cell precursors. No evidence has been found for specific determinants controlling differentiations of limbs or eyes or any other organs or cell types. These pathways are directed by later processes, which arise only during gastrulation and subsequent embryogenesis.

Egg Shells

In many but not all eggs a shell is deposited on their outer surface before they are shed or laid (Fig. 6.11). The shell protects the egg from mechanical damage and dehydration. Insect eggs have a shell consisting of specialized proteins that is referred to as the chorion. Chorion proteins are synthesized in follicle cells that surround the egg. They are secreted around the egg and structured into the chorion. In *Drosophila* there is very little time for the formation of the chorion before the eggs are layed. There are only a few copies of the genes for chorion proteins in the *Drosophila* genome and they cannot support a sufficiently high rate of transcription to generate enough chorion mRNA for optimal chorion production. Just as a mechanism evolved to amplify the genes for rRNA in oocytes, a mechanism evolved in follicle cells to specifically amplify the region containing the chorion genes. The process of replication of this region does not involve rolling circle replication but the more normal bidirectional growth of new DNA strands (Fig. 6.12). Replication starts near the chorion genes and proceeds for some distance in both directions. Multiple initiations of replication result in an onionskin structure in which the central portion is present in five to ten copies while flanking regions are present as progressively fewer copies per cell (Fig. 6.13). There are two X-linked genes that are required for amplification of the region (*7F*1-2) that includes two chorion genes, *S*36-1 and *S*38-1. Since the two controlling genes map at distant loci (12*A*–*D* and 5*D*–6*C*), their products must act in *trans*. Mutations in either of these genes result in female sterility due to underproduction of all major chorion proteins. They appear to single out a 100 kb region of DNA near *7F* for preferential replication in follicle cells. At 66*D* on the third chromosome there is another cluster of chorion genes. Four chorion proteins are encoded within a 11.6 kd DNA segment that is amplified 60-fold specifically in terminally differentiating follicle cells. Overreplication of this segment depends on the presence of a 3.8 kd sequence that acts in *cis* to control bidirectional replication in follicle cells. Thus, there are two well-characterized cases in which the genome is different in specialized differentiated cells: rRNA in oocytes and chorion genes in follicle cells. These are the only cases of specific sequence amplication in embryos or adults out of several dozen that have been carefully investigated. A few more such cases might show up as other genes are analyzed in other cells, but it is clearly a rare strategy used in differentiation.

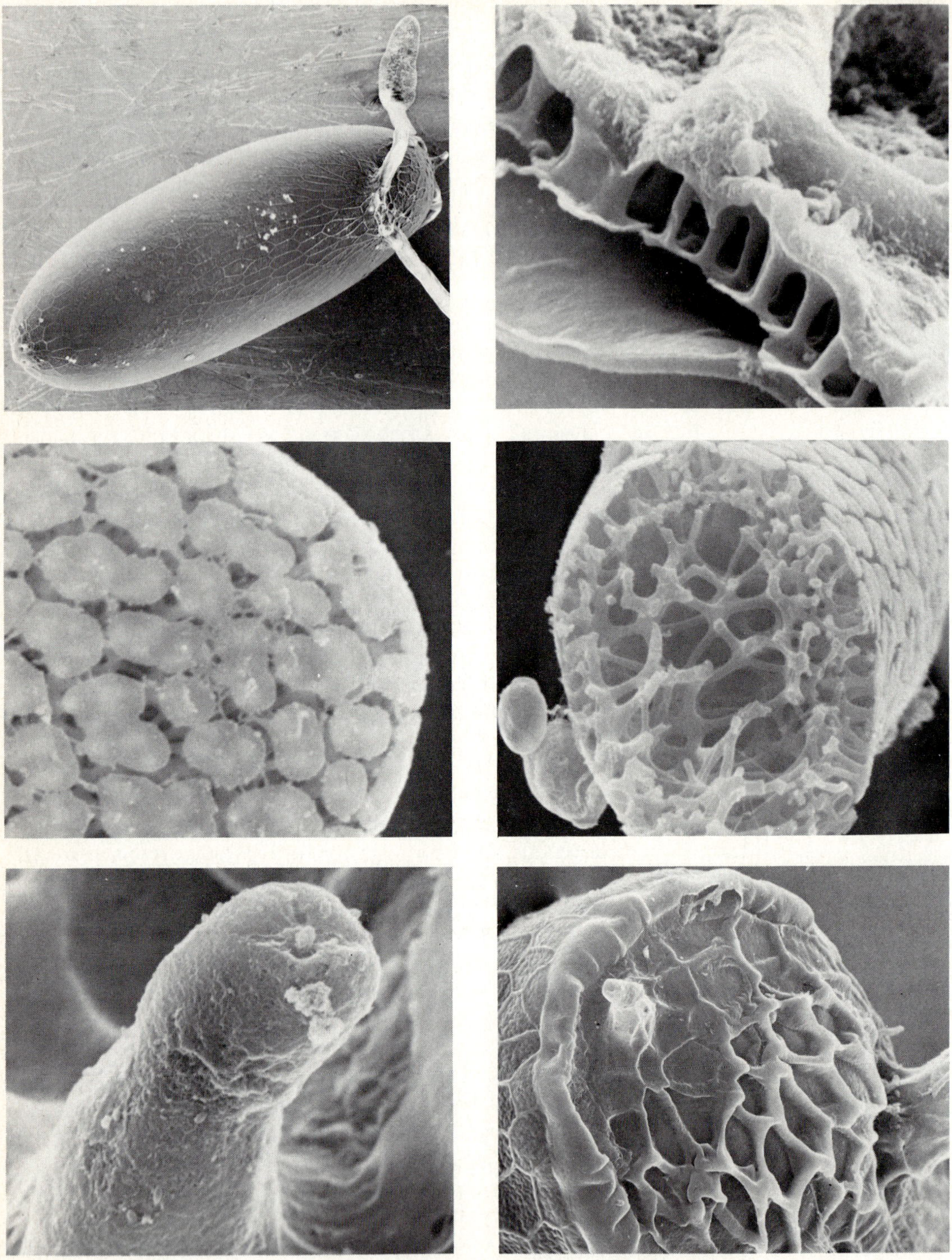

Figure 6.11. The chorion of a *Drosophila* egg. The outlines of the follicle cells that deposited the chorion on the egg can be seen. The micropyle can be seen at the right-hand side of the egg and in the enlargement. Just before fertilization, a guard cell in the micropyle retracts to allow sperm to enter.

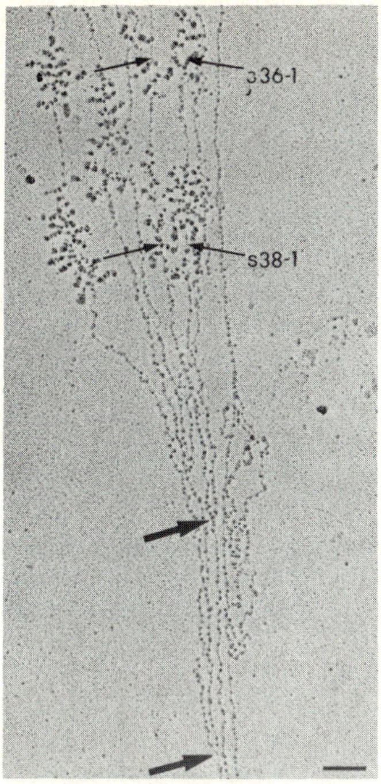

Figure 6.12. Amplification of chorion genes. Chromatin of differentiating *Drosophila* follicle cells has a region of multiforked structures that result from repeated bidirectional replication. Two chorion genes, *s*36-1 and *s*38-1, are transcribed from this region. Replication forks are indicated by arrows.

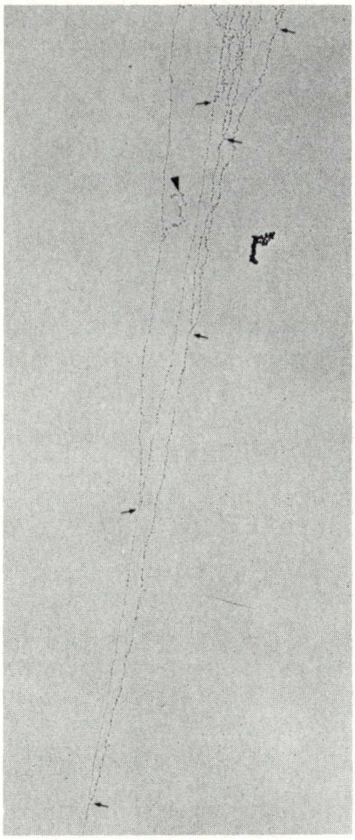

(A)

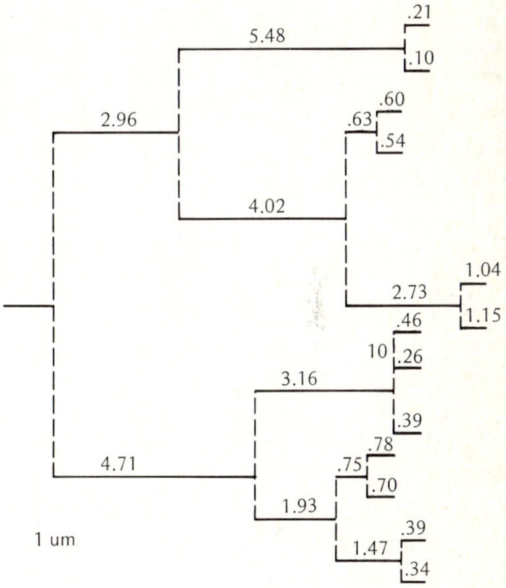

(B)

Figure 6.13. Bifurcation patterns. *(A)* Arrows indicate replication forks. These do not occur symmetrically along sister chromatin strands. The region is at least 10 μm long. *(B)* The lengths between forks are presented on the drawing.

Summary

Differentiation of sperm in males and eggs in females occurs in the gonads. Sperm are produced from stem cells continuously throughout adult life. Meiosis reduces the genetic complement to haploid in spermatids. Most vertebrates further differentiate sperm into high motile small cells with an optimum chance of finding and penetrating an egg of the same species. Oogenesis generates eggs characterized by massive reserves used following fertilization. Yolk is often donated from liver or nurse cells. In amphibians the genes coding for large ribosomal RNA are specifically amplified in oocytes so as to direct a high rate of rRNA synthesis. In *Drosophila* the follicle cells surround the egg in a chorion. The genes coding for chorion proteins are specifically amplified in follicle cells to direct a high rate of synthesis of the egg shell components.

SPECIFIC EXAMPLES

1. Spores

While the passage of generation to generation proceeds from fertilized egg to adult to egg in animals and higher plants, in many organisms the mature form grows directly from a germinated spore. Some spore-formers are exclusively or predominantly haploid, such as bacteria and many fungi and amoebae. In these species the spores are formed from the cells produced by normal mitotic divisions. Other species are predominantly diploid or have alternating cycles of haploid and diploid growth. The Bryophyta plants such as mosses, liverworts, and hornworts produce both sperm and eggs as well as spores (Fig. 6.14). Many vascular plants such as ferns also produce spores.

Spores are small, dry single cells protected by a heavy wall. They are resistant to drought, fire, and famine. Spores of some plants that were stored several thousand years ago have been germinated recently and the plants are growing well. This strategy to overcome adversity is exceptionally successful.

The spore-forming bacteria, such as *Bacillus,* respond to nutritional deprivation by expressing a new set of spore-specific genes. These genes require RNA polymerase sigma factors that differ from the sigma factor used to initiate vegetative gene transcription. It is not yet clear whether or not this change in the RNA polymerase subunits will apply to sporulation genes in other organisms. During sporulation, *Bacilli* accumulate the unusual amino acid diaminopimelate and form a forespore within the mother cell. The spore has a specialized cell wall and is resistant to a variety of environmental insults that would kill vegetative bacteria.

Most fungi produce haploid spores that disperse, germinate, and give rise to new hyphal growth before a new cycle of sporulation is induced. In the common breadmold, *Neurospora crassa,* the spores, referred to as conidia, are held aloft on aerial hyphae. These are extended about once a day in

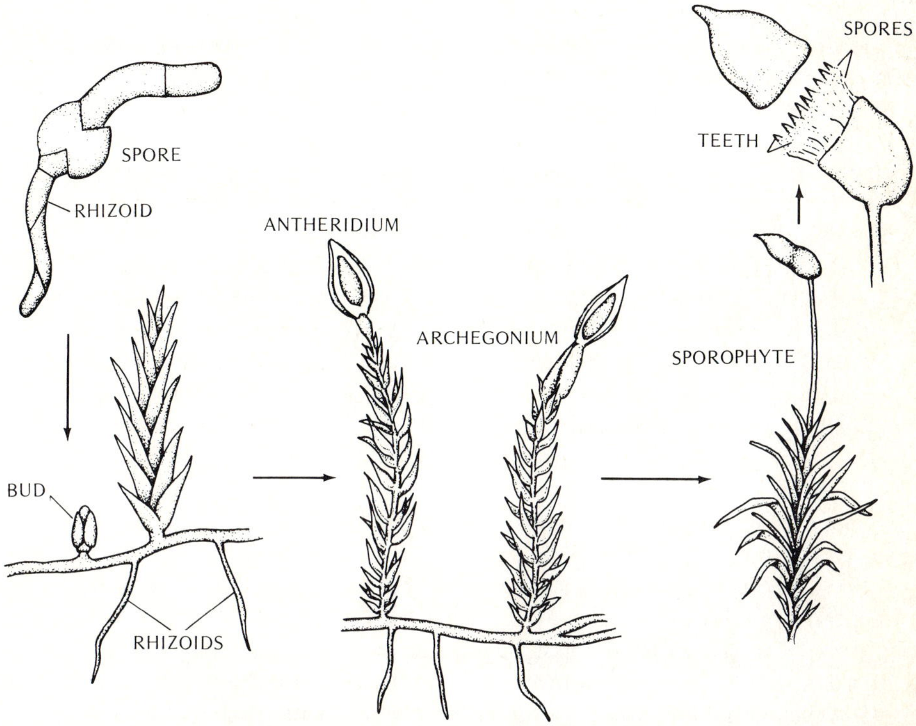

Figure 6.14. Life cycle of a moss. Spores germinate and send out processes. Rhizoids root the moss while protonema spread and send up buds. The shoots develop male or female sex organs (antheridia or archegonia, respectively). A fertilized egg from an archegonium puts up a sporophyte. At the tip of the sporophyte a new generation of spores differentiate within the sporangial cap. This structure eventually bursts, flinging spores around.

a circadian rhythm that is entrained by light. A light–dark regimen results in conidiation in the hours just before dawn when the humidity is high in the natural environment. During conidiation nuclei migrate through the hyphae to the apices, where they are separately encased in a small but tough cell wall. The slightest breeze will disperse the conidia. The spores of mosses and ferns are similar to conidia. Of course, the shapes of the plants that bear them vary enormously in size and shape, but they all grow from small unicellular spores.

Many soil amoebae, such as *Dictyostelium,* differentiate into spores. These nonphotosynthetic cells have properties common to both plants and animals. They ingest yeast or bacteria by phagocytosis but also synthesize cellulose during terminal differentiation. The first signal toward spore formation is the lack of an available food source. The cells respond by producing the enzymes and receptor proteins necessary for chemotaxis to each other. As is described in detail in Chapter 11, aggregates of up to 10^5 cells then integrate their differentiation to produce fruiting bodies in which some of the cells vacuolize and produce a cellulose stalk several millimeters in height on which the spores are held. At the top, each cell secretes a cellulose wall around itself and goes dormant as a spore.

A spore has need of energy, carbon, and nitrogen reserves to use when it germinates. In *Dictyostelium* spores, as in many other spores the nonreducing disaccharide, trehalose, and the amino acid, glutamate, serve as reserves. Trehalose also stabilizes membranes so that they are not disrupted by the dehydration that is common in spores. Dormancy is ensured by removal of almost all free water in spores. Under these conditions all metabolism stops and hydrolysis of labile bonds is kept to a minimum. Moreover, many proteins in spores are stabilized by cross-linking with disulfide bonds.

Germination usually results when spores are rehydrated. In some cases, a germination inhibitor is secreted as the cells encapsulate into spores. The germination inhibitor blocks activation of spores while they are still held together. When dispersed, the germination inhibitor, a low-molecular-weight metabolite, is diluted away. The spores swell due to the uptake of water and burst out of their spore coats. Subsisting initially on their reserves, the newly emerged cells search the new locale for nutrients.

For seedless organisms the formation of spores provides a way to avoid intemperate periods by waiting for a change as well as by dispersing to a new place. Since sporulation is fast and simple compared with the differentiation of sperm and eggs, it is a wonder that sexual propagation evolved at all. However, other environments are more stable and the time and energy needed to produce and fuse eggs and sperm are well worth it. The opportunities opened by meiotic shuffling of genes from two parents to produce gametes able to combine with other genetic assortments have clearly had selective advantage in the radiation of animals and higher plants. The sporogenous organisms have an advantage under certain conditions in the production of large numbers of spores that can rapidly proliferate in ephemeral conditions.

2. Meiosis and Maturation

All sexual organisms exhibit an alteration of generations during which the genome is present first as a haploid, then as a diploid, then as a haploid again. Usually one or the other generation is greatly emphasized: the diploid in higher plants and animals, the haploid in fungi and mosses. The edible marine alga *Ulva stenophylla,* however, forms large leaves as both a haploid and a diploid (Fig. 6.15).

Haploid spores of *Ulva* swim about and then settle to the bottom, where they grow into green leaves referred to as gametophytes because they produce gametes. In *Ulva* these are genetically one of two mating types. Gametes of opposite mating types look exactly alike but have surface properties that distinguish them. Only gametes of different mating types will fuse to form a diploid zygote. The zygote grows into a green leaf referred to as a sporophyte because it produces spores. All of the cell divisions in the stages of the life cycle described so far result from mitotic divisions in which hap-

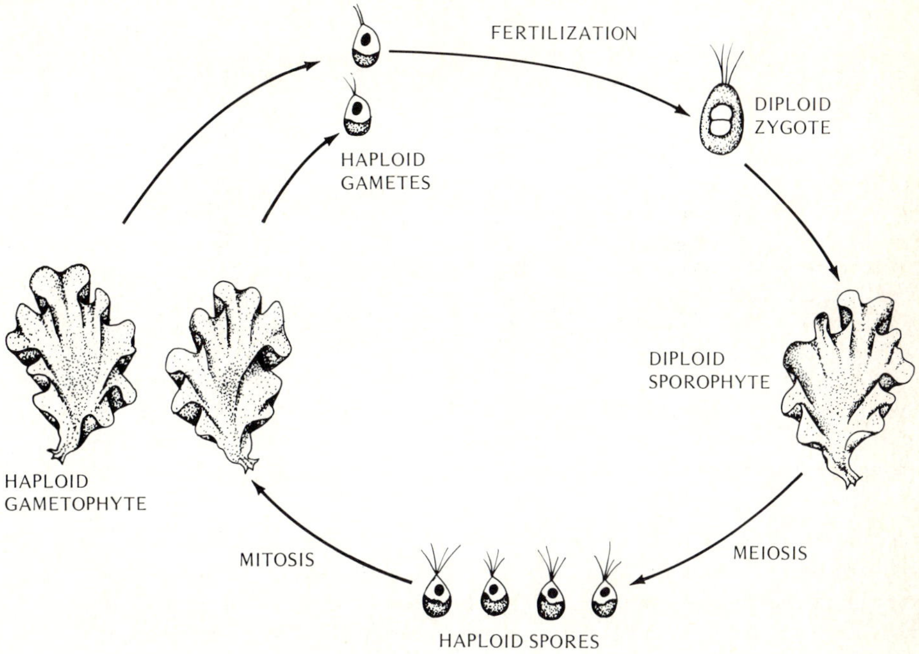

Figure 6.15. Alteration of generations of sea lettuce, *Ulva stenophylla.* Haploid gametophyte and diploid sporophyte adults are almost indistinguishable. Both have large edible green leaves. Spores grow into adult gametophytes that produce the gametes by mitosis. Meiosis in the germ cells of sporophytes give rise to haploid spores that are either male or female.

loid cells give rise to haploid progeny and diploid cells give rise to diploid progeny. However, the next step produces haploid spores from the diploid sporophyte and requires reduction of the chromosome content.

Two meiotic divisions of the diploid cell give rise to four haploid spores. Before the first division each DNA double strand in the chromosomes replicates, producing tetrads. Homologous chromosomes pair at the metaphase plate and then separate at anaphase (Fig. 6.16). Unlike a mitotic division, in which one copy of each chromosome goes to each of the poles, the first meiotic division takes both copies of one chromosome to a pole and both copies of the homologous chromosomes to the other pole. The progeny cells of the first meiotic division have the diploid amount of DNA but each chromosome is homozygous except in a few places. The choice of maternal or paternal (mating type *a* or mating type *b*) is independent for each chromosome and so there is a mixture of parental genes. The DNA in these primary meiotic cells does not replicate before the second meiotic division, which therefore produces haploid cells, each with a random assortment of chromosomes.

The diploid generation in the breadmold *Neurospora* is greatly reduced and can be avoided altogether. However, when conidia of one mating type

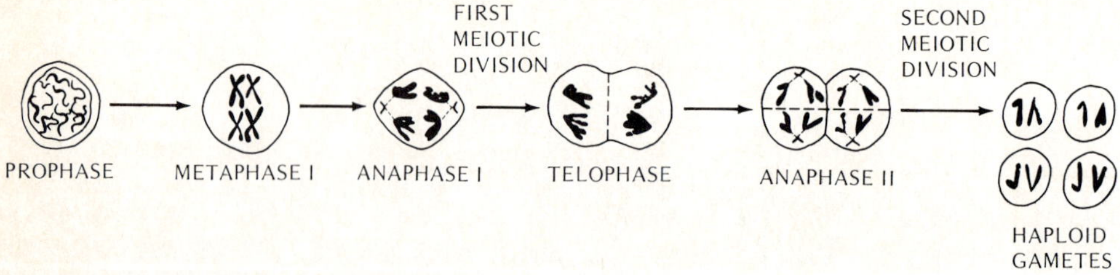

Figure 6.16. Chromosomal segregation during meiosis. As the chromosomes condense during meiotic prophase chiasmata form between homologues. By metaphase I recombining chromosomes appear in X shapes. Homologues separate in anaphase I such that following the first meiotic division, the daughter cells are essentially homozygous containing two identical copies of each chromosome except where recombination of each chromosome has occurred. No replication of the chromosomes occurs before anaphase II when the copies separate. The second meiotic division produces four haploid gametes.

fall into the protoperithecium of another mating type, the nuclei fuse to form a diploid zygote. Without more ado, the diploid nucleus enters into meiosis and gives rise to four haploid spores. These divide once mitotically to give rise to eight haploid ascospores that are released and can germinate (Fig. 6.17).

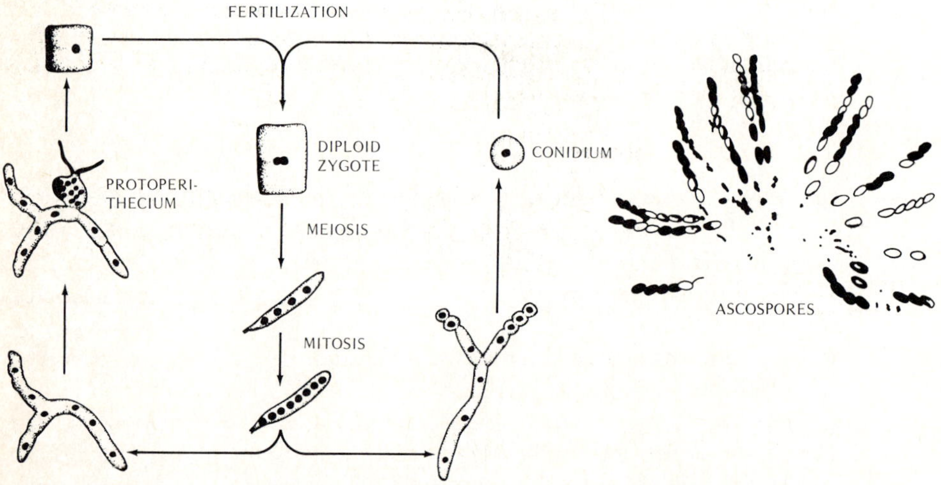

Figure 6.17. Meiotic segregation in *Neurospora*. The mycelium grows as a haploid but differentiates sexual structures referred to as protoperithecia. Fertilization within these structures gives rise to a diploid zygote. Two meiotic divisions produce four haploid cells with an assortment of genes from the two parental cells. They are held in a row within an ascus. They mature into ascospores after a single mitotic division. If the mating strains have differently pigmented spores, the segregation of the genes controlling this phenotype can be analyzed in the ascus. Notice that spore color is always segregated in pairs of cells, since the last division is of haploid cells. Equal numbers of pigmented and unpigmented ascospores indicate that a single Mendelian gene is responsible for this phenotype.

As the chromosomes condense and pair in preparation for the first meiotic division, homologus regions of the paired chromosomes synapse under the control of specialized structures that appear only during the differentiation of meiotic cells. Chiasmata can be seen to form at these points. They are visible manifestations of genetic recombinations in which the DNA double strand of each chromosome at the synapse is broken and joined to the DNA of the homologous chromosome. In this way portions of chromosomes from mating type *A* can be exchanged for the homologous portion of the chromosome from mating type *a*. This specialized mechanism for high-frequency recombination occurs only in differentiated meiotic cells and results in further scrambling of parental genes to give cells with new assortments of each allele. Selection can then favor the most adaptive complement.

In animals, sperm and eggs are haploid cells produced by meiotoc divisions of spermatogonia and oocytes, respectively (Fig. 6.18). The four products generated by two meiotic divisions of a spermatogonial cell form

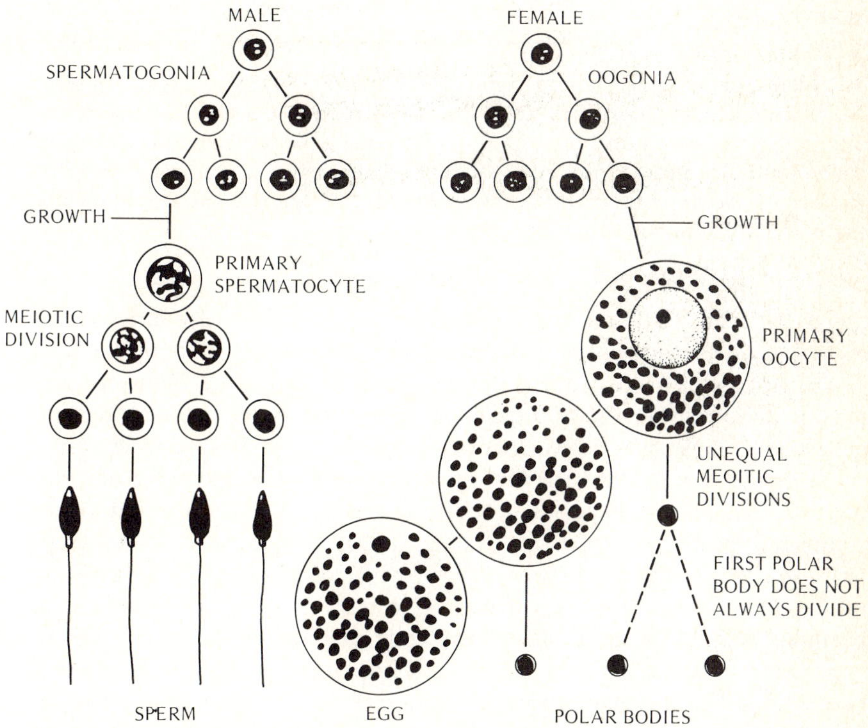

Figure 6.18. Meiosis during gametogenesis. Primary spermatocytes and oocytes are generated by mitotic divisions of the germ cells. Two meiotic divisions produce spermatids and an egg and polar bodies. The meiotic divisions of primary oocytes are highly unequal with almost all the cytoplasm kept for the egg. Polar bodies later degenerate. Spermatids differentiate into spermatozoa without further division. Fertilization can occur in primary or secondary oocytes as well as in eggs depending on the species involved.

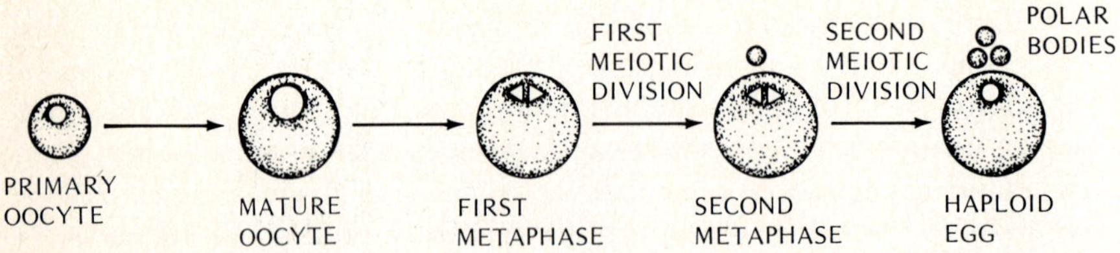

Primary Oocyte	First Metaphase	Second Metaphase	Haploid Egg
Flatworms	Cone worms	Amphibians	Coelenterates
Roundworms	Molluscs	Mammals	Echinoids
Polychaete worms	Insects		
Clams			
Dogs			

Figure 6.19. Stages in egg maturation at which fertilization occurs. In some worms and other species, sperm penetrates young primary oocytes. The oocyte nucleus undergoes two meiotic divisions, producing polar bodies before the pronuclei fuse to form the diploid zygote nucleus. Fertilization occurs at later stages in differentiation of oocytes in other organisms. In humans the egg is at the second meiotic metaphase when sperm penetrate.

functional sperm. The meiotic divisions of an oocyte are very different. At each division one nucleus is budded off with a minimum of cytoplasm as a polar body and the other retains the great bulk of the oocyte. Thus, the two meiotic divisions give rise to a single large egg and either two or three tiny polar bodies, depending on whether or not the first polar body divides again.

Oogenesis occurs early in life, yet the eggs are used only following sexual maturity. In many species development of eggs is arrested before the completion of the meiotic divisions (Fig. 6.19). In humans the first meiotic prophase starts in the third month of embryonic development and then stops until puberty when the hypothalamus secretes a gonadotropin-releasing hormone (Gn-RF) that stimulates the anterior pituitary to release follicle-stimulating hormone (FSH) and luteinizing hormone (LH). These hormones stimulate the maturation of the ovaries, which then secrete two female sex hormones, estrogen and progesterone. During the menstrual cycle these hormones rise and fall periodically. FSH and LH stimulate the growth of follicles surrounding primary oocytes that have yet to divide meiotically. Hormonal feedback loops lead to a sharp peak of LH in the middle of the monthly cycle of humans and the most-developed follicle bursts and releases an egg (Fig. 6.20). During this process the oocyte completes the first meiotic division and forms a polar body. The oocyte then proceeds to the metaphase of the second meiotic division, at which point it again is arrested. Fol-

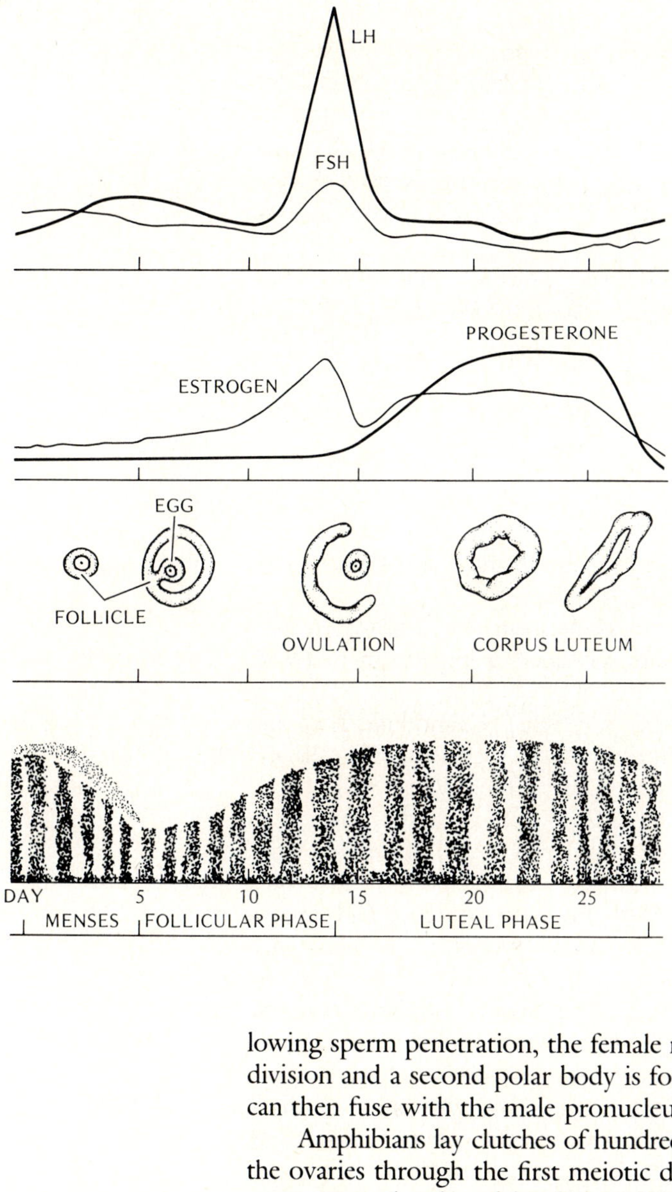

Figure 6.20. Menstrual cycle in humans. The anterior pituitary produces lutenizing hormone (LH) and follicle-stimulating hormone (FSH) in a cyclic pattern repeated every 28 days. These hormones interact with ovarian tissue to regulate the rise and fall of estrogen and progesterone levels. Each month a follicle swells following menses. Two weeks later an egg is released at ovulation. The uterine lining is sloughed during the period of menses and enlarges throughout the rest of the cycle.

lowing sperm penetration, the female nucleus undergoes the second meiotic division and a second polar body is formed. The haploid female pronucleus can then fuse with the male pronucleus to generate the diploid zygote.

Amphibians lay clutches of hundreds of eggs. The oocytes develop within the ovaries through the first meiotic division and then arrest. At spawning, progesterone is released by the ovarian tissue and stimulates maturation. The oocytes leave prophase arrest and enter meiotic reductive division. The large oocyte nucleus, referred to as the germinal vesicle, breaks down and the rate of protein synthesis increases severalfold. The primary oocytes have progesterone-binding proteins of 110 kd on their surfaces that induce the cells to enter the second meiotic division and prepare themselves for fertilization. This is one of the few cases in which a steroid hormone (progesterone) acts on the surface rather than entering the cells and affecting nuclear activity.

When oocytes of the South African clawed toad, *Xenopus laevis,* are treated with progesterone, there is a rapid release of bound calcium ions within the cytoplasm. The Ca^{++} ions associate with calmodulin and together they stimulate protein kinase activity. There is some evidence that this protein kinase accounts for the activity of a maturation promotion factor (MPF) found in progesterone-treated oocytes. MPF is a protein with autocatalytic activity that may be the phosphorylated form of a protein kinase. Autophosphorylation may account for its autocatalytic properties while phosphorylation of other proteins may trigger maturation. Germinal vesicle breakdown (GVBD) liberates components into the cytoplasm. Some of these play essential roles in the first cleavage of the egg.

In echinoids, such as sea urchins, females produce eggs continuously during their season. Oogenesis proceeds without stop through meiosis to produce eggs with a single haploid pronucleus that are shed into the surrounding water.

Related Readings

Chandley, A., Hotta, Y., and Stern, H. (1977). Biochemical analysis of meiosis in the male mouse. Chromosoma *62,* 243–253.

De Cicco, D., and Spradling, A. (1984). Localization of a *cis*-acting element responsible for the developmentally regulated amplification of *Drosophila* chorion genes. Cell *38,* 45–54.

Feinberg, J., Pariset, C., and Weinman, S. (1985). Calmodulin level and cAMP-dependent protein kinase activity in rat spermatogenic cells and hormonal control of spermatogenesis. Devel. Biol. *108,* 179–184.

Gerhart, J., Wu, M., and Kirschner, M. (1984). Cell cycle dynamics of an M-phase-specific cytoplasmic factor in *Xenopus laevis* oocytes and eggs. J. Cell Biol. *98,* 1247–1255.

Groudine, M., and Conkin, K. (1985). Chromatin structure and *de novo* methylation of sperm DNA: implications for activation of the paternal genome. Science *228,* 1061–1068.

Jost, J. P., Geiser, M., and Seldran, M. (1985). Specific modulation of the transcription of cloned avian vitellogenin II gene by estradiol-receptor complex *in vitro.* Proc. Natl. Acad. Sci. *82,* 988–991.

Kitada, K., and Omura, T. (1984). Genetic control of meiosis in rice *Oryza sativa* L: III. Effects of *ds* genes on genetic recombination. Genetics *108,* 697–706.

Newport, J., and Kirschner, M. (1984). Regulation of the cell cycle during early *Xenopus* development. Cell *37,* 731–742.

Orr, W., Komitopoulou, K., and Kafatos, F. C. (1984). Mutants suppressing in *trans* chorion gene amplification in *Drosophila.* Proc. Natl. Acad. Sci. *81,* 3773–3777.

Osheim, Y., and Miller, O. L. (1983). Novel amplification and transcriptional activity of chorion genes in *Drosophila melanogaster* follicle cells. Cell *33,* 543–553.

Raff, E. (1984). Genetics of microtubule systems. J. Cell Biol. *99,* 1–10.

Tamarkin, L., Baird, C., and Almeida, O. (1985). Melatonin: A coordinating signal for mammalian reproduction. Science *227,* 714–720.

Tambes, R., and Shapiro, B. (1985). Metabolite channeling: a phosphoryl creatine shuttle to mediate high energy transport between sperm mitochondrion and tail. Cell *41,* 325–334.

Wassarman, P., Schultz, R., and Letourneau, G. (1979). Protein synthesis during meiotic maturation of mouse oocytes *in vitro.* Synthesis and phosphorylation of a protein localized in the germinal vesicle. Devel. Biol. *69,* 94–107.

Wasserman, W., Houle, J., and Samuel, D. (1984). The maturation response of stage IV, V, an VI *Xenopus* oocytes to progesterone stimulation *in vitro.* Devel. Biol. *105,* 315–324.

Waters, S., Distel, R., and Hecht, N. (1985). Mouse testes contain two size classes of actin mRNA that are differentially expressed during spermatogenesis. Mol. Cell Biol. *5,* 1649–1654.

White, R., Perriman, N., and Gehring, W. (1984). Differentiation markers in *Drosophila* ovary. J. Embryol. Exp. Morph. *84,* 275–286.

Wolgemuth, D., Celenza, J., Bundman, D., and Dunbar, B. (1984). Formation of the rabbit zona pellucida and its relationship to ovarian follicle development. Devel. Biol. *106,* 1–14.

Fertilization

CHAPTER 7

Although the entry of sperm pronuclei into eggs at fertilization has been studied in many organisms, none has been studied in such detail as that of sea urchins such as *Strongylocentrotus purpuratus* and other echinoderms. To a large extent this is due to the fact that one can repeatedly get large amounts of sperm and eggs from sea urchins and fertilization takes place outside the body such as on a microscope slide. While the details of fertilization may differ in other species, it is likely that the processes are similar to those in sea urchins. The emphasis may be shifted from one strategy to another, depending on the particular requirements of the organism. Unless otherwise stated, the following descriptions and analyses have been carried out in sea urchins and related echinoderms.

The Acrosome Reaction

As the sperm approach the egg, they encounter a jelly that surrounds and extends far beyond the surface of the egg. This egg jelly is composed of sialoglycoproteins and sulfated fucose glycoproteins in a tenuous gel. While traversing this region, the bed of G actin explosively polymerizes into filamentous actin (F actin) and extends the acrosomal vesicle into an elongated tubule. This prepares the sperm to adhere and penetrate the egg (Fig. 7.1). An encounter with jelly is sensed by a specific 210 kd protein present on

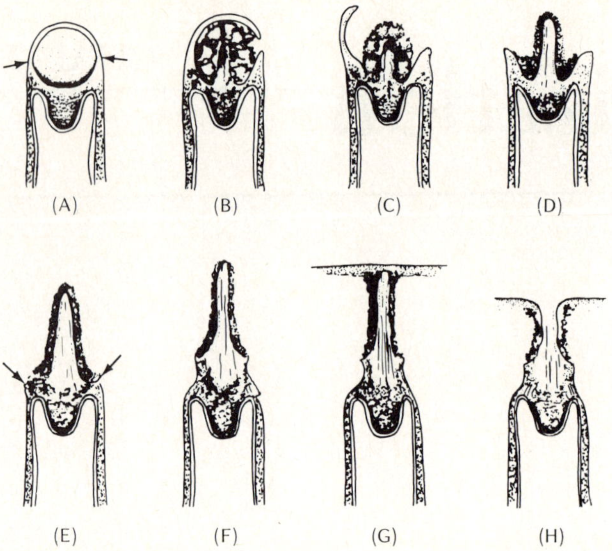

Figure 7.1. Sea urchin acrosome reaction. *(A)* The acrosomal membrane lies beneath the sperm membrane. *(B)* Following activation the two membranes fuse and the subacrosomal actin polymerizes into filaments that distend the acrosomal process. *(C)* The fused membranes fold back, exposing the contents of the acrosome. *(D)* The overlying membrane is lost. *(E)* The acrosomal membrane joins the sperm membrane only at the junction marked by arrows. *(F)* The acrosomal tubule elongates further as microfilaments continue to push from beneath. *(G)* The gametes bind to each other. *(H)* Gamete membranes fuse.

the surface of sea urchin sperm that responds by affecting the ion permeability of the sperm.

The acrosomal reaction of sperm can be triggered in the absence of eggs by the addition of the glycoprotein carrying sulfated fucose polymer of egg jelly. However, the solution bathing the sperm must contain Ca^{++} ions for the glycoprotein to elicit a response. When sperm are suspended in a salt solution approximately that of seawater but lacking Ca^{++}, no acrosomal reaction occurs following addition of the fucose polymer. It has been shown that addition of egg jelly gives rise to a rapid uptake of Ca^{++} and Na^{+} ions and a concomitant secretion of H^{+} and K^{+} ions. The presence of egg jelly in the environment is recognized by a specific glycoprotein on the surface of sperm over the acrosome. If this glycoprotein is covered by a monoclonal antibody that binds to it, jelly no longer triggers the uptake of Ca^{++} nor the release of acid by the sperm and the acrosome reaction does not take place.

The inductive signal for polymerization of the subacrosomal actin appears to be the alkalinity of the sperm cytoplasm resulting from secretion of H^{+} ions. When the subacrosomal actin is freed from the sperm of echinoderms at the pH of seawater, it remains in an insoluble form associated with two proteins of 230,000 and 250,000 daltons that appear to keep the actin from polymerizing. These proteins are dissociated from actin at pH 8.0.

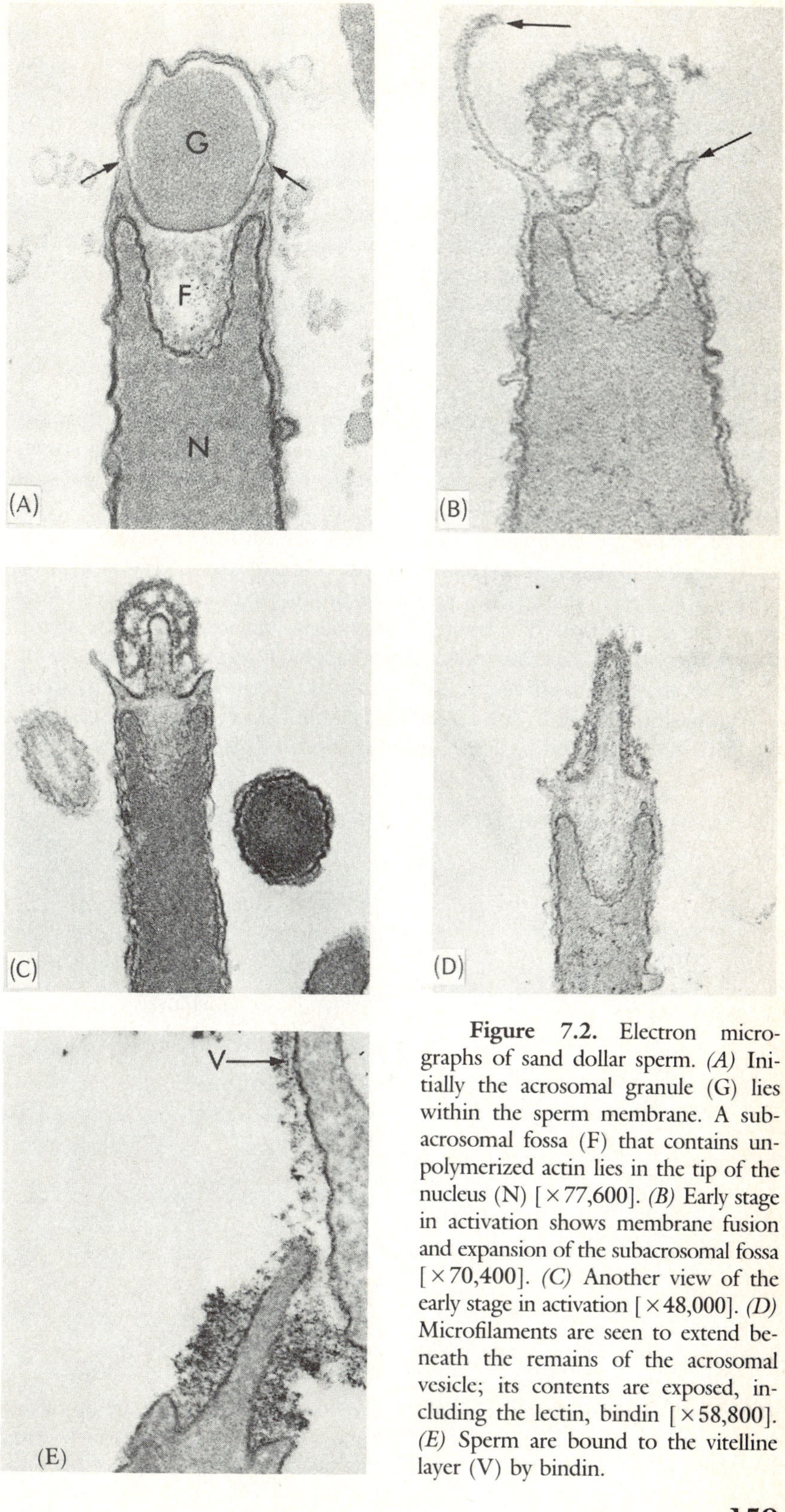

Figure 7.2. Electron micrographs of sand dollar sperm. *(A)* Initially the acrosomal granule (G) lies within the sperm membrane. A subacrosomal fossa (F) that contains unpolymerized actin lies in the tip of the nucleus (N) [×77,600]. *(B)* Early stage in activation shows membrane fusion and expansion of the subacrosomal fossa [×70,400]. *(C)* Another view of the early stage in activation [×48,000]. *(D)* Microfilaments are seen to extend beneath the remains of the acrosomal vesicle; its contents are exposed, including the lectin, bindin [×58,800]. *(E)* Sperm are bound to the vitelline layer (V) by bindin.

The requirements for the sulfated fucose glycoprotein can be bypassed by addition of an ionophore (A23187). This small molecule embeds itself in the membrane of the sperm and allows exchange of Ca^{++} and Na^{+} ions for H^{+}. Addition of A23187 in normal seawater triggers the acrosome reaction. However, if the internal pH of the sperm is kept low by acidification of the seawater, the ionophore has no effect.

These results have suggested that the actin-binding proteins (230 and 250 kd) keep the actin in an unpolymerized state. When activated by egg jelly or the ionophore, H^{+} ions rush out in exchange for entering Na^{+} ions. The internal pH rises and the actin-binding proteins dissociate to liberate unbound G actin, which is sufficiently concentrated to spontaneously polymerize into microfilaments and thrust the acrosomal process forward.

Similar mechanisms extend the acrosomal tubule in sperm of other organisms, but they do not always involve polymerization of actin. In sperm of the horseshoe crab, *Limulus,* a rearrangement of the microfilaments from a coil to a parallel array extends the acrosomal process.

The membrane of the acrosomal vesicle fuses with the overlying sperm membrane and liberates its contents near the point of fusion (Fig. 7.2). The acrosomal vesicle in mammals as well as sea urchins contains a protease that acts during fertilization. The protease may allow penetration of the sperm through the vitelline layer by hydrolzing this glycoprotein covering.

When the acrosomal vesicle fuses with the sperm membrane its inner surface becomes exposed. The acrosomal membrane of sea urchins carries a carbohydrate-binding protein referred to as bindin. It is a well-characterized

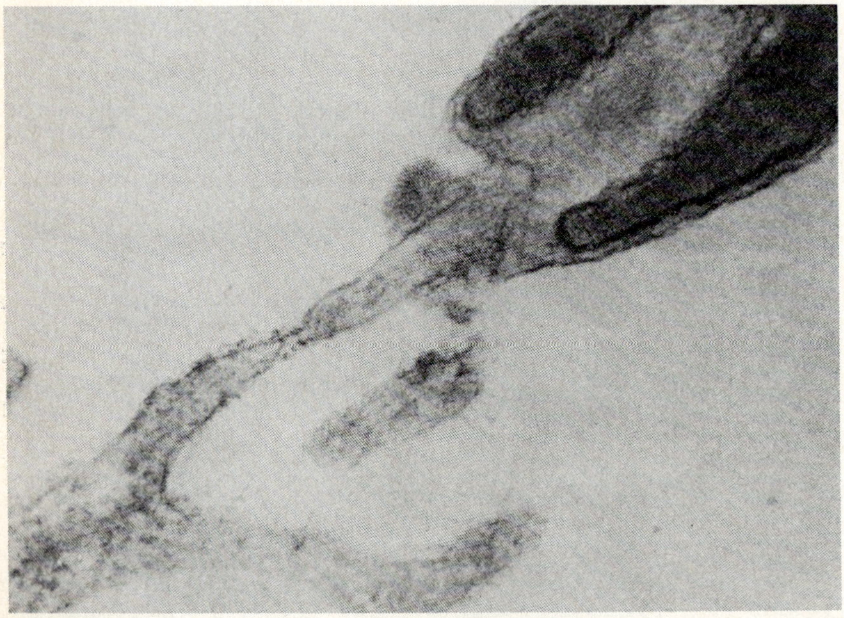

Figure 7.3. Moment of fusion in sea urchin sperm and egg. The tip of the sperm touches a microvillus on the egg surface and the membranes fuse. A cytoplasmic bridge will form through which the sperm nucleus enters the egg [×54,600].

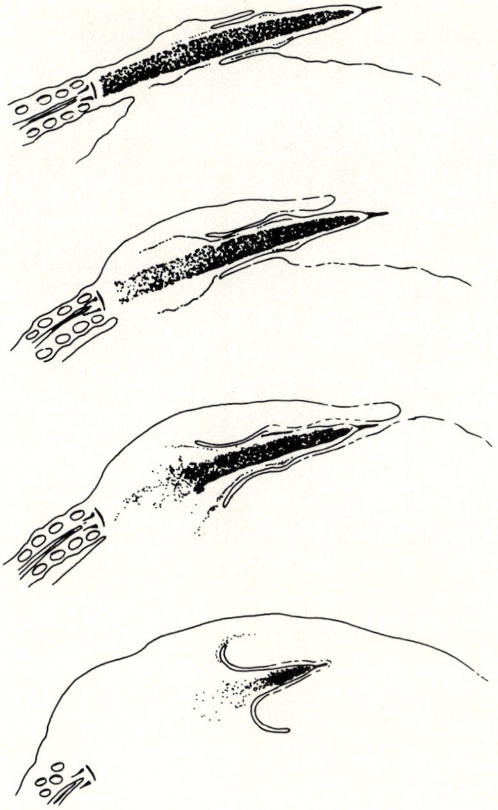

Figure 7.4. Sperm entry in rabbits. The sperm nucleus fuses near its base with the egg and is then engulfed by the rising egg membrane. The whole sperm is drawn into the egg.

protein of 30,500 daltons that can be thought of as a lectin since it specifically binds to carbohydrates. Bindin has strong affinity to a unique glycoprotein on the surface of sea urchin eggs and appears to hold the sperm to the vitelline layer.

The juxtaposed sperm and egg membranes then fuse, allowing the sperm pronucleus to enter the egg cytoplasm (Fig. 7.3). The point of fusion varies between sperm of different species; in sea urchins it is on the acrosomal membrane while in mammals it is toward the base of the head of sperm (Fig. 7.4). The cytoplasmic connection widens until the nucleus is free to migrate into the egg, where it is directed toward the female pronucleus.

Blocks to Polyspermy

Sperm pronuclei add their haploid complement to that of the egg pronucleus to form the diploid zygote. However, if more than a single sperm pronucleus participated in zygote formation, problems of genetic balance would occur later in development. If more than a single sperm enters eggs of most organisms, this condition of polyspermy results in gross defects later in cleavage. In eggs of some species several sperm enter; however, all but

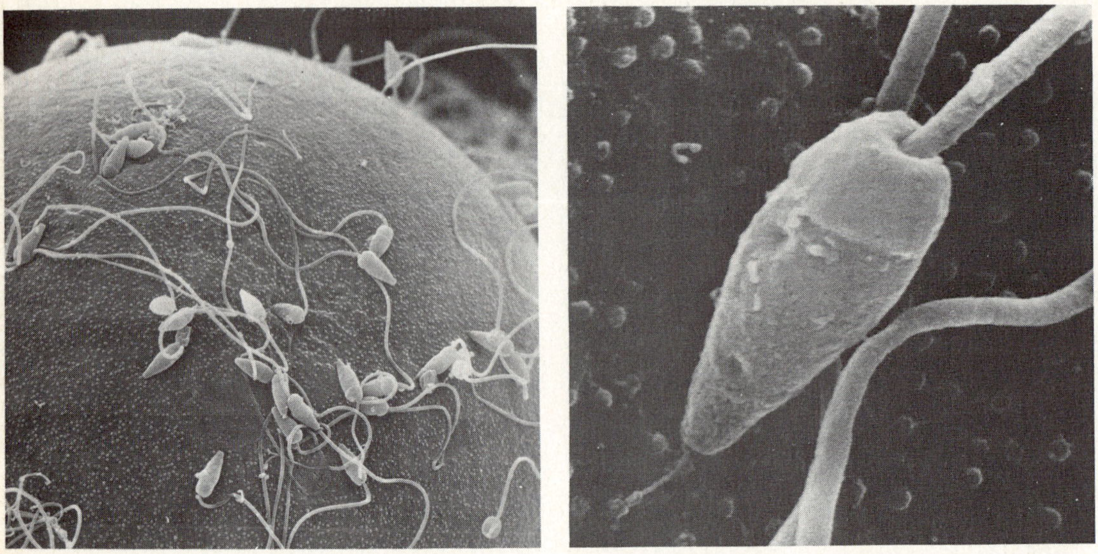

(A) (B)

Figure 7.5. *(A)* Hundreds of sperm bind to a sea urchin egg when the concentration of sperm in the seawater is high. *(B)* Each sperm binds at its tip.

one degenerate. In most cases only a single sperm is allowed to enter before the egg surface is changed to block fusion of any latecomers (Fig. 7.5). There are two major blocks to polyspermy: (1) a fast incomplete electrical block and (2) a slower complete physical block.

The resting egg pumps out sodium before fertilization so that the internal concentration is much lower than the surrounding environment. Ionic balance is maintained by a high K^+ ion concentration within the egg. Fusion of the sperm membrane with the egg membrane opens Na^+ ion channels and allows sodium to enter rapidly. The electrical potential across the egg membrane is initially negative (-70 mV in sea urchin eggs), but fertilization results in rapid depolarization for a period of a few minutes. During this period sperm fusion is blocked. This rapid electrical block to polyspermy occurs in many eggs, including those of echinoderms, echiuroid worms, and anuran amphibians, but not those of mammals or fish. It appears that the danger of near-simultaneous fertilization is not great in mammals. If the membrane potential is maintained at its original negative voltage by a voltage clamp, then sea urchin eggs will become polyspermic under normal conditions of fertilization. Clearly, this ionic defense plays a significant role to ensure that each egg is fertilized by a single sperm.

During the period of electrical block to sperm fusion, a slower mechanical block to sperm is raised. Just beneath the surface membrane of sea urchin eggs lies a layer of cortical granules. These membrane-enclosed vesicles contain various enzymes and structural proteins. Starting at the point of sperm entry, the cortical granules fuse with the egg membrane and liberate their contents. The egg is contained within a glycoprotein covering, the vitelline

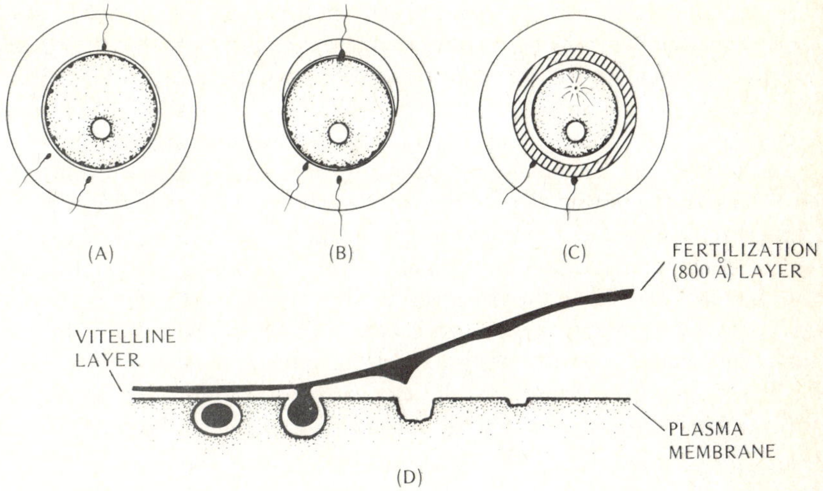

Figure 7.6. Cortical reaction in sea urchin eggs. *(A)* Sperm first cross the jelly layer *(outermost circle)* and are activated. *(B)* The top sperm has fused and initiated the cortical reaction, which lifts the vitelline layer *(second circle)*. *(C)* The vitelline layer has been converted into the fertilization layer and asters extend from the sperm pronucleus. *(D)* Cortical granules lie just under the egg plasma membrane. Exocytosis following fertilization allows their contents to alter the vitelline layer and convert it to the fertilization layer.

layer. The contents of the cortical granules are trapped between the vitelline layer and the egg surface following the cortical reaction. Among the enzymes released is a protease, which frees the vitelline layer from its attachment points to the egg membrane and allows it to expand. A peroxidase is also released, which cross-links the proteins in the vitelline layer, adding rigidity to this structure. Starting about 25 seconds after sperm entry and taking a few minutes, the cortical reaction sweeps around the egg surface. The vitelline layer is raised and hardened to form the sperm-impermeable fertilization envelope (Fig. 7.6).

Fertilization of mammalian eggs also triggers a cortical reaction in which cortical granules release their contents into the space surrounding the egg causing the zona pellucida to be altered such that it is an effective mechanical block to entry of late-arriving sperm. In sea urchins and mammals the protective layers enclose the developing embryo until late in the blastula stage when the embryo secretes a protease and hatches out.

The initial trigger for the cortical reaction appears to be a rise in the free Ca^{++} ion concentration to a level above 1 μM. Before addition of sperm, the internal free Ca^{++} level is considerably below this level in eggs although there is a large amount of calcium bound up in the internal membrane system, the endoplasmic reticulum. Starting at the point of sperm entry, Ca^{++} is released and triggers exocytosis of the cortical granule contents. Free Ca^{++} also stimulates further calcium release in adjacent regions. This calcium-stimulated calcium release propagates through the eggs of deuterostome organisms at about 10 μm/second and reaches the opposite pole of small eggs in under a minute while taking 5 or more minutes to get there in large eggs. In protostome organisms the calcium is taken up from the surrounding water rather than being released from internal stores. Protostome and deuterostome organisms are distinguished by their mechanisms of gastrulation.

Sea urchins, like mammals, are deuterostomes and do not depend on external calcium for propagation of the cortical reaction. A local rise in Ca^{++} ion concentration at one point on the egg surface generated by pricking in a calcium-containing solution will initiate a wave of cortical granule breakdown. The free Ca^{++} can be visualized if the eggs are preloaded with aequorin, a protein that emits light when it binds free Ca^{++}. Addition of sperm to such eggs results in a band of light, which can be seen to travel from the point of entry of the sperm across the egg to the opposite pole (Fig. 7.7). The band indicates the autocatalytic release of free Ca^{++}. Fusion of the

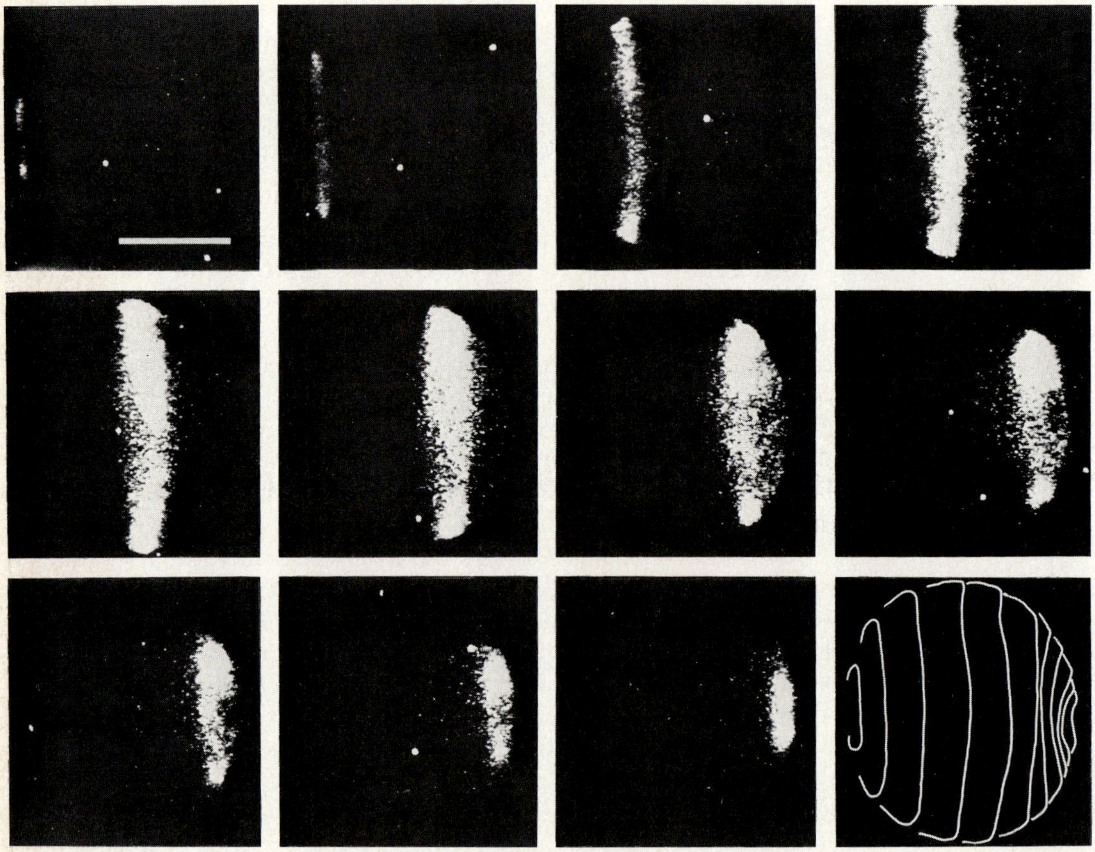

Figure 7.7. A wave of calcium release passes through a fish egg (medaka) starting from the point of sperm penetration. This egg was injected with the calcium-dependent light-releasing protein, aequorin, before fertilization. The band of light is drawn at 30-second intervals. The scale marker indicates 0.5 mm.

membranes surrounding the cortical granules with the surface membrane appears to depend on free Ca^{++} ions.

The mechanism leading to release of bound Ca^{++} has recently been further explained. Fusion of sperm with sea urchin eggs apparently introduces the phosphosugar inositol 1,4,5-trisphosphate or liberates it from membranes of the egg by a phospholipase C activity. Inositol trisphosphate has been shown to stimulate the cortical reaction at concentrations of only 10 nM by a Ca^{++}-dependent mechanism. When the inositol triphosphate concentration rises above this threshold, Ca^{++} is released from internal stores and triggers cortical granule exocytosis. It also stimulates further production of inositol trisphosphate from phosphatidyl inositol 4,5-bisphosphate. This autocatalytic loop of inositol trisphosphate–stimulated calcium release and calcium-stimulated inositol trisphosphatase production propagates across the egg rapidly. Similar processes involving inositol trisphosphate have been implicated in the mechanism by which some growth hormones activate mammalian cells.

Activation of Metabolism

Before fertilization sea urchin eggs are fairly quiescent cells. Protein synthesis and energy generation are taking place but at a low rate; RNA synthesis has slowed to a very low level and DNA synthesis has ceased. Within an hour all of this machinery must be activated to support the rapid cell division characteristic of the cleavage stage. Five minutes after fertilization of sea urchin eggs, the rate of protein synthesis increases five- to tenfold. DNA synthesis starts about 30 minutes later and the first cell division occurs about an hour after that.

One minute after insemination of sea urchin eggs there is a release of acid into the surrounding solution. This leaves the internal environment more alkaline, and direct measurement in sea urchin eggs indicates that the internal pH increases from 6.8 to 7.2 at this time. This change in H^{+} ion concentration activates egg metabolism. Eggs treated with a weak base such as ammonia are activated even in the absence of fertilization. Although ammonia treatment results in activation of protein and DNA synthesis, cell division does not occur. Triggering cell division requires the centrioles provided by sperm.

Protein synthesis following fertilization is initially directed by maternal mRNA present in the egg. Although some transcription can be observed within an hour of fertilization in certain embryos, in others, transcription seems to be suppressed while the DNA is rapidly replicated during the early cleavage stages. DNA replication proceeds at a higher rate during early cleavage than at any other time in the life of the organism. The densely packed replication forks might be hindered if genes were in the process of being transcribed into RNA. Later, as the blastula is formed, DNA replication slows and transcription of genes can be seen to take place.

Cytoskeletal Reactions

When the sperm pronucleus enters the egg, it rapidly moves toward the female pronucleus. The cytoskeleton provides both motive force and direction to ensure that the pronuclei can get together and allow pairing of the chromosomes. A considerable number of other cytoplasmic rearrangements also occur in the fertilized egg. In some pigmented eggs such as those of the frog, a massive rearrangement of the pigment relative to the point of sperm entry is readily visible. This can lead to demarcations on the egg surface that are useful in assigning fate-map positions for regions of the egg that will undergo specific differentiations later in development. For instance, an area on the side opposite that at which a sperm entered a frog egg becomes relatively unpigmented and shows up as a gray crescent. In an undisturbed embryo the region of the gray crescent will be subdivided into blastula cells, which will be the first to invaginate during gastrulation. However, if the egg is rotated upside down, the egg components will be redistributed by the force of gravity bearing on the denser regions of the egg. These rearrangements of cytoplasm and yolk undoubtedly pull on the cytoskeleton and result in the blastopore forming at other sites. In some cases such rotation will give rise to two separate blastopores and even two independent early embryos forming from the same fertilized egg. These experiments are described in Chapter 9. While the details of cytoskeletal organization in the zygote are not yet understood, it is clear that the position of the initial cleavage planes and the fate of the blastomeres is very sensitive to their arrangement.

When the pronuclei have migrated into close proximity, they fuse. The chromosomes replicate and line up on a metaphase plate between the centrioles provided by the sperm. Diploid complements are separated and surrounded by newly formed nuclear membrane. Simultaneously the surface membrane constricts, separating them into the first two cells.

Summary

Changes in cellular concentration of ions, principally calcium and protons, orchestrate the acrosome reaction and fusion of sperm and eggs. Sperm pass through a jelly zone on their way to eggs and have specific jelly receptors on their surface that affect their ion permeability. Calcium and sodium ions enter the sperm in exchange for protons. The increase in intracellular pH leads to polymerization of subacrosomal actin that extends the fusing acrosome. In echinoderms such as *Strongylocentrotus purpuratus,* the inner membrane of the acrosomal membrane carries a lectin that binds to a glycoprotein on the egg surface. Fertilization results in an increase in free calcium and pH in the egg. The calcium triggers a cortical reaction that raises the fertilization membrane as a mechanical block to polyspermy. The pH increase triggers metabolic activity and microtubule polymerization in the egg. In eggs of many species, the ion flux results in a fast, although incomplete, electrical block to polyspermy that lasts for a few minutes. In sea urchins the nuclear DNA is replicated and the first division is underway within an hour.

SPECIFIC EXAMPLE

Pollination

Flowering plants (angiosperms) reproduce by the fusion of sperm and eggs, but the generation of gametes and processes of fertilization differ significantly from those in animals. The gametes differentiate in specific structures of the flowers (Fig. 7.8).

The male gametes differentiate within pollen held in the anthers. The female gametes differentiate within ovules held in the ovary. The two come together after pollen falls on the stigma that is connected to the ovary by a style. Pollen of the same species as the stigma triggers release of water from the underlying cells, imbibes it, and germinates (Fig. 7.9). A long pollen tube is extended from the pollen cell that descends the style and carries the sperm nucleus to the ovule. There, the pollen tube releases the haploid sperm

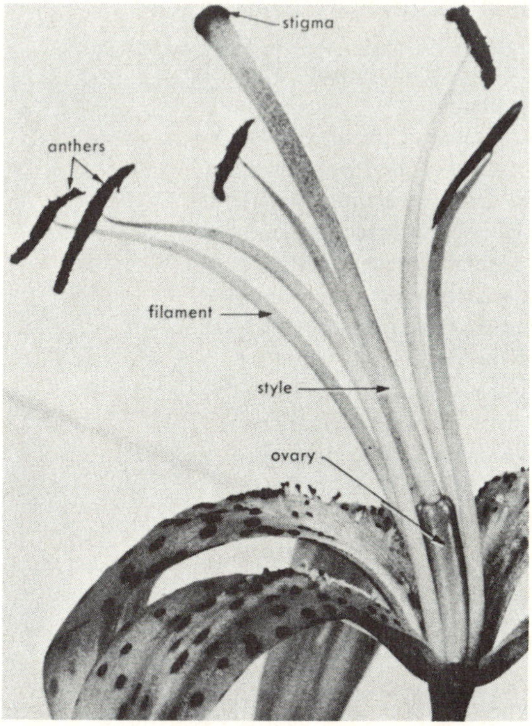

Figure 7.8. Reproductive organs of a flower. Filaments have anthers at their ends, where pollen is made and released. Pollination requires pollen to land on the stigma and extend a tube down the style to the ovary, where the eggs are held.

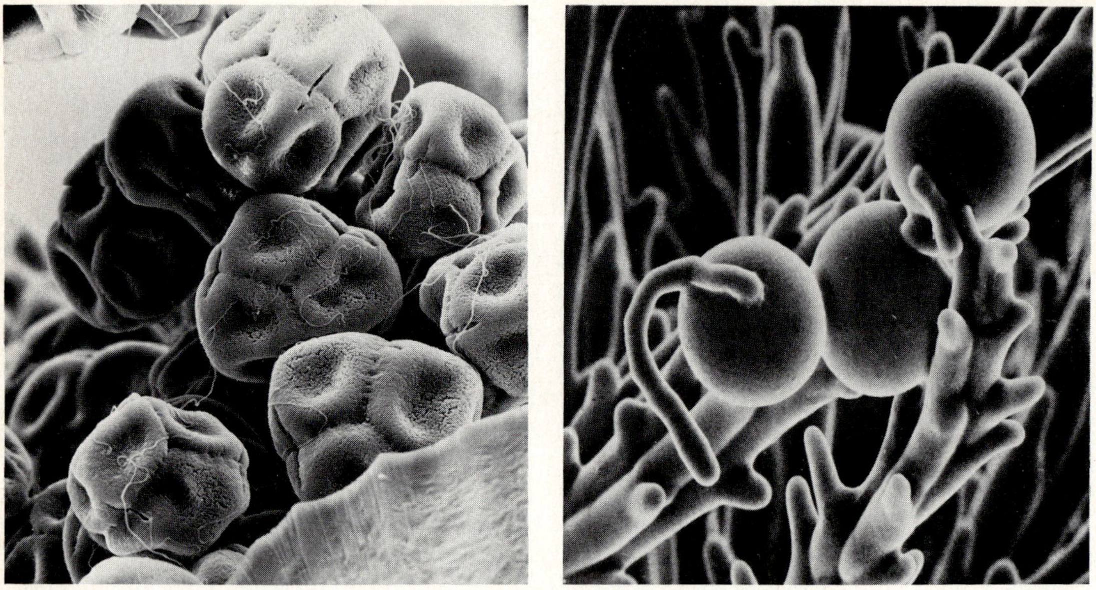

(A) (B)

Figure 7.9. Pollen germination. Pollen tubes emerge soon after pollen encounters the stigma surface. *(A) Castaos bippinatus* and *(B) Phalaris.*

that fertilize the haploid egg as well as adding a male nucleus to the binucleate endosperm cell. The embryo is nurtured by the trisomic (two female genomes, one male genome) endosperm during early development. This double fertilization is characteristic of all flowering plants.

Flowers undoubtedly evolved from leaflike structures of primitive vascular plants. Typical flowers are composed of four whorls of modified leaves: (1) sepals, (2) petals, (3) stamens, and (4) carpels. These structures are attached to a modified stem referred to as a receptacle. Sepals protect the outside of the flower while it is differentiating and are usually green. Petals are large and colorful in those plants that depend on insects or birds for dissemination of pollen. Stamens consist of filaments supporting the anthers in which pollen is formed. Carpels are the central whorl of modified floral leaves at the base of which the ovary differentiates. The stigma is the opening at the top of the style that leads down to the ovary. In many flowers several carpels fuse; in others, each carpel remains as an individual structure. Both single and fused carpels are referred to in the classical literature as pistils. There are many specialized botanical terms for flowering parts.

Peas in a pea pod are seeds that developed from fertilized eggs at the fused margins of carpels (Fig. 7.10). The arrangement of the eggs is reminiscent of spore formation along the margin of leaves in many species of ferns. In flowering plants the leaf margins have fused to form the protective ovary.

Within the ovary, multiple ovules grow as dome-shaped masses of cells

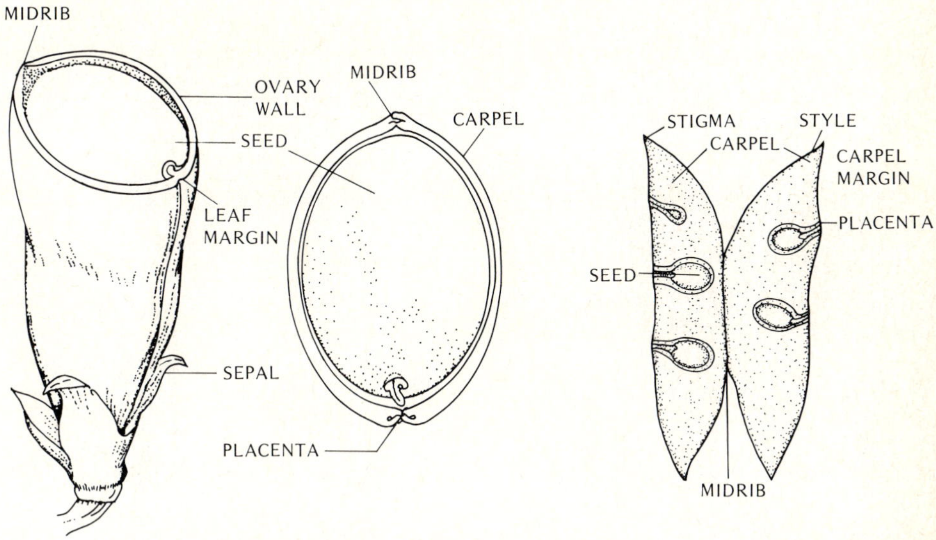

Figure 7.10. Pea pod. The ovary wall is fused at the leaf margin, where the seeds attach to the placenta. On the other side there is a midrib. The ovary wall is formed from carpels. Sepals protect the base.

attached to the ovary wall by a structure referred to as the placenta. At the center of this mass can be seen a megaspore mother cell that is considerably larger than most cells in the flower. In a series of nuclear divisions, nuclear migrations, and cellularizations, it will give rise to the egg and the endosperm cells (Fig. 7.11).

The large mother cell undergoes two meiotic divisions to give rise to four haploid megaspores. Usually, three disintegrate and their contents are reabsorbed. All subsequent nuclei of the embryo sac are derived from a single haploid nucleus. The nucleus of the surviving megaspore divides three times to produce eight haploid nuclei. Two of these nuclei occupy the large central region that will become the endosperm cell. The other six nuclei become enclosed in cellular membranes, but only the one nearest the micropyle opening of the ovule will become the egg. The other haploid cells of the ovule (antipodals and synergids) usually degenerate shortly after fertilization.

Pollen differentiates from pollen mother cells in the anthers at the end of the stamens. Each mother cell undergoes two meiotic divisions to give rise to four haploid cells referred to as spores. The nucleus of each spore divides mitotically, giving rise to two haploid nuclei in each pollen grain (Fig. 7.12). One of them serves as the tube nucleus while the other becomes surrounded by cellular membrane and is referred to as the generative cell. Following germination, the generative cell divides mitotically to give rise to two sperm cells (Fig. 7.13).

Some plants are predominantly self-pollinating, with the pollen falling directly on adjacent stigma. Others are cross-pollinating, with the pollen of

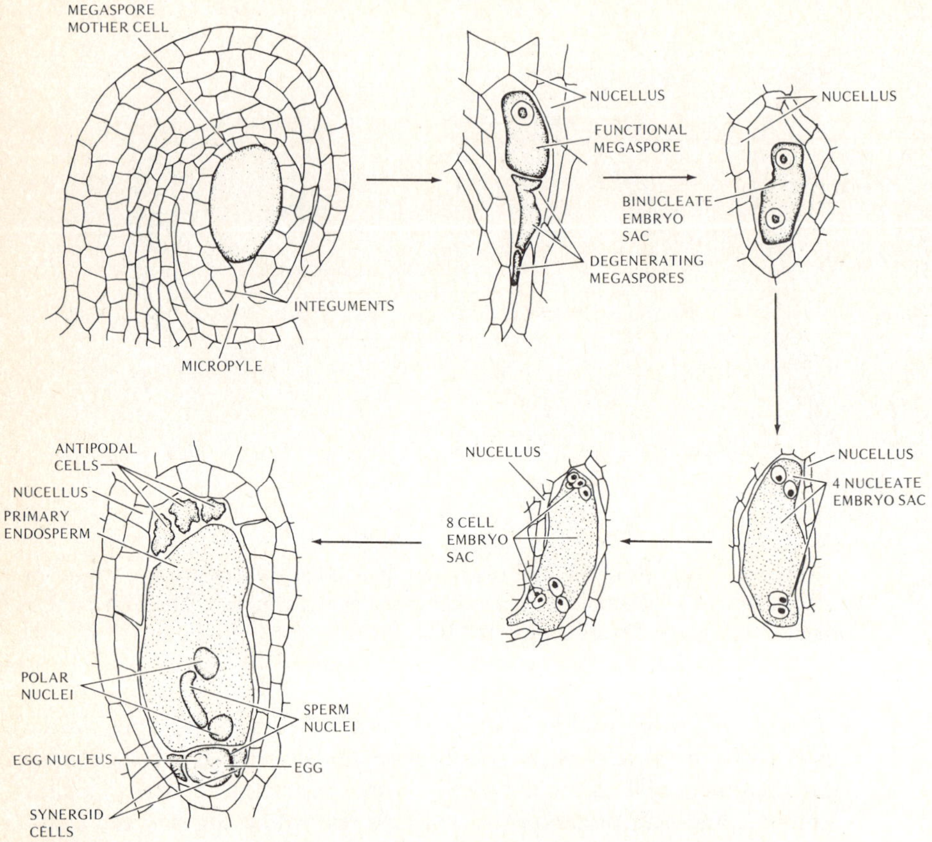

Figure 7.11. Embryo sac differentiation. Within an ovary a large megaspore mother cell is protected by integuments except at the micropyle. The megaspore mother cell divides meiotically to give rise to four megaspores, three of which degenerate. The megaspore nucleus divides as the nucellus expands. After the next nuclear division, pairs of nuclei go to opposite ends of the embryo sac. The embryo sac divides into one large cell and several smaller ones. Mature embryo sac contains antipodal cells that will degenerate following fertilization, a large endosperm cell with two haploid nuclei, and synergid cells surrounding the egg that lies over the micropyle. Pollen donate nuclei to both the endosperm cell and the egg.

one plant fertilizing a separate plant. Dispersal of gametes in vascular plants cannot depend on the mechanisms that had evolved in aquatic predecessors or in land animals because flowering plants grow in the air and are not motile. Solutions have been found in the use of wind, insects, birds, and even bats as pollen vectors.

Grasses, ragweeds, and poplars produce enormous quantities of pollen and rely on wind-borne pollination. The stigmas in many cases expose a large surface to catch the blowing pollen. In catkin-bearing trees such as cottonwoods, alders, and birches, the pollen is released before the leaves unfold so their dispersal is not hindered. Plants that produce colorful flowers usually

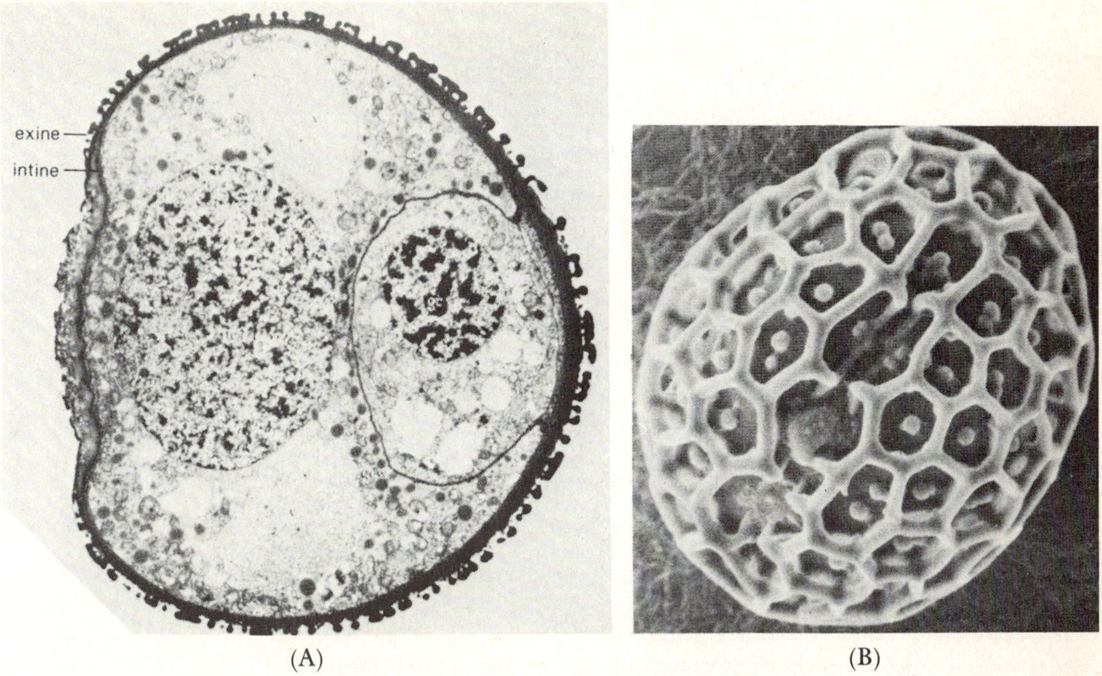

Figure 7.12. Pollen. *(A)* The thick double-layered wall protects two nuclei: one for the tube cell and one to fertilize an egg, in this case a blood lily. *(B)* The surface of pollen of the angiosperm *Eranthemum* is highly sculptured and intricately molded.

rely on living vectors to transport the pollen. The mutual evolution of specific plants with specific vectors has provided an enormous variety of naturalistic tales. For example, a species of orchid *(Ophrys)* mimics a female wasp to attract male wasps. In the process of attempted copulation, pollination is effected. No matter how the right pollen reaches the appropriate stigma, it can then fertilize the plant.

Flowers are open to all sorts of pollen but usually trigger germination only in those of the same species. There is now some evidence that the necessary recognition system relies on cell-surface carbohydrates on the surface of the stigma and specific proteins on the surface of the pollen that recognize unique carbohydrate structures. Compatible pollen take up water from the stigma cells, swell and germinate. A pollen tube descends the style carrying the sperm cells toward the ovary. In some cases the pollen tube must enzymatically dissolve the tissue of the style to gain passage. It is not known exactly how the outgrowth of the tube is directed to the ovary, but chemotactic response to ionic gradients in the style is a likely possibility. The pollen tube enters an ovule via the micropyle and liberates two sperm nuclei. One fertilizes the egg and the other joins the two haploid female nuclei of the endosperm cell (Fig. 7.14).

Fertilization not only initiates development of the plant embryo but also

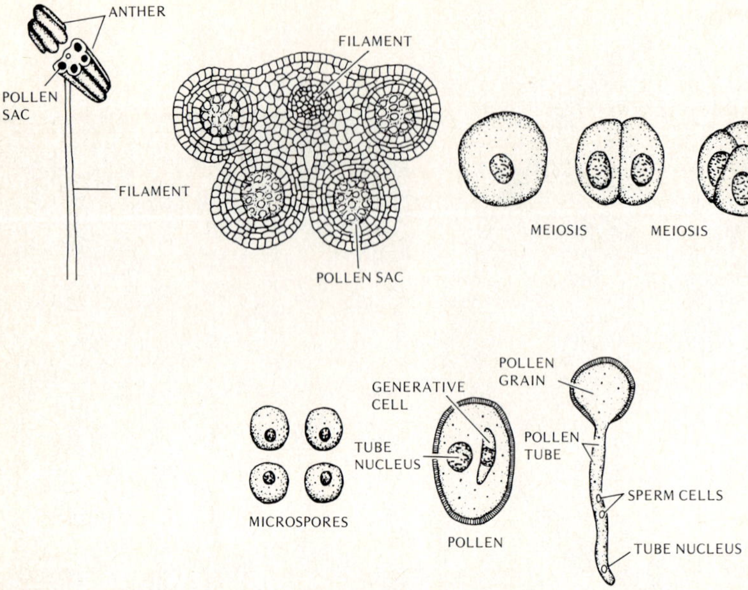

Figure 7.13. Differentiation of pollen. Within pollen sacs of anthers, pollen mother cells divide meiotically to generate four haploid microspores. These divide mitotically to give rise to pollen grains that carry the generative cells. Upon germination the generative cell divides into two sperm cells still within the pollen tube. The pollen tube nucleus is usually found at the tip of the elongating tube.

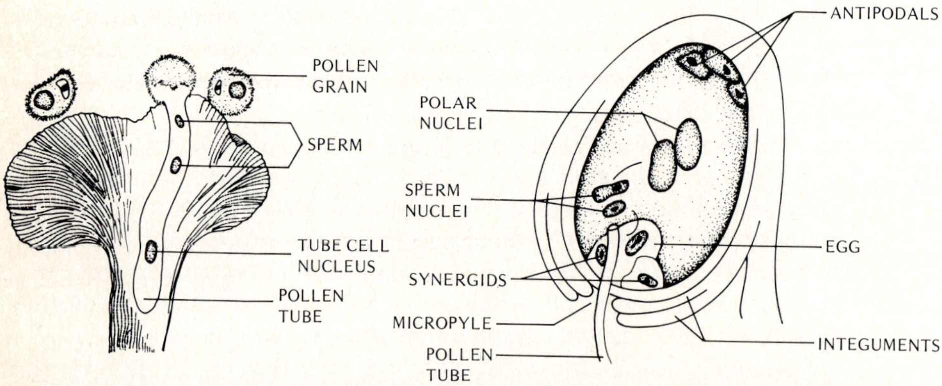

Figure 7.14. Fertilization of flowers. The pollen tube carries two sperm cells down to the ovary where it passes between the integuments and through the micropyle until reaching the egg sac. Two identical sperm nuclei are released from the end of the pollen tube. One enters the egg and fuses with the egg nucleus to form the new diploid zygote. The other enters the endosperm mother cell and fuses with the two maternal polar nuclei to form the triploid endosperm.

affects the growth of some floral parts to form a fruit and the withering of other parts that have served their function.

Very little can be said about the molecular mechanisms that direct egg or pollen differentiation or the biochemical consequences of fertilization. To a large extent this is due to the lack of sufficient homogeneous material as well as the inaccessible location at which fertilization occurs. As ultramicromolecular biological techniques are increasingly turned toward plants, this situation may change.

Related Readings

Alexandraki, D., and Ruderman, J. (1985). Expression of α- and β-tubulin genes during development of sea urchin embryos. Devel. Biol. *109,* 436–451.

Anderson, M., Hoggart, R., and Clarke, A. (1983). The possible role of lectins in mediating plant cell—cell interactions in "Chemical taxonomy, molecular biology and function of lectins" edited Goldstein, I., and Etzler, M. Alan Liss, N.Y.

Austin, C. (1978). Patterns in metazoan fertilization. Curr. Top. Dev. Biol. *12,* 1–9.

Cline, C., Schatten, C., Balczon, R., and Schatten, G. (1983). Actin-mediated surface motility during sea urchin fertilization. Cell Motil. *3,* 513–524.

Dubé, F., Schmidt, T., Johnson, C., and Epel, D. (1985). The hierarchy of requirements for an elevated intracellular pH during early development of sea urchin embryos. Cell *40,* 657–666.

Ferrari, T., Bruns, D., and Wallace D. (1981). Isolation of a plant glycoprotein involved in control of intercellular recognition. Plant Physiol. *67,* 270–277.

Florman, H., and Wassarman, P. (1985). O-linked oligosaccharides of mouse egg ZP3 account for its sperm receptor activity. Cell *41,* 313–324.

Gilkey, J., Jaffe, L., Ridgeway, E., and Reynolds, G. (1978). A free calcium wave traverses the activating egg of the medaka, *Oryzias latipes.* J. Cell Biol. *75,* 448–466.

Glabe, C. (1985). Interaction of the sperm adhesion protein, bindin, with phospholipid vesicles. J. Biol. Chem. *100,* 794–799.

Gould-Somero, M., and Jaffe, L. A. (1984). Control of cell fusion at fertilization by membrane potential. In Cell fusion: gene transfer and transformation, Beers, R., and Basset E. (ed.). Raven Press, New York.

Gunderson, G., and Shapiro, B. (1984). Sperm surface proteins persist after fertilization. J. Cell Biol. *99,* 1343–1353.

Podell, S., and Vacquier, V. (1984). Wheat germ agglutinin blocks the acrosome reaction in *Strongylocentrotus purpuratus* sperm by binding a 210,000 mol. wt. membrane protein. J. Cell Biol. *99,* 1598–1604.

Schatten, G., and Hulser, D. (1983). Timing the early events during sea urchin fertilization. Devel. Biol. *100,* 244–245.

Schatten, G., Maul, G., Schatten, H., Chaly, N., Simerly, C., Balczon, R., and Brown, D. (1985). Nuclear lamins and peripheral nuclear antigens during fertilization and embryogenesis in mice and sea urchins. Proc. Natl'l. Acad. Sci. *82,* 4727–4731.

Swann, K., and Whitaker, M. (1985). Stimulation of the Na/H exchanger of sea urchin eggs by phorbol ester. Nature *314,* 274–277.

Vacquier, V. (1979). The interactions of sea urchin gametes during fertilization. Am. Zool. *19,* 839–849.

Webb, D., and Nuccitelli, R. (1985). Fertilization potential and electrical properties of the *Xenopus laevis* egg. Devel. Biol. *107,* 395–406.

Whitaker, M., and Irvine, R. (1984). Inositol 1,4,5-trisphosphate microinjection activates sea urchin eggs. Nature *312,* 636–639.

CHAPTER 8

Cleavage

The first few divisions of most eggs occur orthogonally; that is, the second cleavage is 90° from the plane of the first cleavage. Since the egg is three-dimensional, there are two possible planes orthogonal to the first one. The third division is orthogonal to the first two, usually bisecting the more yolky vegetal hemisphere from the less yolky animal hemisphere. The fourth division, which is parallel to the previous (third) cleavage plane, divides both hemispheres (Fig. 8.1).

There are many variations on this general plan that depend on the total amount of yolk in the egg. Moreover, there are embryos in which the first few cleavage planes are not orthogonal to each other, the most common of which are the spiral cleaving embryos discussed at the end of this chapter and in Chapter 16. Asymmetry in all three spatial axes is clear by the third division in spiral cleavage. Such embryos are also characterized as mosaic in that killing or removing a blastomere results in an embryo that is missing the structures that normally differentiate from the wounded piece. This may be an expression of the asymmetry established during the first few cell divisions in spiral cleaving eggs.

Cleavage in Echinoderm Embryos

In sea urchin embryos the first division separates the egg into two equal halves, from the animal pole to the vegetal pole. The yolk is equally distrib-

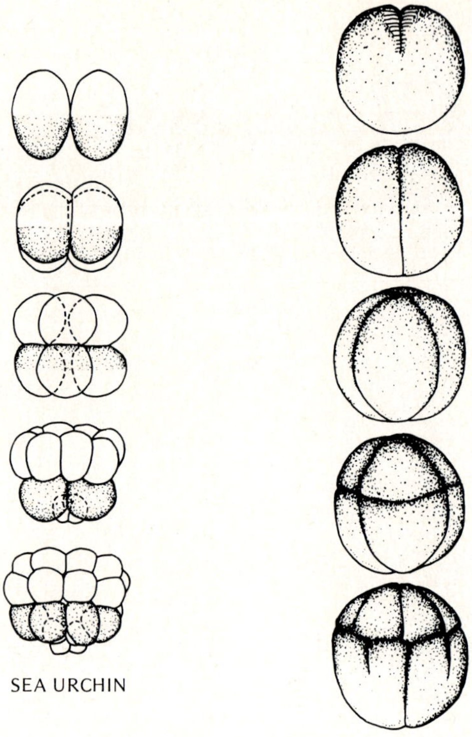

Figure 8.1. Holoblastic cleavage in echinoderms and amphibians. The first three divisions of these eggs divide up the whole cell. The first two divisions start at the animal pole and end at the vegetal pole, producing four equal-sized blastomeres. The third division is at right angles to the first two cleavage planes and separates the animal hemisphere from the vegetal hemisphere.

uted in each of the two blastomeres. If these blastomeres are separated, each will develop into a normal but half-sized embryo. The second cleavage is parallel to the first but offset 90°. It also goes from the animal to the vegetal pole. If the four equal blastomeres are separated at this stage, they will each develop into a normal but quarter-sized embryo. Clearly, no essential determinants have been sequestered in specific cells by the four-cell stages. This type of egg is referred to as regulative.

The third division in sea urchins is near the equator between the poles. The cells in the vegetal half contain considerably more yolk than those in the animal half. If the blastomeres are separated at this stage, the animal hemisphere cells develop in quite a different way from the vegetal hemisphere cells. The difference may be a consequence of the amount of yolk.

The fourth division bisects the animal cells into eight approximately equal-sized cells but divides the vegetal cells into two quite different-sized cells, the macromeres and the micromeres. The four small micromeres lie at the vegetal pole. Later in development they will give rise to the first cells to invaginate the blastula. These cells become the primary mesenchymal cells responsible for laying down the skeleton of the pluteus larva.

Control of the orientation and positions of the cleavage planes is central to the early organization of the embryo. Each cleavage appears to direct the orientation of the next. However, cell division is not essential for this process. When cell division is inhibited by treatment of certain eggs with ultraviolet (UV) light or hypotonic solutions, the orientation of the mitotic spindles still follows the normal sequence of orthogonal shifts at each nuclear division. Thus, the controlling events appear tied to the metaphase plate.

Cleavage in Amphibian Embryos

Frog eggs are much larger and contain far more yolk than sea urchin eggs. The yolk is heavy and concentrates in the vegetal hemisphere. Sperm usually penetrates near the equator separating the animal and vegetal hemisphere. The first cell division bisects the egg starting at the animal pole and traveling toward the vegetal pole. The division furrow forms more slowly in regions high in yolk, and the division of the first two cells slows noticeably as it nears the vegetal pole. The second division is parallel and orthogonal to the first, again starting at the animal pole and proceeding at a diminishing rate to the vegetal pole. The third division separates the animal hemisphere from the vegetal hemisphere. Thereafter, cell divisions proceed more rapidly in the animal hemisphere than in the vegetal hemisphere, so that within a short time there are many more animal cells than vegetal cells. The animal cells are also much smaller than the yolky vegetal ones (Fig. 8.2).

Time-lapse microcinematography of fertilized frog eggs has shown that there are rhythmical contractions of the whole egg superimposed on the

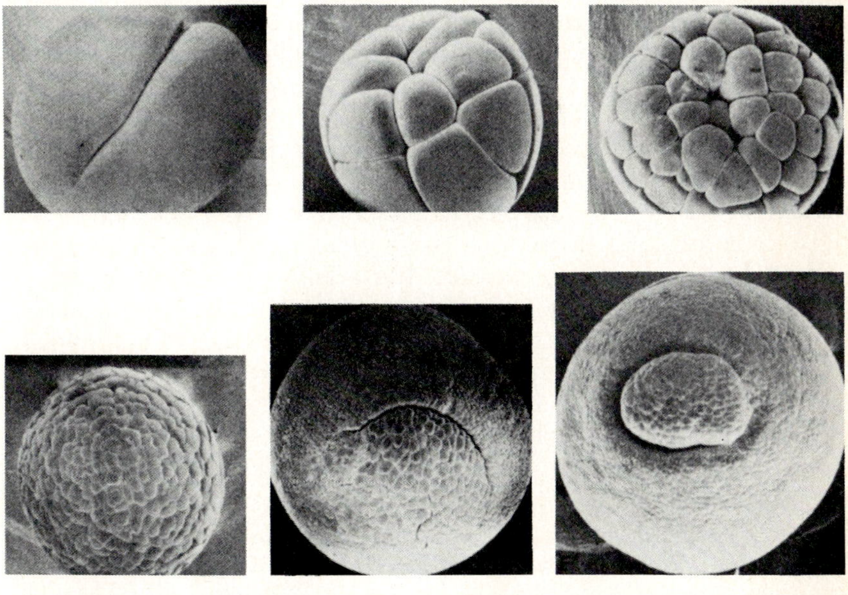

Figure 8.2. Cleavage of a frog egg. The first cleavage in fertilized eggs of *Rana pipiens* starts as a groove at the animal pole. After four divisions there are 16 blastomeres. Within the next five hours the whole egg is divided into several thousand cells.

movements generated by the cell divisions themselves. These contractions may be involved in the timing of events as well as the positioning of components within the subdividing embryo.

The cells formed by cleavage of *Xenopus* zygotes are held together and coupled by cytoplasmic bridges. Small molecules freely diffuse between adjacent cells via these bridges and may integrate metabolic functions. Following the mid-blastula transition (MBT), cell divisions are complete but gap junctions are inserted between cells of the animal hemisphere that permit continued communication between these cells. A specific blastomere can be uncoupled from its neighbors by injection of antibodies specific to the 54 kd gap junction protein. When one of the cells in the animal hemisphere at the 16-cell stage is uncoupled, it affects subsequent neural differentiation of the cells derived from the treated blastomere. This result seems to indicate that intercellular communication through gap junctions may be important for the specification of developmental fate.

DNA and protein synthesis proceed at a high rate in fertilized frog eggs but RNA transcription is not measurable until after several cell divisions. In embryos of the South African clawed toad, *Xenopus,* the first 12 divisions are rapid and synchronous, occurring about every 40 minutes. Thereafter the cleavage period slows to about 2 hours and synchrony is lost. RNA synthesis cannot be observed before the twelfth division but then increases rapidly. This shift in cell division and transcriptional activity has been called the mid-blastula transition (MBT). During the first 12 divisions the number of nuclei and the amount of nuclear DNA increases about 4000-fold. It appears that the MBT results from titration of a component by the increasing number of nuclei. Transcription can be stimulated prematurely by injecting exogenous DNA equivalent to about 4000 nuclei (24 ng) into eggs shortly after fertilization. The source of DNA makes no difference and can even be that of bacterial plasmids. The injected DNA is packaged into structures closely resembling nuclei within the egg. Thus, the MBT trigger could be either the amount of DNA or the amount of nuclear membrane in the embryo. It has been directly shown that the MBT does not depend on the number of cell divisions or the time following fertilization but only on the ratio of nuclei to cytoplasm. Transcription of specific genes introduced into the egg is controlled by the repressing substance just as endogenous transcription is. This mechanism may function to ensure rapid DNA replication unhindered by transcriptional machinery during the early cell divisions of the embryo. During later blastula stages transcription of certain genes essential for effective gastrulation is permitted following titration of the repressing substance by the exponentially increasing nuclear material.

Cleavage in Teleost Fish, Reptiles, and Birds

Although cleavage of the eggs of some primitive fish is complete and development proceeds essentially in the pattern found in amphibians, cleav-

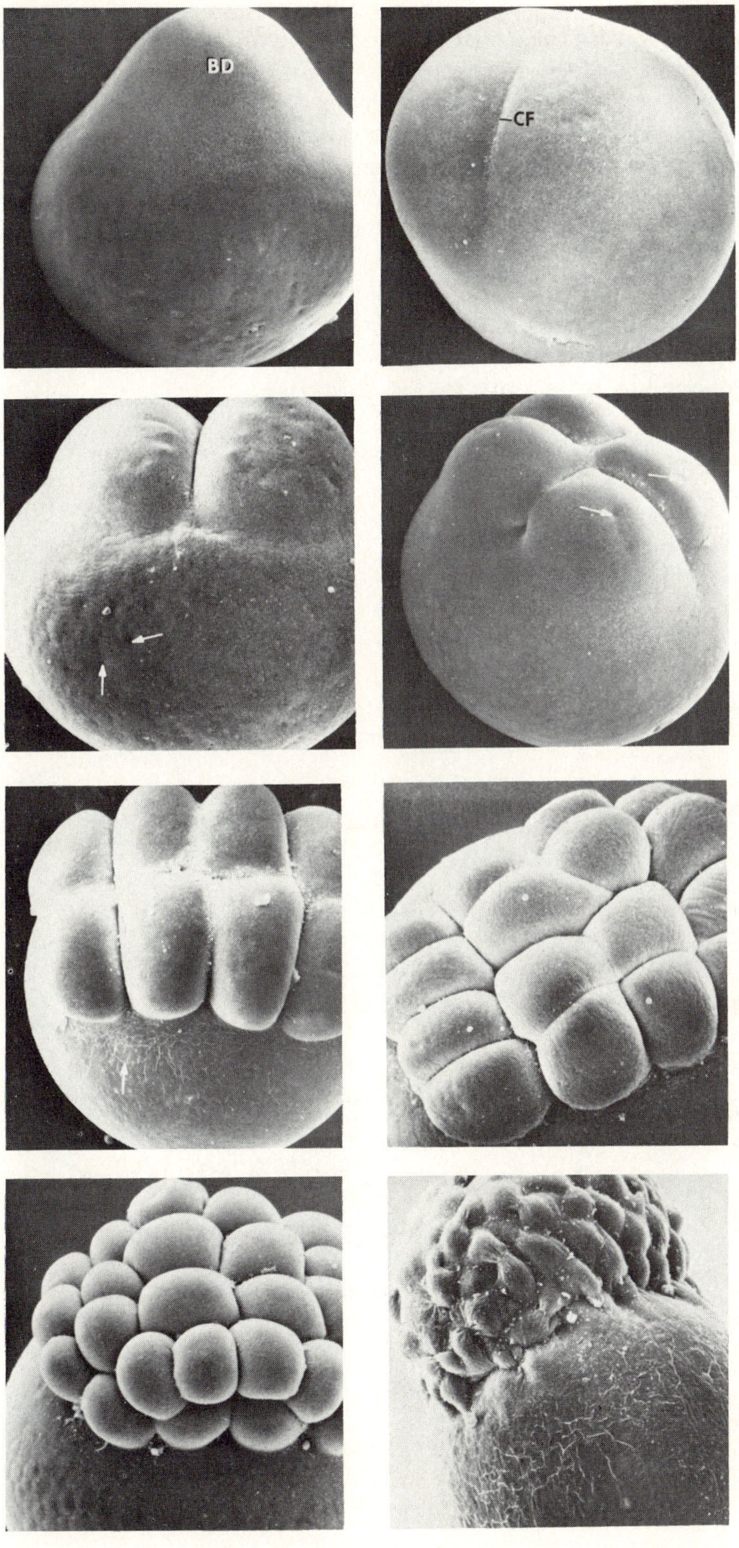

Figure 8.3. Meroblastic cleavage in zebra fish. Cytoplasm protrudes (BD) near the point of sperm entry and it is then bisected by the first cleavage furrow (CF). Cleavage is incomplete and does not divide the yolky vegetal portion of the embryo. The second cleavage is orthogonal to the first, producing four blastomeres. Two more divisions produce 8 and 16 cell embryos. By the 32-cell stage the blastodisc protrudes noticeably. Cleavage continues in the blastodisk without affecting the yolk.

age of the larger eggs of bony fish (Teleostei) is partial and resembles that of the still larger reptilian and avian eggs. Large eggs that have accumulated a lot of yolk can support the development of larger embryos, but the very mass of yolk has startling effects on the early stages of embryogenesis. Yolk platelets take up almost all the volume in some eggs, leaving the other cellular components such as ribosomes, mitochondria, enzymes, and the nucleus in a relatively small cytoplasmic cap on one side. Fusion of the male and female pronuclei occurs in this cytoplasmic cap. Following the first nuclear division, membrane is formed from the surface to separate them but does not surround the nuclei completely. Further nuclear divisions give rise to more nuclei, each of which is walled off from its neighbors but still open to the rest of the egg below. Since these cleavages do not divide the whole egg into discrete cells, they are referred to as meroblastic (Fig. 8.3). Eggs

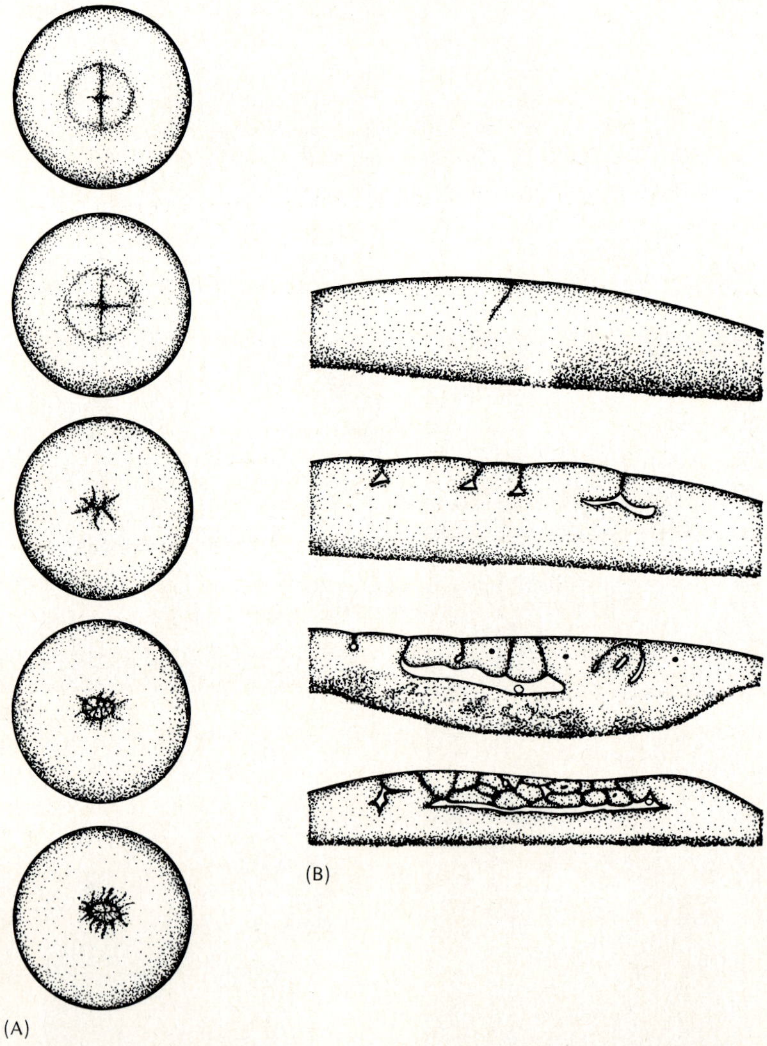

Figure 8.4. Cleavage of a chicken egg: *(A)* surface view; *(B)* cross-section of germinal disk. Cytoplasm is limited to a relatively small germinal disk on top of the yolk. The first two cleavage furrows are orthogonal, but division is incomplete: the cells are open to the yolk at the bottom. Walls continue to be put down at odd angles to produce the blastodisk. Cells near the center of the blastodisk become fully enclosed in membrane and lift off the yolk. Further out the walls of future cells continue to divide up the cytoplasm.

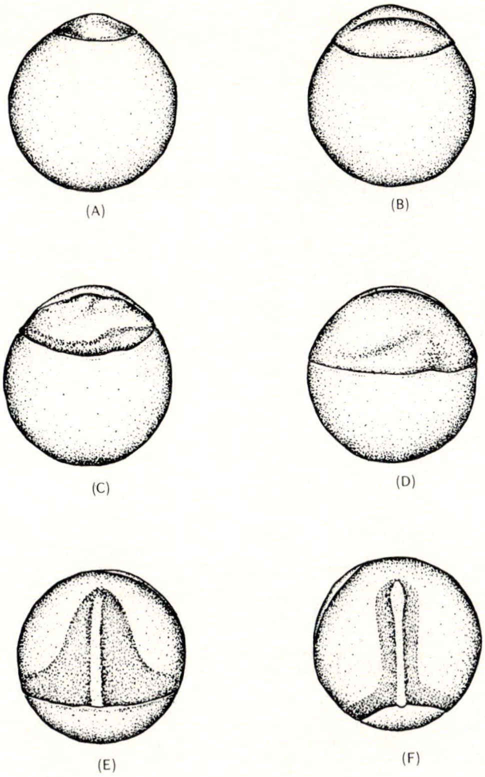

Figure 8.5. Epiboly in a teleost fish, *Fundulus. (A)* The blastoderm forms a cap on the yolk sphere. *(B)* The blastoderm starts to spread around the yolk by expansion of the cells. *(C)* As epiboly continues, gastrulation begins at the edge of the blastoderm. *(D)* The embryonic shield is visible. *(E)* The axis of the embryo forms. *(F)* Epiboly finally results in the embryonic cells surrounding the yolk.

in which the divisions divide the whole egg are said to have holoblastic cleavage.

Meroblastic cleavage in a fertilized chicken egg results in a disk of embryonic nuclei partially surrounded by cell membranes. After a few dozen nuclei have formed, cell membranes close off under the nuclei making intact cells. The nuclei at the periphery are still open at the bottom while the central disk of cellular blastomeres continues to get larger. In this way a flat disk of cells forms on the egg (Fig. 8.4).

Especially in eggs of some fish and reptiles, there is a period of expansion of the early cells of the embryonic disk. The initially round cells elongate and spread as a sheet over the surface of the egg until the noncellular yolky portion is engulfed by the cellular blastomeres. This process, termed epiboly, is rapid and dramatic in some embryos. It looks as if the whole egg is being swallowed by the sheet of cells (Fig. 8.5).

Birds produce eggs with the most yolk. Early cleavage in fertilized eggs results in the formation of blastomeres in the cytoplasmic cap. Those in the center of the growing disk become fully enclosed in cellular membranes. The cleavage planes are not precisely ordered, and so a jumble of cells is made that is several cells thick at the center of the disk. At the edges, nuclear division proceeds with the formation of partially enclosed cells. The yolk is tapped later by special tissues that develop following gastrulation.

Mammalian Early Development

Although the evolutionary ancestors of mammals were probably similar to modern-day reptiles and developed from yolky eggs with meroblastic division, a radical change in embryogenesis of mammals has favored the production of small eggs with much reduced yolk content. Cleavage in mammalian eggs is holoblastic, dividing the whole egg into a number of small cells (Fig. 8.6). The first division occurs about 24 hours following fertilization. The next three divisions occur at 12-hour intervals. During these early cleavages in mammalian embryos, there is a requirement for expression of both maternal and paternal genes, unlike the situation in echinoderm or amphibian embryos.

The radical change that allowed mammals to reproduce from small holoblastic eggs was the evolution of placental nutrition of the embryo. Development of the embryo within the duct system of the mother has evolved several times, so that some fish, amphibians, and reptiles have live births. In mammals this has been followed by mechanisms that allow the developing embryo to gain nutrients from the maternal circulation. When mammalian embryos have divided into only a hundred cells, the resulting balls of cells invade the uterine wall. The uterus is especially adapted in mammals to ac-

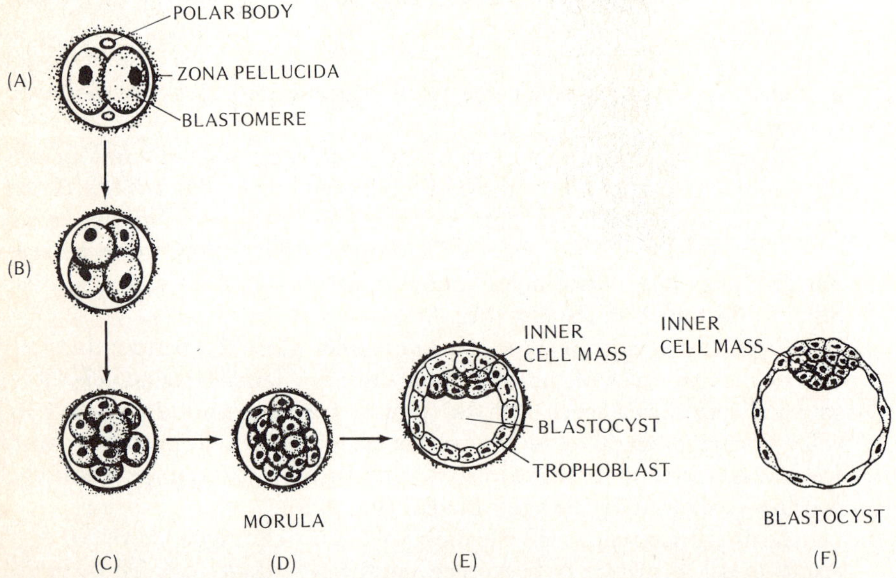

Figure 8.6. Cleavage of a human embryo. *(A)* Polar bodies are formed before the first cleavage. *(B)* The second holoblastic cleavage produces four equal-sized cells. *(C)* Cleavage becomes asynchronous after the third division. *(D)* The morula is still within the zona pellucida. *(E)* An inner cell mass can be recognized in cross section. *(F)* After hatching from the zona pellucida, the blastocyst cavity expands as a result of accumulation of fluid.

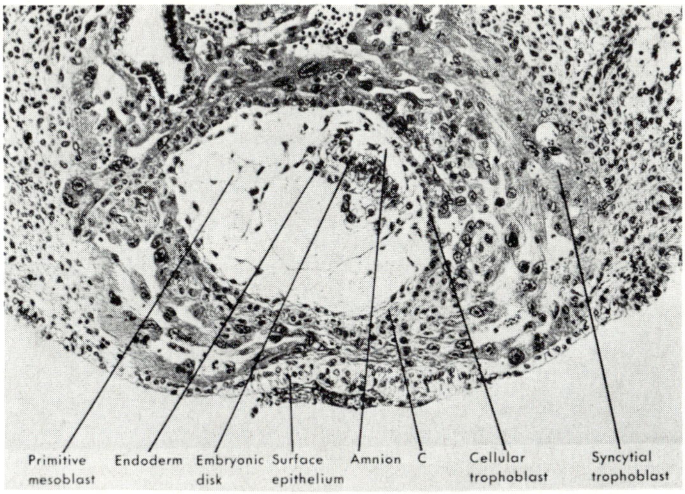

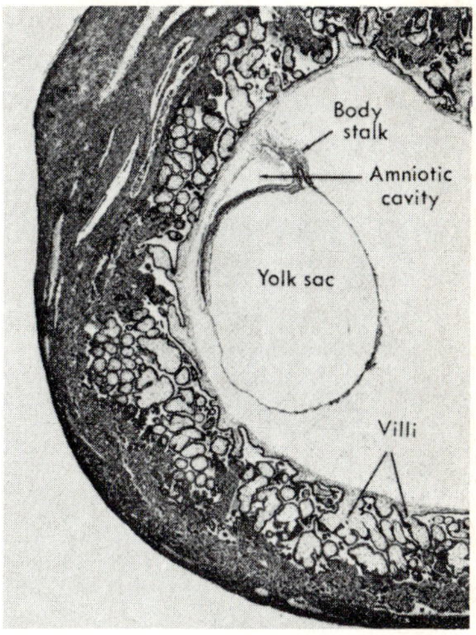

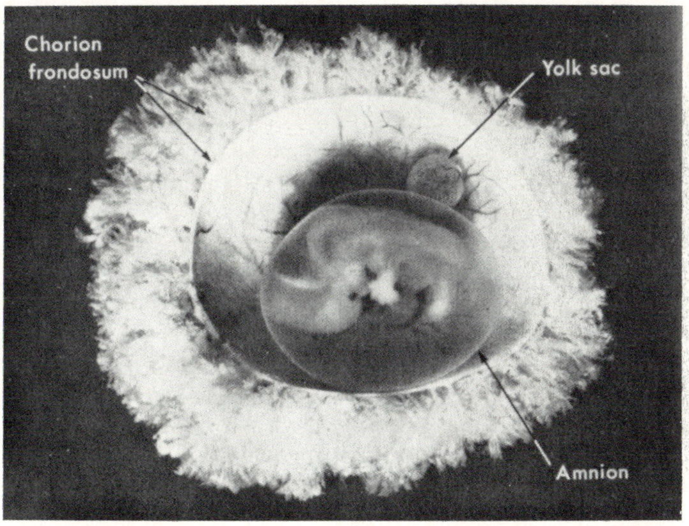

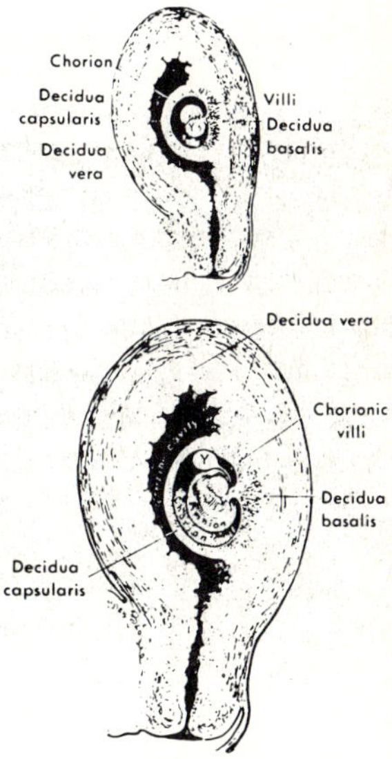

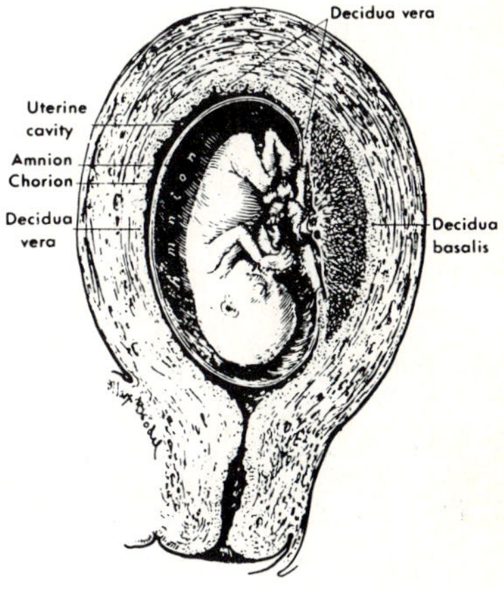

Figure 8.7. Placental and extraembryonic membranes in a human embryo. The body of the embryo is attached to the trophoblast by a stalk. The amniotic membrane lifts off the embryonic plate and surrounds the embryo. The yolk sac membrane arose as endodermal cells spread to cover the blastocyst cavity and then separated from the trophoderm when the extraembryonic coelon expanded. The trophoderm proliferates as chorionic tissue. The placenta and decidua grow throughout gestation.

cept the embryo and surround it with tissue richly supplied with blood. The embryonic blastula sends out processes that interdigitate with the capillary bed of the uterine walls. Nutrients are passed from the maternal bloodstream to the embryonic cells, allowing the embryo to grow and freeing it from dependence on yolk reserves present initially in the egg. The placenta grows by proliferation of both maternal and embryonic tissue and sustains the life of the embryo until birth (Fig. 8.7).

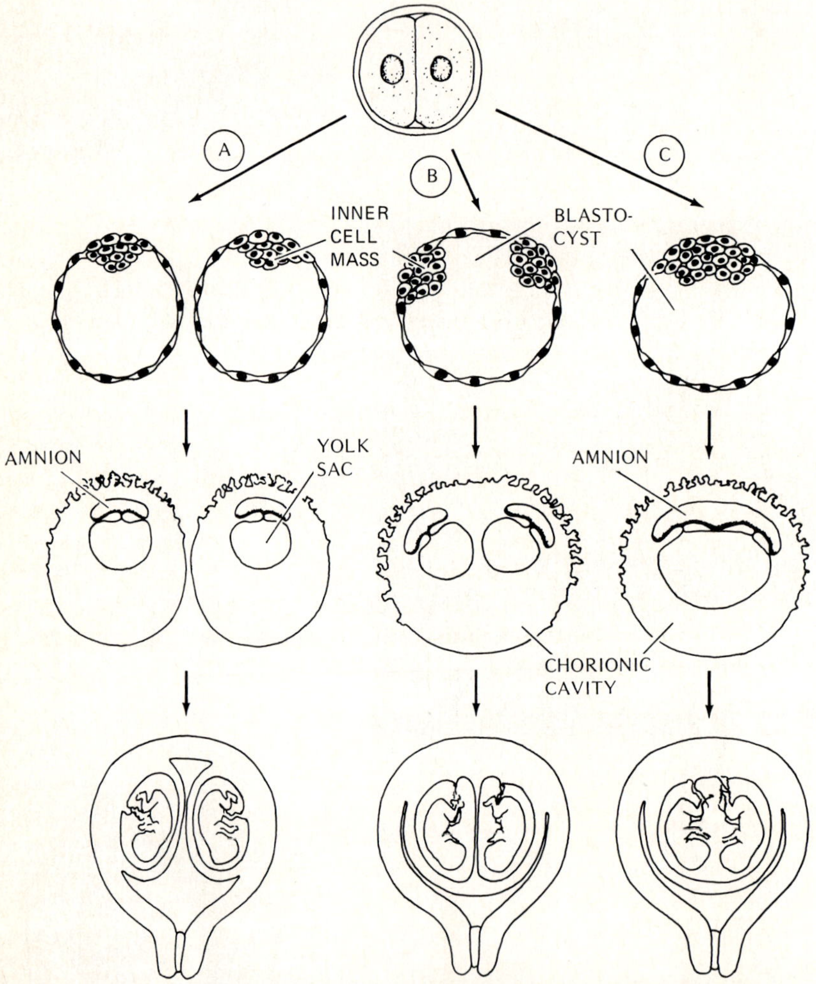

Figure 8.8. Monozygotic twins in humans. *(A)* After the first cleavage, the two cells may separate fully. Each twin develops from a separate blastocyst and has its own amnion, chorion, and placenta. *(B)* A single blastocyst may form two inner cell masses. The resulting twins share a common chorion and placenta, but each has its own amnion. *(C)* Two primitive streaks may form within a single embryonic plate. Both twins share a common chorion, placenta, and amnion. If a single primitive streak splits longitudinally, the twins will be conjoined.

Twinning can occur at several stages in mammalian embryogenesis (Fig. 8.8). Often the first two cells fail to adhere, and each develops into a normal child served by its own placenta. If two inner cell masses are formed, then the twins will share a single placenta. If twinning occurs at a yet later stage, conjoined embryos will develop into Siamese twins. The fact that a single inner cell mass can generate two essentially normal embryos clearly indicates that there has been no irreversible determination of developmental potential in these cells even by the time the embryonic disk is formed.

The Blastocoel

After six cell divisions there are 64 cells in a blastula. In both holoblastic and meroblastic embryos, a cavity is formed as the cells in the center of the mass dissociate from each other and fluid fills the space. This liquid-filled cavity is referred to as the blastocoel. The detailed geometry of the blastocoel varies considerably in embryos of different organisms.

In echinoderms such as sea urchins, the early divisions produce cells of about the same size. The blastocoel begins to be visible by about the sixth division and is enlarged during the next few divisions. By the time there are a few hundred cells, the blastula resembles a hollow ball surrounding the blastocoel.

In amphibians such as frogs, the early divisions generate more cells in the animal hemisphere than in the yolky vegetal hemisphere. The blastocoel forms above the large cells of the vegetal hemisphere as a result of pumping sodium ions into the intercellular space. The increase in osmotic pressure causes water to flow into the blastocoel. When well formed, it extends under the cells of the animal hemisphere (Fig. 8.9).

Eggs that undergo only partial cleavage, such as those of fish, reptiles, and birds, form a blastocoel only between the blastodermal cells in the em-

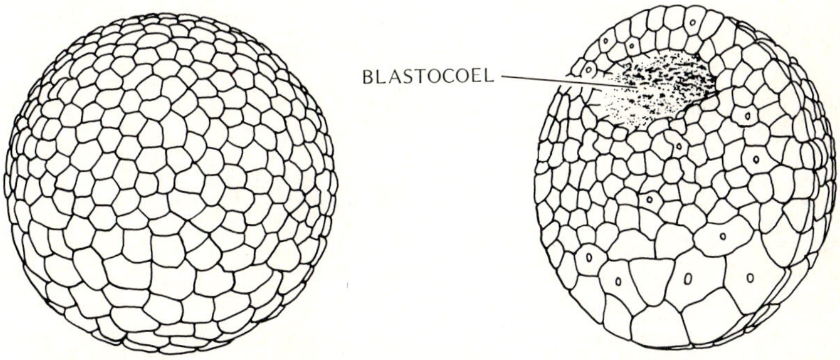

Figure 8.9. Amphibian blastocoel. The outer surface shows no sign of the blastocoel, but when a blastula is cut in half the cavity is clearly seen in the animal hemisphere.

bryonic disk. Holoblastic cleavage in these embryos results in a blastodisk several cells thick. The blastocoel separates the outermost layer from underlying cells that overlie the large mass of yolk. The blastocoel expands laterally as the disk expands until it resembles a flattened blastocoel of holoblastic embryos (Fig. 8.10).

Blastocoel formation in mammalian embryos almost seems to take place twice. Initially, the blastula forms much as in other holoblastic embryos. A central cavity is generated that transforms the close-packed cells of the morula into the hollow ball of the blastocyst. While the blastocyst is in the process of embedding in the uterine wall, a small group of cells within the blastocoel reiterates many of the steps seen in blastula formation of meroblastic embryos. The mammalian embryo develops from this inner cell mass. It expands into the blastocoel and is connected to the outer layer of cells only by an embryonic stalk. The outer layer, termed the trophectoderm, will develop into placental tissue that will provide a source of nutrients for the embryo proper.

As the inner cell mass grows it divides into several layers. The embryonic disk separates an overlying amnionic cavity from the underlying primitive yolk sac. The yolk sac contains no yolk but is referred to as such because of its homology to the yolk sac in meroblastic embryos. Cells in the embryonic disk continue to proliferate until they form a sheet several cells thick. Then a cavity forms separating the outermost layer of cells from those beneath it. This cavity is directly homologous to the blastocoel in meroblastic embryos. This is why in mammalian embryogenesis we have to think about blastocoel formation at two separate stages. Development in the mammalian embryonic disk then proceeds much as it does in chick embryos except that the mother continuously provides nutrients in mammals while the yolk provides nutrients in chicks.

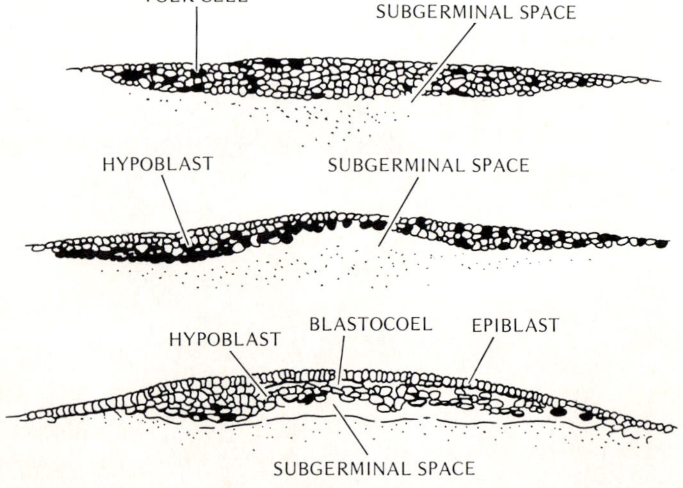

Figure 8.10. Avian blastocoel. The blastoderm rises slightly off the yolk, leaving a subgerminal space. However, the blastocoel forms between the outer epiblast cells and the larger hypoblast cells over the subgerminal space.

Gene Expression at the Blastula Stage

Rapid cleavage increases the number of nuclei exponentially. The total number of copies of each gene doubles at each cell division. By the time there are several hundred cells, expression of specific genes can be measured. A large number of genes are expressed in the later stages of blastula formation. Some of these appear to be necessary for the next stage in embryogenesis—gastrulation. There are several lines of evidence that support this statement.

Treatment of fertilized sea urchin eggs with a drug, actinomycin D, that blocks transcription has little or no effect on the cleavage pattern in blastulae but blocks gastrulation. This may be because the drug blocks expression of genes essential for gastrulation but it could also be due to unknown side effects of the drug.

In some cases conditions can be found in which the eggs of one species are fertilized by the sperm of another species. Early cleavage stages almost always proceed normally and a blastocoel is formed. Then embryogenesis stops abruptly. These results have been interpreted to mean that the maternal and paternal genomes are incompatible in these interspecies crosses and cannot direct normal gastrulation. If this is so, it also means that up until that stage it did not matter very much if the genomes were compatible or not. Blastula formation in many organisms seems to proceed on maternal mRNA laid down in the egg during oogenesis. Only toward the later stages of blastula formation is there a requirement for expression of the embryo's own genes.

By the time a blastula is well formed there are marked inhomogeneites in the cells. Some have more yolk than others, some were formed before others, some are large, and some are small. Although direct evidence has been difficult to get, it is likely that some genes are turned on only in specific cells. Likewise, some genes are turned off only in specific cells. These alterations in gene expression will ultimately come to change the physiological functions of the cells and they will start to diverge down the pathways of tissue differentiation.

Already in the late blastula stage certain cells have started to specialize. A small number of cells will leave the outer layer of cells and, either individually or in sheets, will invade the blastocoel. This invasion signals the start of the next stage, gastrulation.

Summary

Holoblastic division of small, relatively yolk-free eggs divides them into approximately equal-sized blastomeres whereas meroblastic division of large yolky eggs is initially incomplete and later gives rise to small embryonic cells and a large mass of yolk. The first few cleavage planes are at right angles to each other in most cases. In echinoderms and amphibians holoblastic division results in a ball of cells surrounding a central cavity. In fish and birds meroblastic division produces a mass of cells that rises off the underlying yolk and then separates into epiblast and hypoblast layers on either side of the blastocoel. In mammals holoblastic cleavage produces a ball of cells surrounding a central cavity, the blastocyst cavity. Within that cavity a small group of about a dozen cells, referred to as the inner cell mass, develops into the embryo. The surface layer of cells differentiates as extraembryonic tissues. Transcription of genes accelerates as the number of nuclei increases exponentially.

SPECIFIC EXAMPLES

1. Role of the Centriole

Cell cleavage passes through a plane defined by the position of the centrioles (Fig. 8.11). Spindle fibers composed of microtubules connect centrioles on opposite sides of the metaphase plate to the condensed chromosomes. As the chromosomes move apart, one complement goes toward one centriole and the other complement goes toward the other centriole. The spindle fibers extend radially from each pole and appear to hold up the cell surface except at the middle, over the metaphase plate. Cell surface membrane indents over the metaphase plate and then proceeds to make a division furrow across the cell, separating the centrioles and their attached chromosomes. In this way each daughter cell receives a full complement of chromosomes and a pair of centrioles. These replicate and separate before the next division.

Centrioles are electron-dense bodies that can be recognized in electronmicrographs. They function as microtubule-organizing centers (MTOC). Eggs of echinoderms such as sea urchins have an abundant supply of tubulin, but

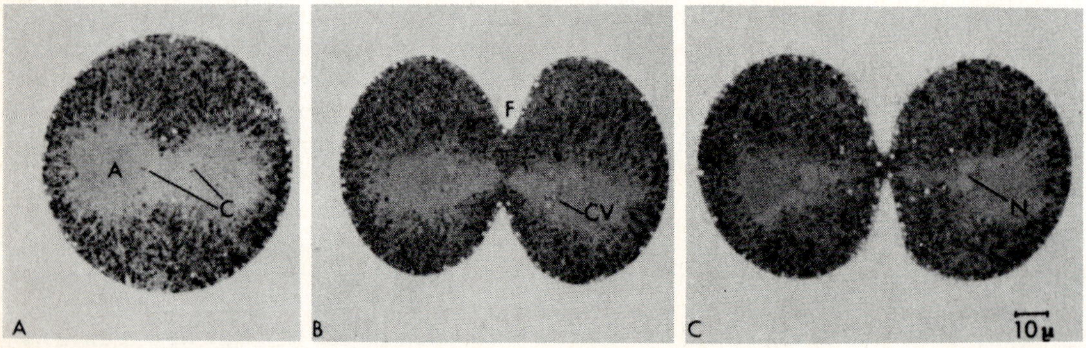

Figure 8.11. First cleavage in sea urchins. *(A)* Relatively yolk-free cytoplasm is organized by the mitotic apparatus. *(B, C)* The cleavage furrow (F) cuts between daughter nuclei (N).

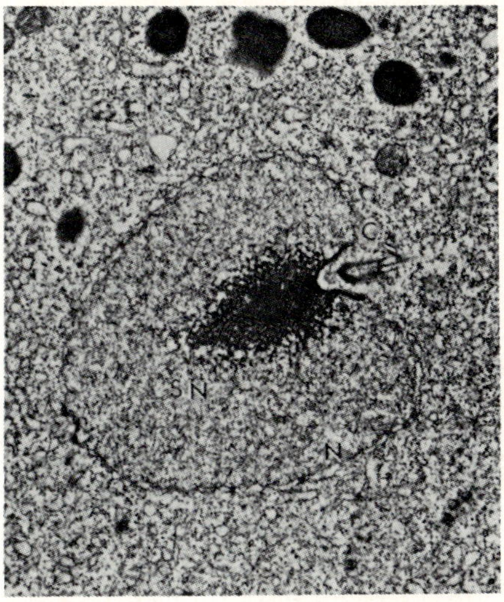

Figure 8.12. Sperm centriole. Shortly after penetration in the sea urchin *Arbacia punctulata,* the sperm nucleus swells. Just behind it is the pair of sperm centrioles (C).

there are no assembled microtubules. The unfertilized egg lacks the ionic environment and the nucleating center that would permit polymerization. The fertilizing sperm contributes a pair of centrioles that form the seed around which the microtubules grow to give rise to the sperm aster (Fig. 8.12). At some point in the maturation of oocytes the egg centrioles are destroyed, but it is not clear at exactly what point. This leaves the egg ready but waiting for the sperm centrioles.

Microtubule assembly is dependent not only on an MTOC but also on an alkaline pH. Sperm entry has significant local consequences to intracellular pH and calcium release. The pair of sperm centrioles that enter behind the nucleus provides the MTOC on which polymerization of tubulin is initiated.

Sperm fusion results in a propagative explosion of sequestered egg calcium stores, which transiently increases the available calcium throughout the fertilized egg cytoplasm. After this there is an increase in intracellular pH throughout the entire cytoplasm. Microtubules rapidly elongate from the sperm centrioles and push the sperm nucleus toward the center of the egg. The microtubules continue to extend and soon connect the egg nucleus with the sperm nucleus. The egg nucleus migrates rapidly (15 μm/minute) toward the center of the sperm aster. This movement is independent of the function of microfilaments because it is not inhibited by cytochalasin but is directly dependent on the function of microtubules since it is sensitive to microtubule inhibitors such as colcemid, vinblastin, or taxol. The egg nucleus moves toward the center of the sperm aster by attachment of the connecting microtubules to the nuclear membrane. Microtubules are undergoing net disassembly near the sperm centriole. Inhibitors of microtubule disassembly, such as taxol, block this motion. About 8 minutes after the fusion of sperm and egg of the sea urchin *Lytechinus variegatus,* the pronuclei come to lie side by side at the center of the egg.

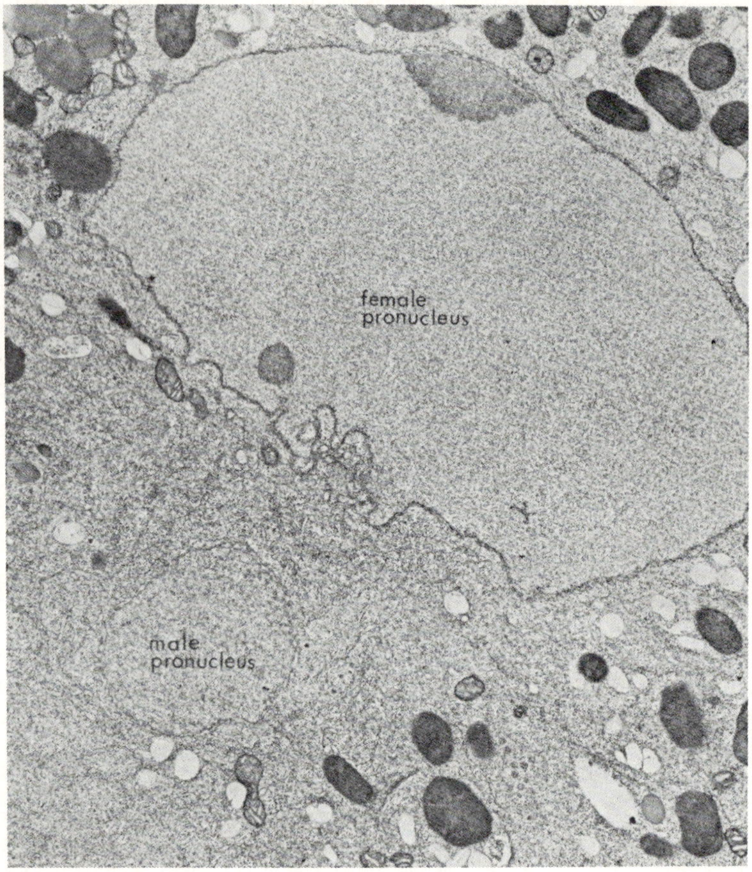

Figure 8.13. Pronuclei meeting. The large female and smaller male pronuclei of the sea urchin *Arbacia punctulata* have been brought together and will soon fuse.

The adjacent pronuclei then move fairly slowly (3 μm/minute) toward the center of the egg (Fig. 8.13). The sperm aster continues to enlarge and the centrioles first begin to separate in a direction perpendicular to the direction of centration. This separation establishes the plane of the first cleavage.

When the pronuclei reach a position near the center of the egg, they fuse to form the first diploid nucleus of the individual. At the same time the microtubules of the astral rays disassemble. Following syngamy, a bipolar apparatus extends from the centrioles that are now on opposite sides of the diploid nucleus. At prophase of the first mitotic division, the nuclear envelope breaks down (Fig. 8.14). The mitotic apparatus then forms between the centrioles and engages the chromosomes at the metaphase plate. This assembly/disassembly cycle of microtubules may be the consequence of ionic changes in the zygote. Nuclei are re-formed as the chromosomes leave the metaphase plate and the spindle is disassembled. The first cleavage then proceeds through the metaphase plate giving rise to the first two cells of the embryo. The anterior/posterior axis is established and the path of events leading to subsequent divisions is entrained. These movements during the first half hour of embryogenesis are shown schematically in Figure 8.14.

The entry of the sperm centrioles modifies the local ion environment

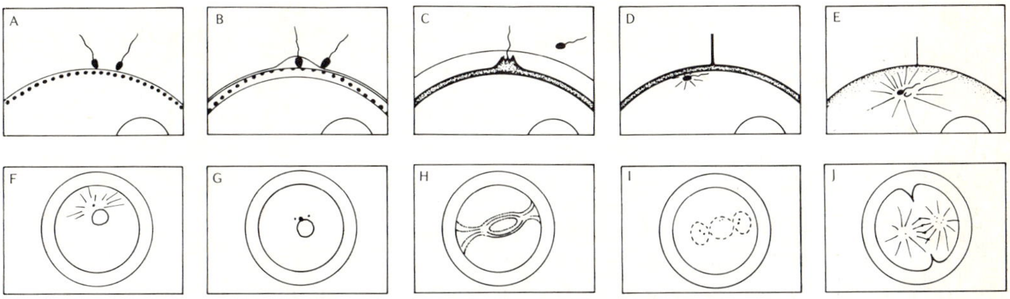

Figure 8.14. Schematic view of sea urchin fertilization. *(A)* Sperm attach to the egg surface. *(B)* The fertilization layer elevates. *(C)* Fertilization cone forms around the erect immotile sperm. *(D)* Sperm glides along the egg cortex. *(E)* The sperm aster radiates and moves the male pronucleus toward the center of the egg. *(F)* At lower magnification the aster can be seen connecting the male and female pronuclei. *(G)* Fusion of the pronuclei (syngamy) occurs near the center of the egg. *(H)* A microtubular streak deforms the nucleus. *(I)* The streak disassembles prior to nuclear breakdown. *(J)* Cleavage is perpendicular to the axis of the mitotic apparatus.

and provides an MTOC for microtubule assembly. The elongating microtubules push on the egg cortex and the sperm nucleus so that it moves deeper into the egg. Microtubule assembly continues until the sperm aster reaches the egg nucleus. The two pronuclei are drawn together and then centered in the egg by the function of microtubules radiating from the sperm centriole. The pathway taken by the adjacent pronuclei defines the plane of the first division. The separated centrioles subsequently determine the orientation of the metaphase plate and the topological position of the daughter nuclei that will be separated by the first division.

If the pathway of migration of adjacent pronuclei is perturbed by inhomogeneities in the fertilized egg, such as localized concentrations of yolk platelets, the first cleavage plane and the embryonic axis will be affected. Asymmetric cleavage or spiral cleavage such as occurs in the eggs of many species may be the consequence of such inhomogeneities.

2. Spiral Cleavage in Molluscs

Eggs of the elephant tusk mollusc, *Dentalium,* range colorfully from green to red in different individuals. The pigment is distributed in regions of the egg loaded with yolk granules. At both the animal and vegetal poles there are unpigmented regions of cytoplasm relatively free of yolk. The first few cleavages of *Dentalium* are markedly asymmetric (Fig. 8.15). Following fertilization, the bulk of the yolk-free cytoplasm at the vegetal pole is blebbed out into a lobe. The first division passes to one side of the stalk that connects the lobe to the embryo so that it is now attached to one (CD) but not the other (AB) cell. Before the next cleavage, the lobe is retracted into the CD cell and then blebbed out again. The second division is spiral and leaves the polar lobe connected only to the D cell.

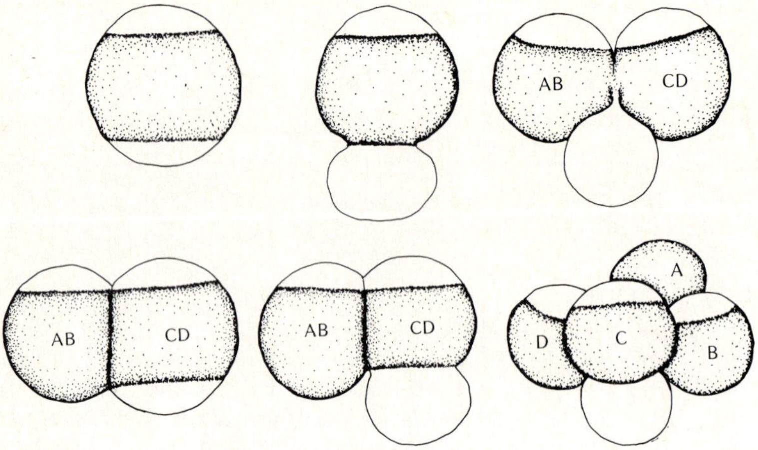

Figure 8.15. Spiral cleavage of the mollusc, *Dentalium*. A fertilized *Dentalium* egg constricts a lobe to one side of the first cleavage plane. After the first cleavage, the lobe is retracted but is reformed before the second cleavage, which occurs at different angles in the different blastomeres. After the second division, the lobe is connected to cell D.

The yolk-free material in the polar lobe is essential for differentiation of the mesoderm rudiment in the larva. If the polar lobe is nipped off at the time of either the first or second cleavage, the larva that develops lacks the mesoderm rudiment. No nuclear material was removed in this experiment; only the relatively yolk-free cytoplasm was excised, leaving the D cell with

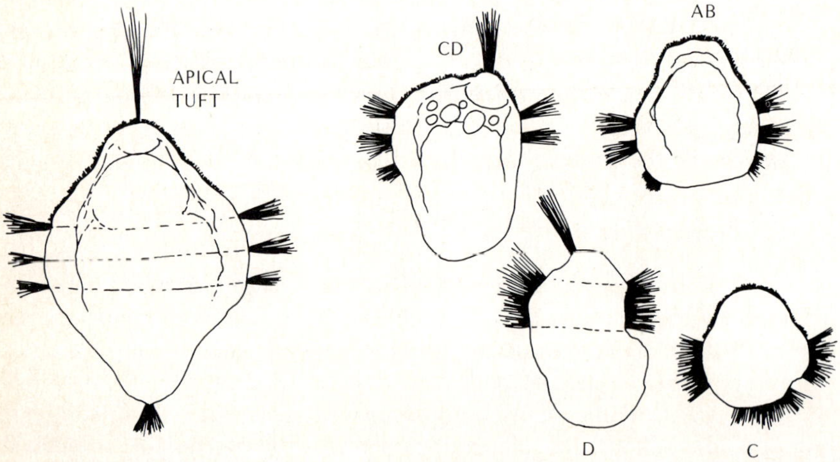

Figure 8.16. Partial development of isolated *Dentalium* blastomeres. Normal larvae have apical tufts as well as lateral tufts. Larvae that develop from single blastomeres isolated from two- or four-cell embryos form apical tufts if they came from the blastomeres with a polar lobe (CD or D). The other blastomeres (AB or C) develop into larvae that lack this tuft.

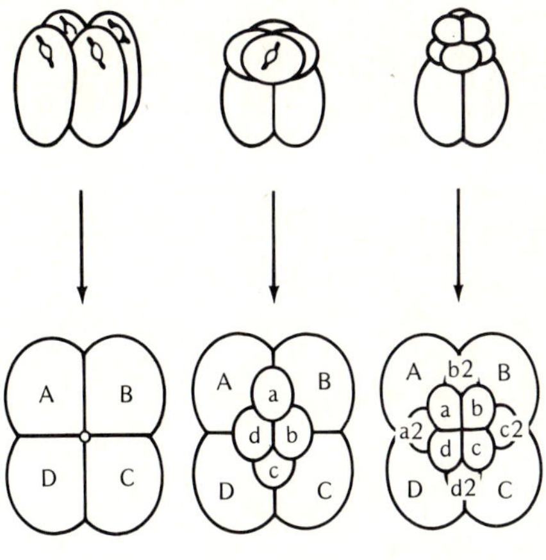

Figure 8.17. Spiral cleavage in a snail. The orientation of the mitotic spindle as well as the cleavage planes clearly show sinestral spiral cleavage. Two views of the four, eight and twelve cell embryos of *Lymnaea peregra* are diagrammed.

a yolkier composition than normal. Somehow this results in the lack of proper differentiation. Similar conclusions can be drawn from separating the blastomeres at the two- or four-cell stage and allowing development to proceed in isolation. Cells that carry the polar lobe (cell CD and later cell D) develop into normal although smaller larvae complete with an apical tuff. Cells that lack the polar lobe cytoplasm develop into defective larvae lacking the rudiment of the mesoderm (Fig. 8.16).

These results have been interpreted as indicating a requirement for vegetal pole material in the development of the mesoderm. The most striking characteristic of the vegetal pole cytoplasm is its lack of yolk granules. Embryos from which the polar lobe has been removed end up with a higher than usual yolk content, and it may be this characteristic that derails the normal train of developmental events.

The freshwater snail, *Lymnaea peregra,* has either right-handed (dextral) or left-handed (sinistral) twists to its shells. The pattern of cleavage in *Lymnaea* embryos is also spiral, and even at the first division a handedness can be seen. They cleave in either a dextral or sinistral manner. By the third cleavage the asymmetry is easily observed (Fig. 8.17).

The decision as to right- or left-hand cleavage that ultimately results in snails with differently twisted shells is determined by a single gene, *D*. Mutations in this gene result in left-handed cleavage that leads to left-handed snails. The product of the *D* gene must be present in the egg before fertilization for dextral cleavage to occur. Thus, all eggs from homozygous sinistral females *(d/d)* cleave in a sinistral fashion whether they are fertilized by a sperm carrying a wild-type gene *(D)* or a mutant gene *(d)*. *D/d* embryos from a *d/d* mother cleave in a sinistral manner. Eggs from a *D/D* or a *D/d* mother cleave in a dextral manner irrespective of the paternal *D* gene. This is a strict maternal inheritance dependent on a single dominant gene, *D*.

By micromanipulation a small amount of cytoplasm can be withdrawn from an egg of a *D/D* mother and injected into the egg of a *d/d* mother. If this is done six hours before the first cleavage, the divisions will follow the dextral pattern. Injection of as little as 3 percent of the cytoplasm reorients the cleavage planes within six hours. These results suggest that the *D* gene product acts catalytically on a determining process. Alternatively, the mechanism that determines handedness is delicately poised and can be easily shifted to the right by a small amount of the product of the *D* gene. It will be very interesting to see what this product is. A protein that interacts with the components of the cytoskeleton is a good guess.

Related Readings

Freeman, G., and Lundelius, J. (1982). The developmental genetics of dextrality and sinistrality in the gastropod *Lymnaea peregra.* Willhelm Roux Arch. *191,* 69–83.

Gurdon, J., Mohun, T., Fairman, S., and Brennan, S. (1985). All components required for the eventual activation of muscle-specific actin genes are localized in the subequitorial region of an uncleaved amphibian egg. Proc. Natl. Acad. Sci. *82,* 139–143.

Jeffrey, W. (1984). Spatial distribution of messenger RNA in the cytoskeletal framework of ascidian eggs. Devel. Biol. *103,* 482–492.

Jeffrey, W., and Meier, S. (1983) A yellow crescent cytoskeletal domain in ascidian eggs and its role in early development. Devel. Biol. *96,* 125–143.

Maller, J., Poccia, D., Nishioka, D., Kidd, P., Gerhart, J., and Hartman, H. (1976). Spindle formation and cleavage in *Xenopus* eggs injected with centriole-containing fractions from sperm. Exp. Cell Res. *99,* 285–294.

McGrath, J., and Solter, D. (1984). Inability of mouse blastomere nuclei transferred to enucleated zygotes to support development *in vitro.* Science *226,* 1317–1319.

Mitchison, T., and Kirschner, M. (1984). Dynamic instability of microtubule growth. Nature *312,* 237–241.

Mohun, T., Maxson, R., Gormezano, G., and Kedes, L. (1985). Differential regulation of individual late histone genes during development of the sea urchin (Strongylocentrotus purpuratus). Devel. Biol. *108,* 491–502.

Newport, J., and Kirschner, M. (1982). A major developmental transition in early *Xenopus* embryos. Cell *30,* 675–696.

Pittman, D., and Ernst, S. (1984). Developmental time, cell lineage, and environment regulate the newly synthesized proteins in sea urchin embryos. Devel. Biol. *106,* 236–242.

Raven, C. (1966). Morphogenesis: the analysis of mulluscan development. Pergamon Press, Oxford.

Schatten, G., Schatten, H., Bestor, T., and Balczon, R. (1982). Taxol inhibits the nuclear movements during fertilization and induces asters in unfertilized sea urchin eggs. J. Cell Biol. *94,* 455–465.

Speksnijder, J., Mulder, M., Dohmen, M., Hage, W., and Bluemink, J. (1985). Animal-vegetal polarity in the plasma membrane of a molluscan egg: a quantitative freeze-fracture study. Devel. Biol. *108,* 38–48.

Suprynowicz, F., and Mazia, D. (1985). Fluctuation of the Ca^{++}-sequestering activity of permeabilized sea urchin embryos during the cell cycle. Proc. Nat'l Acad. Sci. *82,* 2389–2393.

Ubbels, G., Hara, K., Koster, C., and Kirschner, M. (1983). Evidence for a functional role of the cytoskeleton in determination of the dorsoventral axis in *Xenopus laevis* eggs. J. Embryol. Exp. Morph. *77,* 15–37.

Warner, A., Guthrie, S., and Gilula, N. (1984). Antibodies to gap-junction protein selectively disrupt junctional communication in the early amphibian embryo. Nature *311,* 127–131.

Wiley, L. (1984). Cavitation in the mouse pre-implantation embryo: Na/K-ATPase and the origin of nascent blastocoel fluid. Devl. Biol. *105,* 330–342.

CHAPTER 9

Gastrulation

Gastrulation is characterized by the movement of cells into the blastocoel (Fig. 9.1). As a consequence the embryo becomes a complex multilayered structure since some cells remain on the surface while others enter the interior. The external cells are referred to as ectodermal and will give rise to specific tissues such as skin and nerves. The inner cells are referred to as endodermal and will form the gut and many glands. Between these so-called germ layers are other cells referred to as mesodermal. They will give rise to muscles and many other organs.

The mechanisms of invagination of cells into the blastocoel have diverged considerably during evolution and it is best to consider the distinct classes separately before drawing homologies between them.

Sea Urchin Gastrulation

When the blastula has divided into about a thousand cells, a few cells at the vegetal pole detach from the surface of the hollow ball and invade the blastocoel. These primary mesenchymal cells are derived from the four micromeres that were formed at the vegetal pole at the fourth division of sea urchin eggs. They will give rise to the skeletal structures of the sea urchin larva (Fig. 9.2).

The molecular and biochemical changes that result in invasiveness of

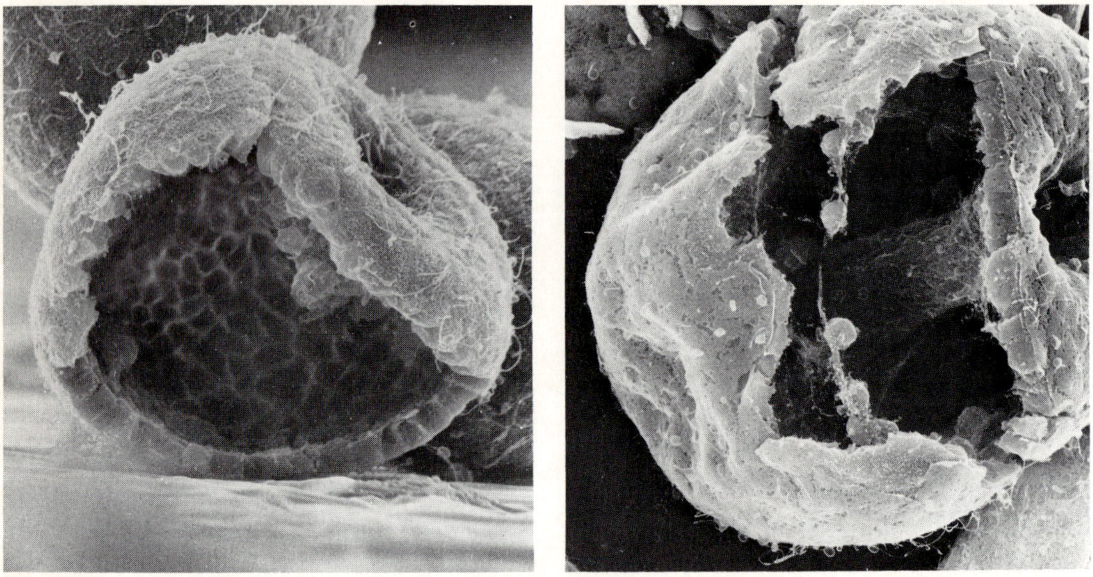

Figure 9.1. Early gastrula of a sea urchin. Embryos were fractured so that primary mesenchymal cells can be seen in the blastocoel. Indentation at the blastopore proceeds across the blastocoel.

the primary mesenchymal cells are unknown and can only be guessed at from the gross morphological characteristics of early gastrulation. The cells that form the surface of the hollow ball all seem to be strongly adherent to each other. Both the inner and outer surfaces are smooth as a result of the presence of extracellular layers covering them. The primary mesenchymal cells appear to lose the adhesions that kept them in the surface layer of cells as they migrate to the interior. They must also breach the inner surface covering to reach the blastocoel. Once inside, they can be seen to move about unhindered and do not appear to adhere strongly to the inner surface or to each other. Thus, a change in the adhesive mechanisms on the surface of these cells seems implicated. A better understanding of cellular adhesive mechanisms in general might direct attention to the molecular changes associated with this first step in gastrulation.

Shortly after the primary mesenchymal cells have invaded the blastocoel a well-defined blastopore can be seen. The blastopore of sea urchin embryos starts as a flattening of the surface at the vegetal pole. Then the blastopore dimples as the whole sheet of cells starts to turn inward. These morphological changes suggest that contractile processes are functioning on or in the cells at the vegetal pole. There are no obvious changes in the adhesion of these cells. The sheet of cells stays intact. But on the inner face, the invaginating cells extend long processes that wave about in the blastocoel. They appear to scan the inner surface of the embryo until they attach quite firmly to positions near the opposite (animal) pole. These strands emanating from the blastoporal cells come under tension and may help to draw the invaginating cells across the cavity. At least they seem to guide the cells across the

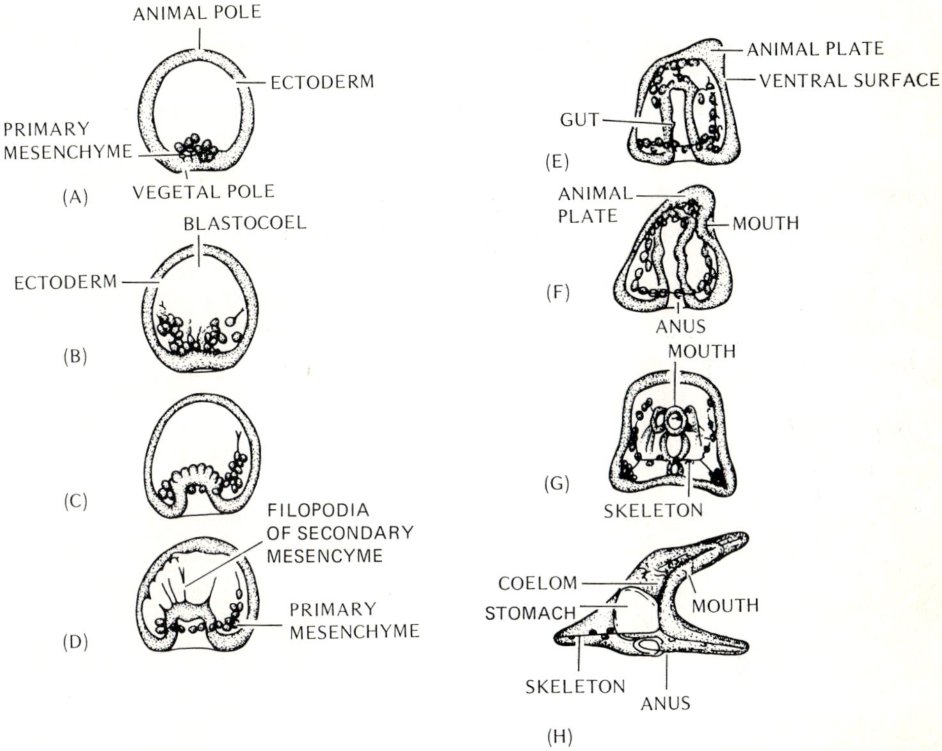

Figure 9.2. Gastrulation in sea urchins. *(A)* First the primary mesenchymal cells enter the blastocoel at the vegetal pole. *(B)* Then the blastopore dimples and *(C)* extends inward. *(D)* Filopods extend from the invaginating cells and *(E)* guide the archenteron to the opposite pole. *(F)* The mouth will later fuse to form the larval gut. *(G)* Primary mesenchymal cells secrete spicules to form the skeleton. *(H)* Arms extended and mouth open, the pluteus larva is fully developed and ready to feed.

cavity. When the invagination has proceeded across the blastocoel, the gastrula looks like a soft balloon into which one has stuck a finger.

The blastopore itself is an opening rather than an organ of distinctive cells. Cells on the sides of a blastula move continuously through the blastopore to become interior endodermal tissue. Some conditions have been found which permit the formation of a blastopore but lead to movement outward rather than inward. This exogastrulation occurs when certain salts such as lithium are added to the seawater. Perhaps the osmotic state of the blastocoel fluid relative to the surrounding seawater helps direct normal gastrulation. Exogastrulation is also observed when embryos are dissociated at the 8 or 16-cell stage into animal and vegetal blastomeres. Blastulae derived from the macromeres of the vegetal hemisphere almost always exogastrulate. However, if a few micromeres are added back to the macromeres, the ensuing blastulae undergo quite normal gastrulation. These experiments indicate that the presence of primary mesenchymal cells derived from the micromeres plays a directive role in gastrulation of embryos derived from the yolky halves of eggs.

The cavity formed by entry of the sheet of cells is referred to as the archenteron. It is open to the outside at the blastopore and will function as the alimentary canal in the larva. A second opening results from fusion of the far end of archenteron with the wall of the gastrula. The second opening becomes the mouth of the larva, allowing one-way traffic along the alimentary canal. The second opening is quite far from the blastopore in sea urchin embryos as in well as in the embryos of many other organisms including all vertebrates. Based on these features, these phyla are referred to as deuterostomes. In other organisms such as molluscs and insects the first opening becomes the mouth or the mechanisms of gastrulation are sufficiently different from that in deuterostomes that they are referred to as protostomes. The evolutionary divergence of protostomes and deuterostomes appears to be very ancient. Since that time, further evolution has been specific to one class or the other with no crossover.

The larger eggs of some primitive chordates make blastulae very similar to those of sea urchins, but the cells contain a higher complement of yolk. Gastrulation of the protochordate *Amphioxus* starts as a flattening at the vegetal pole that reaches halfway across the diameter of the embryo. This flat sheet bends into the blastocoel in such a way that the embryo looks like a soft balloon into which one has stuck a fist. The blastopore then contracts

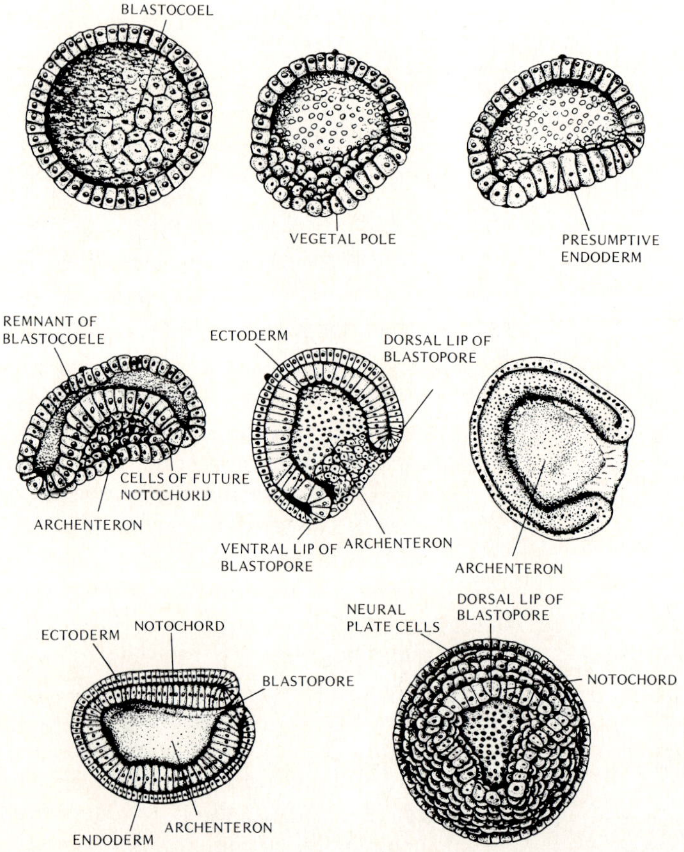

Figure 9.3. Gastrulation in *Amphioxus*. Sections through embryos of this simple chordate show the topography of invagination of the endoderm followed by closing of the blastopore. A notochord forms between the ectoderm and the mesoderm on the dorsal side.

until its diameter is only a small percentage of that of the gastrula. Now the *Amphioxus* gastrula looks much like that of a sea urchin (Fig. 9.3).

Amphibian Gastrulation

The yolky eggs of amphibians form blastulae in which the blastocoel lies over the large vegetal cells and is domed over by a layer of animal cells. The blastopore forms on one side below the separation of animal and vegetal hemispheres. In most embryos the blastopore forms on the side opposite the point of sperm entry. In these embryos fertilization or pricking with a fine needle induces rotation of the cortex relative to the inner components of the egg. In pigmented eggs the internal pigment granules are drawn toward the site of sperm penetration leaving a relatively unpigmented region, the gray crescent, on the opposite side (Fig. 9.4). Later in development the blastopore will form just below the gray crescent. The eggs of many amphibians are not highly pigmented and no gray crescent can be seen. However, the dorsal lip of the blastopore also forms opposite the point of sperm entry in these embryos, due to the redistribution of internal components relative to the surface layer. This process establishes the future dorsal/ventral polarity.

The internal components can be artificially moved by rolling over a fertilized amphibian egg and keeping it upside down. This procedure alters the position of the blastopore at gastrulation. Amphibian eggs are inhomogeneous with the yolk stratified and concentrated by its difference in density. Changing the direction of the force of gravity on the yolk platelets redistributes them and alters the position of invagination at gastrulation. Furthermore, conditions that stabilize microtubules such as D_2O or taxol, block the redistribution that normally occurs upon sperm entry and result in si-

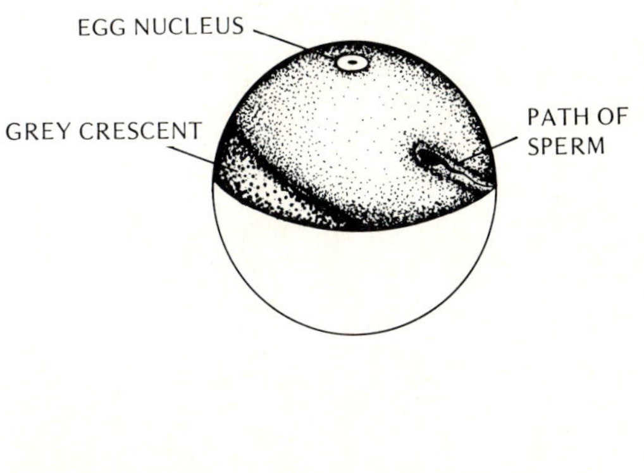

Figure 9.4. Formation of a grey crescent in amphibian eggs. The animal hemisphere in many amphibians is pigmented. Sperm entry initiates a rotation of the outer layer (cortex) relative to the insides. This leaves a relatively less pigmented crescent on the side opposite the sperm entry point.

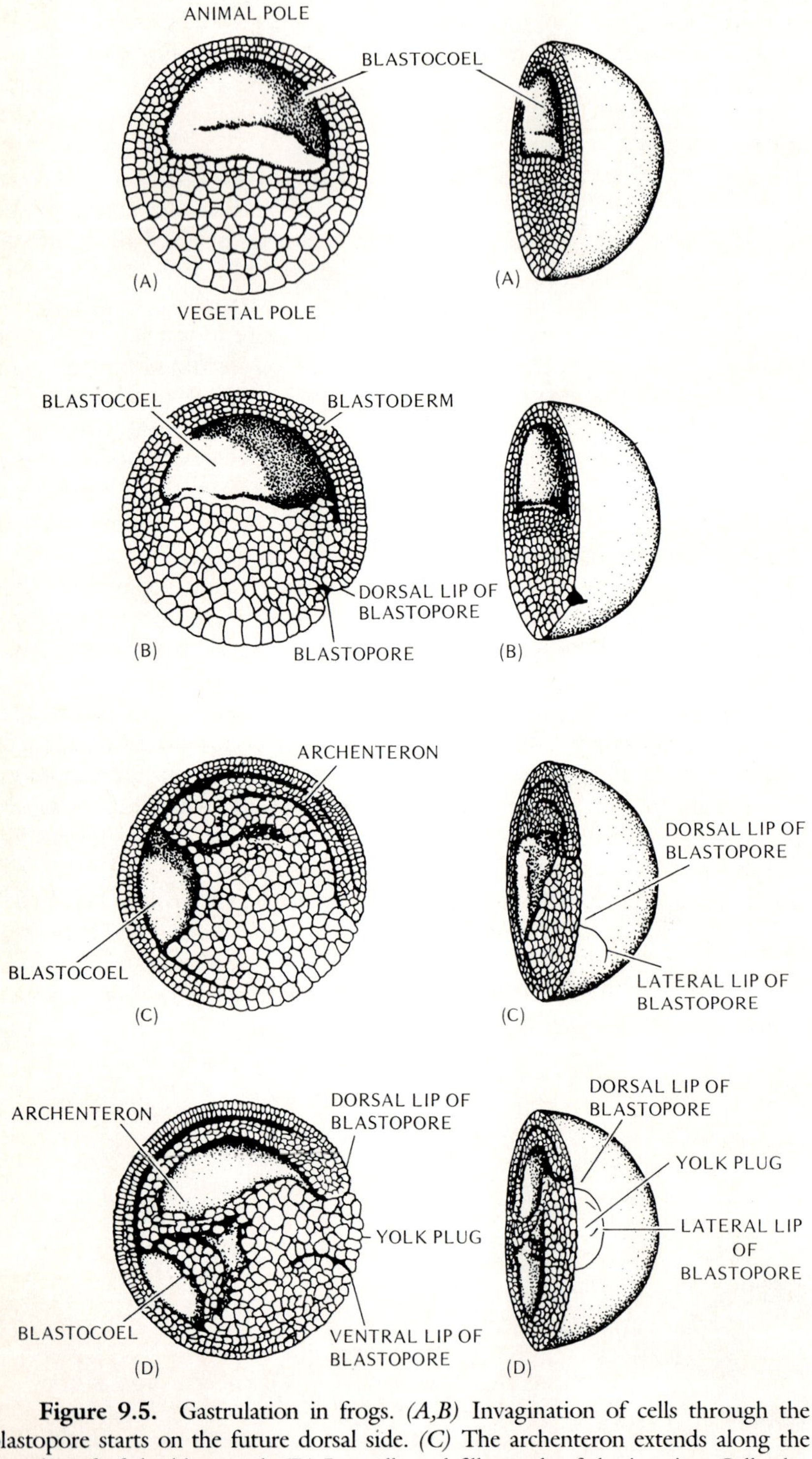

Figure 9.5. Gastrulation in frogs. *(A,B)* Invagination of cells through the blastopore starts on the future dorsal side. *(C)* The archenteron extends along the dorsal roof of the blastocoel. *(D)* It swells and fills much of the interior. Cells also invaginate over the lateral and ventral lips of the blastopore, leaving a yolk plug.

multaneous invagination through a blastopore that extends almost all the way around the blastula. Embryos treated in this way end up with far more cephalic tissue than normal embryos and construct multiple eyes. Conditions that depolymerize microtubules, such as high pressure or cold, result in greatly decreased cephalic differentiation.

The blastopore of frog embryos is an elongated lip through which cells move into the interior (Fig. 9.5). The blastopore initially forms as a groove just below the equator and then spreads laterally down toward the vegetal pole. The lips meet close to the vegetal pole where ventral tissues will differentiate later. Migration of cells into the interior is much less extensive at the ventral lip than at the dorsal lip. As in sea urchins, the blastopore is an opening and not a stable tissue; cells continue to move over the blastopore lips into the interior throughout gastrulation. As gastrulation proceeds, the blastocoel cavity is replaced by the cavity of the archenteron which is open to the outside and will develop into the alimentary canal.

When the blastopore first forms, cells at the lips take up unusual elongated shapes. On the outside they are strongly attached to each other by tight junctions and desmosomes. On the inside the cells break adhesive connections with each other and extend upwards toward the blastocoel. The attached ends near the outer surface are constricted into thin necks while most of the cellular contents appear squeezed into the bulbulous inner ends (Fig. 9.6). The cells take on the appearance of bottles and are referred to as bottle cells. Running down the neck of bottle cells are microtubules and microfilaments that appear to hold the cells in these elongated shapes and may be involved in the bulk movement of their contents into the blastocoel.

The contraction of the outer end of bottle cells to form necks results in the indentation of the surface that can be seen from without. These invaginating cells continue to invade the blastocoel, turning toward the roof of the animal hemisphere shortly after entering. The invasion proceeds as the sheet of cells flows over the blastoporal lips. Other cells are released from the sheets to make up the mesodermal tissues.

Cells near the blastopore appear to have an innate ability to invade tissues. If a small group of cells are taken from near the blastopore and placed

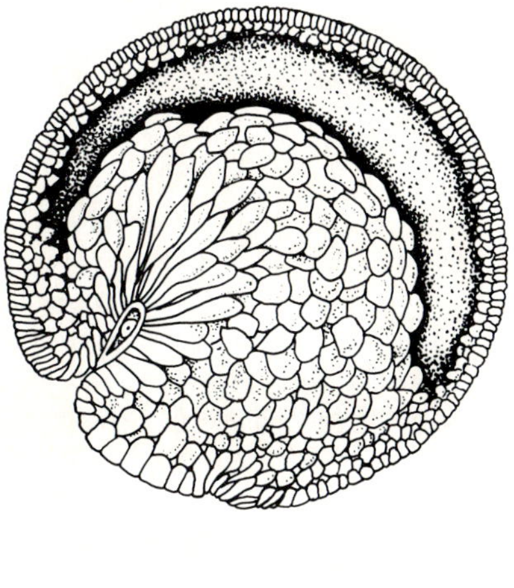

Figure 9.6. Bottle cells. Invaginating cells of amphibian gastrulae are constricted at their bases and bulbous at their tips and so have been called bottle cells. They are seen at both the dorsal and ventral lips of the blastopore.

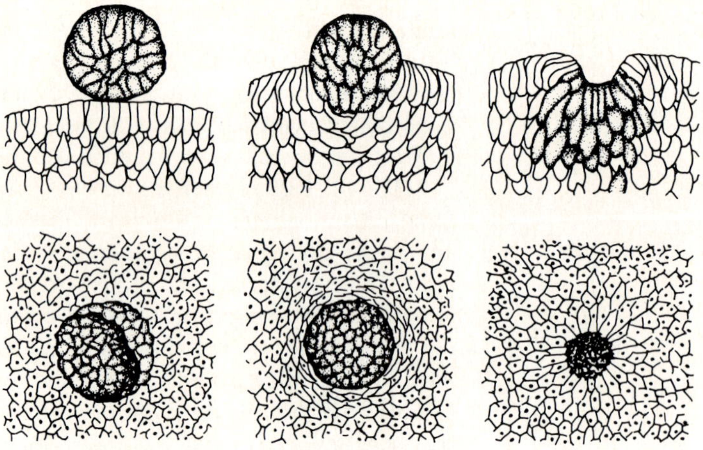

Figure 9.7. Invagination of cells from the lips of an amphibian blastopore as seen in cross-section and from above. A small group of cells at the blastopore was dissected out and freed of extracellular matrix. A ball of these cells was deposited on a sheet of endodermal cells. The ectodermal ball of cells rapidly sank beneath the surface and gave rise to bottle cells and a subsequent indentation of the sheet.

on endodermal tissue they invade the endodermal tissue (Fig. 9.7). Ectodermal cells taken from other positions on the embryo will not do this. Instead, they will spread to cover the surface. The blastoporal cells seem to have quite different surface properties that lead to invagination of tissues. Cells at the dorsal lip move over the roof of the blastocoel and migrate much further than cells that enter at the ventral lip. This difference in migration capacity is established earlier during cleavage perhaps as a consequence of the initial cortical rotation triggered by sperm entry.

The sequence of development in frogs is so regular that once a sperm and egg have fused, the rest of embryogenesis proceeds with each event determining the next. Sperm can enter anywhere in the animal hemisphere, although most often the successful sperm fuses near the equator. The gray crescent then forms opposite the sperm entry point and cleavage of the egg

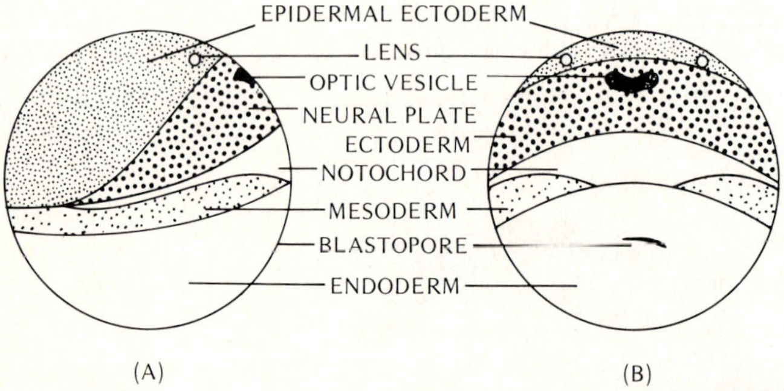

Figure 9.8. Frog fate map. Whether viewed from the side *(A)* or the front *(B)*, a fate map can be drawn on the surface of an amphibian egg as soon as a sperm enters. The contents of the egg will be divided into cells that will follow invariant paths to epidermal, neural, mesodermal, or endodermal differentiations. Even the positions that will give rise to the lens and optic vesicles can be reliably predicted. The blastopore will form opposite the sperm entry point. Cells that will form dorsal mesoderm come to lie in a deep layer of the blastula beneath the dorsal lip of the blastopore.

gives rise to the blastula. The blastopore forms among those cells that come to incorporate the egg material near the gray crescent. They invaginate and will later differentiate into the foregut. It is all so regular that one can predict what structures will arise from different portions of a fertilized egg (Fig. 9.8). The anterior/posterior axis is established in relationship to the distribution of yolk platelets while the dorsal/ventral axis is established by cortical rotation in response to sperm entry. These inhomogeneities control the timing and positioning of the gastrulation processes.

A fate map indicates the tissues that will be derived from different portions of a fertilized egg in the normal course of events but does not indicate when this fate is determined. Normally, ventral tissues are formed on the side where the sperm enters. However, if one or two of the large vegetal blastomeres of the presumptive ventral side is replaced at the 64-cell stage by similar vegetal blastomeres from the dorsal side, the overlying cells of the animal hemisphere will form a second dorsal lip and proceed to form a second axis (Fig. 9.9). Cells, which by fate mapping would have been epidermal ectoderm, can change at this stage to give rise to neural and mesodermal tissues when signaled by dorsal vegetal cells. Clearly, their fate is not yet irreversibly set.

When a dorsal piece of a blastula is cut out just after gastrulation has started and exchanged with a piece of a similar size elsewhere on another blastula, the grafted piece will differentiate in a manner that is normal for its new position (Fig 9.10). That is, the cells regulate to their new surroundings. When one carries out the same experiment a little later when gastrulation is almost completed, quite different results are observed. For instance, cells that are dorsal to the center of the blastopore normally differentiate into the neural plate and give rise to the brain and spinal column. When cut out after gastrulation and inserted into a region that is expected

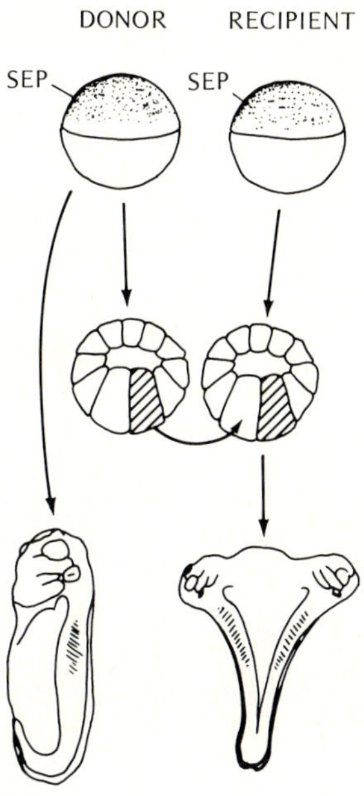

Figure 9.9. Induction of a second axis in *Xenopus* embryos. At the 64-cell stage a vegetal cell on the side opposite that of the sperm entry point (SEP) can be removed and used to replace a similar vegetal cell on the same side as the SEP of a recipient embryo. This dorsal vegetal cell *(cross-hatched)* induces invagination of recipient animal cells of the presumptive ventral side and differentiation of dorsal tissues. Transplantation of ventral vegetal cells to the same position had no consequences.

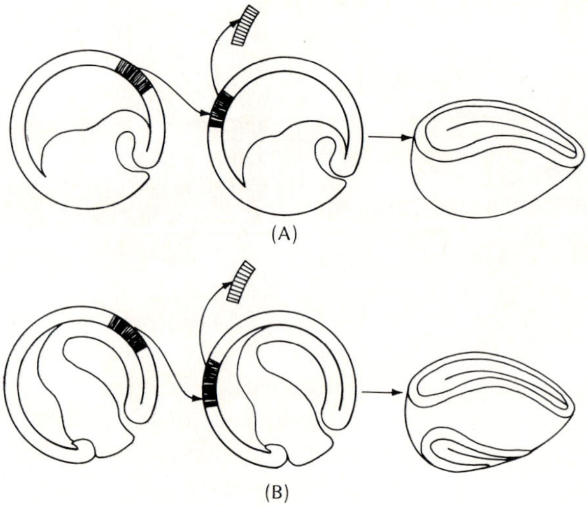

Figure 9.10. Regulation and determination in the newt. *(A)* Early in gastrulation transplanted neural ectoderm tissue regulates to fit in with differentiation of the host when put in a region fated to be epidermis, that is, it differentiates as epidermis rather than neural tissue. *(B)* Later in gastrulation the same operation results in differentiation of a second neural plate. At this stage the neural ectoderm is determined to differentiate as neural tissue.

to make lateral ectoderm, the grafted cells nevertheless differentiate into a neural plate. Although on the flank, they round up to make a neural tube. Thus, their fate has been irreversibly determined after gastrulation.

Transplantation of cells from the dorsal lip of a blastopore into the blastocoel of an early gastrula will often induce a second embryo (Fig. 9.11). These results show that even after invagination through a blastopore has started and cellular cleavage is almost completed, host cells can be redirected by altering the normal position of cells fated to be notochord. Regulation of fate at this late a stage in embryogenesis is common in vertebrates but usually not found in invertebrates.

Transplantation of the dorsal lip tissues was observed to organize a secondary axis about 60 years ago by Hilde Mangold in the laboratory of Hans

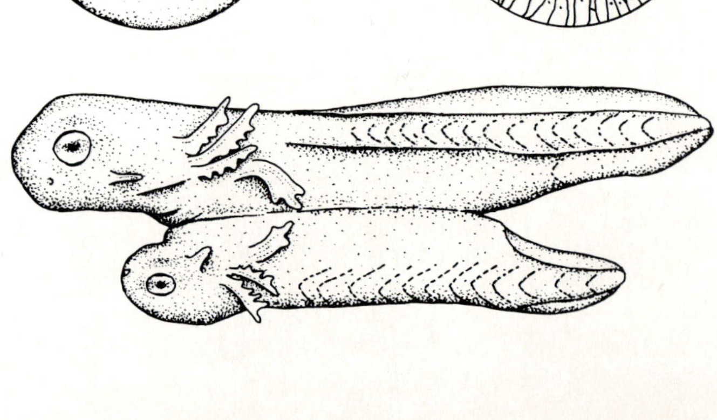

Figure 9.11. Dorsal lip of the blastopore. Cells taken from the dorsal lip of a newt embryo will induce cells of a recipient embryo to form a second axis when inserted into the blastocoel on the side opposite the host blastopore. Only some of the neural and mesodermal cells of the second embryo are derived from the donor tissues; others are host cells that have been redirected. Use of donor and host embryos with different nuclear markers allows the origins of the cells to be recognized with confidence.

Spemann. During the last half-century many attempts have been made to determine what is special about dorsal lip cells that results in secondary embryonic axis induction. The answer is that there is nothing very special about these cells. Many other treatments can be found that give secondary embryonic axis induction. Implantation of heat-killed cells, staining with certain dyes, and injection of inert substances, such as ash, can all induce secondary embryonic axes to one extent or another. It has been suggested that the insult itself sets off the chain of events resulting in a new axis.

The question of what establishes the dorsal/ventral polarity of the amphibian embryo might be better answered if we knew how vegetal blastomeres interacted with animal blastomeres back at the 64-cell stage. A few vegetal cells on the side opposite that of the sperm entry point (SEP) have the ability to allow nearby animal blastomeres to form dorsal rather than ventral tissue. The vegetal blastomeres do not themselves contribute progeny to the dorsal axis but develop as endoderm of the gut. Therefore, they must be signaling adjacent animal blastomeres to follow a pathway, leading through a blastopore, to dorsal fates. Since only those vegetal blastomeres on the presumptive dorsal side and not those on the presumptive lateral or ventral sides have the ability to induce a complete second axis, a dorsal/ventral polarity is clearly established in 64-cell embryos but can be changed by experimental manipulation. Only the few dorsal vegetal blastomeres seem to be set in their ways at this stage. These special cells receive axis-inducing properties during cortical rotation and cleavage of the egg cytoplasm long before transcription is activated in the embryo.

Since multiple axes can be induced shortly after fertilization by redistributing cellular components stratified by gravity, it looks as if in the normal course of events the blastula is formed in an unstable state where gastrulation at almost any point is easily set in motion. Normally, a single blastopore is initiated below the gray crescent and gives rise to a single archenteron. However, a variety of manipulations can trigger the formation of multiple embryonic axes.

The Notochord

Later in frog gastrulation the dorsal roof of the archenteron and overlying mesenchymal cells condense into a relatively tightly packed column of cells that will become the notochord. The notochord is used for structural support in the larvae of some chordates, but in frogs it disappears as a discrete structure later in development. However, in the early development of all vertebrates it carries out an essential step in the causal pathway of embryogenesis—it induces overlying ectoderm to form the neural plate (Fig. 9.12).

The cells of the notochord primordium become progressively compacted and surrounded by an extracellular sheath that contains collagen. At

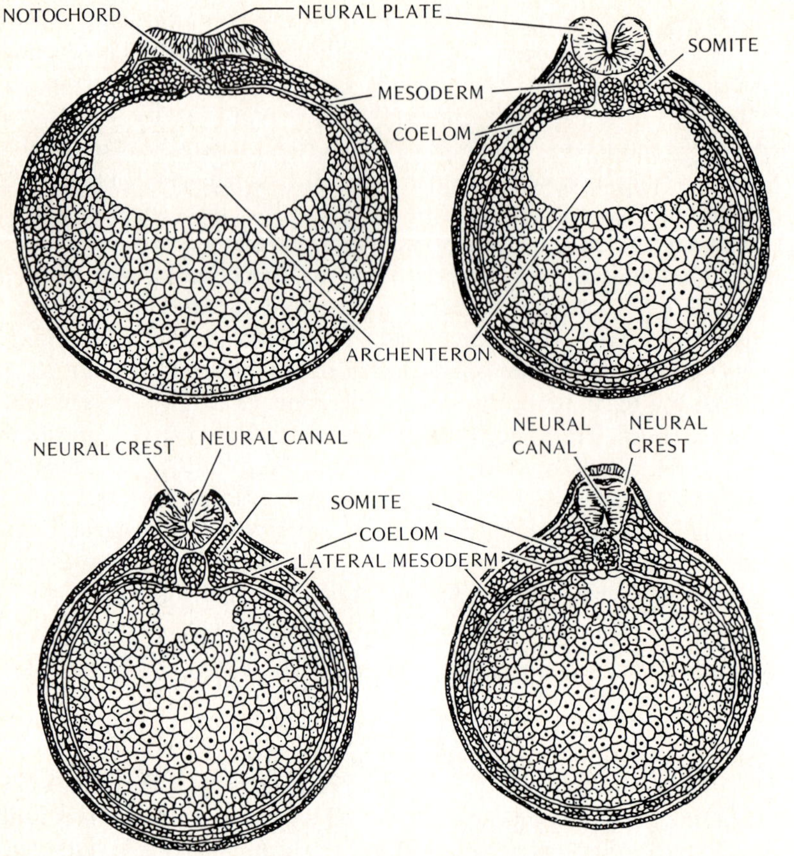

Figure 9.12. Amphibian notochord. Mesodermal cells on the dorsal midline condense into a notochord. A neural plate forms in the overlying ectoderm and somites condense in the lateral mesoderm.

this stage the notochord can be cut out and moved relatively intact to a new position in another host embryo. As long as the notochord is placed underneath an ectodermal layer, it will induce these cells to form a neural plate. The artificially induced neural plate will round up and form a secondary neural tube. Thus, a new embryonic axis has been formed by transplantation of the notochord. The induction is very rapid and dramatic at the morphological level, and the initial cause can be clearly assigned to the notochord. Other tissues taken from a gastrula do not induce the formation of a neural plate. Is the notochord putting out diffusing substances that instruct nearby ectoderm to form a neural plate or is it the mechanical consequence of the close presence of a notochord, or both? Unfortunately, the size of the notochord presents too little tissue for present methods of biochemical and molecular analysis. However, it is strongly felt by many that an understanding of the causal interactions between notochord tissue and overlying ectoderm tissue would be a major advance in the analysis of embryogenesis.

Gastrulation in Chick Embryos

The blastocoel of a chick embryo is a flattened space between the outer layer of cells and the underlying cells that rest on the great mass of yolk. The blastodisk is shaped somewhat like a shield with a broad anterior section and a narrower posterior section. Gastrulation starts at the posterior

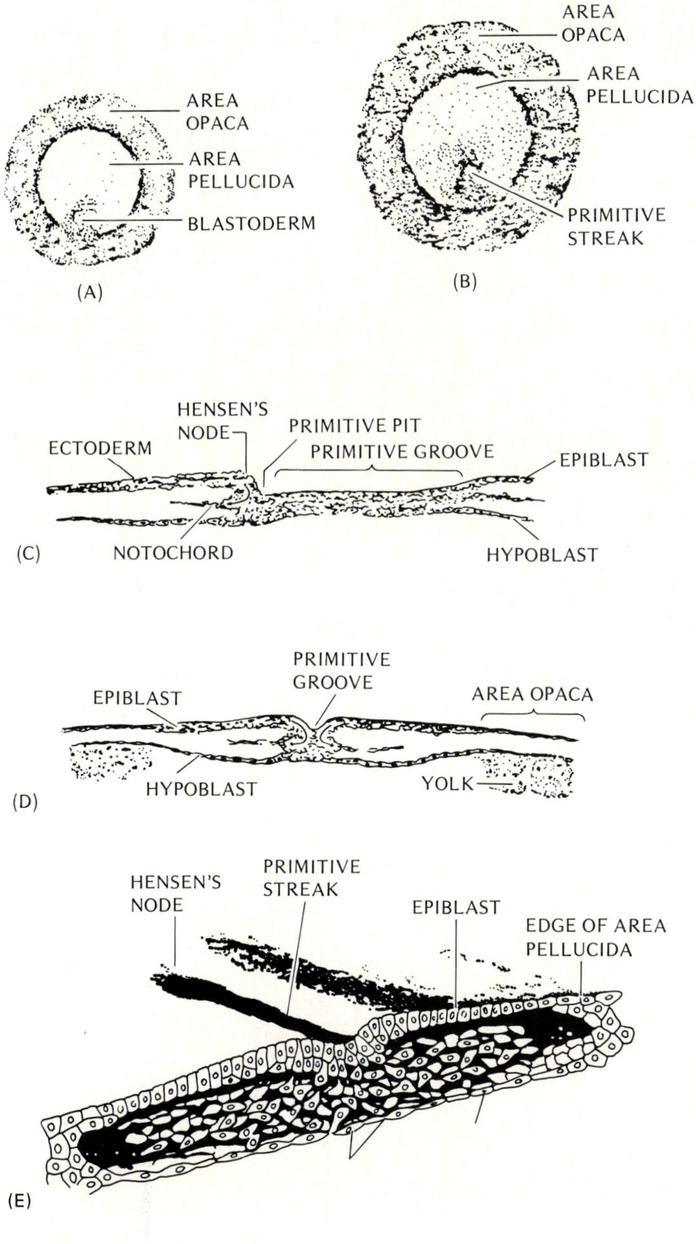

Figure 9.13. Gastrulation in chicks. *(A)* After 3 hours incubation, the cells thicken in a quadrant of the blastoderm. The area opaca contains cells not yet fully enclosed by membranes. *(B)* By 10 hours the accumulated cells converge to form the primitive streak. *(C)* Seen in section, a 17-hour-old chick embryo has a primitive groove ending at Hensen's node. Cells invaginate and spread between the ectodermal and endodermal layer. *(D)* A transverse section at this stage shows that cells move laterally from the primitive groove as well. *(E)* Migrating cells fill most of the blastocoel. They appear to drop off the epiblast at the primitive streak.

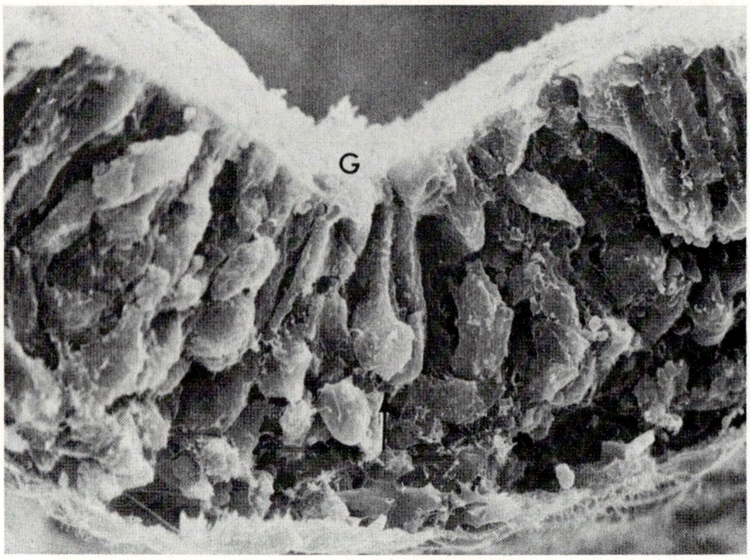

Figure 9.14. Chick bottle cells. Gastrulation in chicks also results in the formation of bottle cells at the invagination groove (G). (SEM ×2100).

end and advances as a streak toward the anterior. Cells move from the sides of the shield to the midline, where they drop into the blastocoel and then migrate laterally and toward the anterior (Fig. 9.13). As the primitive streak indents, cells along the midline take on the bottle shape that was first noticed in amphibian gastrulae (Fig. 9.14). The mechanism of formation of the primitive streak in chick embryos may be similar to that which forms the blastopore in frog embryos. All along the primitive streak cells lose their adhesion to the ectodermal layer and enter the blastocoel as loosely attached cells. Some of them reassemble under the midline to form the notochord between the underlying endodermal layer and overlying ectodermal layer.

Seen from outside, the indentation of the primitive streak proceeds to a point about two-thirds the way across the blastodisk. Movement of the

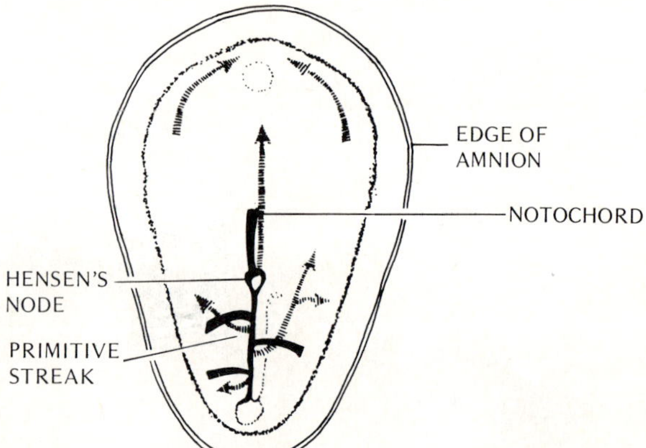

Figure 9.15. Movement through the primitive streak. Hensen's node marks the anterior end of the primitive streak. Cells enter the blastocoel at the streak and move laterally and to the anterior. The notochord forms from cells that move directly anterior along the midline. Growth of the blastodisk ahead of Hensen's node expands the size of the embryo.

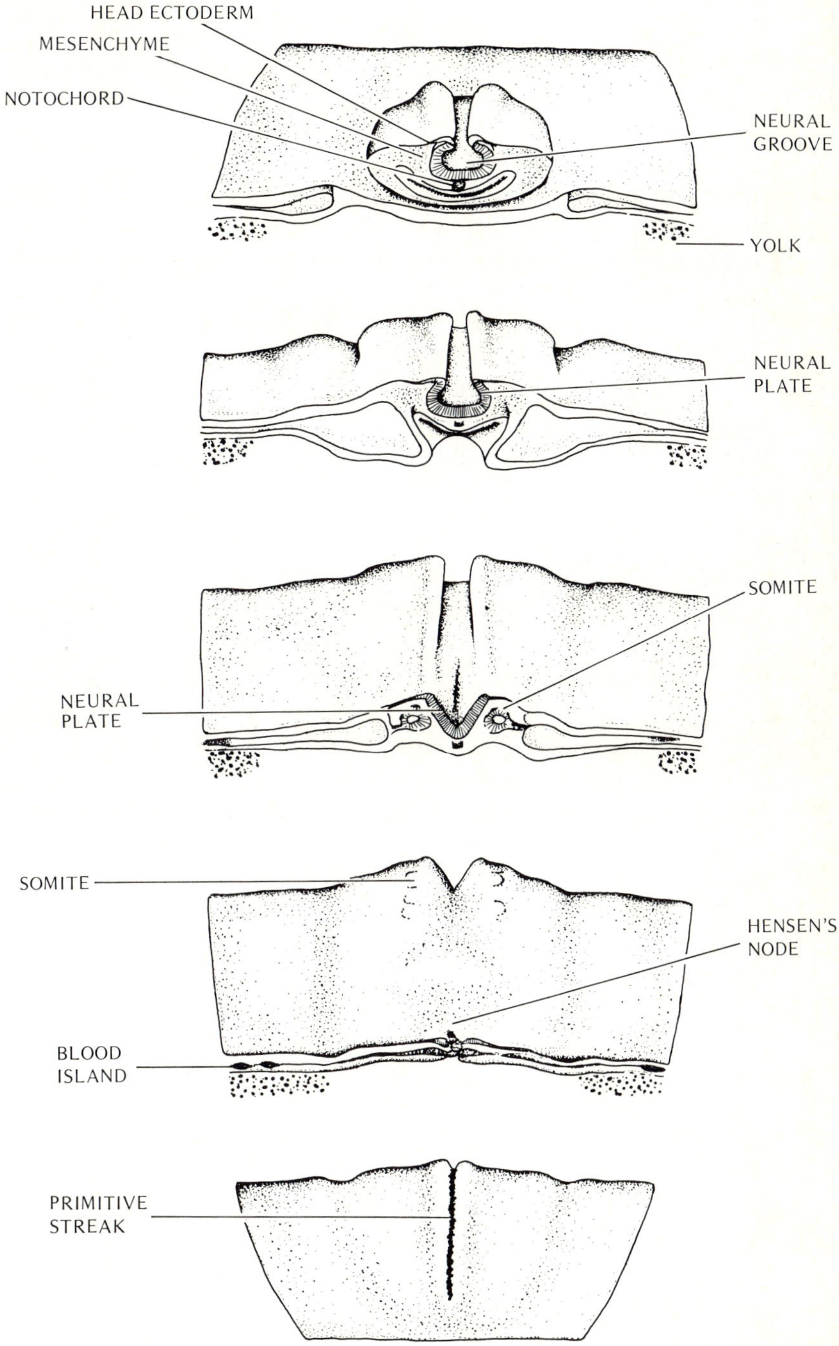

Figure 9.16. One-day-old chick embryo. Sections drawn at various positions along the embryonic axis show tissue differentiation more advanced in the anterior than in the posterior at one day of development. Hensen's node is near the posterior now since most of the growth of the embryonic tissue occurred ahead of it. Gastrulation is almost complete by this stage.

streak then appears to reverse and proceed from anterior to posterior. In fact, the whole embryo is growing rapidly using the nutrients derived from the yolk. At the anterior end of the primitive streak there is a thickening of cells just before the regressive movement is seen. Cells in this thickening, referred to as Hensen's node, become separated from the ectodermal cells just ahead of them by growth of embryonic tissue. As the embryonic axis elongates, this separation increases until Hensen's node lies near the posterior of the much larger embryo. Unless care is taken to determine the frame of reference, Hensen's node appears to be traveling rapidly toward the posterior of the embryo (Fig. 9.15).

The new embryonic tissue that grows just ahead of Hensen's node is directed by the underlying notochord formed during gastrulation. Almost as soon as the notochord appears the ectodermal tissue over the midline differentiates into a neural plate that soon forms the neural tube (Fig. 9.16). Mesodermal cells that entered anterior of Hensen's node during gastrulation also proliferate and then condense into clearly visible blocks of cells called somites, one on each side of the midline. As Hensen's node becomes relatively closer to the posterior of the elongating embryo, more and more pairs of somites condense from the mesodermal cells. The first somites formed are those of the anterior, and somite formation proceeds in a very regular manner toward the posterior. Many other tissues are differentiating even while the pairs of somites are forming. These include the optic vesicles of the brain and the heart primordium. Stages in chick development use the number of easily countable pairs of somites as a reference.

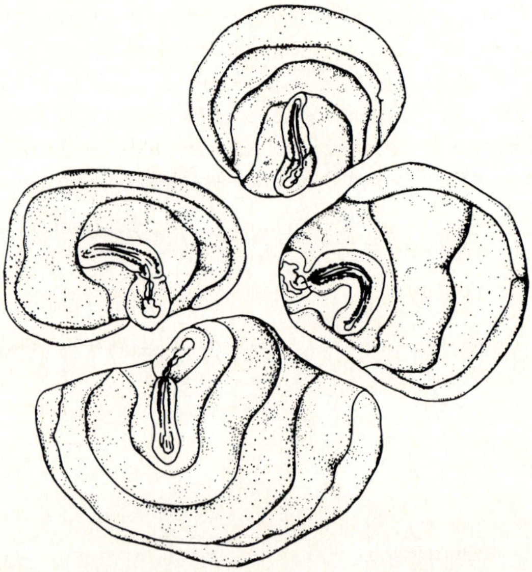

Figure 9.17. Quartered chick blastoderm develops four embryonic axes. A blastoderm was cut with a fine glass needle before a primitive streak had formed. Each piece formed a primitive streak, gastrulated, and differentiated somites.

Although the primitive streak normally forms at a single position in the narrow end of the blastodisk and elongates as a streak toward the broader end, gastrulation can occur from any part of the blastodisk. Cutting a blastodisk into four approximately equal pieces before the formation of the primitive streak results in four independent primitive streaks. In fact, each quarter develops into a separate early embryo complete with optic vesicles, spinal cord, and flanking somites. The heads of these small embryos all point toward the center of the quartered blastodisk (Fig. 9.17).

This fairly simple experiment demonstrates that cells of the early blastodisk can differentiate into almost any tissue under the influence of a primitive streak. The primitive streak forms near the growing periphery of the blastodisk and elongates toward the center, where the cells formed earlier and the blastocoel is more definitely formed. Once a primitive streak has formed it inhibits the formation of secondary streaks within a continuous ectodermal layer. However, if the ectodermal layer is surgically cut into quadrants, each piece forms a primitive streak, undergoes gastrulation, and develops normally at least up to the 20-somite stage.

Gastrulation in Mammalian Embryos

The evolution of mammals from organisms similar to reptiles has left its traces in the steps of early embryogenesis. The large yolky eggs of reptiles undergo meroblastic cleavage to form a blastodisk much as in chick embryogenesis. Gastrulation proceeds through a primitive streak into the flattened blastocoel. Mammalian gastrulation also proceeds through a primitive streak in which cells drop off an ectodermal layer into a thin space above an endodermal layer of cells. However, mammalian embryos have very little yolk and cleavage is holoblastic. Just looking at the blastula we would expect gastrulation to proceed as in sea urchins or *Amphioxus*. However, the trophodermal tissue of mammalian blastulae forms only the extra embryonic placental tissue. The embryo proper develops from the dozen or so cells of the inner cell mass.

The inner cell mass grows to give rise to a multilayered mass extending into the blastocoel and then delaminates to give rise to an amniotic cavity over the embryonic plate and the empty yolk sac beneath it. Each of these cavities is roofed over by cells derived from the inner cell mass, but these cells will not be included in the individual that develops (Fig. 9.18).

Although the fertilized egg has developed a blastocoel in which cells are positioned, no organized process similar to gastrulation in other organisms directs the inner cells. The inner cell mass flattens to form the embryonic plate. A cavity forms between the ectodermal layer on the side of the amnion and the endodermal layer on the yolk sac side. This cavity is analogous to the blastocoel of chick embryos. A primitive streak can be seen to start near the edge of the embryonic plate in the epiblast layer. Lateral cells

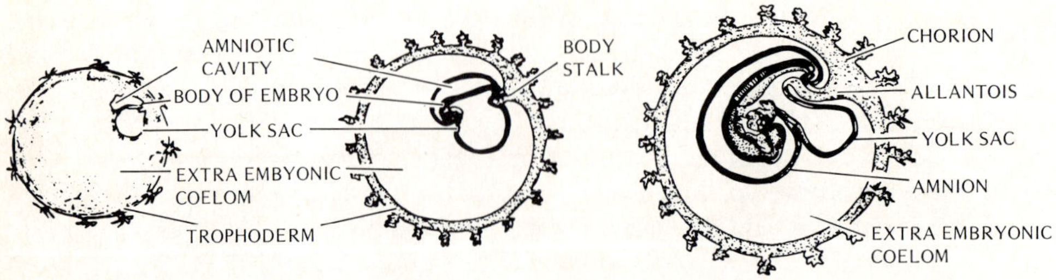

Figure 9.18. Early mammalian embryo. The body of the embryo develops from a two-layered sheet of cells separating the amniotic cavity from the yolk sac. The embryo is suspended in the extra embryonic coelom by the body stalk attached to the trophoderm. A primitive streak forms in the epiblast of the body of the embryo. Gastrulation proceeds through the primitive streak as in avian embryos. A neural tube flanked by somites forms the embryonic axis.

are drawn to the streak as cells drop into the cavity. Once inside, these mesodermal cells migrate laterally and toward the center of the plate. Just as in chick embryos, the primitive streak elongates toward the center and the mesodermal cells condense under the midline to form the notochord. A structure similar to Hensen's node can be seen when the primitive streak is about halfway across the embryonic plate. Further growth occurs predominantly anterior of Hensen's node, soon leaving it in a progressively posterior position. Ectodermal cells overlying the notochord are induced to form a neural plate and lateral mesodermal cells condense into pairs of somites flanking the neural tube, which rounds up between them. The next few stages in mammalian embryos are similar to those in chick embryos, and both clearly are processes that first arose in reptiles, in which almost identical processes of early embryogenesis occur.

Homologies in Gastrulation

Gastrulation takes place after the period of rapid cell division has generated the majority of the cells of an early embryo. It is a stage of dramatic movement of the cells relative to each other. In embryos of all the organisms we have discussed there is a unique position at which cell migration occurs—the blastopore in sea urchins and frogs; the primitive streak in birds and mammals. The position of the blastopore in sea urchins is at the vegetal pole. In frogs the blastopore is as close to the yolky end as is compatible with invagination. It occurs just below the floor of the blastocoel among yolky cells. In birds and mammals the primitive streak first forms near the margin of the blastodisk in proximity to cells that can supply nutrients from metabolism of yolk or from placental sources. Thus, a characteristic of the blastopore is that it is formed from cells well supplied with nutrients.

The blastopore is small and stays localized in sea urchin gastrulae. It is initially large and then closes in the protochordate *Amphioxus*. It spreads to form a circle in frogs, and it extends along the embryonic axis as the primitive streak in birds and mammals. In each case, some cells at the blastopore change their adhesive properties so that they can dissociate from the ectodermal layer and enter the central cavity. Other cells enter the blastocoel in a sheet continuous with the external ectodermal layer of cells in sea urchins and frogs. In birds and mammals the cells invade individually, but the endodermal tissue in these organisms was already laid down in the blastodisk.

Some of the mesodermal cells condense into groups in all cases, but a notochord is formed only in chordates. In sea urchins the primary mesenchymal cells give rise to the skeletal structures of the larvae. In birds and mammals the mesenchymal cells form many structures, including the bilaterally symmetric somites. The somites give rise to the lateral muscles and skeletal elements such as ribs.

Gastrulation transforms the relatively simple blastula and blastodisk structures into complex multilayered embryos. It proceeds in an ordered fashion to establish the polarity of the embryonic axis from the posterior to the anterior ends. Conceptually, there are several other ways to generate a multilayered embryo such as the unlocalized shedding of cells into the blastocoel. But this is never observed in any organism. Perhaps the process of cell migration itself during gastrulation is essential to give positional information to the embryo. The cells that lead through the blastopore will differentiate as anterior structures and will entrain those around them to do likewise. The last cells to invaginate will differentiate as posterior structures. Thus, the temporal polarity of gastrulation is transformed into the spatial polarity of the embryo.

The geometry of gastrulation changed dramatically as larger eggs containing more yolk evolved. From a simple pore, the site of invagination has evolved into a streak. This might seem a radical change, but a transition strategy has been found. The early embryogenesis in an amphibian of the Ecuadoran jungle was recently observed in detail. These hylid frogs lay large eggs abundantly filled with yolk. Cleavage, although slow, is complete, that is, holoblastic as in other amphibians. Gastrulation also proceeds as expected through a blastopore to form the archenteron. However, the embryo develops from a disk of cells left on the dorsal surface following gastrulation and not from the internal cells. The structures formed within the blastocoel remain extraembryonic and only provide nutrients to the embryonic disk on the surface. They function in a manner similar to the extraembryonic tissues of chicks and mammals. Subsequent development of the embryos of these hylid frogs has not yet been studied in detail but appears to proceed much as in chick embryos. A notochord forms beneath the thin outer layer of cells of the disk and is followed by differentiation of a neural tube with flanking somites. The posterior is found near the site of the initial blastopore toward the vegetal pole of the egg. It is likely that further observation will show that a primitive streak directs a second gastrulation among the cells of the embryonic disk. Evolution from this strategy to that used by birds and mammals does not seem such a big step.

Summary

As the blastula forms, certain cells acquire altered adhesive properties. At gastrulation they leave the sheet of cells on the surface and invade the blastocoel. In many species these invasive cells become elongated into bottle shapes before they break loose. Gastrulation in echinoderms is initiated by the invasion of primary mesenchymal cells that are the descendants of 16-cell-stage micromeres. They enter at the vegetal pole and are followed by invagination of the sheet of cells at the vegetal pole. Gastrulation in primitive chordates is topologically similar. In amphibians the invaginating sheet follows the dorsal roof of the blastocoel. Gastrulation in birds and mammals occurs along the length of the primitive groove that stretches from posterior to anterior. Cells move to the streak, sink inwards under the epiblast, and then migrate laterally and to the anterior.

The point of invagination is not set in unfertilized amphibian eggs. However, as soon as the sperm entry point can be seen, it can be predicted with some confidence that the blastopore will form on the opposite side. If the fertilized egg is rolled over or a ventral blastomere is exchanged for a dorsal blastomere in the vegetal hemisphere, a blastopore will form on the same side as the sperm entry point. Cells become progressively determined as embryogenesis proceeds. Once the notochord has formed under the dorsal epithelium, the neural plate is induced and will form the neural tube no matter where it is placed on an early embryo.

SPECIFIC EXAMPLE

Gene Expression at Gastrulation

Several different molecular techniques have been used to try to determine the number of new genes expressed at various stages in development. Since the morphological compexity of an embryo continuously increases during embryogenesis, it was initially thought that the number of genes expressed would have to increase as well. However, it has turned out that only a few specific genes are activated at specific stages of embryogenesis.

Different mRNA molecules might be expected to hybridize to different portions of the genome. Therefore, the proportion of the genome that would form DNA–RNA hybrids with RNA from different stages should indicate the number of genes that are active at any given stage. However, this approach was soon found to have serious problems in eukaryotic organisms because portions of many mRNA molecules were shown to be complementary to many different regions of the DNA. A specific mRNA not only hybridized with the gene that coded for it but also with DNA sequences that do not code for any transcripts as well as with genes that code for related but distinct RNAs (Fig. 9.19). There are repeated sequences of 100 to 1000 bases interspersed with unique sequence DNA throughout all eukaryotic genomes. Some of these sequences are transcribed into RNA. The families

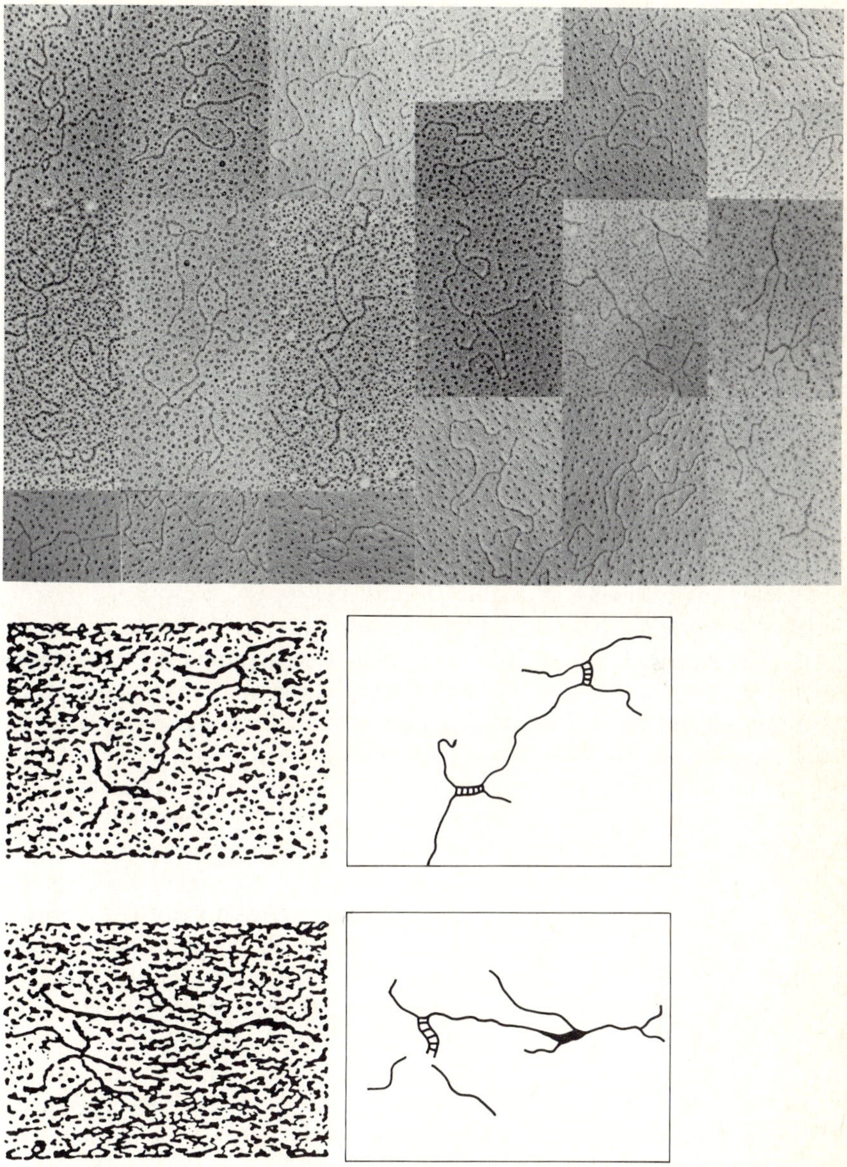

Figure 9.19. Reannealing of repeated sequences. When double-stranded DNA fragments are separated into single strands a few thousand bases long and then allowed to reanneal for a short period of time, double-stranded regions reform by base pairing that links many molecules to each other. The double-strand regions are a few hundred bases long and are separated by single-stranded regions that are about a thousand bases long. The regions that reanneal at low C_0t (initial DNA concentration × time) are present many more times in the genome than the sequences that remain single stranded. Double- and single-strand regions can be seen in electron micrographs when the molecules are coated with protein. The DNA molecules in this micrograph were isolated from the amphibian *Xenopus*. The proportion of repetitive DNA varies in different eukaryotes from 30 to 90 percent. Much of it appears to serve no useful function.

of repeat sequences include transposons that appear to favor their own replication and insertion in novel genetic places (selfish DNA). They also include duplicated portions of genes that are carried along as harmless baggage (junk DNA). Some sequences function in several different genes and evolved from a common ancestral gene (shared exons). These and other problems make the interpretation of bulk hybridization results difficult.

Various techniques have been developed to overcome these problems, but none is fully satisfactory. Genomic DNA can be sheared into fragments only a few hundred bases long to try to separate most of the repeat sequences from the unique regions. Hybridization at low DNA concentration for a short period (low C_0t reannealing) favors hybrid formation among repeated sequences since the probability that such fragments will find each other is greater than for unique sequences. However, the last members of a repeat family will remain with the unhybridized unique sequences. Nevertheless, most repeat sequences become double stranded and can be removed. Remaining single-stranded DNA can be hybridized to mRNA of a specific stage—for instance, gastrulation—and the DNA–RNA hybrids can be isolated. These DNA sequences are enriched in those expressed at a given stage.

Using sea urchin DNA sequences enriched for those expressed at gastrulation to analyze mRNA transcripts at other stages has indicated that most gastrula sequences are expressed during oogenesis and the mRNAs are still present during cleavage and blastula stages. These transcripts persist even after gastrulation and are found in the pluteus. Thus, the global picture of transcription does not shed much light on the subtle changes going on in cells as they diverge at gastrulation. Perhaps it is not surprising that most genes are active throughout embryogenesis, since they would be expected to be needed for nuclear and cellular replication, membrane formation, and other housekeeping functions. A change in expression in only a few genes could have dramatic consequences for subsequent morphogenesis and physiology, but more specific approaches are needed to recognize these determinative genes.

Detailed analysis of specific cloned sequences has provided probes for genes transcribed at unique periods in development. There are at least six genes that code for actin in the sea urchin genome. They can be distinguished by using DNA from the 3′ untranslated portion of the transcription units. These terminal sequences do not code for actin and so have diverged to the point that they will not cross-hybridize. Actin mRNA is at very low abundance during the first 10 hours of development. These mRNAs start to accumulate at the blastula stage (14 hours) and are prevalent following gastrulation (40 hours) (Fig. 9.20).

The CyIIa actin gene is activated at 14 hours in the vegetal plate prior to invagination. Its mRNA accumulates in secondary mesenchymal cells that develop into the larval gut. The CyIIa mRNA cannot be observed in cells other than those of the archenteron. This is the earliest cell-type-specific marker known for the cell lineage leading to gut tissue.

The majority of the actin mRNA is a 1.8 kb transcript from the CyIIIa gene. This gene is expressed in gastrulae and pluteus larvae but not in adult tissues. The CyIIIa mRNA increases about 50-fold during gastrulation and

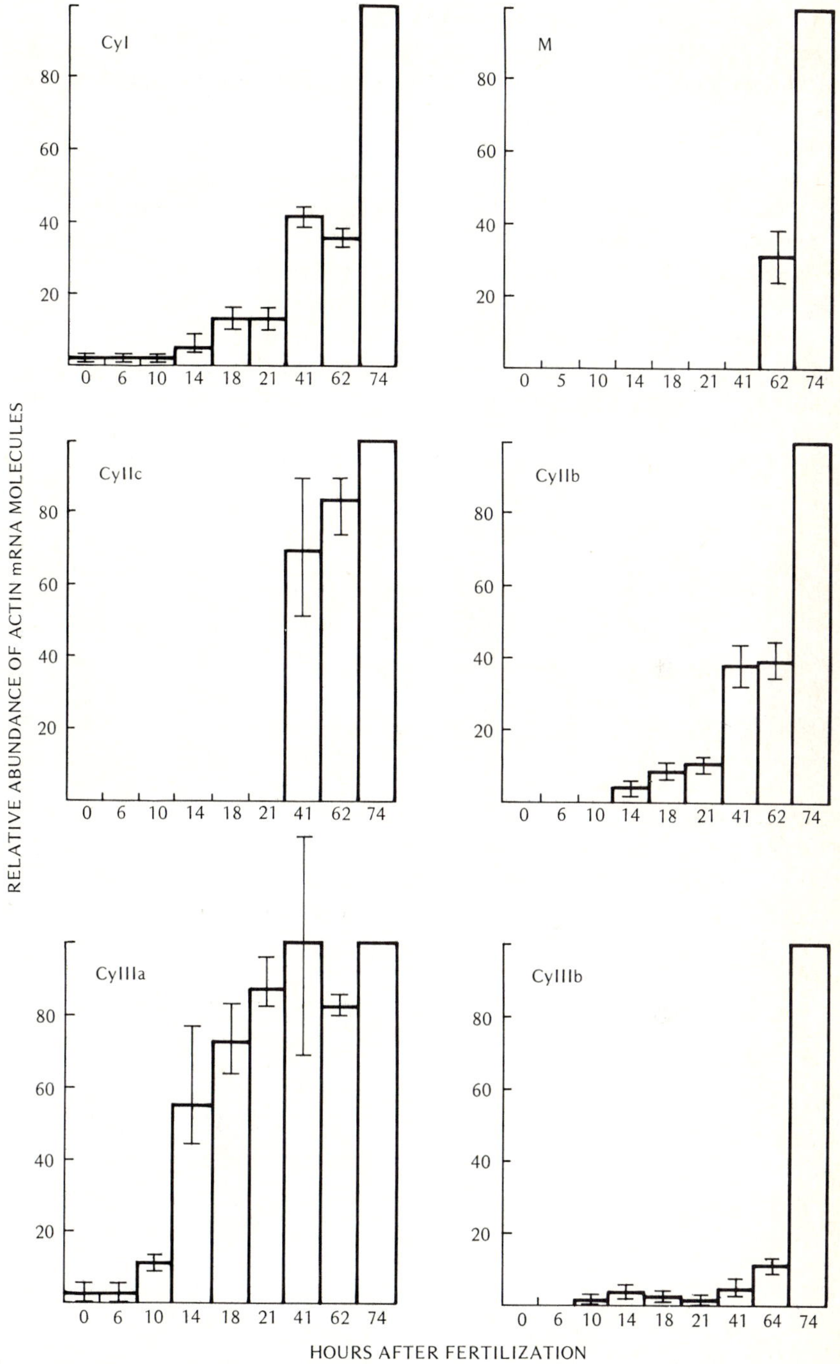

Figure 9.20. Temporal expression of actin genes during sea urchin development. RNA was isolated at various times after fertilization and size separated. Northern blots were probed with labeled sequences from the 3′ untranslated portions of six different actin genes. The amount of probe hybridized at each stage is presented in relation to the maximum level observed for that probe.

is found predominantly in dorsal ectoderm cells. There are about 400 such cells in late blastulae. During gastrulation they change from cuboidal to squamous epithelial cells as the result of contraction of actin containing microfilaments. Contraction of these microfilaments is triggered by elevated intracellular free calcium.

Sea urchin genes coding for calcium-binding proteins of about 15 kd are also specifically expressed in dorsal ectoderm cells at gastrulation. Three closely related genes, termed Spec 1 a, b, and c, can be distinguished by untranslated 3′ regions of their respective mRNAs. The transcripts are 1.5 kb in length and begin to accumulate at the blastula stage in predorsal ectoderm cells. Maximum accumulation occurs late in gastrulation and reaches a level at least 100-fold greater than that in cleavage-stage embryos. The rate of synthesis of the 15 kd proteins they code for increases as a result of the

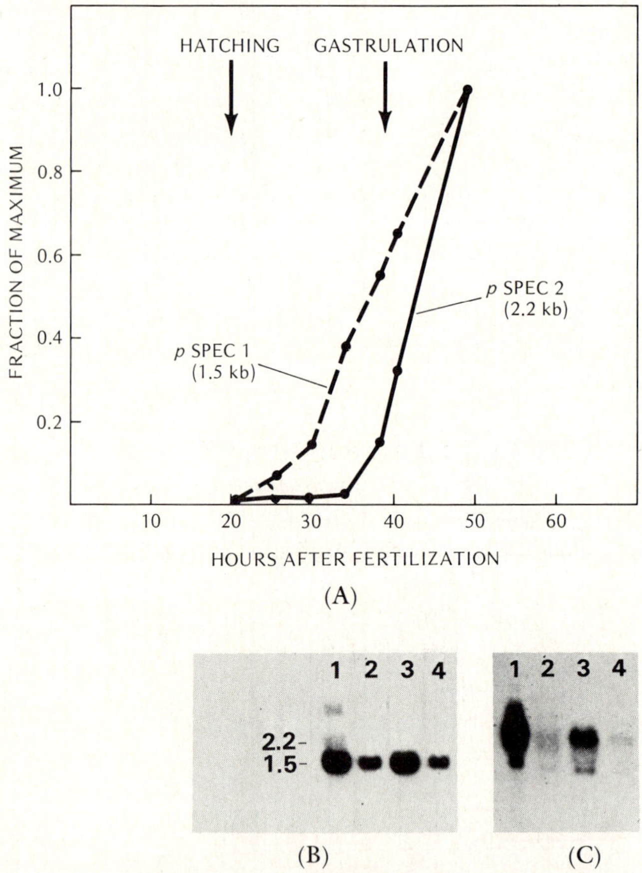

Figure 9.21. *(A)* Accumulation of Spec 1 and Spec 2 mRNA during sea urchin development. RNA of various stages was analyzed by hybridization to labeled clones of genes for the 15-kd calcium-binding proteins. Both Spec 1 *(B)* and Spec 2 *(C)* mRNA accumulate in gastrulae but these genes are not coordinately regulated. The mRNAs were found on Northern blots at higher levels in pluteus and gastrula ectoderm (lanes 1 and 3) than in pluteus or gastrula endoderm/mesoderm (lanes 2 and 4).

appearance of these mRNAs. The levels of Spec 1 mRNAs decrease in the pluteus larvae. A related set of genes, termed Spec 2, are expressed about 10 hours after the Spec 1 genes (Fig. 9.21). Spec 2 genes, of which there are at least three, are transcribed into 2.2 kb mRNAs that also code for calcium-binding proteins of about 15 kd. However, Spec 2 mRNA is only a tenth as abundant as those coded for by Spec 1 genes.

These analyses of specific genes point out the difficulties of recognizing the transcription patterns by global RNA hybridization techniques. The actin mRNAs all cross-hybridize within the protein-coding region and would not be distinguished unless 3′ untranslated sequence probes were used. The same is true of the Spec genes. However, using specific probes, it is clear that none of the actin genes are coordinately expressed with other members of the family and that Spec 1 genes are regulated independently of Spec 2 genes. All of these genes are preferentially activated in dorsal ectoderm cells and may play roles in the specialization of these cells for the morphological changes they undergo at gastrulation as well as later during metamorphosis. Only as more developmentally regulated genes are analyzed in depth will a clear picture of transcriptional control of embryogenesis come into focus.

There are about 10,000 different mRNAs in the cytoplasm of sea urchin embryos. Only about 5 to 10% of these are cell lineage-specific. Thus, expression of only about 1000 genes, starting at the late blastula stage and continuing until terminal differentiation of the larval tissues, appears to be sufficient for cell-type divergence throughout sea urchin embryogenesis.

Related Readings

Angerer, R., and Davidson, E. (1984). Molecular indices of cell lineage specification in sea urchin embryos. Science *226,* 1153–1160.

Black, S., and Gerhart, J. (1985). Experimental control of the site of embryonic axis formation in *Xenopus laevis* eggs centrifuged before first cleavage. Devel. Biol. *108,* 310–324.

Boucaut, J., Darribere, T., Poole, T., Aoyama, H., Yamada, K., and Thiery, J. P. (1984). Biologically active synthetic peptides as probes of embryonic development: a competitive peptide inhibitor of fibronectin function inhibits gastrulation in amphibian embryos and neural crest cell migration in avian embryos. J. Cell Biol. *99,* 1822–1830.

Carpenter, C., Bruskin, A., Hardin, P., Keast, M., Angstrom, J., Tyner, A., Brandhorst, B., and Klein, W. (1984). Novel proteins belonging to the troponin C superfamily are encoded by a set of mRNAs in sea urchin embryos. Cell *36,* 663–671.

Carson, D., Farach, M., Earles, D., Decker, G., and Lennarz, W. (1985). A monoclonal antibody inhibits calcium accumulation and skeleton formation in cultured embryonic cells of the sea urchin. Cell *41,* 639–648.

del Pino, E. M., and Elinson, R. P. (1983). A novel developmental pattern for frogs: gastrulation produces an embryonic disk. Nature *306,* 589–590.

Fink, R., and McClay, D. (1985). Three cell recognition changes accompany the ingression of sea urchin primary mesenchyme cells. Devel. Biol. *107,* 66–74.

Gerhart, J., Black, S., and Scharf, S. (1983). Cellular and pancellular organization of the amphibian embryo. Modern Cell Biol. *2,* 483–507.

Gerhart, J., Ubbels, G., Black, S., Hara, K., and Kirschner, M. (1981). A reinvestigation of the role of the grey crescent in axis formation in *Xenopus laevis.* Nature *292,* 511–516.

Gimlich, R., and Gerhart, J. (1984). Early cellular interactions promote embryonic axis formation in *Xenopus laevis.* Devel. Biol. *104,* 117–130.

Gurdon, J., Fairman, S., Mohun, T., and Brennan, S. (1985). Activation of muscle-specific actin genes in *Xenopus* development by an induction between animal and vegetal cells of a blastula. Cell *41,* 913–922.

Holtfreter, J. (1943). Properties and functions of the surface coat in amphibian embryos. J. Exp. Zool. *93,* 251–323.

Holtfreter, J. (1944). A study of the mechanics of gastrulation. J. Exp. Zool. *95,* 171–212.

Keller, R. (1980). The cellular basis of epiboly: an SEM study of deep-cell rearrangements during gastrulation in *Xenopus laevis.* J. Embryol. Exp. Morph. *60,* 201–234.

Malacinski, G. (1983). Sperm penetration and the establishment of the dorsal/ventral polarity of the amphibian egg. In *Biology and fertilization,* C. Metz and A. Monroy (eds.). Academic Press, San Diego.

Shott, R., Lee, J., Britten, R., and Davidson, E. (1984). Differential expression of the actin gene family of *Strongylocentrotus purpuratus.* Devel. Biol. *101,* 295–306.

Slack, J. (1984). *In vitro* development of isolated ectoderm from axolotl gastrulae. J. Embryol. Exp. Morph. *80,* 321–330.

Steinberg, M. (1970). Does differential adhesion govern self-assembly processes in histogenesis? Equilibrium configurations and the emergence of a hierarchy among populations of embryonic cells. J. Exp. Zool. *173,* 395–434.

CHAPTER 10

Neurulation Organogenesis

The mark of all vertebrates is a dorsal nerve cord. It develops from the neural plate induced in the dorsal ectoderm by the notochord. Initially the neural plate looks, as one would expect from its name, like a flat plate slightly raised from the lateral ectodermal sheet. The next stage in the differentiation leading to the spinal cord is the rounding up and fusion of the edges to form a hollow tube (Fig. 10.1).

In the electron microscope, cells in a neural plate do not appear to be very different from surrounding cells. They form a sheet of columnar cells that are tightly bound to each other by tight junctions and desmosomes. At the outer edge there are indications of a concentration of actin microfilaments running across the axis of the body. When the neural plate rounds up, the outer edge of the cells in the midline contracts, giving rise to wedge-shaped cells (Fig. 10.1).

Contraction of the microfilaments plays an essential role in the rounding up of the neural plate. It can be specifically inhibited by treatment with cytochalasin, a drug that binds to microfilaments and blocks their function. Under these conditions the neural plate remains as a flat sheet and the neural tube does not form.

It has been suggested that tension on the array of microfilaments might trigger their contraction. If so, then once a few cells in a neural plate contract, the contraction will spread to adjacent cells and lead to a rounding up of the tube. The localization of the microfilaments to the outer ends of neural

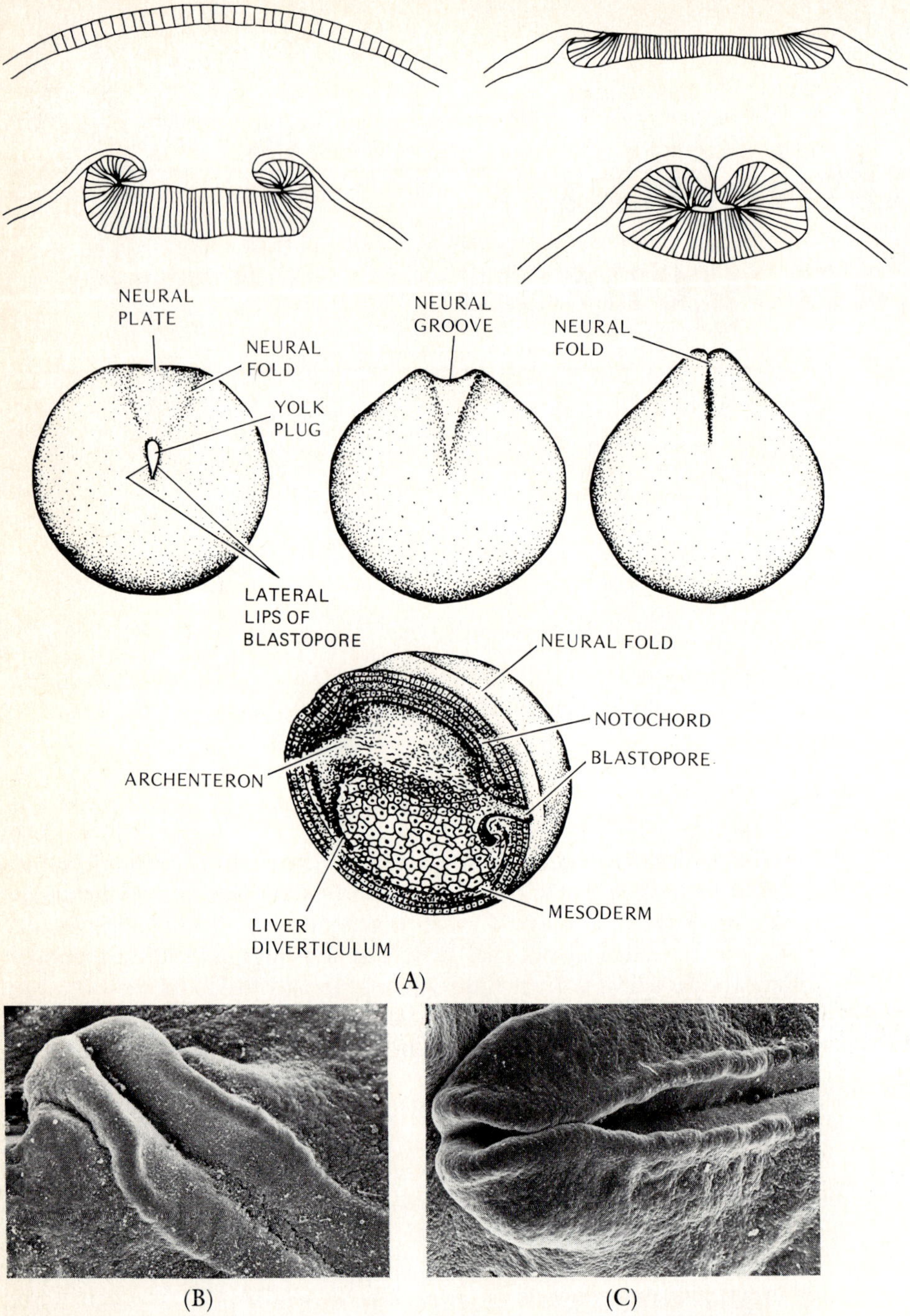

Figure 10.1. Neurulation. *(A)* Cells of the neural plate contract and take on a columnar shape. Further contraction of the external surface curls the plate. The neural plate covers the dorsal surface anterior to the blastopore in amphibian embryos. Neurulation results in clearly observable neural groove and neural fold. In cross section the neural plate can be seen to lie directly above the notochord. *(B)* In the chick embryo the neural groove is raised at the anterior. *(C)* Later the head fold closes.

plate cells ensures that the curling will be outward. This model has been shown mathematically to be able to account for the progressive rolling up of the neural plate starting in the head process and passing toward the posterior. Whatever the mechanism is, it is clear that once a neural plate is formed, it has an innate tendency to roll up independently of surrounding tissue. The neural plate can be excised free of the rest of the embryo and cultured in isolation. Nevertheless, the sheet of cells rounds up and forms a tube.

Neural Crest Cells

When the edges of the tube meet, cells just below the dorsal surface bind to each other and close the tube. A change in extracellular material coating these cells appears to play some role in this fusion of the edges. The neural tube now lies just beneath the dorsal ectoderm surface and runs the length of the embryo's axis. A few cells are left above the fusing edges of the neural tube. These cells, referred to as neural crest cells, have an interesting and complex fate (Fig. 10.2).

Neural crest cells leave their position at the top of the neural tube and migrate extensively throughout the embryo. Even while they are migrating they profilerate, and once they have found an available site of arrest they proliferate further. Many of the tissues of the final embryo are derived from these neural crest cells. Once they leave the neural crest, these cells lose much of the surface glycoprotein, N-CAM, implicated in cell–cell adhesion. This may facilitate movement through surrounding tissue. Later, when in place, the amount of N-CAM on the surface increases again, perhaps to allow adhesion at the new position.

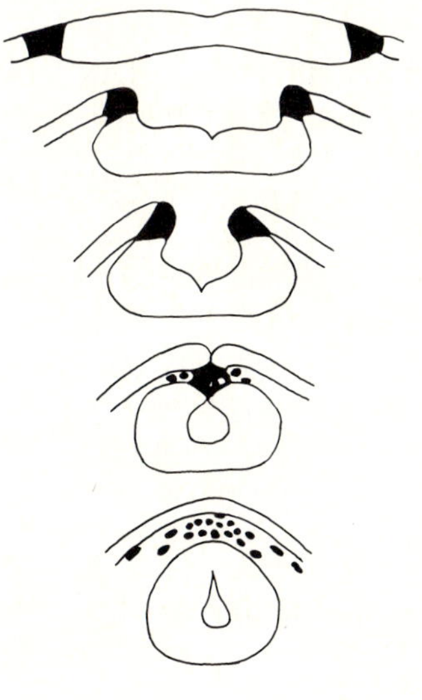

Figure 10.2. Neural crest cells come from the margins of the neural plate. As the edges of the neural tube fuse, the neural crest cells separate from the overlying ectoderm. From the dorsal midline they migrate to positions throughout the head and body.

The pathways of migration are complicated and appear to be determined by the surroundings rather than the ontogeny of the neural crest cells. Thus, if neural cells from the region of the head are put in the trunk region, they migrate to the positions normally occupied by neural crest cells of the trunk. The reverse operation results in trunk neural crest cells migrating to the positions usually filled by neural crest cells from the head. The exact nature of the directions provided by the surrounding tissues is unknown, but extracellular matrices and landmark structures are good guesses.

The list of tissues that are derived at least in part from neural crest cells is surprisingly long. It includes neural structures such as autonomic ganglia, jugular ganglia, facial root ganglia, and Schwann and sheath cells. It also includes many structures of the head including bones, cartilage, striated muscle, and dermis of the face and neck as well as fibroblasts and endothelium of the cornea and melanocytes of the iris in the eye. These tissues do not seem to have much in common except, perhaps, a strong selective advantage to an organism that has recently evolved a spinal cord. Positioning complex sensory organs at the anterior of an organism served by a central nervous system would be advantageous in seeking prey and avoiding obstacles and predators. The facial structures that position the eyes appear to have evolved after the development of a neural tube. Recruiting the neural crest cells to these functions seems a good strategy.

Neural crest cells also give rise to melanocytes of the dermis and epidermis, cells of the carotid body and several glands, adrenomedullary cells, and ciliary muscles. This assortment of differentiated tissues includes those predominantly derived from endoderm, mesoderm, or ectoderm layers of the gastrula. Yet, neural crest cells are all of ectodermal origin. They appear to break all the rules concerning the division of potential among the primary layers established at gastrulation. Neural crest cells may be the exceptions that emphasize the rule or, on the other hand, the rule that internal glands are derived from endoderm, muscles from mesoderm, and neural tissue from ectoderm may not be very meaningful. Evolution may have just used available nearby cells to form a new organ.

Neural crest cells undergo biochemical differentiation once they have reached their sites of arrest. Some ganglion cells are cholinergic and synthesize acetylcholine whereas others are adrenergic and synthesize the appropriate neurotransmitters and receptors. Whether a given neural crest cell will be cholinergic or adrenergic is not determined by its site of origin but only by its final position in the embryo, as has been shown by transplantation experiments. Moreover, differentiation of neural crest cells in culture to cholinergic or adrenergic states can be manipulated by the culture conditions. In fact, it appears that neural crest cells can be simultaneously cholinergic and adrenergic and specialize to one or the other signaling system only later when in place. Likewise, some neural crest cells become melanocytes and express the genes necessary for the synthesis of the dark pigment, melanin. Luckily, this occurs only in the iris and not in the cornea, although neural crest cells go to both tissues. The final position appears to determine the type of differentiation these cells undergo.

Somites

In vertebrate embryos mesodermal cells that flank the neural tube condense into somites at about the time the neural tube closes (Fig. 10.3). The spaces between these cells disappear as the cells become tightly apposed. An extracellular sheath then surrounds each somite, clearly delineating it from the surrounding cells. Although formation of the somites occurs at regular distances starting near the anterior and proceeding toward the posterior, the stimulus does not appear to be passed from head to tail. Barriers can be put

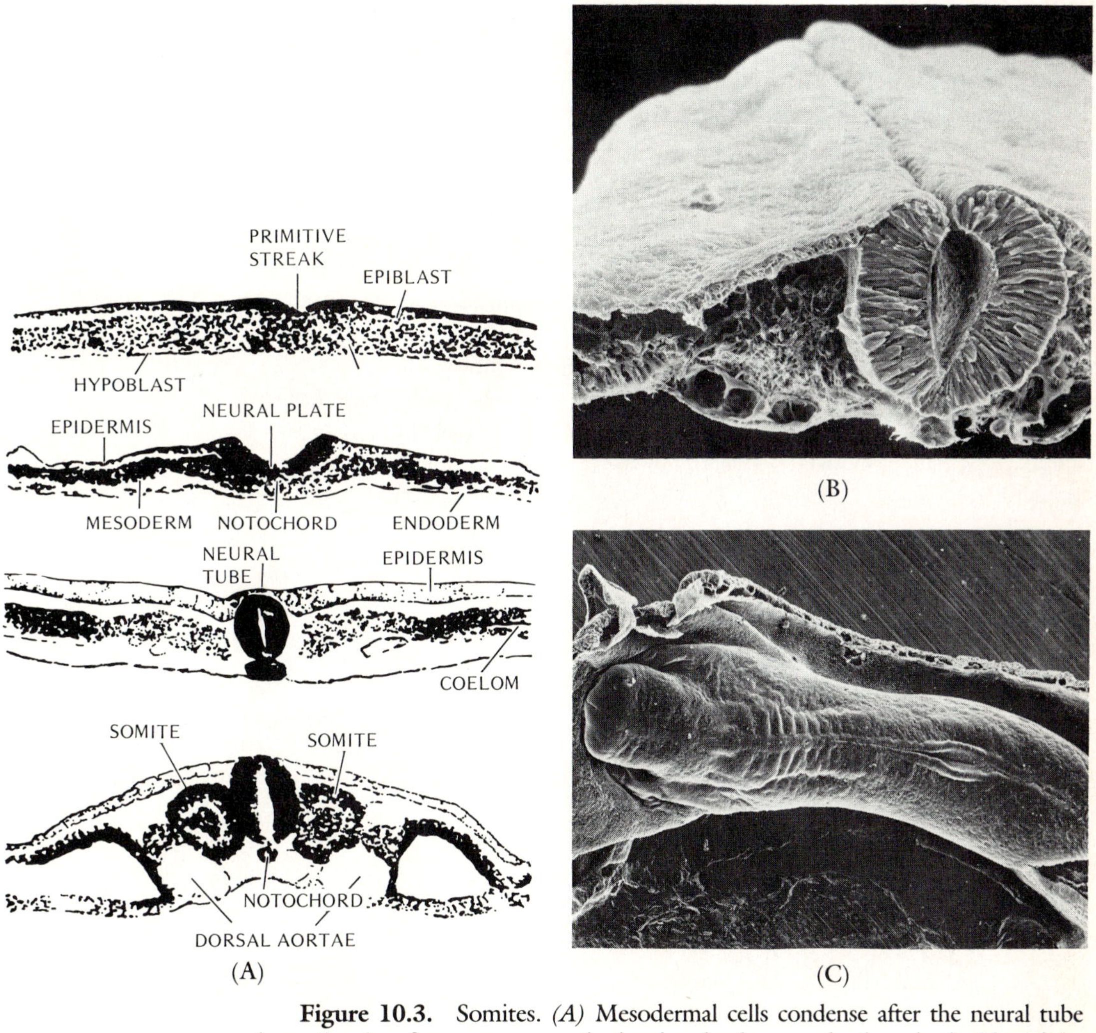

Figure 10.3. Somites. *(A)* Mesodermal cells condense after the neural tube closes. Somites first appear near the head and subsequently along both sides of the dorsal midline of chick embryos. *(B)* A fractured chick embryo shows the neural tube flanked by mesodermal cells. *(C)* Somites can be clearly seen from above when the neural tube is partially closed.

at various positions along the axis of a developing embryo without stopping the progress of somite formation. The barriers can be thin sheets of plastic that should stop the passage of all molecules, large and small. The interpretation of these results is that somite formation occurs at a fixed time following the formation of the neural plate and related structures during gastrulation. It proceeds sooner in the anterior than to the posterior. What mechanism can generate this sort of timing is presently unknown.

Later in development some of the cells of each somite give rise to the vertebrae of the spinal column that surrounds and protects the spinal cord. They also give rise to the ribs and lateral muscles. The regular spacing of bilateral somites has reminded some of the sequential arrangement of organs most dramatically seen in annelids such as earthworms. However, the type of segmentation seen in worms involves regular repeats of complete organ systems and not just the cell condensation seen in somites. As will be discussed in Chapter 15, a great deal is known concerning the mechanism of development of segments, but the general rules and concepts do not appear to have much bearing on the repetitive formation of somites.

Gut and Heart

In both sea urchin and frog embryos the alimentary canal forms directly from the archenteron. The blastopore becomes the anus and a mouth opening forms at the anterior. Of course, in tadpoles, further specializations of this canal have produced a throat, stomach, and intestines by the time it hatches.

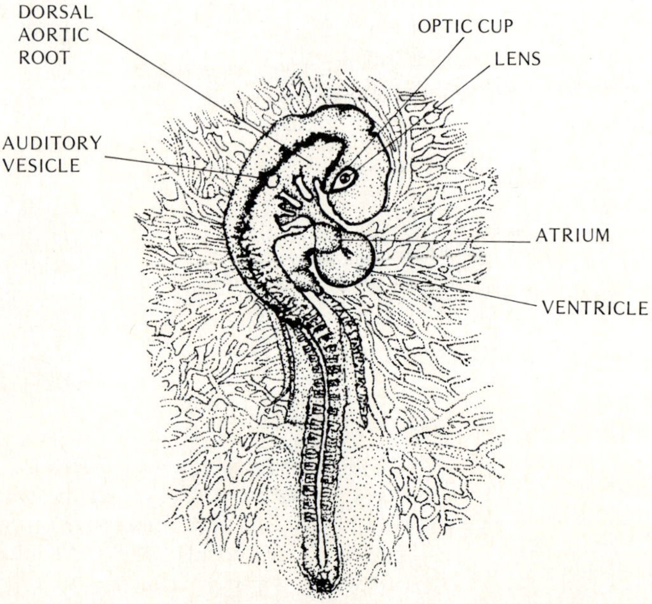

Figure 10.4. The chick embryo lies on top of the yolk. It is fed by an extensive vascular system. Blood is pumped through the capillaries by a well-developed heart. The gut forms as the embryo lifts off the yolk. The embryo is at the 29-somite stage. The neural tube has fully closed and the brain and eyes are developing.

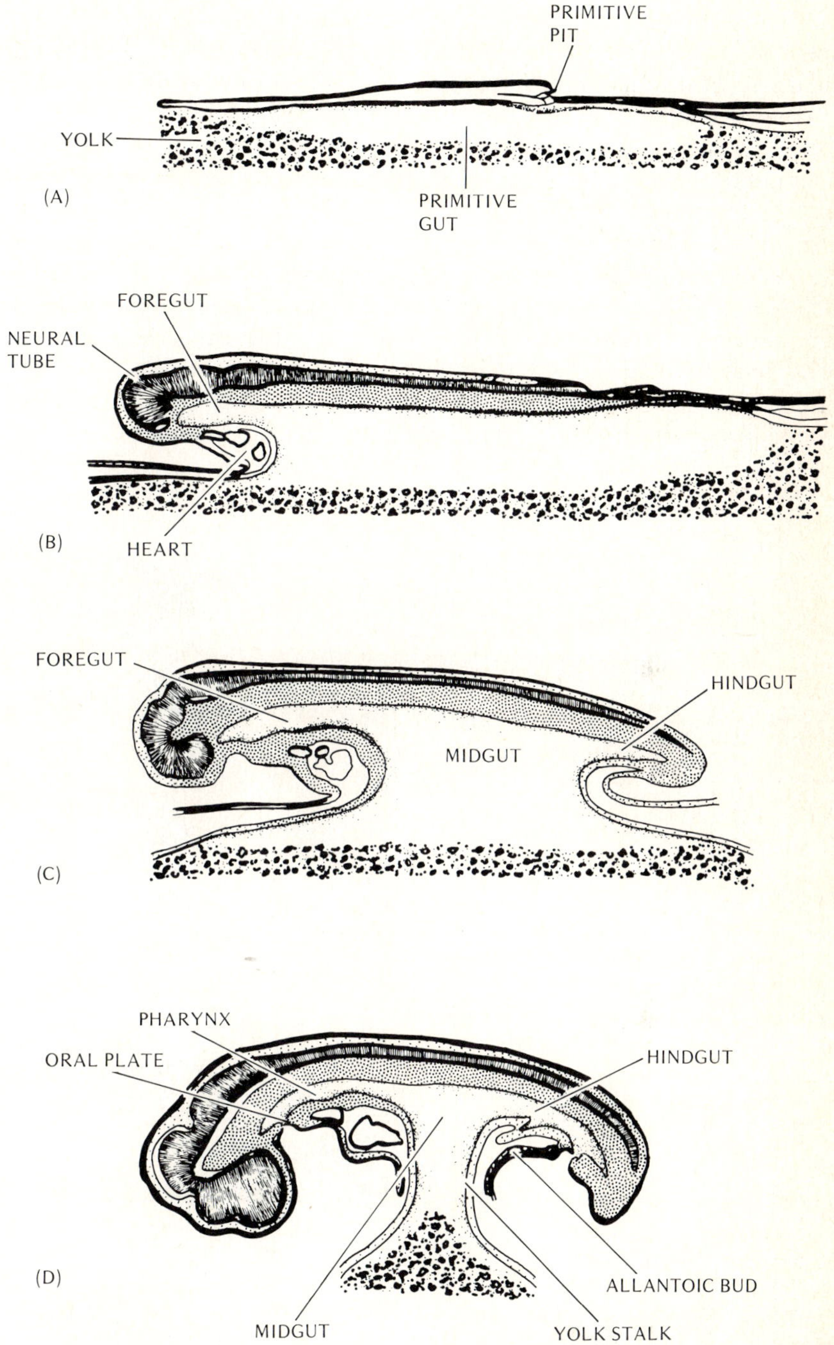

Figure 10.5. Development of the gut in chicks. *(A)* The subgerminal space can be considered a primitive gut in 1-day embryos. *(B)* By the next day the gut is better defined. *(C)* Foregut, midgut, and hindgut are recognizable as the amnion is drawn under the embryo. *(D)* After 3.5 days of development, the chick embryo has risen off the yolk and its gut is connected to the yolk only by the yolk stalk.

In both birds and mammals, no archenteron forms because invagination through the primitive streak does not occur as a sheet of cells. During the period of neurulation, a fluid-filled cavity appears below the avian blastodisk separating it from the yolk. As the embryo grows, it lifts off from the cells surrounding the yolk and progressively pinches off some of this cavity to form the gut (Figs. 10.4 and 10.5). In mammals the embryonic plate lies over an empty yolk sac. But just as in chicks, the embryo rises from this plate, pinching off a portion of the yolk sac to form its gut. Subsequent development molds this cavity to the final architecture of the alimentary canal.

Soon after the closure of the neural tube, spaces lined with endothelial cells can be seen among the mesodermal cells just below each somite. In both chick and mammalian embryos these spaces come together over the midline and fuse. This cavity will develop as the heart and aorta (Fig. 10.6). Morphogenesis of the heart in human embryos was described in Chapter 1.

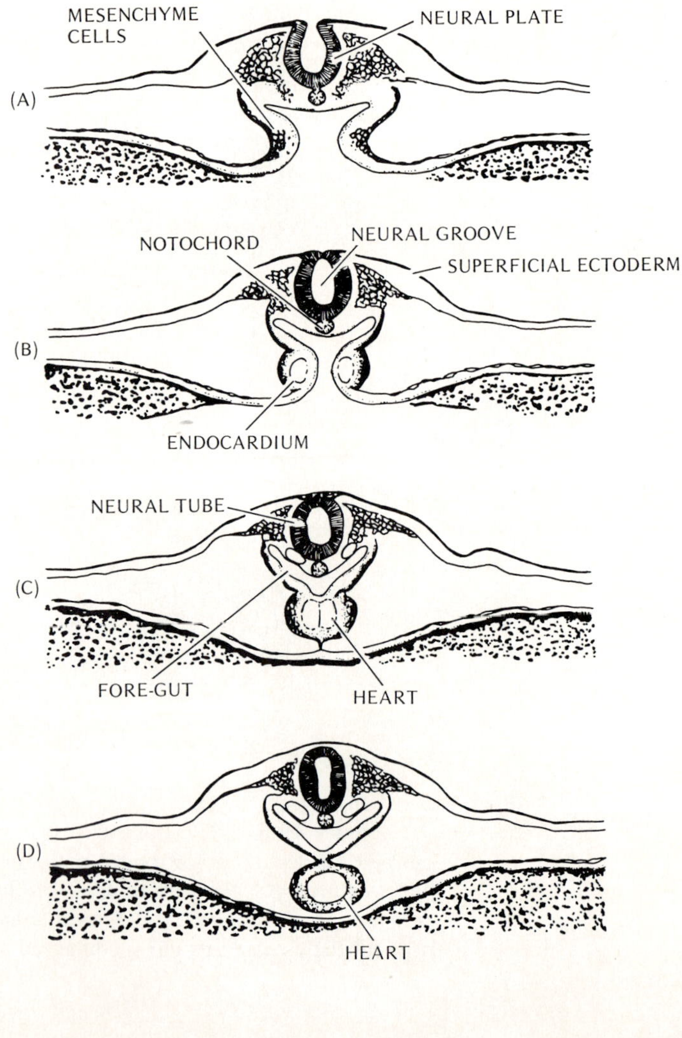

Figure 10.6. Formation of the heart in chick embryos. *(A,B)* At between 25 and 30 hours of incubation, mesenchymal cells on both sides of the midline just behind the head fold form tubes. *(C,D)* They coalesce to form a single tube below the notochord.

The basic body plan is now laid out along the axis of the embryo: a neural tube lies over the notochord, which is flanked by somites; under the notochord is the heart primordium, which, in turn, lies over the primitive gut. Each is surrounded by an extracellular matrix that delineates it and participates in subsequent morphological development. The details of these subsequential steps have adapted to generate the particular body plan of each species.

Eyes and Sex Cells

Eyes provided an enormous advantage to evolving vertebrates (Fig. 10.7). Early in the development of all vertebrates, the neural tube in the head bulges to each side. As the neural tube closes over, these bulges extend all the way

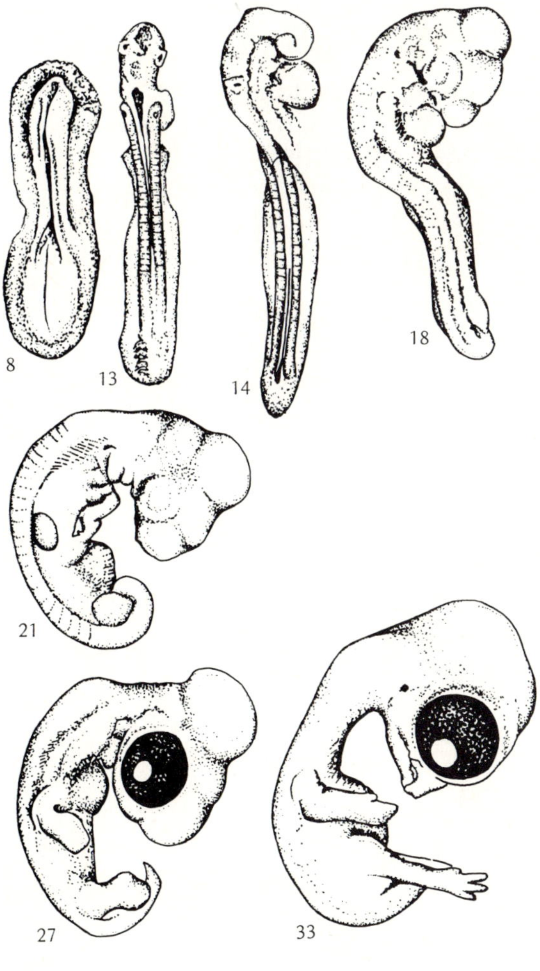

Figure 10.7. Eyes start to be formed even before all the somites are formed. They remain prominent in chick embryos until hatching. The stages in chick embryogenesis are given.

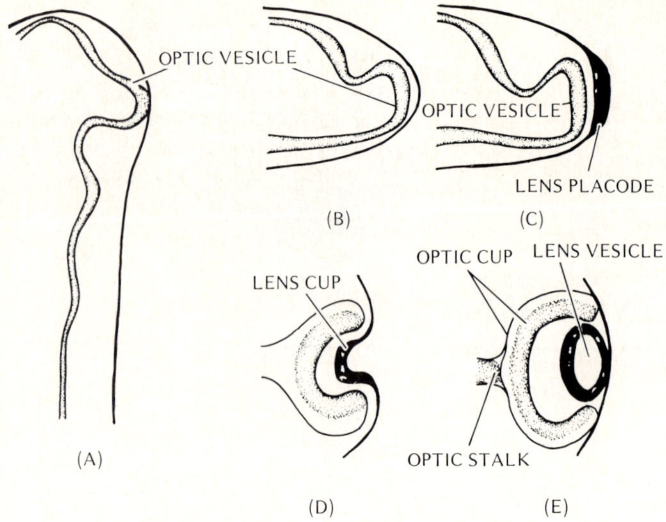

Figure 10.8. Eye formation. *(A,B,C)* Epithelial tissue over the optic vesicle thickens to form the lens placode. *(D)* Together, the optic vesicle and lens cup indent. *(E)* The lens fuses and is covered by the cornea. The optic cup is two layers thick and differentiates into a light-sensitive retina.

to the ectodermal layer of cells. Juxtaposition of these tissues not only induces the overlying ectodermal cells to differentiate into lens cells but also results in cupping of the optic bulge (Fig. 10.8). The optic vesicle that forms is now two layers thick. The cells in the outer layer facing the lens differentiate into light-sensing cells whereas those behind become pigmented. This extension of the brain then undergoes structural specialization to optimize the eye to see and focus on objects.

Perhaps the most important cells in any embryo, from an evolutionary point of view, are the sex cells—sperm and eggs. Although these cells have differentiated for as long as there have been embryos, they are recruited to the gonads from quite different parts of the early embryos of different phyla.

Embryogenesis of sea urchins results in a feeding larve, referred to as a pluteus, which has no sex cells. Only after metamorphosis into the familiar spiny adult are eggs or sperm formed in the gonads. Since most of the adult is derived from mesenchymal cells of the pluteus, the sex cells probably come from those tissues as well.

In frogs the sex cells, also called germ cells, are derived from large yolky cells on the floor of the blastocoel. Following gastrulation the germ cells are in the endodermal tissue of the gut. From there they migrate to the position of the gonads.

In chicks the primordial germ cells are found at the anterior of the blastodisk, well ahead of the head process (Fig. 10.9). They move during subsequent development to the gut, enter the bloodstream, and finally land in the gonads. If the tissue at the anterior of the blastodisk is cut out and removed, embryogenesis appears to proceed smoothly; however, the chicks that hatch are all sterile. They lack sex cells. The signals that guide primary germ cells are not specialized to a given species. Thus, if the anterior margin of the blastodisk of a duck embryo is used to replace that of a chick, the

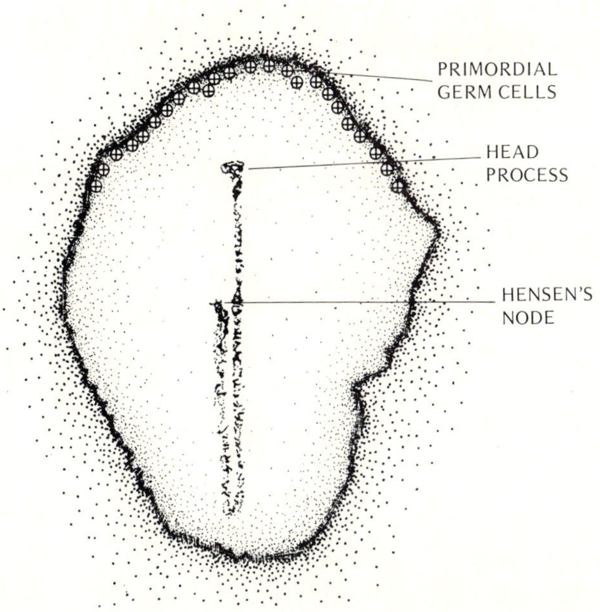

Figure 10.9. Location of primary germ cells in the chick blastoderm. Surgical removal or lethal irradiation of the circled regions at the anterior of the blastoderm results in lack of either sperm or eggs in chick embryos. Later in development, cells from this area migrate to the gonads and differentiate.

embryo develops as a chick but has the sex cells of a duck. There is even evidence that primary germ cells of a mouse introduced into a chick will end up correctly positioned in the gonads.

In mammals germ cells are derived from cells in the posterior of the yolk sac. These cells also leave their initial position late in development and migrate to the gonads. ((Fig. 10.10).

There is no separation of developmental potential of primary germ cells into either male or female. These cells can differentiate into either eggs or

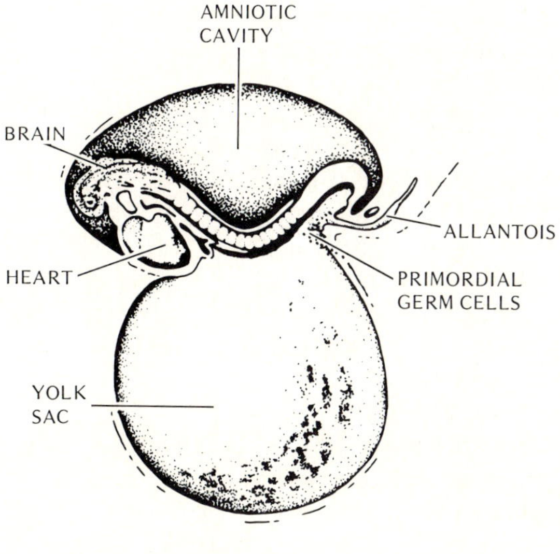

Figure 10.10. Location of primary germ cells in the human embryo. Sex cells can be traced back to the endodermal tissue of the yolk sac in early embryos.

sperm. What determines their subsequent dramatic differences is the genetic makeup of the gonadal tissue. Primary germ cells introduced into a male (XY) will differentiate as sperm whereas those injected into a female (XX) will differentiate as eggs. It appears that the genetic differences of embryos are expressed in different molecules on the surface of the gonadal cells in such a way as to entwine the arriving germ cells in a manner that induces differentiation into either sperm or eggs. The differentiation of sexual organs is described in detail in Chapter 12.

Summary

Contraction of the outer edge of neural plate cells leads to the rounding up of the neural tube. The edges fuse, leaving a few cells at the crest below the epithelial layer. These neural crest cells migrate to the anterior and sides of the embryo and differentiate into a wide variety of cell types depending on the environment in which they land. Lateral mesodermal cells coalesce into slabs on either side of the neural tube, which then segment into somites that will give rise to vertebrae, ribs, and flanking muscles. The cavities that will form the gut and heart become lined with epithelial cells and form tubular structures. At the anterior, the neural tube extends to the sides, where optic vesicles are formed. Close apposition of optic vesicles induces overlying epithelial cells to differentiate into lens cells. The cells that will subsequently differentiate into either sperm or eggs migrate to the gonads, where they are instructed concerning the sex of the individual.

SPECIFIC EXAMPLES

1. Epidermal–Dermal Interaction

Chick embryos develop scales on their feet by 16 days of incubation. The scales on the top of the toes have a different structure and contain different keratin proteins than those of the footpad. These tissues present favorable material to analyze the relationship between morphogenesis and biochemical differentiation in an organ composed of dermis and epidermis.

The skin on the top of the toes is covered with scutate scales that form overlapping plates (Fig. 10.11). The reticulate scales of the footpad are radially symmetric and adapted to provide traction (Fig. 10.11). The major proteins found in scales are the water-insoluble keratins closely related to the subunits that make up intermediate filaments. They are produced in massive amounts in skin and scales and can be characterized by their molecular weight (size) as well as with antibodies specific to different classes of keratin. The epidermis of scutate scales accumulate six α keratins in the range of 50,000 to 60,000 daltons and three β keratins of about 15,000 daltons. The epidermis of reticulate scales accumulate three α keratins but no β ker-

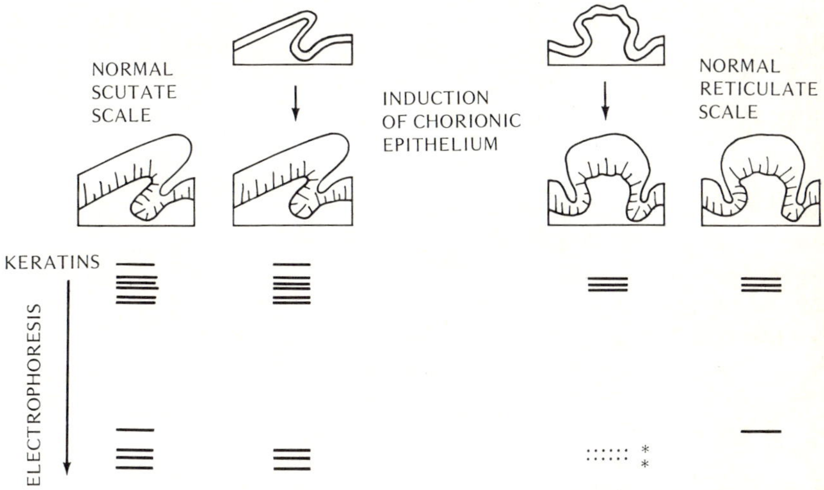

Figure 10.11. Keratins in scutate and reticulate scales. The electrophoretic mobility in SDS-polyacrylamide of water insoluble chicken keratins present in scutate and reticulate scales were compared with those induced in chorionic epithelium by dermis of scutate or reticulate scales. The patterns in the reconstructed tissue were similar to their normal homologs with a few exceptions. The starred β-keratins are present only in cells of the subperiderm.

atins. The α and β keratins can be distinguished by specific antibodies. The different-sized keratins can be separated by solubilizing the proteins of scales with the detergent sodium dodecyl sulfate (SDS) and running them on a slab of polymerized acrylamide. This technique separates the α and β keratins into discrete bands (Fig. 10.11). Thus, both the morphology and the protein synthetic pattern differs between these epidermal cells.

α keratins are found by fluorescent antibody staining throughout the epidermis of reticulate scales but are absent in the underlying dermis. They are present only in the inner epidermal cell layer (the alpha stratum) of scutate scales. β keratins are found only in the outer epidermal cells of scutate scales (beta substratum). This raises the question as to whether this pattern of protein synthesis is determined during morphogenesis of the scale or is directed by the underlying dermal cells following morphogenesis. To answer this question a responsive ectodermal layer of cells that can keratinize but normally does not was needed that could be combined with isolated dermal fragments from scutate and reticulate scales. The chorioallantoic epithelium of the egg proved ideal for this experiment.

Chick embryos are surrounded by several extra embryonic membranes that enclose the fluids of the egg. The chorioallantoic membrane is a highly vascularized tissue that carries out gas exchange for the embryo and segregates nitrogenous wastes. The epithelium of the chorioallantoic membrane is not normally keratinized in the chick egg but, if exposed to air through a crack in the shell, it rapidly stratifies to form glistening keratinized patches that protect the embryo from dehydration. This response demonstrates that the epithelial cells of the chorioallantoic membrane can keratinize under some stimuli.

A brief treatment with the divalent cation chelator EDTA loosens the attachment of epithelial layers from underlying mesenchymal tissues and can be used to isolate the epithelial cells from the chorioallantoic membrane. The experiment involves placing the unkeratinized epidermal layer of cells over a small (2-mm^2) piece of scale dermis, freed of its own epidermis by treatment with EDTA, and then allowing the composite to differentiate on the top of the chorioallantoic membrane of another embryo. The chorioallantoic membrane of an egg can be conveniently exposed by cutting a small window in an egg, and it provides the nutrients needed to sustain the dermal–epidermal sandwich placed on it. The window is then closed with a strip of gauze and the egg kept in a humid chamber. Under these conditions, the piece of chorioallantoic epithelium combined with a scale dermis

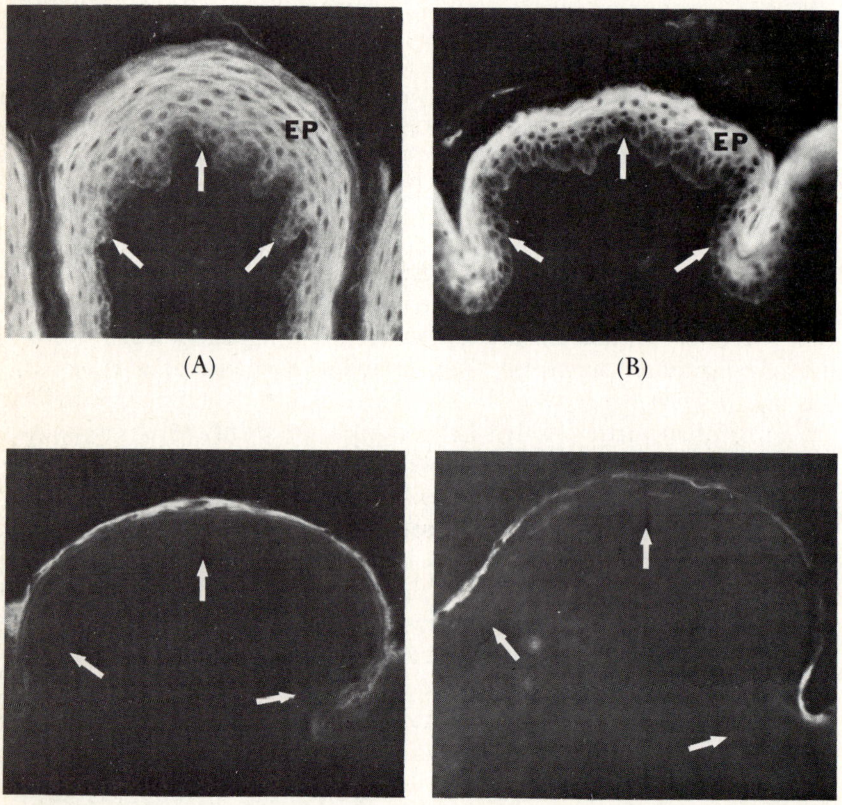

(A) (B)

(C) (D)

Figure 10.12. Keratin localization visualized by bound fluorescent antikeratin antibodies. The arrows indicate the junction of dermal and epidermal (EP) tissue: *(A,C)* reticulate scale; *(B,D)* chorion epithelium induced by reticulate scale. The antibody used in *A* and *B* is specific to α-keratin. The antibody used in *C* and *D* is specific to β-keratin. The strata basal, intermedium, and corneum of the epithelium contain α-keratins whereas only the subperiderm contains β-keratins.

fragment undergoes biochemical differentiation (Fig. 10.11). When in association with dermis of scutate scales, epithelial cells make six α keratins as well as three β keratins. When in association with dermis of reticulate scales, the epithelial cells make only three α keratins. These results clearly show that even after morphogenesis dermal cells are able to direct the pattern of biochemical differentiation in associated naive epidermal cells.

Epithelial cells induced by reticulate scale dermis synthesize a small amount of β keratin. However, immunofluoresence with antibodies specific to β keratin has shown that the β keratin is restricted to a subperiderm layer of cells (Fig. 10.12). This tissue is normally sloughed during development but is retained in the experimental preparation. The scale epidermal cells contained no β keratin. Both normal reticulate scale and the experimental epidermis combined with reticulate dermis contain large amounts of α keratin.

Mutants

There are strains of chickens that carry a mutation called *scaleless.* In these birds development of scutate scales fails to occur and the reticulate scales are abnormally shaped. The mutation primarily affects differentiation of the beta substratum of the scutate scales. Dermis of 16-day embryos isolated from scaleless strains fails to direct the formation of scutate scale epidermis. The mutation impairs dermal differentiation for the ability to direct histogenesis and biochemical differentiation of overlying epidermis. The epidermis of scaleless mutant embryos responds normally to the instructions of normal scutate scale dermis. Thus, this mutation clearly points to a tissue-(dermis-)specific property that is able to direct the pathway of differentiation in adjacent tissues. Unfortunately, the nature of the interaction, be it direct contact or short-range induction, is not yet known.

2. Seeds

Higher plants are dispersed by seeds. This strategy overcomes the problem of immobility of plants and ensures that they can colonize new environments. The seed carries an embryonic plant, referred to as the axis, as well as nutritional reserves inside a protective covering, the seed coat. Seeds vary enormously in size and shape, but the basic plan is similar in all seeds. The difference in size depends mostly on the amount of nutritional reserves in the endosperm. The seeds of some orchids are microscopic and contain essentially no endosperm whereas the seeds of the Seychelles nut palm are almost a meter in diameter (Fig. 10.13).

Seeds develop from fertilized ovules in the ovary. The diploid zygote divides three times to produce eight cells all in a row extending into the seed. The cell nearest the micropyle expands by forming a large vacuole, further inserting the embryo into the endosperm (Fig. 10.14). This terminal cell is referred to as the suspensor. At the other end, the cleavage plane rotates 90° to produce side-by-side cells. Further division alternates in a roughly orthogonal manner to produce a globular cell mass that will one day become the new seedling.

Figure 10.13. The world's largest seed, *Lodoicea seychellarum,* or seychelles nut palm. The "double coconut" takes five years to mature and weighs up to 18 kg. The seed held by the man has been removed from the fruit husk.

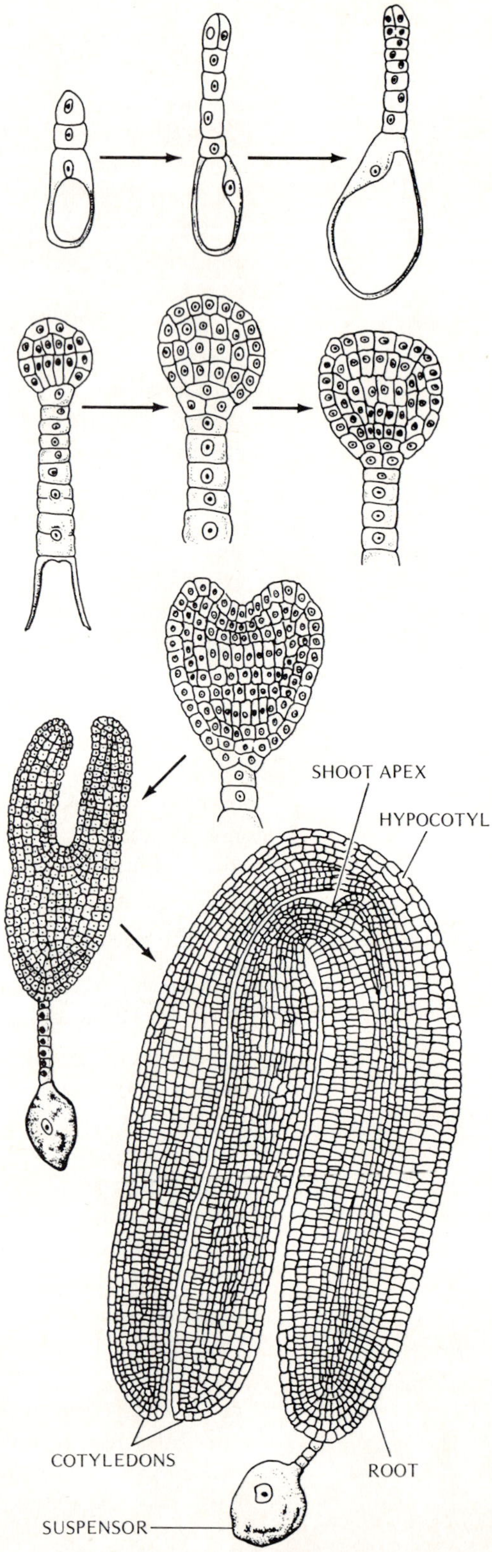

Figure 10.14. Dicot embryo as it grows within the seed. The heart-shaped structure transforms over a period of weeks into the embryo with folded cotyledons.

When the globular mass contains only a few hundred cells, the outer and inner layers can already be seen to differ. At the apex in dicot seeds two projections extend to give the so-called heart structure (Fig. 10.14). These projections elongate over the next week or two to form the embryonic cotyledons that hold the nutritional reserves of the young seedling. Growth of the plant embryo proceeds at the expense of the endosperm and is further nourished by material from the ovary tissue. In many seeds the cotyledons, together with the axis, become longer than the seed itself and so they must fold over. They straighten out at germination.

The grasses and cereals, such as wheat, are monocots and develop one rather than two cotyledons (Fig. 10.15). They also retain far more endosperm tissue, which is utilized following germination. The aleurone layer of surface cells secretes hydrolytic enzymes, such as amylase and protease, into the endosperm at germination and the resulting breakdown products are absorbed into the growing seedling by the single cotyledon, which is referred to as the scutellum in this case.

Hydrogen peroxide (H_2O_2) is a byproduct of normal metabolic processes. Buildup of this compound can lead to cell death. In plants, H_2O_2, acting through a free-radical mechanism, inhibits the enzymes that catalyze CO_2 fixation and so it is imperative to remove it. Hydrogen peroxide is split into water and oxygen by a reaction catalyzed by the large heme-containing enzyme, catalase. In *Zea mays* (corn), there are four isozymes of catalyse that are synthesized at various stages in the tissues of the seeds. Three of the catalyses have been shown to be the products of separate unlinked genes, *cat*-1, *cat*-2, and *cat*-3, whereas the fourth is minor and appears only briefly

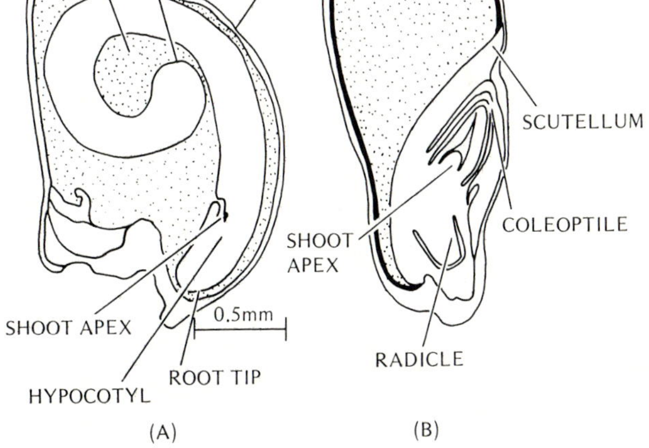

Figure 10.15. Monocot embryo in the seed. A large amount of endosperm is retained for use following germination: *(A)* onion seed; *(B)* wheat seed or grain.

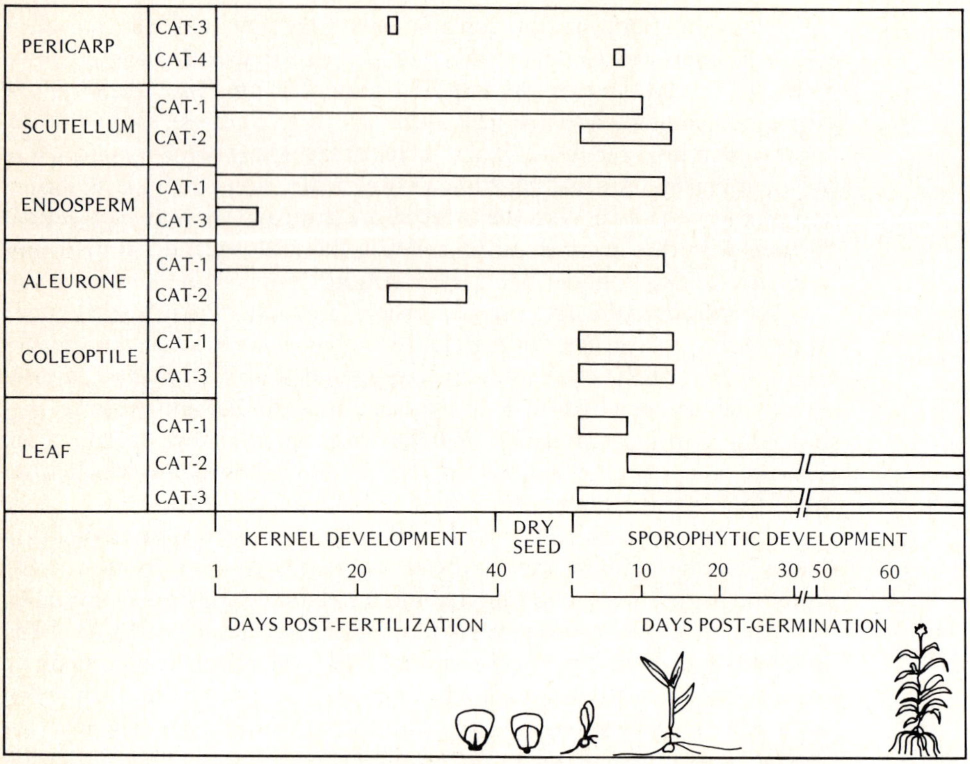

Figure 10.16. Temporal and spatial regulation of catalase. The bars indicate the stages when *cat*1, *cat*2, or *cat*3 are expressed in specific tissues. The seed tissues disappear after germination and leaves appear.

in pericarp tissue following germination. The three major isozymes accumulate at different times in different tissues (Fig. 10.16).

The scutellum contains CAT-1 during kernal development but then replaces it with CAT-2 during the first week following germination. The specific activity of catalase in the scutellum reaches a peak at the third day post-germination about tenfold higher than that found in kernals. In some strains of *Zea mays* the specific activity remains high for the next two weeks while in other strains it drops back down. The difference is due to different alleles of a *trans*-acting locus, *Car*-1. Synthesis of CAT-2 in isolated scutella is dependent on the plant hormone, abscisic acid (ABA), that is normally provided by the plant embryo.

CAT-2 is also present in the new, young leaves. However, synthesis is completely dependent on the presence of light. Dark-grown (etiolated) leaves make no CAT-2. Upon illumination both CAT-2 and CAT-3 isozymes are made.

CAT-1 is present in the scutellum, endosperm, and aleurone tissues of seeds. In the endosperm it is synthesized only in the second week after pollination. Expression of the *cat*-1 gene in the scutellum is regulated by a closely linked locus, *Car*-2. This locus regulates CAT-1 accumulation in a tissue-specific and time-specific manner.

During the first few weeks of seed maturation, the number of cells in the embryo increases rapidly. The endosperm, with its triploid nuclei, also divides up into many cells. Peas take about two months to develop fully. After the first three weeks there is no significant increase in the number of cells in the axis or the cotyledons but the cotyledon cells enlarge three- or fourfold over the next two weeks as they accumulate nutritional reserves including protein in protein bodies and starch. There is also a significant increase in RNA and DNA in the cotyledon cells as they become polyploid. During the last two weeks of maturation, peas dry out, losing about a third of their water content and becoming dormant. They can stay viable for months in this form. As they enter the dormant phase, seeds accumulate large amounts of the plant hormone, abscisic acid (ABA). This hormone is involved in dropping leaves (abscission) of deciduous trees in the fall but is also involved in blocking precocious germination of seeds. Mutants of some plants as well as certain species produce less ABA than necessary or fail to respond to it and, as a result, the seeds germinate while still on the plant (vivipary). This is advantageous for mangroves but usually defeats the purpose of aerial dissemination.

Upon germination, starch is metabolized by amylase. Transcription of the amylase gene is induced by gibberellic acid, a plant hormone produced when the seed takes up water and germinates. However, transcription of the amylase gene does not occur if ABA is still high even if gibberellic acid is also present. In this way ABA keeps the nutritional reserves from being used prematurely. ABA also induces the synthesis of storage proteins such as wheat germ agglutinin in wheat cotyledons.

The cells of cotyledons are crammed full of membrane-enclosed protein

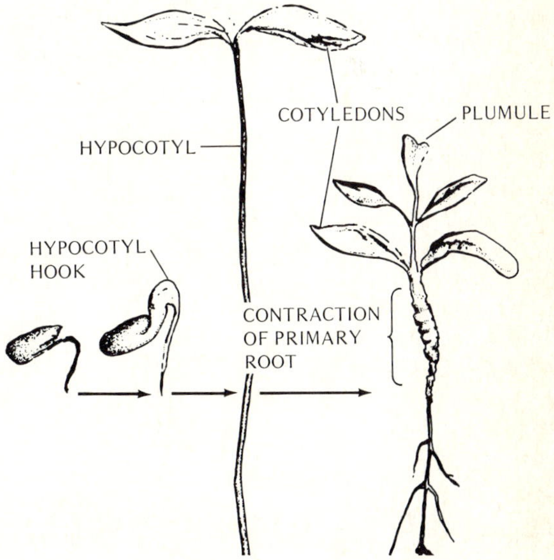

Figure 10.17. Germination of a dicot seed. The seed coat bursts and the hypocotyl hook extends. The primary root goes into the soil and then contracts to bring the plant close to the earth.

bodies. These organelles contain a half-dozen different proteins that accumulate in massive amounts for nutritional storage. These proteins are synthesized on endoplasmic reticulum membranes and sequestered in vesicles. Many are glycoproteins and their processing pathways have been described in detail. Considerable evolutionary conservation of the primary structure of storage protein is found in seeds of different plants. The mRNA coding for storage proteins has been measured during maturation of soybean seeds and found to increase by a factor of 100,000 between the early cotyledon stage and midmaturation. At peak these mRNA molecules make up 10 percent of the total poly-A^+ RNA. In peas mRNAs coding for five different storage proteins were followed. Each had a characteristic period during which it reached a maximum and then declined, indicating that there is a temporal sequence for protein accumulation and that each gene is independently regulated.

Germination of seeds follows uptake of water, referred to as imbibition. In some cases a cold period or high heat are also required. The embryonic root, or radicle, expands and bursts through the weakened seed coat. Protein, carbohydrate, and fat in the storage tissues are rapidly used up during the first week as the result of the synthesis and release of hydrolytic enzymes into the stores. The root rapidly elongates and puts out hairs while the leaf primordia unfold (Fig. 10.17). In an appropriate place the seedling will thrive.

3. Development of Sand Dollars: A Pictorial Review

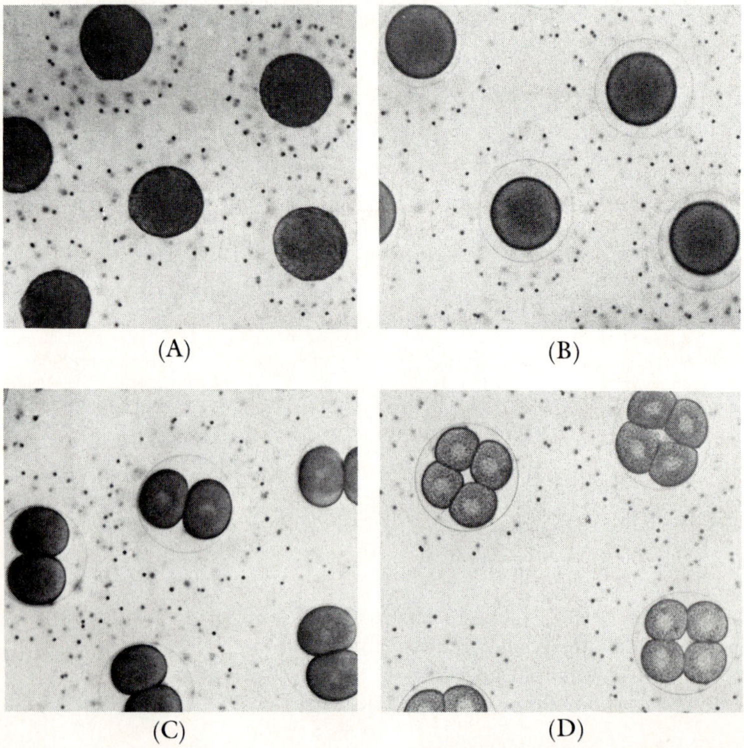

Figure 10.18. Cleavage of *Dendraster*. *(A)* Sand dollar eggs are 120 μm in diameter and are surrounded by a jelly coat. Imbedded in the jelly are pigmented cells. *(B)* Two minutes postfertilization the fertilization layer rises to form a permanent block to polyspermy. *(C)* At one hour the zygote has divided into two blastomeres in these holoblastic embryos. *(D)* At one and a half hours four cells have resulted from the second cleavage, which is orthogonal to the first. Each of the four cells could form a complete embryo if separated from the other cells.

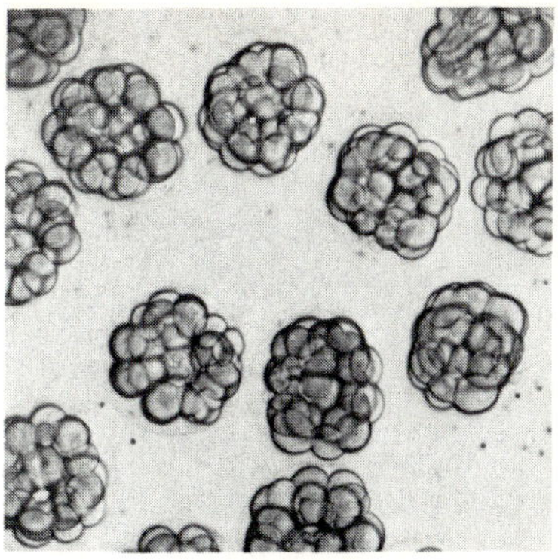

(A)

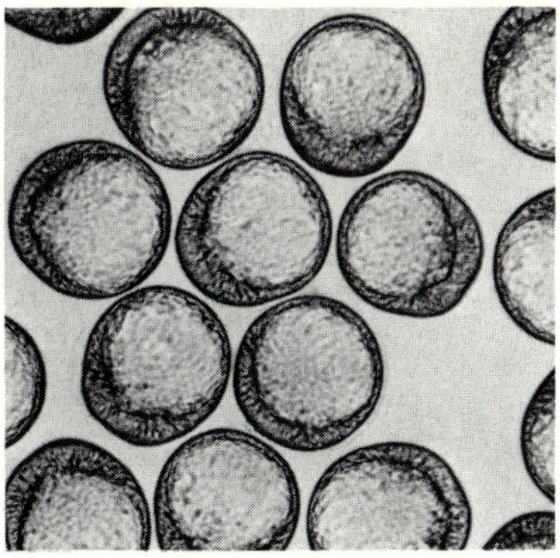

(B)

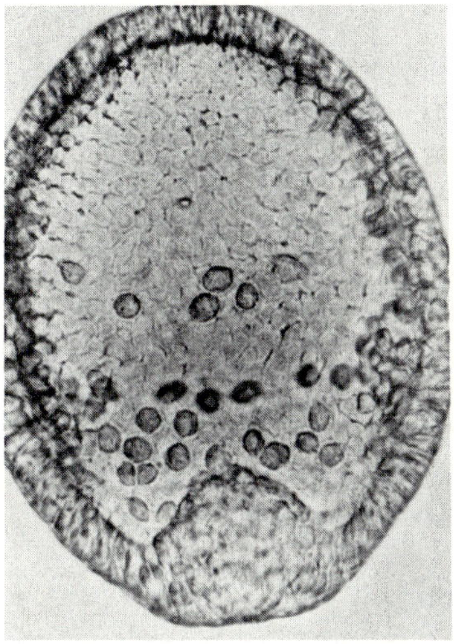

(C)

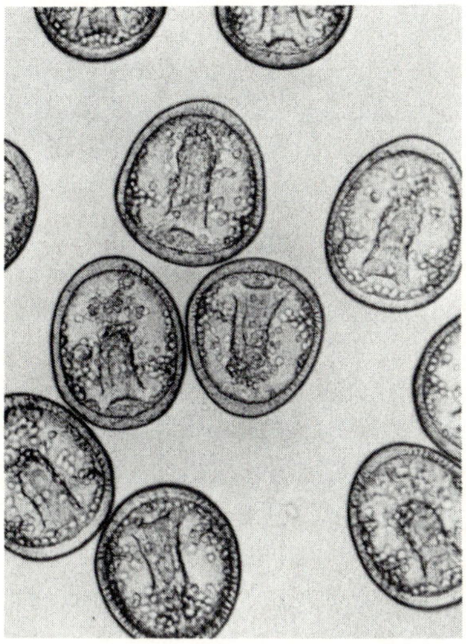

(D)

Figure 10.19. Gastrulation of *Dendraster*. *(A)* At three hours the fourth division has resulted in blastomeres of different sizes: micromeres, macromeres, and mesomeres. *(B)* At seven hours the blastulae have hatched out of the fertilization layer using a specific protease. *(C)* At ten hours the primary mesenchymal cells derived from the micromeres enter into the blastocoel. *(D)* At fourteen hours gastrulation occurs as cells move through the blastopore. New genes are activated at this stage.

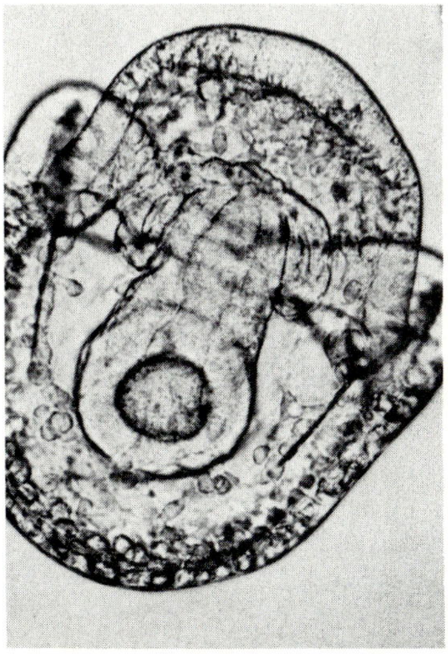

(A)

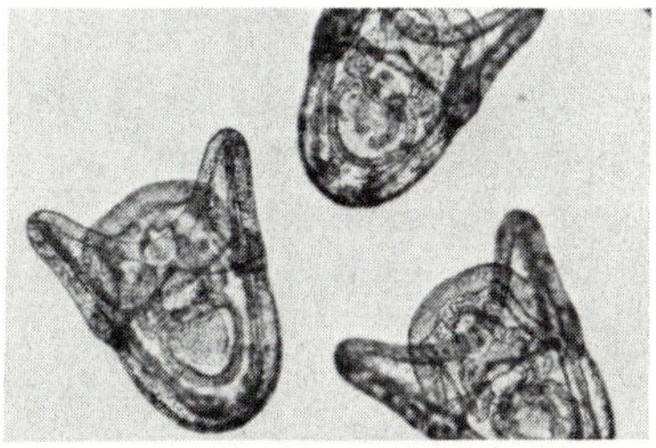

(B)

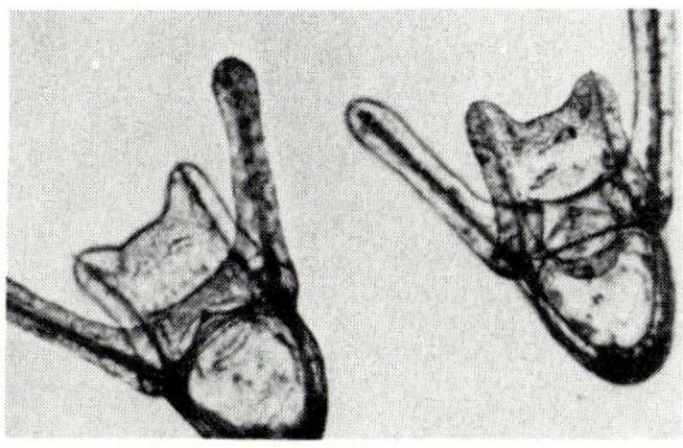

(C)

Figure 10.20. Organogenesis in *Dendraster*. *(A)* At twenty-three hours the archenteron has crossed the blastocoel and will fuse with the ectoderm to form the mouth in these deuterosomes. Spicules are formed by the primary mesenchymal cells. The circular opening in the middle is the larval anus. *(B)* At twenty-eight hours the gut is well formed. Gut enzymes including β-glucanase and α-amylase increase more than a hundredfold during the next day. *(C)* At thirty-five hours embryogenesis is complete and the larvae can feed and grow.

Related Readings

Chandlee, J., and Scandalios, J. (1984). Regulation of Cat1 gene expression in the scutellum of maize during early sporophytic development. Proc. Natl. Acad. Sci. *81,* 4903–4907.

Chandler, P., Spencer, D., Randall, P., and Higgins, T. (1984). Influence of sulfur nutrition on developmental patterns of some major seed proteins and their mRNAs. Plant Physiol. *75,* 651–657.

Chrispeels, M., Vitale, A., and Staswick, P. (1984). Gene expression and synthesis of phytohemagglutinin in the embryonic axis of developing *Phaseolus vulgaris* seeds. Plant Physiol. *76,* 791–796.

Gans, C., and Northcutt, G. (1983). Neural crest and the origin of vertebrates: a new head. Science *220,* 268–274.

Goodman, C., Bastiani, M., Doe, C., du Lac, S., Helfand, S., Kuwada, J., and Thomas, J. (1984). Cell recognition during neuronal growth. Science *225,* 1271–1279.

Jacobsen, J., and Beach, L. (1985). Control of transcription of α-amylase and rRNA genes in barley aleurone protoplasts by gibberellic and abscisic acid. Nature, *316,* 275–277.

Langridge, P., Eibel, H., Brown, J., and Feix, G. (1984). Transcription from maize storage protein gene promotors in yeast. EMBO J. *3,* 2467–2471.

Le Douarin, N. (1984). Cell migrations in embryos. Cell *38,* 353–360.

Lofberg, J., Nynas-McCoy, A., Olsson, C., Jonsson, L., and Perris, R. (1985). Stimulation of initial neural crest cell migration in the axolot embryo by tissue grafts and extracellular matrix transplanted on microcarriers. Devel. Biol. *107,* 442–457.

Marchuk, D., McCrohan, S., and Fuchs, E. (1985). Complete sequence of a gene encoding human type I keratin: sequences homologous to enhancer elements in the regulatory region of the gene. Proc. Natl. Acad. Sci. *82,* 1609–1613.

Rafalski, A., Scheets, K., Metzler, M., Peterson, D., Hedgcoth, C., and Soll, D. (1984). Developmentally regulated plant genes: the nucleotide sequence of a wheat gliadin genomic clone. EMBO J. *3,* 1397–1403.

Sawyer, R., O'Guin, G., and Knapp, L. (1984). Avian scale development. Devel. Biol. *101,* 8–18.

Schoenwolf, G., and Franks, M. (1984). Quantitative analyses of changes in cell shapes during bending of the avian neural plate. Devel. Biol. *105,* 257–272.

Schweizer, J., Kinjo, M., Furstenberger, G., and Winter, H. (1984). Sequential expression of mRNA-encoded keratin sets in neonatal mouse epidermis: basal cells with properties of terminally differentiating cells. Cell *37,* 159–170.

Sengupta-Gopalan, C., Reichert, N., Banker, R., Hall, T., and Kemp, J. (1985). Developmentally regulated expression of the bean β-phaseolin gene in tobacco seed. Proc. Nat'l. Acad. Sci. *82,* 3320–3324.

Schoenwolf, G. (1985). Shaping and bending of the avian neuroepithelium: morphometric analyses. Devel. Biol. *109,* 127–139.

Thiery, J-P., Delouvée, A., Grumet, M., and Edelman, G. (1985). Initial appearance and regional distribution of the neuron-glia cell adhesion molecule in the chick embryo. J. Cell Biol. *100,* 442–456.

van Straaten, H., Thors, F., Wiertz-Hoessels, L., Hekking, J., and Drukker, J. (1985). Effect of notochord implant on early morphogenesis of the neural tube and neuroblasts: histometrical and histological results. Devel. Biol. *110,* 247–254.

Vitale, A., and Chrispeels, M. (1984). Transient N-acetylglucosamine in the biosynthesis of phytohemagglutinin: attachment in the Golgi apparatus and removal in protein bodies. J. Cell Biol. *99,* 133–140.

Study Questions
Part II

1. Give two cases of differential gene amplification during development. Give the cell type and the genes involved.
2. Where in the chick blastoderm do the primary germ cells first appear? What is the evidence for their location?
3. Give two mechanisms that block polyspermy in echinoderm eggs.
4. What happens when the membrane potential of a sea urchin egg is maintained by a voltage clamp at (a) -5 mV and (b) -70 mV and then exposed to sea urchin sperm at 10^6/ml?
5. Describe two mechanisms that function to adhere cells to each other in a tissue.
6. What is the evidence that the male and female pronuclei approach each other in a fertilized egg as a result of microtubule-generated force?
7. What would you expect to happen to echinoderm sperm suspended in Ca^{++}-containing seawater at pH 6.2 when an ionophore that allows passage of Ca^{++} and H^+ is added?
8. a. What effect does actinomycin D, which blocks RNA synthesis, have on protein synthesis and morphogenesis in fertilized sea urchin eggs?
   b. What can you conclude from this fact?
9. What is the mechanism that results in a dramatic increase in transcription rate of RNA at midblastula stage in *Xenopus*? What is the experimental evidence supporting this conclusion?
10. What is the experimental finding that shows that microfilaments play an active role in neurulation?
11. Define:
    a. acrosome
    b. spermatocytes
    c. archenteron
    d. inner cell mass
    e. determination
    f. blastula
    g. cortical granule
    h. gray crescent
    i. meroblastic cleavage
    j. bottle cells
    k. notochord
    l. nurse cell
    m. stem cell
    n. somite
    o. fate map
    p. primary streak
12. Give four distinct tissues derived from neural crest cells.

13. Give, in order of appearance, eight discrete steps that occur in a temporal sequence during embryogenesis of frogs and are necessary before crystalins accumulate in a specific cell type.
14. What advantages do spore-forming organisms have?
15. What is the advantage of a sexual cycle?
16. Name two cell types found in vertebrates that undergo meiotic division.
17. At what stages in oogenesis in different animals does sperm penetration occur?
18. How does progesterone stimulate the maturation of amphibian oocytes?
19. How do pollen cells recognize stigma of the same species?
20. Where are pollen formed?
21. How do sperm centrioles affect pronuclear migration?
22. What are the consequences of removing the polar lobe of the mollusc *Dentalium* at the time of first cleavage?
23. What is the evidence that the product of a specific gene (D) is necessary during oogenesis for dextral cleavage in the snail *Lymnaea*?
24. Why is it thought that microtubules play a regulatory role in establishing the position of the blastopore in amphibian embryos?
25. What is the earliest difference in the dorsal/ventral axis that can be recognized in *Xenopus* embryos?
26. What are some of the difficulties in measuring the amount of the genome transcribed at various stages when hybridization of mRNA to total genomic DNA is used?
27. What techniques have been used to measure specific actin transcripts during sea urchin development?
28. What is the evidence that two genes coding for 15 kd calcium-binding proteins are independently regulated in sea urchin development?
29. Why is the chick chorioallantoic membrane a useful epithelial tissue for analysis of keratin induction by dermal tissues?
30. How is the starch that is accumulated in the endosperm utilized by the seedling of a monocot?
31. Give two families of related genes in monocots in which there is a temporal sequence of expression of the individual members.

PART III

Chosen Systems

Biology does not always advance in clearly logical ways. Sometimes an organism is chosen for intensive study for one reason and then turns out to be amenable to analysis in quite another way. This has been the case for the nematode *Caenorhabditis elegans,* which was chosen for neurobiological reasons and then found to be excellent for developmental studies. The fruit fly, *Drosophila melanogaster,* was chosen for genetic mapping studies and only later used to work out mechanisms and processes of development. The soil amoeba, *Dictyostelium discoideum,* was studied because it seemed so bizarre and then was found to face many of the same problems that embryonic cells face early in development. The advantages of these three systems for detailed genetic analyses have attracted studies on a variety of developmental problems. Moreover, the relative simplicity of these organisms encourages a comprehensive approach to the study of their development.

The other systems chosen in this section concern basic problems of sexual dimorphism, life and death, and the construction of arms and legs. The hormonal control of sexual differentiation presents an especially clear case of directed specialization and also has medical relevance. Form is often generated by removal of tissues by cell death. Thus, selective cytolysis is as important as selective growth. Both function in establishing the structure of sexual organs as well as limbs. The clear pattern of structure in limbs has allowed the consequences of various microsurgical manipulations to be assessed. The results have been analyzed in terms of several intriguing models. These should be kept in mind when thinking of developmental processes throughout embryogenesis.

Here in these terminally differentiating tissues, the role of specific genes and processes can be clearly defined. The mechanisms that were discussed in Part I come into play in unique and specialized ways. By going into a developmental system in some depth, one can assess the relative importance of a given process. The answers are not all in yet, but that just increases appreciation of the available facts. During the coming years more and more will be known about these and other systems.

CHAPTER 11

Developmental Processes in *Dictyostelium*

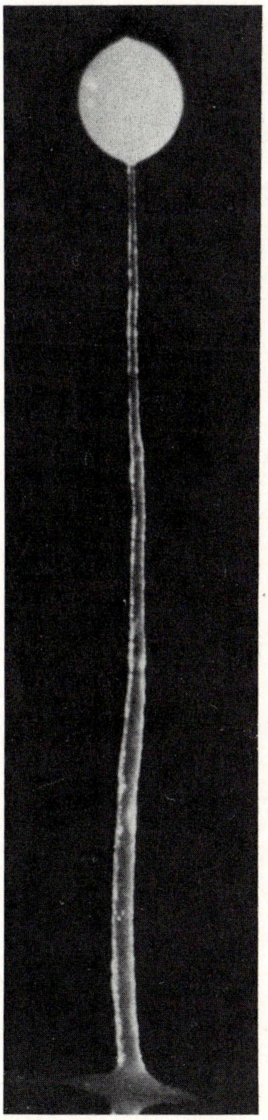

Dictyostelium (Fig. 11.1) solves many of the problems that occur in early embryogenesis: how to integrate a large number of cells so that they proceed through temporal stages in synchrony; how to generate different cell types in the proper proportions and in the proper positions; and how to construct optimally functional multicellular forms. This eukaryotic organism has received considerable attention because it develops into two very distinct terminal cell types: spores and stalk cells. Both cell types are biochemically, as well as morphologically, very different from the amoebae present at the start of development (Fig. 11.2).

The basic problem of how a group of apparently identical cells can diverge to form two distinct tissue types is presented in extreme simplicity. Because very large numbers of cells (up to 10^{11} at a time) can be induced to develop synchonously, many high-resolution biochemical analyses have been carried out on *Dictyostelium*.

These soil amoebae can be grown in the laboratory on defined medium. Because they grow by binary fission, the population increases exponentially. When development is induced in an exponentially growing culture, initially all the cells are identical with the exception that they are

Figure 11.1 Fruiting body of *Dictyostelium discoideum*. A ball of about 10^5 spores, referred to as a sorus, is held up on a cellular stalk about 2 mm high. The spores are dispersed when the fruiting body is disturbed.

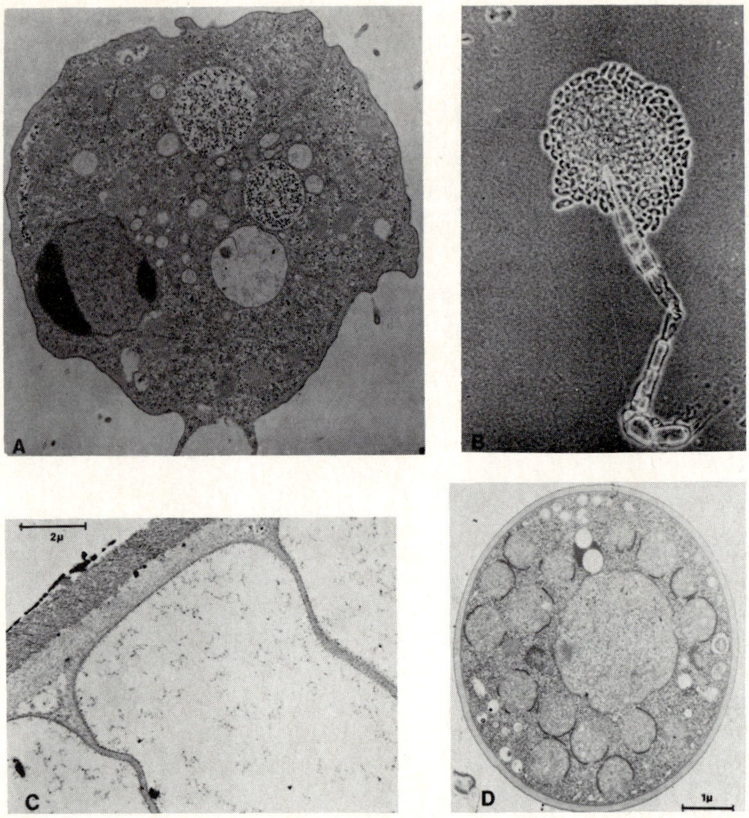

Figure 11.2 Cells of *Dictyostelium discoideum.* *(A)* At the start of development the cells are unspecialized motile amoebae with prominent nucleoli visible within the nucleus (transmission electron micrograph [TEM]). *(B)* Development leads to the formation of a fruiting body with spores held up on a stalk packed with stalk cells. This is an exceptionally small fruiting body, which usually consists of 10^4 to 10^5 cells (phase contrast micrograph). *(C)* Stalk cells vacuolize within the stalk tube, which consists of aligned cellulose fibers. These cells are dead (TEM). *(D)* Spores are enclosed in an extracellular spore coat. They have condensed down by losing water. The internal membranes are studded with ribosomes. Upon germination the cell will swell and break out of the spore coat (TEM).

distributed randomly throughout the cell cycle; that is, some will be in S phase and others in G2. Since all cells arrest in G1 phase shortly after the initiation of development and proceed to terminal differentiation in the absence of further division, differentiation is uniform among billions of cells.

Development of *Dictyostelium* is normally induced by the replacement of nutrient medium by a buffered salts solution; the withdrawal of exogenous amino acids triggers the cells to stop cell growth and start the biochemical differentiations that will ultimately lead to formation of spore and stalk cells. Because development occurs in the absence of growth, one need not worry that a change in proportions of a given cell type might be due to differential growth rates, a process that plays a major role in the development of many tissues in embryos. In *Dictyostelium* a change in cell types indicates conversion of one cell type to another.

Within a few hours following the initiation of development the previously solitary amoebae become sociable and aggregate into groups of about 10^5 cells (Fig. 11.3). This is about the same number of cells found in a gastrula, a fly, or a hydra. Once in an aggregate, the cells differentiate into two distinct cell types: pre-spore and pre-stalk cells. The pre-stalk cells are found predominantly at the anterior of the elongating slug-shaped organism while the pre-spore cells are in the posterior. As their names imply, pre-spore cells later differentiate into spores whereas pre-stalk cells differentiate into stalk

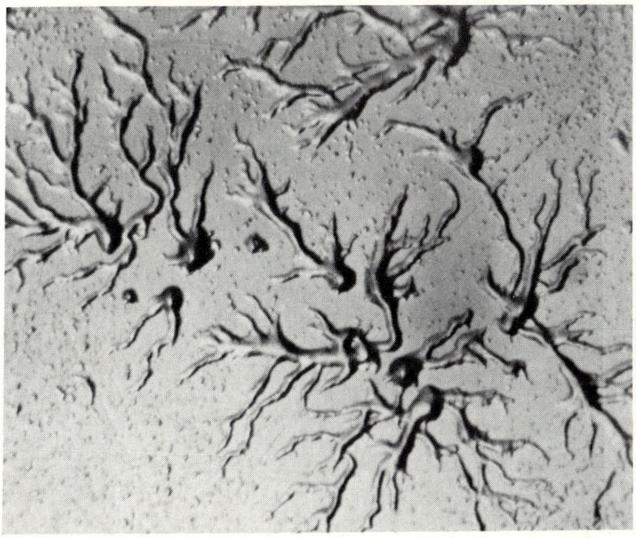

Figure 11.3 Streaming aggregates. *Dictyostelium* cells are chemotactically attracted to cAMP, which they secrete. A relay system can pass the signal over distances of several centimeters. About 10^5 cells converge on an aggregation center along streams.

cells. Since about 80 percent of the total number of cells in an aggregate form spores, about 80 percent of the slug cells are pre-spore cells. The remainder are pre-stalk cells. Thus, we have a beautifully simple fate map in slugs with the anterior 20 percent distined to be stalk cells and the posterior 80 percent destined to be spore cells (Fig. 11.4).

After 18 hours of development, terminal differentiation into spores and stalk cells can occur. The relative positions of the anterior and posterior cells

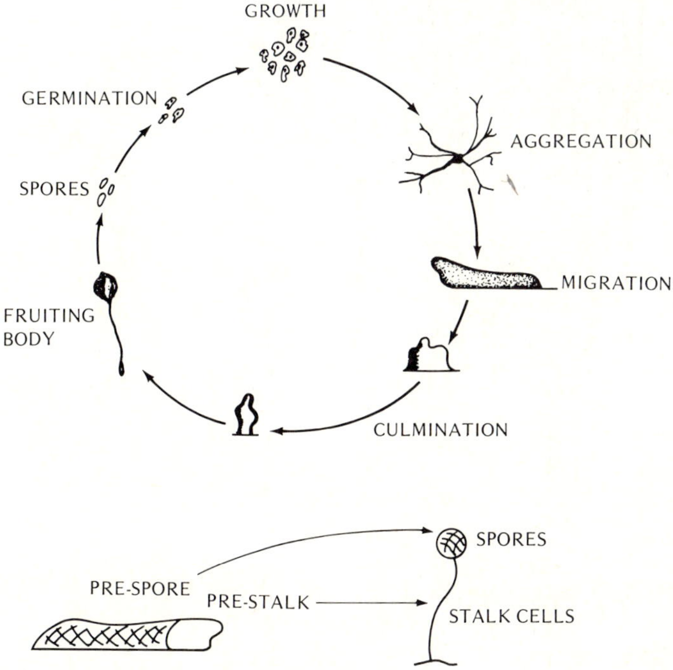

Figure 11.4 Life cycle and fate map of *Dictyostelium.* Amoebae aggregate after 8 hours of development, form slugs in the next 8 hours, and culminate to form fruiting bodies by 24 hours of development. Spores take 4 hours to germinate. The amoebae that emerge can grow with a doubling time of 4 hours. Cells at the anterior of a slug differentiate into stalk cells while those in the posterior three-quarters encapsulate into spores. The pre-spore and pre-stalk cells in slugs can be distinguished by several biochemical specializations.

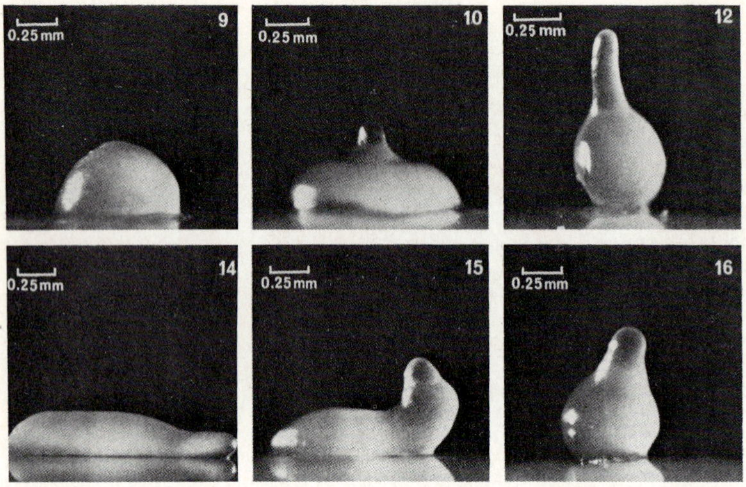

(A)

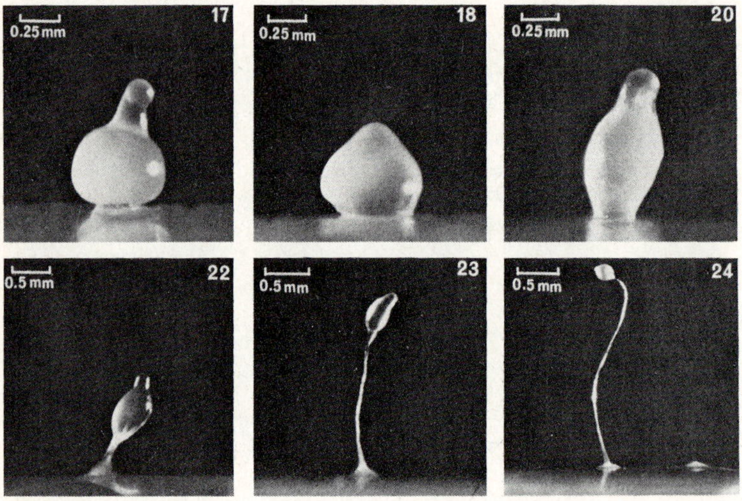

(B)

Figure 11.5 Morphogenesis of *Dictyostelium discoideum*. Nine hours after the initiation of development, tight aggregates are covered with an extracellular matrix that acts as a sheath. The shape of the aggregate changes every hour, forming a tip at 10 hours, a neck at 12 hours, a slug at 14 hours and so on. The last three photographs (at 22, 23, 24 hours) are enlarged half as much as the others to show all of the 2-mm fruiting body.

are reversed by the movement of pre-stalk cells through the pre-spore cells behind them. The pre-stalk cells enter the top of the stalk in a movement compared to a fountain running backwards. The pre-spore cells are lifted on the extending stalk until they are held aloft. Then they encapsulate into dormant spores. From that height above the support their chances of good dispersal are increased. This whole developmental process takes 24 hours and is so regular that you can set your watch by it. The extreme synchrony allows the biochemical steps to be carefully monitored at hourly intervals (Fig. 11.5).

Chemotaxis

For initially dispersed cells to make an aggregate, they must have a signaling system to attract each other. During the first few hours of development they synthesize a chemotactic relay system and become responsive to the attractant. A variety of different compounds including metabolites and small peptides have evolved as the chemotactic agents in species closely related to *Dictyostelium.* In the most intensively studies species, *D. discoideum,* the chemoattractant is cyclic AMP (cAMP).

At the start of the development amoebae of *D. discoideum* synthesize only very low levels of cAMP, and the enzyme responsible for its synthesis from ATP, adenyl cyclase, is undetectable. Within a few hours the specific activity of adenyl cyclase increases to reach a peak at 6 hours of development. The specific activity then gradually decreases throughout the remainder of development (Fig. 11.6).

Since the increase in specific activity of adenyl cyclase can be inhibited by either cycloheximide, an inhibitor of protein synthesis, or actinomycin D, an inhibitor of RNA synthesis, it is likely that adenyl cyclase accumulates as the result of the developmental activation of its gene and translation of the resulting adenyl cyclase mRNA. However, direct proof that the adenyl cyclase gene is induced must await isolation of a cloned segment carrying the gene.

The measured amount of adenyl cyclase activity is sufficient to account for the measured rate of cAMP synthesized in aggregating cells. To activate the enzyme the cells must be stimulated by a pulse of cAMP flowing over

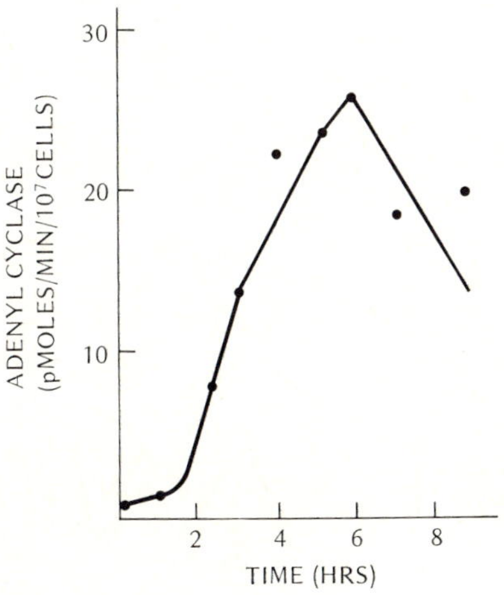

Figure 11.6 Developmental kinetics of adenyl cyclase. The specific activity of adenyl cyclase was measured at hourly intervals following the initiation of development in *D. discoideum.*

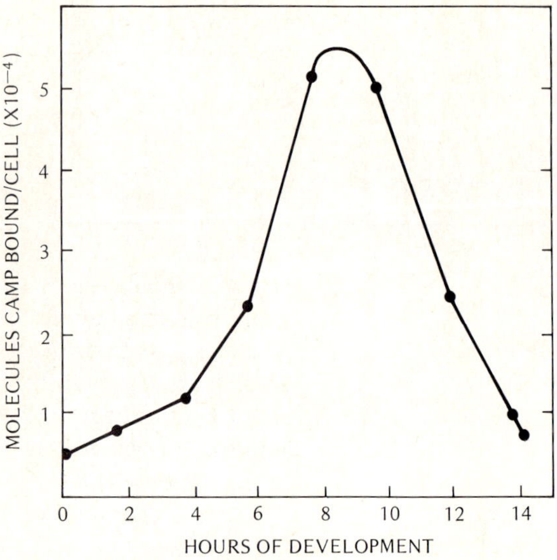

Figure 11.7 cAMP-binding sites. The ability of intact cells to bind cAMP was measured every two hours during early development of *D. discoideum.* The concentration of cAMP that the cells were suspended in was 40 n*M*.

them. This is what happens normally in aggregating cells that are near other cells signaling them. This stimulation of adenyl cyclase activity by exogenous cAMP results in a relay of the signal because each cell secretes cAMP to about a tenfold higher level than it is exposed to. The process is similar to the microwave relay system that periodically, over a continent, receives a signal from the East, amplifies it, and passes it on to the West. It also shows that developing cells of *Dictyostelium* have excitable membranes. Not only is adenyl cyclase inserted into the membrane but a specific cAMP receptor protein of 41,000 daltons is also added to the surface. The signal, cAMP, does not enter the cell but triggers a response in which another molecule, cGMP, is used as the internal relay. The surface cAMP-binding protein is absent from cells during the first few hours of development but is rapidly inserted into the membrane to reach a peak at 8 hours (Fig. 11.7). It is removed after aggregation is over.

This same 41 kd cAMP-binding protein also appears to be involved in directing the movement of each cell toward the region of highest cAMP concentration. In the absence of a gradient in cAMP, cellular movement is random. However, when cAMP is added from a given direction the cells rapidly move by pseudopodal action toward the highest level of cAMP. Chemotactic movement is also mediated internally by stimulation of guanyl cyclase. Mutations that result in perturbation of cGMP metabolism have dramatic effects on the pattern of chemotactic cell movement. It is not known exactly how cGMP controls cellular movement; however, Ca^{++} ions are strongly implicated. Within seconds of chemotactic stimulation of cells with cAMP the internal Ca^{++} concentration increases. In association with the calcium-binding protein, calmodulin, Ca^{++} inactivates a protein kinase that phosphorylates myosin heavy chains. Myosin then becomes activated by dephosphorylation and forms thick filaments just under the cell surface. Contraction at the rear forces extension of pseudopods at the front. The whole process of cAMP stimulation of cAMP synthesis and secretion as well as chemotactic movement results in periodic lurches of the cells toward a central collecting point where the density of cells is greatest. The details of the biochemistry, physiology, and diffusion of cAMP are so well worked out in this system that a variety of behavioral aspects can all be predicted by exact mathematical equations. This may be only one aspect of morphogenesis, but it is satisfying to have it described mathematically.

In the wild as well as in the laboratory, *Dictyostelium* is able to aggregate whether it is present in either a dense or a sparse population. Yet the potential to saturate the environment with cAMP in dense populations must be balanced with the sensitivity to low levels of cAMP found in sparse populations. This is achieved by a regulated enzyme system consisting of a cAMP phosphodiesterase and a specific protein that binds the phosphodiesterase and effectively inhibits it. Under conditions in which the cAMP in the environment is low both proteins are made and secreted between hours 2 and 12 of development. The activity of phosphodiesterase is kept low by the inhibitor. But, under conditions in which the cAMP in the environment is high, synthesis of the inhibitor is repressed and the phosphodiesterase is fully active. It can then reduce the cAMP levels between bursts of synthesis and secretion such that the noise level of cAMP does not drown out the gradient required for chemotactic movement. The relative rates of synthesis of both cAMP phosphodiesterase and its inhibitor correlate with the levels of their respective mRNAs. Thus, transcription of both genes is developmentally induced while that of the inhibitor is also subject to repression by exogenous cAMP.

Control of a few genes, adenyl cyclase, cAMP-binding protein, cAMP phosphodiesterase, and its inhibitor, appear, at least superficially, to account for the dramatic change in behavior seen in the first few hours of development of *Dictyostelium.* These few changes give some insight into how a cell can become both excitable and responsive.

Adhesion and Multicellularity

Just getting the cells together is not sufficient to make a multicellular organism. The cells must stick to each other. At the start of development the amoebae are not mutually adhesive, but by 8 hours of development they have changed their surface properties so that they form tight clumps when they encounter each other. One of the molecules responsible for cell–cell adhesion is a glycoprotein of about 80,000 daltons (gp80). It is synthesized and inserted into the membrance between 6 and 12 hours. Monovalent antibodies, including a monoclonal one, reactive with gp80 can block cell–cell adhesion. Moreover, mutations in a specific gene, *modB,* that interfere with the addition of the carbohydrate side groups to gp80 significantly reduce cell–cell adhesion. Although gp80 may not be the only component involved in the developmentally regulated adhesion mechanism of *Dictyostelium,* it does appear to play an essential role.

These results indicate that carbohydrates carried by gp80 play an essential role in the adhesion system that functions between 8 and 14 hours of development in *Dictyostelium.* As the cells transform into migrating slugs, another adhesive system is brought into play. Adhesion late in development is even stronger than that in aggregates as judged by the ability of dissociated cells to form clumps when shaken in the presence of the chelator of divalent cations, EDTA. Monoclonal antibodies to gp80 that block adhesion of cells developed for 12 hours do not block adhesion of cells developed for 18 hours. However, antibodies to surface antigens present on 18-hour cells can block adhesion late in development. It is not yet clear exactly what molecules elicit the late adhesion, but a glycoprotein of 95 kd is a prime candidate that is presently being analyzed.

These detailed studies have pointed out that a developing system can pass through a series of stages of adhesion using different systems adapted to the changing needs of the organism. There are indications that cell–cell adhesive systems in chick embryos go through similar stages, although different molecular mechanisms are used.

As with many developing tissues, aggregates of *Dictyostelium* become ensheathed with a covering of extracellular matrix. A thin layer consisting of protein and carbohydrate surrounds the group of 10^5 cells from the aggregation stage until the end of development. The sheath is reinforced with cellulose fibers and is sufficiently strong to exclude any latecomers and to keep all the cells together during the migration of the slug.

As the slug migrates, new sheath is produced, which is stretched at the front. They move over the sheath, which is then left behind like a collapsed sausage casing. Polarity of movement of the slug is ensured by the localization of distensible sheath at the front. The sheath effectively blocks movement out the sides or backwards. Extracellular matrices may serve the same function for cells of metazoan tissues. Movement in *Dictyostelium* is both phototactic and thermotactic, leading the slug to the light at the top of the soil.

Movement requires traction as well as motive force. One of the mechanisms that may attach cells to the substratum is provided by the molecule discoidin I. This 26 kd protein has a cell-binding domain similar to the one on fibronectin and binds to the surface of *Dictyostelium* cells. It also has affinity to carbohydrates and thus may also provide traction by associating with polysaccharides of the extracellular matrix and surface sheath. It is synthesized from a small family of three related genes during aggregation, stored in excretory vesicles, and secreted when aggregates are formed. Discoidin is then found between the cells and on the inner surface of the sheath. Mutant strains that fail to produce discoidin are significantly impaired in cell-substratum attachment, as would be expected. Unlike wild-type cells, these strains cannot stream together under a variety of conditions.

The sheath of *Dictyostelium* is also a barrier to diffusion except at the anterior where movement distends it. Thus, small molecules diffuse in and out far more readily at the anterior. We can predict that this will result in gradients of small molecules that might be used to direct differentiation dependent on the position of the cells in the tissue. Once again, it is quite possible that extracellular layers surrounding various tissues in metazoans may also regulate the concentration of diffusable molecules so as to result in spatial inhomogeneities. These gradients, in turn, could trigger specific inductions to lead to further cell type specializations.

Cell-Type Specific Differentiations

One of the dramatic differentiations one can see late in the development of *Dictyostelium* is the production of spore coats that encapsulate each spore (Fig. 11.8). Upon germination, an amoeba emerges from the spore coat, which is left behind like an empty shell. The spore coats can then be conveniently collected and analyzed. It turns out that they contain only three major proteins, termed SP96, SP70, and SP60. Monospecific antibodies have been raised to these spore coat proteins and used to tag their appearance during development. The spore coat proteins are first synthesized when the aggregates become surrounded by sheath. They are packaged in specialized vesicles, called pre-spore vesicles, that are found only in pre-spore cells. When the mRNA for SP96 was analyzed using a cloned sequence, it was found that it accumulates only in pre-spore cells and is virtually absent in pre-stalk cells (Fig. 11.9). These results indicate that the gene for SP96 is induced in pre-spore cells while remaining repressed in pre-stalk cells. SP96 is synthesized and accumulated in preparation for the rapid production of a spore coat around each spore during culmination.

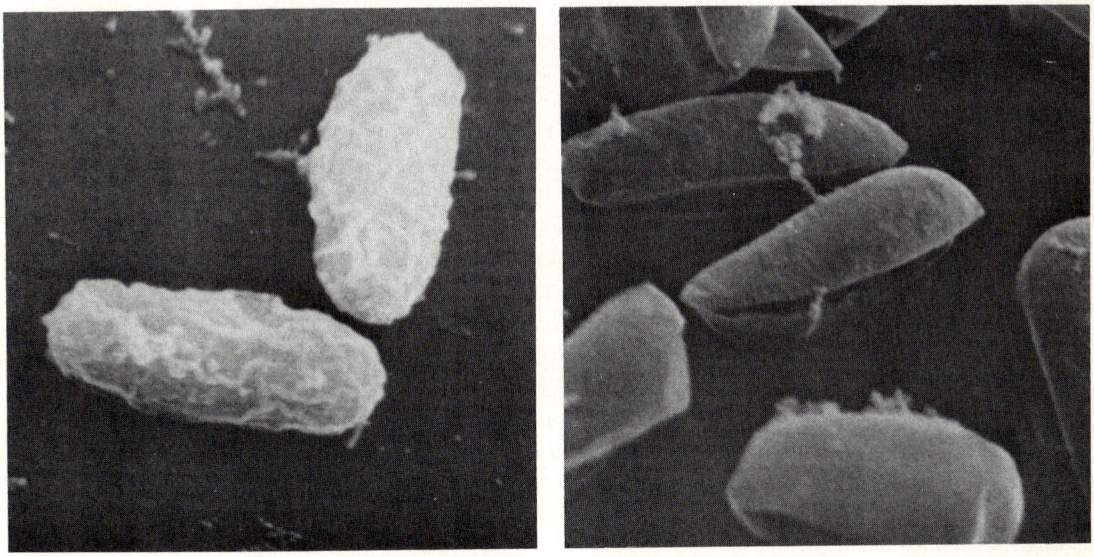

(A) (B)

Figure 11.8 Spores and spore coats. *(A)* The wrinkled surface of spores attests to their shrunken, dehydrated state. After imbibing water, they swell and amoebae emerge. *(B)* The spore coats are stretched and split longitudinally. They are left behind as empty shells.

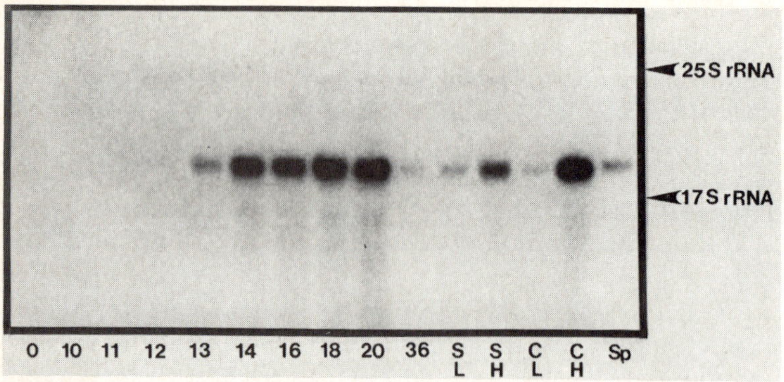

Figure 11.9. Temporal and spatial expression of SP96 mRNA. A Northern blot of RNA extracted from cells at various times (hours) of development was probed with labeled cDNA of the SP96 spore coat protein gene. A 2.2-kb mRNA appears first at 13 hours. Pre-stalk (SL, CL) and pre-spore (SH, CH) cells were isolated from migrating slugs (S) and culminants (C). SP96 mRNA is far more prevalent in pre-spore cells. The ribosomal 25*S* and 17*S* RNA provided size markers.

The spore coat proteins are not the only ones synthesized preferentially in pre-spore cells. Separation of the proteins synthesized during the slug stage on 2D gels has revealed another 15 or so abundant proteins that are synthesized in pre-spore cells but not in pre-stalk cells. Synthesis of about half of these proteins continues throughout the remainder of development and probably produces components necessary for the specialized requirements of spores. The synthesis of the other half of these abundant proteins stops before terminal cytodifferentiation of spores is seen and either produces the components necessary for pre-spore functions or results in sufficient accumulation of the proteins for their function in spores (Fig. 11.10).

One of the abundant pre-spore specific proteins is of particular interest because it is synthesized early in all cells but is preferentially shut off later in development in pre-stalk cells. This points out that cell type divergence can depend equally well on preferential repression of a common gene as on preferential induction of a gene.

In contrast to the pre-spore cells, pre-stalk cells synthesize very few abundant proteins preferentially until the start of terminal cytodifferentiation when they swell and lay down the thick cellulose walls characteristic of stalk cells. During stalk formation about a dozen abundant proteins are made exclusively in stalk cells. So far, nothing can be said about the rare proteins that might be synthesized exclusively in one cell type or the other.

Shortly after the cells have formed multicellular aggregates there are changes in the enzymatic systems that modify proteins. A cAMP-dependent protein kinase appears that adds phosphate to serine and threonine groups on a variety of proteins. Phosphorylation can dramatically change the function of a protein so that new processes occur. At about the same time the mechanisms of trimming and modifying the carbohydrate side chains attached to asparaginyl groups in glycoproteins also change. Post-translational

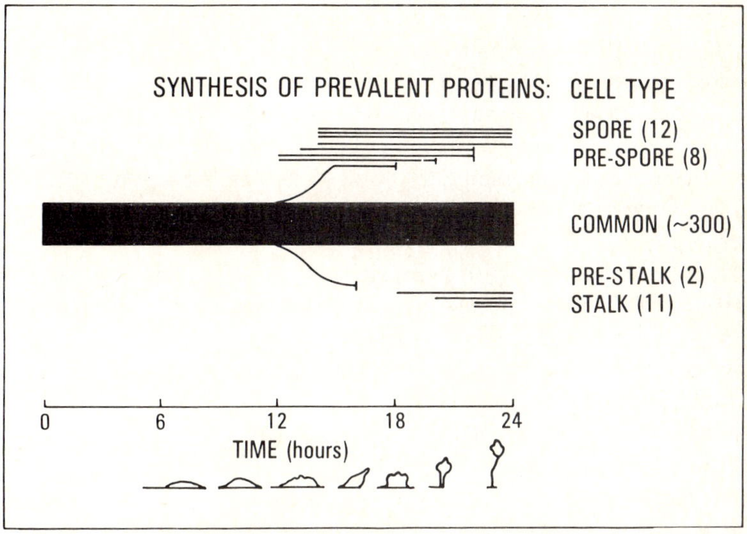

Figure 11.10 Cell type specialization. The synthesis and accumulation of over 300 proteins was measured at 2-hour intervals during development. Most proteins were common to both pre-spore and pre-stalk cells, but a few dozen were cell-type specific. Pre-spore and spore-specific proteins appear following formation of aggregates. Pre-stalk and stalk-specific proteins do not appear until culmination. A few cell-type specific proteins are first synthesized in all cells and then preferentially repressed in one cell type or the other.

modification of several lysosomal enzymes follows a different pathway in prestalk cells than that which is followed in pre-spore cells. Although we do not know all the consequences of these modifications to the functioning of the cells, differentiation of the modification systems themselves is an effective way to simultaneously change the role of a whole series of gene products served by these systems.

Another dozen or so genes preferentially expressed in either pre-spore or pre-stalk cells have been recognized in studies on cloned sequences expressed in slug cells. Some of these genes are expressed at relatively low rates and may code for less abundant proteins. Continued expression of many of these genes requires the conditions found within the slug, that is, high cell density and extracellular cAMP. These appear to be necessary exogenous signals for the expression of the cell-type specific genes.

If groups of cells are taken from the posterior of a slug and induced to undergo terminal differentiation on their own, they will rapidly form spores and a few stalk cells. Groups of cells taken from the anterior will differentiate almost exclusively into stalk cells and form few if any spores. The ability to encapsulate into spores is found only if one takes cells about 25 percent back from the front. This is also the position at which one finds cells that have accumulated the spore coat proteins SP96 and SP70. It appears that the anterior cells have not prepared to encapsulate into spores. But if one allows groups of anterior cells 24 or more hours to regulate before undergoing terminal differentiation, then the normal proportion of cells (80 percent) will encapsulate into spores. During this period of regulation, cells in the isolated groups of anterior cells will synthesize spore coat proteins among other things. Thus, it appears that in unmolested slugs a signal comes from the posterior that specifically represses expression of these genes in anterior cells. Alternatively, the change in topology that occurs upon isolation of anterior fragments gives rise to the conditions necessary for induction of these genes.

The proportion of cells that accumulate spore coat proteins is constant at about 80 percent in slugs that differ in total number of cells by a factor of 10 or more. It is sufficiently constant among individual slugs that it is highly unlikely that the proportion of pre-spore cells was set by a random differentiation mechanism. More likely, a fieldwide signaling system regulates the proportions of pre-spore and pre-stalk cells. The cells apparently can sense the total number of cells in the field, in this case the whole slug, and adapt in a manner that ensures the proper proportions.

Synthesis of the proteins characteristic of pre-spore cells starts at about 14 hours of development whereas synthesis of their mRNAs starts an hour or so earlier. At this time in development cells have entered an aggregate and been surrounded by the extracellular sheath. Cells at the top of the mound or at the center might be in an environment different from the rest of the cells and express certain genes preferentially. The differences in environment might then be stabilized by the polarized movement of the cells relative to the sheath during subsequent migration of the slug. The nature of the components in the environment that give rise to differential gene expression is still not clear, although cAMP and NH_3 are interesting candidates.

Terminal Differentiation

Development of *Dictyostelium* leads to dormant spores held aloft on cellulose stalks filled with dead, vacuolated cells. From these the spores can be dispersed by wind, blowing leaves, or low-flying insects. The spores are small, almost completely dehydrated cells, whereas stalk cells have allowed water to fill a central vacuole that swells to four or five times the volume of the initial amoebae. Thus, one difference between spores and stalk cells is that the former pump most of the water out of the cell while the later allow water to enter. Both cells make large amounts of cellulose during terminal differentiation; spores put it into the spore coat while stalk cells put it into the stalk and their own surrounding cellulose cases. Although the stalk cells die in the process of terminal differentiation, spore cells keep a considerable amount of reserves for use in germination. A major form of these energy reserves is the disaccharide trehalose. All the cells in the slug, both pre-spore and pre-stalk cells, accumulate trehalose but stalk cells convert it for the synthesis of cellulose during terminal differentiation. The enzyme trehalase is active in stalk cells during terminal differentiation, whereas it is active in spore

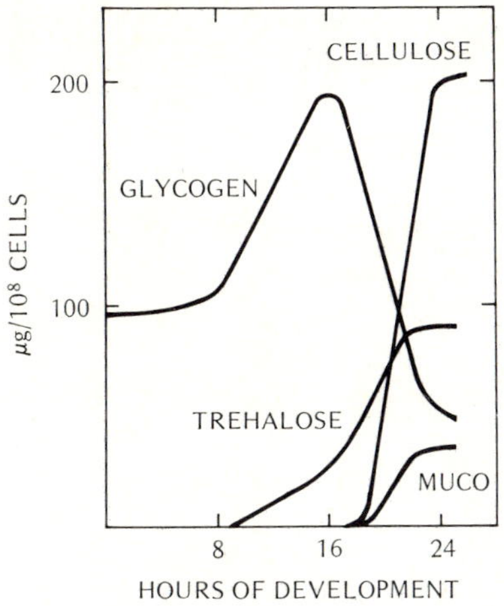

Figure 11.11 Carbohydrate changes during development. Glycogen accumulates until culmination when it is converted into cellulose, trehalose, and mucopolysaccharide. Trehalose starts to accumulate in aggregates.

cells only following germination. Thus, simple control of trehalase activity results in cells with quite different functions.

The spore coat also contains a galactose-rich mucopolysaccharide. It is synthesized by an enzyme, UDP galactose polysaccharide transferase, that is synthesized only in pre-spore cells. Thus, the localization of this enzyme to

Reaction	Enzyme
1.	Glycogen synthetase
2.	Amylase
3.	Maltase
4.	Glucokinase
5.	Glycogen phosphorylase
6.	Phosphoglucomutase
7.	UDPGlc pyrophosphorylase
8.	Trehalose-6-P synthetase
9.	UDPGal epimerase
10.	UDPGal polysaccharide transferase
11.	Cellulose synthetase

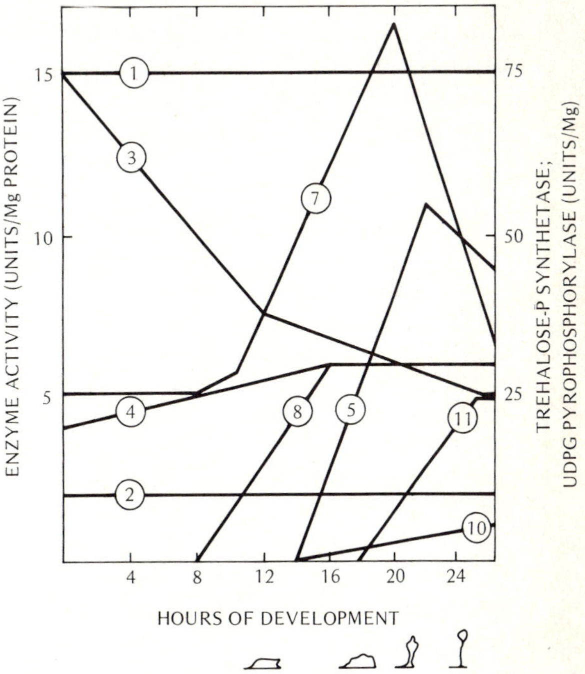

Figure 11.12 Developmental kinetics of pertinent enzymes. The specific activities of the enzymes were measured throughout development. The numbers refer to the enzymes in the table. One unit is defined as the amount of enzyme necessary to produce 1 μmole per minute at 25° when the substrate is saturating.

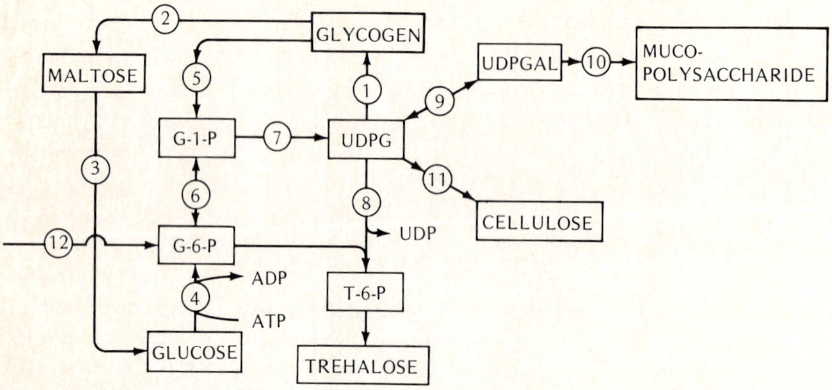

Figure 11.13 Metabolic pathways leading to various carbohydrates. The enzymes are numbered in the same way as those in Figure 11.12. Gluconeogenesis produces glucose-6-phosphate (reaction 12). Reactions 6 and 9 are reversible whereas all the others are essentially irreversible for biochemical reasons.

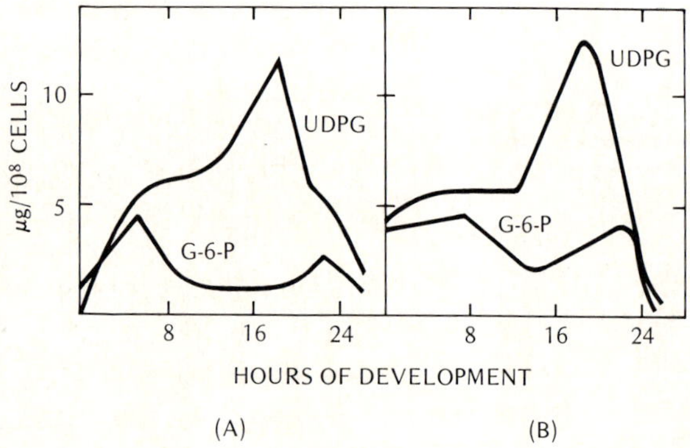

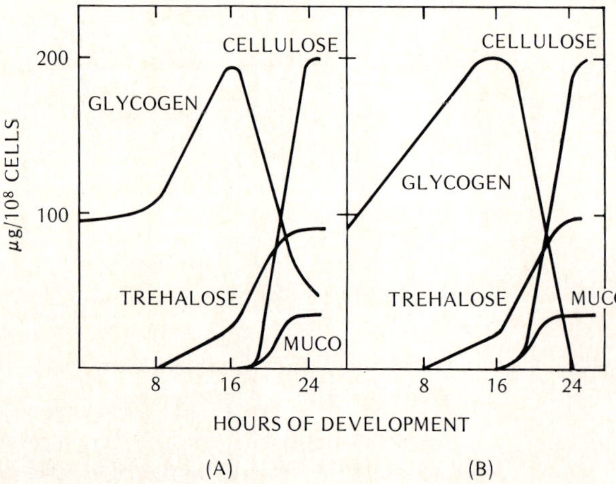

Figure 11.14 Measured and calculated changes in metabolites and polysaccharides. *(A)* Measured values can be compared to *(B)* calculated values for UDP glucose, glucose-6-phosphate, glycogen, trehalose, cellulose, and mucopolysaccharide. The calculated values were arrived at by using the specific activity of each enzyme measured throughout development and the *in vitro* measured Km of each enzyme. The substrates for trehalose synthesis (UDPG and G6P) are both limiting and this was taken account in the equation for this enzyme. The similarity of measured and calculated values indicates that the metabolic map and *in vitro* assays closely approximate the *in vivo* conditions.

a specific cell type prepares them for subsequent spore differentiation. Like the other pre-spore–specific proteins, UDP galactose polysaccharide transferase first appears after about 14 hours of development and is localized in posterior cells.

This analysis has pointed out the dramatic changes in carbohydrates that accompany development in *Dictyostelium*. Glycogen accumulates up until terminal differentiation and then is rapidly converted into cellulose, trehalose, and mucopolysaccharide (Fig. 11.11). Most of the enzymes involved in glycogen metabolism and synthesis of the other carbohydrates have been measured at hourly intervals throughout development. Striking changes are seen in some but not all of these enzymes (Fig. 11.12). The pattern of temporal changes in these enzymes in some cases accounts for the change in various carbohydrates but in other cases the consequences are not so obvious. If one looks at the pathway leading to and from the various carbohydrates, one sees that certain intermediates such as the hexose phosphates and UDP glucose are common to the different pathways (Fig. 11.13).

The flux down each pathway will depend not only on the relative activity of the competing enzymes but also on the relative affinity of the enzymes for their substrates. In each case the Km of the enzymes and changes in their specific activity were measured and the data put into a computer that was programmed with the information in Figures 11.11 and 11.12. At 1 minute intervals of development, changes in the intermediary metabolites and polysaccharides were calculated. The results (Fig. 11.14) showed that the measured changes in enzyme activities actually do account for the observed changes in both small molecules and macromolecules.

Thus, changes in only a dozen or so enzymes can account for the dramatic conversion of specific carbohydrates at the various stages of differentiation in *Dictyostelium*. It does not take changes in the expression of many genes to alter fairly radically the physiology of a cell. Of course, fine tuning may require the regulation of many more genes.

The Number of Genes in *Dictyostelium*

The genome of *Dictyostelium* contains about 50,000 kilobase pairs (5×10^4 kb) of DNA. If all of this DNA coded for mRNA there could be up to 20,000 genes. However, about half of the DNA is present as sequences that are wastelands containing only *A* and *T* that are not transcribed. A variety of lines of evidence indicate that growing cells express only about 5,000 genes. Many of these genes are also expressed throughout development. The number of genes that can mutate to affect some aspect of morphogenesis yet not affect the growth of the cells is found to be about 300. On the other hand, direct nucleic acid analysis has indicated that enough DNA is transcribed exclusively during development that there might be several thousand developmental genes. Either most of these transcripts play subtle supportive roles during morphogenesis or transcriptional control is not perfectly efficient.

The fact that only a small proportion of the total genome appears to play a selectively advantageous role in the individual (~25 percent in the case of *Dictyostelium*) is common among all multicellular organisms. It comes about as the consequence of the small proportion of total energy expenditure necessary for replication of a genome filled with dispensible sequences. Movement of cells takes far, far more energy. The extra DNA is very light baggage to carry around. The dispensible sequences appear to arise from parasitic replication units (selfish DNA) and the very nature of genetic information. Duplications of sequences that occur by rare mistakes in replication or genetic exchange are seldom deleterious. Deletions that could remove dispensible sequences are exceedingly dangerous. If the deletion includes even a small portion of an essential gene or its regulatory element, the individual will be eliminated from the population. Only when the amount of dispensible DNA presents a target size that results in the rate of acceptable deletions being equal to the rate of duplication will the size of the genome stabilize. This analysis predicts that all metozoans will have considerably more DNA in their genomes than actually required to code for their genes. This has been found to be the case in a wide variety of organisms.

Causal Sequences

Vertebrate embryos gastrulate before they form a neural plate and neurulate before they form optic vesicles. The statement that gastrulation is necessary for eye formation appears simplistically obvious. Events that occur in a temporal sequence are often linked causally. But at the detailed level this need not be so. Consider a group of events (A, B, C, D, etc.) that are all triggered by the same signal but are expressed after lags of various times (Fig. 11.15). These four events are not causally connected to each other. If A does not occur, B, C, and D still take place. The arrangement might involve either multiple timers or a single timer. In either case, the events in the temporal sequence are independent.

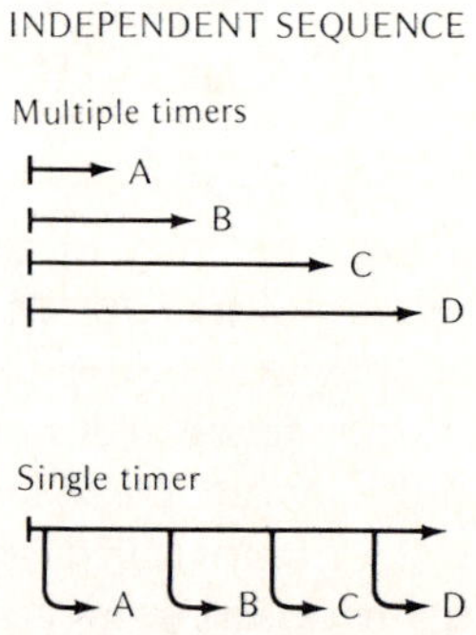

Figure 11.15 Events unlinked by causal relationships may occur in a temporal order independent of each other. Somite formation in a sequence from the anterior to the posterior of a chick embryo is an example of an independent sequence; blocking formation of one somite does not stop the formation of subsequent somites toward the posterior.

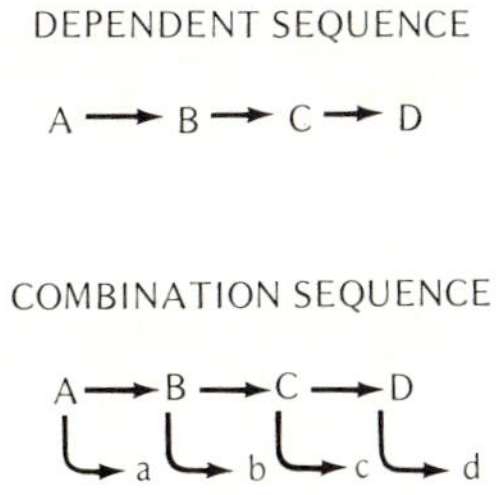

Figure 11.16 Causally linked events in which the last member is dependent on all prior ones. Synthesis of crystallins in lens cells induced by optic vesicles derived from the neural plate over the notochord formed during gastrulation is an example of a dependent sequence. A block in any one of these steps, blocks all subsequent events. Some events *(lower case letters)* may be triggered by causally linked events but are not themselves required for subsequent events. The formation of neural crest cells is not required for crystallin synthesis in the lens, for instance, yet both are dependent on neurulation.

In a dependent sequence, such as that of gastrulation → neurulation → optic vesicles → eyes, each event is causally dependent on the preceding events (Fig. 11.16).

One of the ways to determine the causal connections between events is to specifically perturb one and measure the others. Mutations can serve as "magic bullets," knocking out a single gene activity. For instance, mutations that inactivate the enzyme, UDPG pyrophosphorylase, which is responsible for the synthesis of UDPG, result in a block in terminal differentiation of *Dictyostelium*. UDPG is the precursor used in cellulose synthesis and no cellulose is made in these mutants. The pre-spore cells do not encapsulate and the pre-stalk cells do not make a stalk. Clearly there is a dependent relationship between the expression of the gene for UDPG phyrophosphorylase and fruiting body formation.

Several other genes are expressed at the same stage of development as the gene for UDPG pyrophosphorylase, including genes for several lysosomal hydrolases. Yet mutations in these developmentally regulated genes do not stop morphogenesis. Therefore, these events are part of an independent sequence. The hydrolases may be necessary for fine tuning the selective advantage of the organism under some conditions, but they are clearly dispensible for the gross aspects of morphogenesis under laboratory conditions.

Other mutants have been analyzed that were selected because gross morphogenesis was arrested at some stage. In some of these mutant strains development proceeds through the slug stage and then stops; in others the cells fail even to aggregate. By looking at a dozen developmentally regulated enzymes, several membrane proteins, and selected cloned sequences, it has become apparent that there is a central dependent sequence of events that controls the expression of these marker genes (Fig. 11.17). Mutations affecting step C block the expression of genes dependent on C as well as that of *all* genes controlled by subsequent steps in the dependent pathway.

The nature of the central events is not yet known in molecular detail; they are defined only by the mutations in the stains analyzed. Moreover, this is a bare sketch of the complexity of the system; only a small proportion of the 300 or so developmental genes were analyzed. Reality most likely includes branch points and convergent steps as well as subdivisions of the sequence.

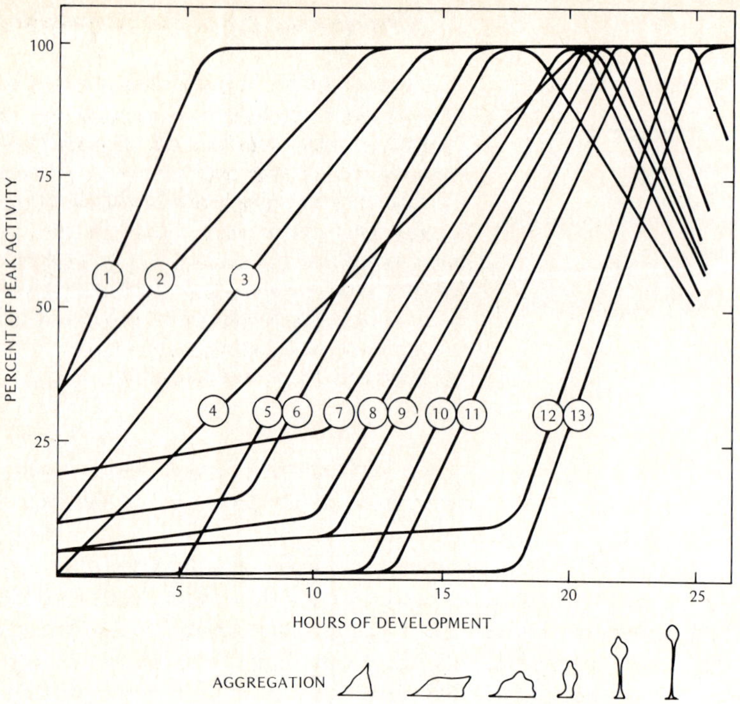

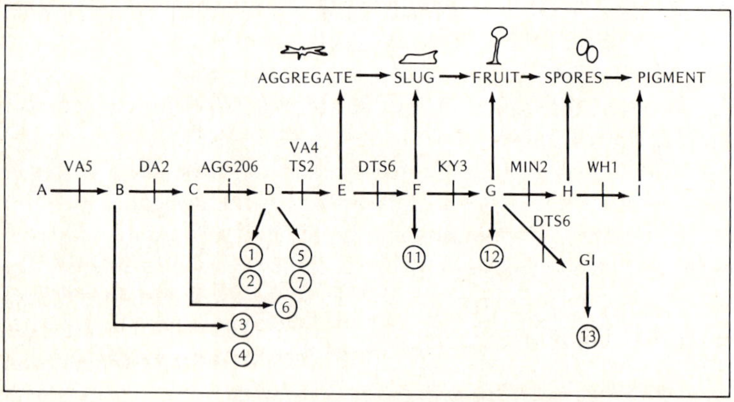

Figure 11.17 Temporally regulated genes controlled by the dependent sequence of development in *Dictyostelium*. Thirteen enzymes have been used as markers for temporal control. Direct genetic and molecular evidence for over half of these enzymes shows that they are products of unique genes. Indirect evidence for the others strongly indicates they are unique gene products. The specific activities of these enzymes increase at various times during development, indicating at least five different temporal stages. Analysis of these gene products has been carried out in ten mutant strains. In some strains (VA5, DA2, Agg206) only a few of the early marker enzymes accumulate and none of the later ones appear. In other strains (VA4, TS2, DTS6, KY3) later markers are turned on but not the culmination-specific markers, alkaline phosphatase (12) or β-glucosidase (13). The mutation in strain DTS6 is temperature sensitive and affects two steps depending on the temperature regimen. These and other results can be interpreted to indicate the presence of a central dependent pathway that controls independent events marked by the accumulation of these gene products.

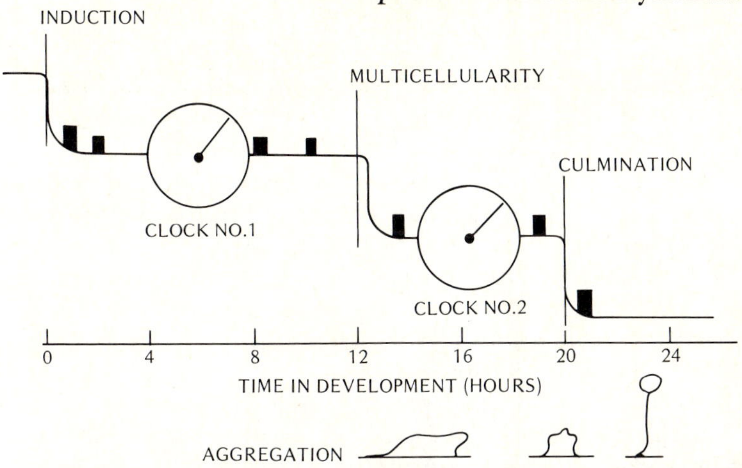

Figure 11.18 Conceptual cascade of temporal regulation in *Dictyostelium*. Three major external signals regulate gene expression: (1) Induction of development occurs where nutrients are removed. (2) Multicellularity is required for late gene expression. (3) Culmination is triggered when the pNH_3 drops below a threshold. Some genes are expressed immediately following induction while others are delayed 8 to 10 hours. Therefore, a timing mechanism (clock 1) is required. Likewise, some late genes are expressed immediately following multicellularity at 12 hours of development while others are delayed 4 to 6 hours. A second timing mechanism (clock 2) is required. The solid blocks indicate the initiation of expression of defined genes.

The development of *Dictyostelium* also involves external signals (Fig. 11.18). There is a clear dependence on acquiring the multicellular state found in an aggregate before expression of late genes can be induced. In some cases this external signal can be bypassed by raising the internal cAMP concentration of the cells by addition of high levels of cAMP. There is also a signal that indicates when a slug can initiate terminal differentiation. External ammonia appears to play a role at this stage. Developing cells secrete ammonia as a by-product of amino acid catabolism. As the protein reserves dwindle, ammonia production declines. Finally, when the ammonia levels drop below a threshold, processes necessary for terminal differentiation into spores and stalk cells can take place. Similar signal systems and dependent sequences may occur in the development of other organisms.

Cellular interactions signal passage through at least five stages of differentiation in *Dictyostelium* (Fig. 11.18): as the density of cells reaches a threshold, certain early genes are induced; depletion of nutrients triggers the second set of genes including that for the cAMP receptor; cells can then respond to pulses of cAMP by activating a third set of genes and forming multicellular aggregates; this induces a fourth set of genes that prepare the cells for terminal differentiation; slugs can remain in this penultimate stage for extended periods of time, but, when the ammonia level drops, the final set of genes is expressed and cytodifferentiation of spores and stalk cells proceeds. At the same time as this temporal sequence is unfolding, a spatial pattern of cell-type specific gene expression is imposed so that posterior cells differentiate as pre-spore cells while anterior cells become pre-stalk cells in preparation for forming fruiting bodies.

Summary

There is a temporal and spatial sequence of biochemical and morphological events during the differentiation of *Dictyostelium* amoebae into the spores and stalk cells of the terminal fruiting body. An underlying causal sequence has been defined by the behavior of a series of mutant strains. Aggregation of cells during early development requires the expression of about 100 developmentally regulated genes. Changes in a half-dozen enzymes and membrane proteins accounts for the newly acquired chemotactic and adhesive properties of the cells. Changes in another dozen or so genes differentiate two cell types in the aggregate, pre-spore cells and pre-stalk cells. The proportions of these cells are regulated so as to be constant over a wide range of sizes and can adapt to the experimental removal of one cell type or the other. Carbohydrate metabolism is modified in both cell types by changes in the relative activity of five to ten key enzymes, so that glycogen is accumulated for eventual conversion to cellulose during formation of the fruiting body. Trehalose and glutamate are stored in spores as reserves for use upon germination.

SPECIFIC EXAMPLE

Dedifferentiation

For many years it was thought that dedifferentiation occurred only in exceptional cases. The classic example was Wolffian lens regeneration in the newt *Triturus*. In the late nineteenth century, Colucci and Wolff found that removal of the lens from the eyes of larval or adult newts was followed by dedifferentiation of the cells in the dorsal rim of the iris and subsequent redifferentiation of these cells to form new lenses. The iris cells are pigmented in the newt and thereby considered differentiated. During dedifferentiation the pigment is lost, and during redifferentiation the genes for cystallins are activated. This regenerative process is not a response to injury of the tissue, since the iris was not cut. It occurs in response to the removal of the lens. Presumably, the lens normally inhibits this response in adjacent epithelial cells of the iris. It was subsequently found that the pigmented epithelium can regenerate a retina as well, as long as a small remnant of retina is left in the injured eye. Thus, epithelial cells can change their function to take on neural roles as well. This process of regeneration occurs in embryos of chicks or mice even when the pigmented epithelium is from one and the retinal fragment is from the other. The demonstration that single pith cells derived from stems of tobacco plants can give rise to complete new plants with a multitude of differentiated tissues (Chapter 1) extended the case for dedifferentiation followed by redifferentiation along new lines.

There are many cell types that fail to dedifferentiate. These include red blood cells and epidermal cells. The former lack a nucleus in mammals and so no longer have the genes for any function; the latter are almost completely filled with keratin and are dead or dying. However, differentiated muscle cells that have accumulated large amounts of actin and myosin can, under some conditions, return to the proliferative phase of myoblasts before once again specializing as contractile cells. Liver cells continue to grow even while carrying out specific digestive functions. The mesenchymal cells at an amputated limb stump lose all sign of specialization and grow to form a blastema. Subsequently, the blastema in some organisms redifferentiates to form the muscle and bones of a regenerated limb. In many cases, the definition of differentiation is not clear, so that dedifferentiation is not recognized. In *Dictyostelium* a variety of well-defined molecular and physiological attributes characterize the stages of differentiation and can be analyzed under conditions of dedifferentiation.

Development of *Dictyostelium* is initiated when the cells are deprived of nutrients. If nutrients are added back at any time prior to terminal differentiation, the cells stop further development and start to grow within about 6 hours. If these cells are taken back out of nutrients after 6 hours and allowed to develop again, they start over from the beginning. All developmental experience has been erased. In fact, if developing cells are refed for only 2 hours, it takes them as long to aggregate following reinitiation of development as it does cells that have been growing for generations. But if developing cells are refed for less than an hour and a half, they reaggregate almost immediately when permitted to develop again. All of the essential physiological functions such as chemotaxis, cAMP relay, and cell–cell adhesiveness are retained for 90 minutes in the presence of nutrients. Any longer under dedifferentiation conditions result in erasure of a critical component necessary for reaggregation.

The first differentiated function to go is the ability to relay the chemotactic signal. At 90 minutes into dedifferentiation, the cells abruptly stop secretion of cAMP in response to a cAMP signal and so they fail to attract surrounding cells. They still retain cAMP binding sites on their surfaces and can respond chemotactically to cAMP but they cannot relay the signal. Shortly thereafter, they lose the ability to respond chemotactically to cAMP because they lose their surface binding sites. They have begun to dedifferentiate.

Nevertheless, dedifferentiating cells that have lost responsiveness to cAMP retain the developmentally acquired ability to adhere to other cells in the presence of EDTA. Only after 6 hours of dedifferentiation is this ability lost. Greater resolution of the sequence of events during dedifferentiation can be reached when critical components are measured at the molecular level. There are several lines of evidence that indicate that a specific glycoprotein, gp80, plays an essential role in the developmentally acquired adhesiveness of the cells. This molecule can be measured by counting the amount of specific anti-gp80 antibody bound to extracts of the cell. Cells at various stages of differentiation and dedifferentiation are broken and their complete complement of proteins solublized with sodium dodecyl sulfate (SDS). The extracts are put in wells on top of a polyacrylamide gel and electrophoretically separated for several hours. Following separation, proteins are transferred to nitrocellulose paper by horizontal electrophoresis. Nitrocellulose has the property of binding almost any protein. The piece of nitrocellulose paper

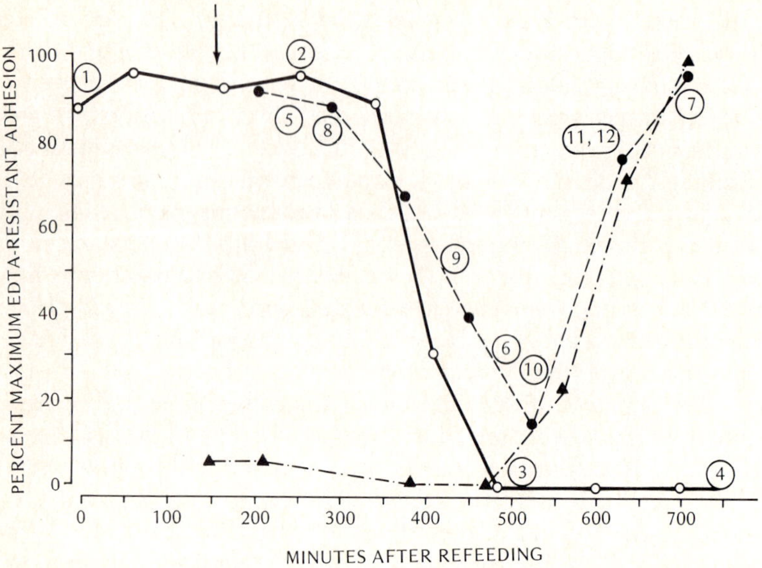

(A)

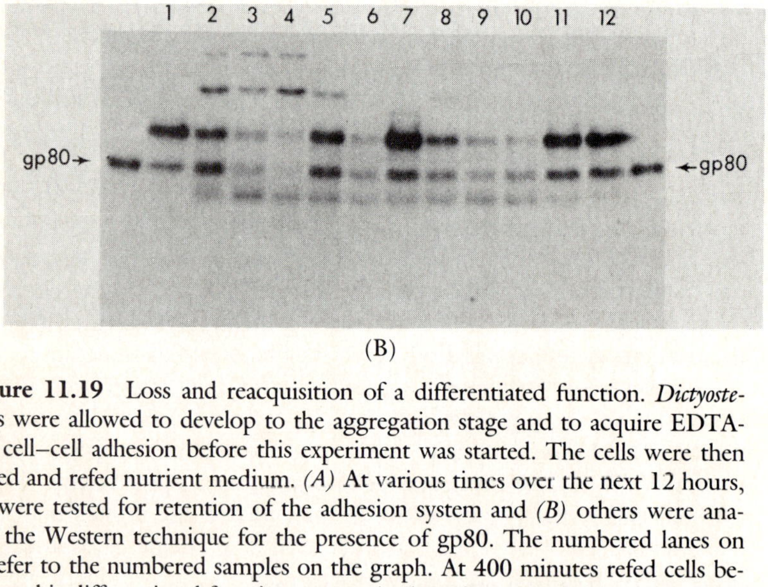

(B)

Figure 11.19 Loss and reacquisition of a differentiated function. *Dictyostelium* cells were allowed to develop to the aggregation stage and to acquire EDTA-resistant cell–cell adhesion before this experiment was started. The cells were then dissociated and refed nutrient medium. *(A)* At various times over the next 12 hours, samples were tested for retention of the adhesion system and *(B)* others were analyzed by the Western technique for the presence of gp80. The numbered lanes on the gel refer to the numbered samples on the graph. At 400 minutes refed cells began to lose this differentiated function (o–o). At 500 (3) and 700 (4) minutes they were nonadhesive and had lost gp80. The antiserum also recognizes gp95 and gp140 on the Western blot. Half the cells that had been refed were induced (I) to redifferentiate at 150 minutes. These cells lost adhesion by 500 minutes (6) (10) but reacquired it by 700 minutes (7) (11) (12) as they redifferentiated (•–•). gp80 was lost and reacquired in parallel. Another culture of the cells that was first induced to develop at 150 minutes began to become adhesive in the presence of EDTA at 500 minutes and was fully adhesive at 700 minutes (▲). These cells rippled (R) at 550 minutes.

with bound *Dictyostelium* proteins is then suspended in a solution of ovalbumin or other nonspecific protein to fill all the remaining binding sites on the paper. The nitrocellulose is then suspended in a solution of specific antibodies, which bind to those proteins that carry certain immunological determinants. The bound antibodies are quantitated by adding radioactively labeled (^{125}I) Protein A of *Staphylococcus aureus*, which fortuitously has a high affinity to the constant region of all IgG antibodies. Autoradiography of the nitrocellulose gives the position on the gel of the proteins recognized by the antibodies and also indicates the amount of the specific molecule in the extract. This immunological technique is referred to as a Western blot in parallel to Southern blots of DNA and Northern blots of RNA on nitrocellulose paper.

Using the Western blot technique it was found that gp80 initially continues to accumulate following dedifferentiation of *Dictyostelium* cells that had developed for 9 hours and acquired EDTA-resistant adhesion (Fig. 11.19). By 8 hours of dedifferentiation these cells have completely lost EDTA-resistant adhesion and the amount of gp80 in the cells has decreased dramatically (Fig. 11.19). Cells allowed to dedifferentiate for 150 minutes and then allowed to develop again, nevertheless, lose adhesiveness before they reacquire it 8 hours later. They also lose gp80 and then reaccumulate it in parallel with adhesiveness.

A mutant strain of *Dictyostelium* has been found that loses the ability to respond to cAMP at the normal time during dedifferentiation but surprisingly retains EDTA-resistant adhesiveness for at least 10 hours. By that time the cells are actively growing. These mutant cells also retain gp80 in their surface membranes for at least 10 hours. The phenotype of this mutant points out that dedifferentiation in *Dictyostelium* is a carefully regulated process controlled by specific genes and that individual components are individually regulated.

Related Readings

Cardelli, J., Knecht, D., Wunderlich, R., and Dimond, R. (1985). Major changes in gene expression occur during at least four stages of development of *Dictyostelium discoideum*. Devel. Biol. *110,* 147–156.

Chisholm, R., Barklis, E., and Lodish, H. (1984). Mechanism of sequential induction of cell-type specific mRNAs in *Dictyostelium* differentiation. Nature *310,* 67–69.

Devreotes, P., and Sherring, J. (1985). Kinetics and concentration dependence of reversible cAMP-induced modification of the surface receptor in *Dictyostelium*. J. Biol. Chem. *260,* 6378–6384.

Dowds, B., and Loomis, W. F. (1984). Cloning and expression of a cDNA that comprises part of a gene coding for a spore coat protein of *Dictyostelium discoideum*. Mol. Cell. Biol. *4,* 2273–2278.

Gerisch, G. (1982). Chemotaxis in *Dictyostelium*. Annu. Rev. Physiol. *49,* 535–552.

Gross, J., Town, C., Brookman, J., Jermyn, K., Peacy, M., and Kay, R. (1981). Cell patterning in *Dictyostelium.* Phil. Trans. Roy. Soc. London *295,* 497–508.

Kopachik, W., Dhokia, B., and Kay, R. (1985). Selective induction of stalk-cell-specific proteins in *Dictyostelium.* Differentiation *28,* 209–216.

Loomis, W. F. (1975). *Dictyostelium discoideum:* a developmental system. Academic Press, New York.

Loomis, W. F. (1982). The development of *Dictyostelium.* Academic Press, New York.

Loomis, W. F., and Thomas, S. (1976). Kinetic analysis of biochemical differentiation in Dictyostelium discoideum. J. Biol. Chem. *251,* 6252–6258.

Martiel, J., and Goldbeter, A. (1985). Autonomous chaotic behavior of the slime mould *Dictyostelium discoideum* predicted by a model for cyclic AMP signalling. Nature *313,* 590–592.

Mehdy, M., and Firtel, R. (1985). A secreted factor and cyclic AMP jointly regulate cell-type specific gene expression in *Dictyostelium discoideum.* Mol. Cell Biol. *5,* 705–713.

Raper, K. (1940). Pseudoplasmodium formation and organization in *Dictyostelium discoideum.* J. Elisha Mitchell Sci. Soc. *56,* 241–282.

Soll, D., and Mitchell, L. (1982). Differentiation and dedifferentiation can function simultaneously and independently in the same cells in *Dictyostelium discoideum.* Devel. Biol. *91,* 183–190.

Springer, W., Cooper, D., and Barondes, S. (1985). Discoidin I is implicated in cell substratum attachment and ordered cell migration in *Dictyostelium discoideum.* Cell *39,* 557–564.

Tasaka, M., Noce, T., and Takeuchi, I. (1983). Pre-stalk and pre-spore differentiation in *Dictyostelium* as detected by cell type-specific monoclonal antibodies. Proc. Natl. Acad. Sci. *80,* 5340–5344.

Yamada, T. (1967). Cellular and subcellular events in Wolffian lens regeneration. Curr. Top. Dev. Biol. *2,* 247.

Yoshida, M., Stadler, J., Bertholdt, R., and Gerisch, G. (1984). Wheat germ agglutnin binds to contact site A glycoprotein of *Dictyostelium discoideum* and inhibits EDTA-stable adhesion. EMBO J. *33,* 2663–2670.

Yumura, S., and Fukui, Y. (1985). Reversible cyclic-AMP-dependent change in distribution of myosin thick filaments in *Dictyostelium.* Nature *314,* 194–196.

CHAPTER 12

Sex Differentiation

At the earliest stage at which they can be recognized, the primary germ cells in mammals are on the floor of the yolk sac. From there they migrate to the genital ridges that develop just above the hindgut. In females the germ cells become completely surrounded by follicle cells. Later in development these germ cells will grow enormously to become oocytes. In males the germ cells associate with cells of the genital ridge in a topological configuration similar to that of testicular tubules (Fig. 12.1). In this configuration the germ cells will differentiate into spermatogonia.

The choice of differentiation into oocytes or spermatogonia is open to germ cells whether they carry the XX (female) or XY (male) configuration of chromosomes. Their fate is determined only by their association with the genital ridge tissue. This is a clear case of cellular interaction determining the pathway of differentiation. Male embryos (XY) form genital ridge cells with the ability to direct differentiation to spermatogonia whereas female embryos (XX) form genital ridge cells with the ability to direct oocyte differentiation. Germ cells of either XX or XY embryos respond to these signals.

What are the signals? There is some evidence that a specific protein or family of proteins referred to as the H-Y antigen plays a role in the association of germ cells with the tissue of the genital ridge. H-Y antigen, as the name implies, was first recognized by specific antibodies that react far more strongly with male embryonic tissue than with female embryonic tissue. H-Y antigen appears to be the product of a gene on the X chromosome that is found in both males and females. However, a gene on the Y chromo-

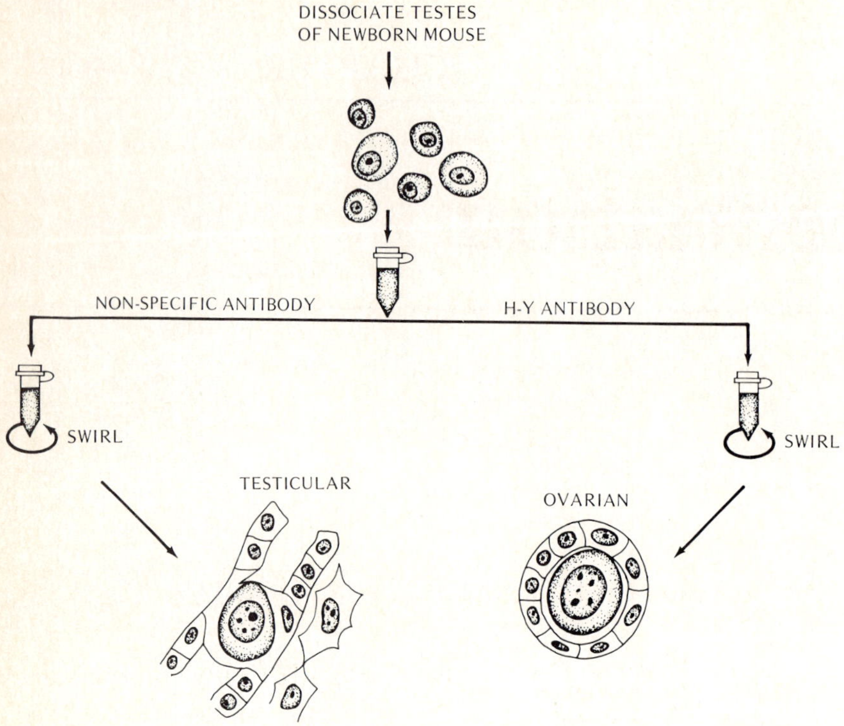

Figure 12.1 H-Y antigen on male mouse testicular cells. Testes of newborn mice were removed and dissociated into germ cells and somatic cells with trypsin. If these were gently mixed together in rotation culture in the absence of any antibody or in the presence of a non-specific antibody, they reassociated in a tubular fashion reminiscent of the original arrangement in testes. However, if they were first coated with antibody to H-Y antigens and then mixed, they reassociated in a different fashion with the somatic cells surrounding the germ cells, an arrangement seen in ovaries.

some, which is found only in males, is needed to activate the H-Y gene and thus it is expressed only in male (XY) embryos. Evidence that the H-Y antigen is controlling in males comes from the use of the antibodies specific to H-Y antigen. When dissociated cells of newborn male mouse testes are coated with the antibody, they reassociate in a configuration reminiscent of that seen in ovaries: the germ cells are completely surrounded by follicular cells. In the absence of added antibody, the testes cells reassociate in the tubular configuration from which they come. These results have been interpreted to imply that H-Y antigen on the surface of genital ridge cells of male embryos leads to the male, tubular, configuration and directs all subsequent differentiation of germ cells down the pathway to spermatogonia. There are conflicting observations in mice that do not support the H-Y antigen hypothesis. Nevertheless, it is clear that the genital ridges of XX individuals direct germ cells to differentiate into oocytes whereas those of XY individuals direct germ cells to differentiate into spermatogonia irrespective of their sex chromosome complement.

The plasticity of differentiation of germ cells is dramatically illustrated in some species of fish. All sheepshead fish start out life as small pink females. As they grow large they become black males. Initially, the germ cells differentiate into eggs whereas later in life they differentiate into sperm. In the sea bass, females are kept as such by the presence of a male. The visual stimulus provided by a male ensures that the hormonal balance will result in differentiation of the germ cells into eggs. When the male is removed or eaten, the females will rapidly change into males. The one that transforms first then represses all the others. In that individual, the germ cells now differentiate into sperm. Although the final form of sperm and eggs differs about as much as any two differentiated cells, germ cells clearly are poised to follow either pathway. The decision is regulated by the hormonal balance in the fish.

Hormonal Control of Sex Differentiation

Sperm and eggs are formed within the body and require ducts to get out. In mammals the genital ducts differentiate from primitive excretory ducts of the embryo: the oviduct develops from the Müllerian duct (sometimes called the paramesonsphros) while the vas deferens develops from the Wolffian duct (also called the mesonephros). Early embryos of both sexes make both of these duct systems but, later, under hormonal control, one or the other degenerates (Fig. 12.2).

In males of mammalian organisms, the gonads produce the steroid testosterone, which then circulates in the bloodstream. The presence of testosterone directs transformation of the Wolffian duct into the vas deferens and keeps it from degenerating. Female embryos produce less testosterone and in them the Wolffian duct spontaneously degenerates. Male gonads (testes) also produce a small peptide hormone, called X, that directs the degeneration of the Müllerian duct so that the female organs are not made. In females, hormone X is not made and the Müllerian duct spontaneously differentiates into the oviduct.

This hormonal control can be easily demonstrated by removing the testes from male rabbit embryos. Neither testosterone nor hormone X is produced. The Wolffian duct degenerates and the Müllerian duct develops into an oviduct in these otherwise male rabbits. If male embryos from which the testes have been removed are treated with testosterone, the Wolffian duct develops into the vas deferens normally but the Müllerian duct fails to degenerate and develops into an oviduct since no hormone X was present in these embryos. The rabbits end up as intersexes. The same result has occurred in humans when pregnant mothers carrying female embryos were treated with testosterone (Fig. 12.3).

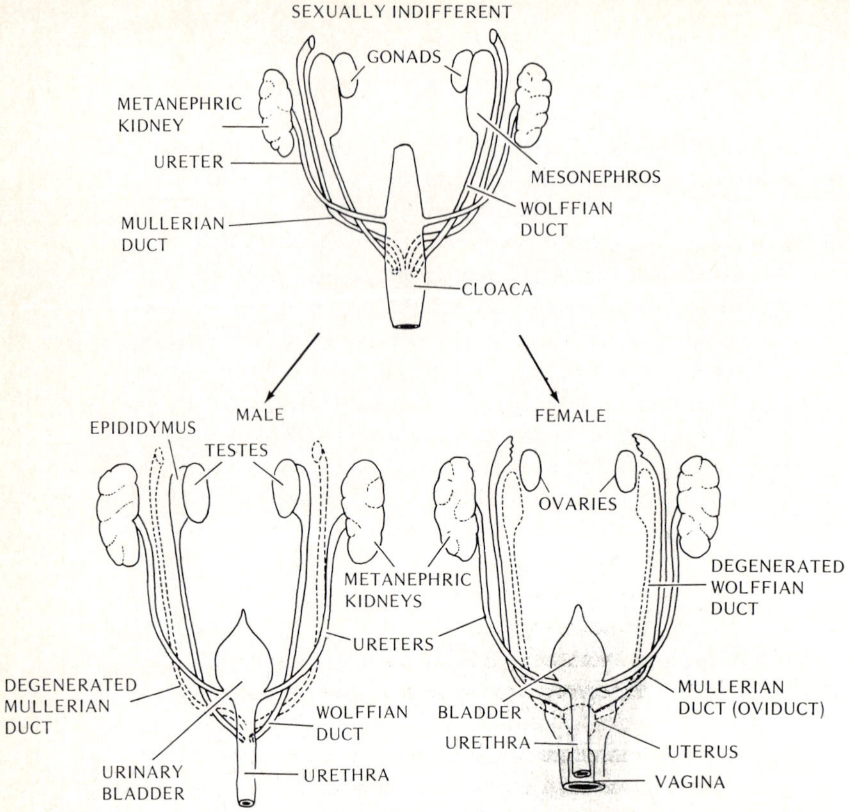

Figure 12.2 Differentiation of male and female duct systems from the androgenous ducts of early mammalian embryos. Male gonads (testes) produce testosterone and hormone X that instruct the Wolffian ducts to differentiate into vas deferens connecting the testes to the urethra and signal death to the cells of the Müllerian ducts. Female gonads make neither of these hormones and so the Wolffian ducts degenerate while the Müllerian ducts differentiate into oviducts that carry eggs from the ovaries to the uterus. In embryos of both sexes the metanephric kidney is connected to the bladder by the ureters.

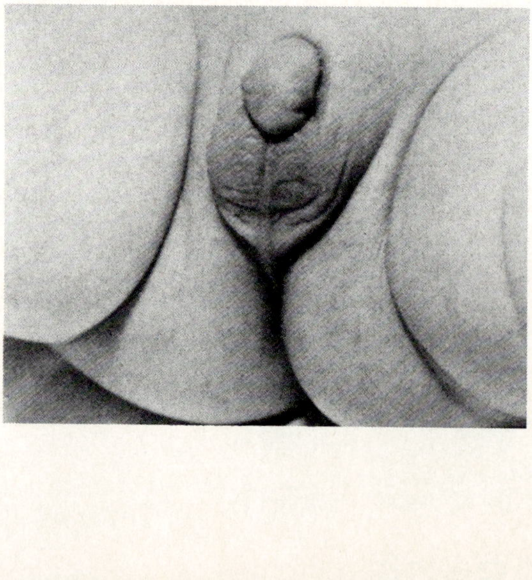

Figure 12.3 Female human infant that developed in the presence of ethyltestosterone. The drug was prescribed to arrest hair loss of the mother during the last seven months of gestation. Such treatment is no longer used with pregnant women. The infant was strongly masculinized. The labia are fused to form a scrotum and the clitoris is enlarged into a small penis.

Secondary sexual characteristics such as facial hair and breast development in humans are also controlled by hormonal balances; the relative levels of testosterone and estrogen regulate the male or female pattern at puberty. In chickens the expression of the vitellogenin genes that produce yolk components are controlled by estrogen. In laying hens the genes are highly active in liver cells whereas they are inactive in the liver cells of roosters. However, estrogen injections will dramatically induce these genes in adult male chickens.

Normal differentiation of the sexual organs is not essential for the viability of the individual. As a consequence, many patients with abnormal or ambiguous genitalia have been studied. One class of pseudohermaphroditism results from genetic defects in the enzymes necessary for synthesis of glucocorticoid hormones. The anterior pituitary tries to compensate for the low level of glucocorticoids by secreting abnormally high levels of ACTH (corticotropin), a hormone that stimulates the synthesis of glucocorticoids. As a result of the high levels of ACTH, the adrenals enlarge and produce high levels of the glucocorticoid precursors. These are converted into testosterone in the adrenals, which normally make very little. Females with this defect in the enzyme necessary for glucocorticoid synthesis develop a normal female reproductive tract internally, since there was no production of hormone X. However, they exhibit various degrees of masculinization of the external genitalia. The clitoris develops as a rudimentary phallus; the labia become hypertrophied and may even fuse to form a scrotum. Treatment of

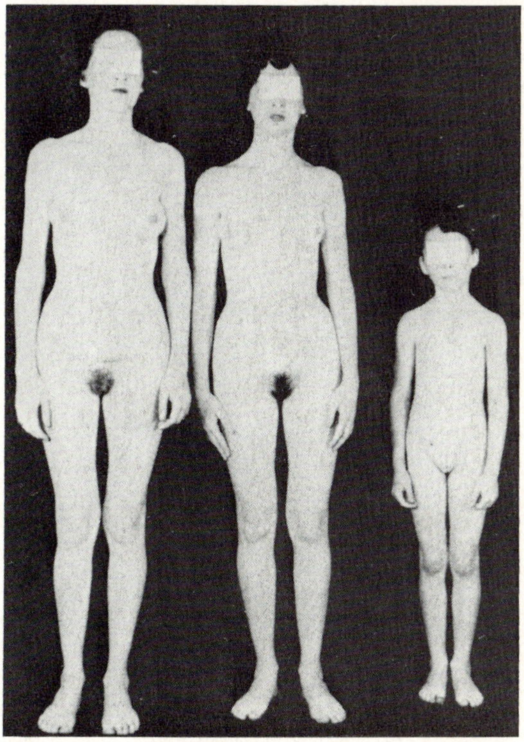

Figure 12.4 Genetically male (XY) humans with testicular feminization. These sisters carry the *Tfm* mutation that results in nonfunctional testosterone-receptor and resulting feminine secondary sexual characteristics. They each had testes that failed to descend and shallow, blind vaginas. Breast and hip development was normal at puberty.

such individuals soon after birth with glucocorticoids can successfully correct most of the subsequent problems and allow normal development into fertile females. The administered glucocorticoids repress ACTH production in the anterior pituitary and the adrenals are no longer stimulated to synthesize testosterone.

Another group of patients includes normal adolescent females who come to the doctor only because menstruation appears to be delayed following puberty. It is found that their vaginas terminate in a deadend; that is, they are not connected to a uterus. These individuals carry 46 chromosomes (XY), making them genetically male. Yet they have feminine body contours with well-developed breasts (Fig. 12.4). It turns out that they carry a mutation affecting the testosterone-receptor protein. While their testes have been producing testosterone normally throughout development, none of their tissues have been able to respond as a consequence of the loss of testosterone-binding activity. The Wolffian ducts degenerated because they could not recognize the testosterone. The Müllerian ducts also degenerated under the influence of hormone X produced by the testes. These patients, referred to as androgen insensitive (AIS), carry mutations in a gene *Tfm* essential for the testosterone receptor. They are left with none of the ducts of the reproductive tract but have the secondary sexual characteristics of females. They usually live their lives as normal, although infertile, females.

Genetic Control of Sex Differentiation

Differentiation of reproductive tissues in the male mode is dependent on the presence of a Y chromosome in humans. This may be due to the requirement for the Y-linked gene that is essential for activity of the H-Y gene on the X chromosome. Normal males are XY. Individuals carrying two Y chromosomes (XYY) are somewhat taller but otherwise indistinguishable. Those with two X chromosomes and a Y chromosome (XXY) develop as males but express Klinefelter's syndrome with various abnormalities.

Individuals lacking a Y chromosome develop as females. Their genetic complement may include one, two, or three X chromosomes. XO individuals with a single X chromosome express Turner's syndrome with specific abnormalities. Those with two Xs (XX) are normal females. Those with three Xs (XXX) have only minor problems.

To a certain extent the insensitivity to differences in the number of X chromosomes is due to the fact that all but one of these chromosomes is inactivated early in embryogenesis. In triple X individuals, two of the X chromosomes are condensed into Barr bodies, leaving a single active X. In normal XX individuals one or the other X is inactivated, and in XO individuals the single X remains active. The inactivation and condensation of an X chromosome occurs when the inner cell mass consists of a dozen or so cells. It is a random process in mammals wherein the paternal or maternal chromosome is inactivated. But once inactivated in a cell of the inner cell mass, the same X remains inactivated in all the progeny cells derived from that inner cell mass cell. As a consequence, adult females are mosaics with some cells using paternal X-linked genes and others using maternal X-linked genes.

Once inactivated, a given X chromosome is almost never reactivated. This type of genetic differentiation is stable through hundreds of cell generations. The only procedure that has been found to result in reactivation is treatment of the cells with 5-azacytidine, a compound known to interfere with the methylation pattern of DNA (Chapter 4). These results have been interpreted as indicating that methylation of certain GC positions on the X chromosome are essential signals for inactivation and condensation of the region of the X chromosome. Since the maintenance methylase preferentially methylates hemimethylated base pairs in newly replicated DNA, once a given X chromosome becomes methylated and thereby inactivated, it passes this state on to all subsequent progeny cells.

Sex in Worms

Some nematodes, such as *Caenorhabditis elegans,* are normally hermaphrodites producing both sperm and eggs. They have an XX type of genome. Rare nondisjunctions can result in XO individuals that develop as males. They produce sperm and specialized appendages to donate the sperm to hermaphrodites.

Rare mutants have been found in which XX nematodes differentiate as males. Other mutants have been found in which XO nematodes differentiate as hermaphrodites. At least seven different genetic loci have been recognized that control differentiation of the gonads and somatic tissue to determine the sex of the worm (Fig. 12.5). The ratio of X chromosomes to au-

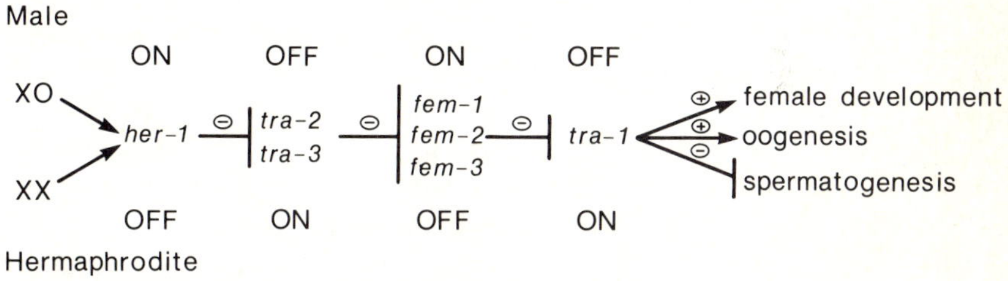

Figure 12.5 Genes that control sexual differentiation in *C. elegans.* Nematodes that carry only a single X chromosome (XO) are males. Those that carry two X chromosomes (XX) are hermaphrodites. The *her*-1 gene is necessary to measure the number of X chromosomes relative to the number of autosomes (A). If the ratio is low, as in XO worms, *her*-1 represses *tra*-2 and *tra*-3, and so they are off. The *fem* genes are active in XO males and repress the *tra*-1 gene. The product of the *tra*-1 induces differentiation of female tissues and oogenesis. Since *tra*-1 is off in XO individuals, female differentiation does not occur. Spermatogenesis then proceeds. In XX hermaphrodites the higher X/A ratio results in *her*-1 being inactive. As a consequence, *tra*-2 and *tra*-3 are on and repress the *fem* genes. Therefore, *tra*-1 is on and female differentiation can proceed while spermatogenesis is repressed.

tosomes (non-X chromosomes) is recognized by the *her*-1 gene. In XO individuals, the *her*-1 gene is on and represses genes *tra*-2 and *tra*-3. In XX individuals, *her*-1 is off and so *tra*-2 and *tra*-3 are not repressed, that is, they are on. The products of *tra*-2 and *tra*-3 in XX hermaphrodites repress the three *fem* genes. In XO males the *tra*-2 and *tra*-3 genes are off, and so the *fem* genes can be on. The *fem* genes repress the *tra*-1 gene and so it is off in males. In XX hermaphrodites the *tra*-1 is on, and its product can induce oogenesis and female development. It also represses the male mode of differentiation of the gonad and somatic tissue. In XO males *tra*-1 is off, and male differentiation is not repressed while female differentiation is not activated.

This cascade of interaction allows fine tuning of sexual dimorphism so as to allow a little maleness in hermaphrodites and complete maleness in males. Genetics of nematodes is sufficiently refined that these relationships of genes could be worked out by constructing double mutants carrying different pairs of mutant genes and observing the consequences to sexual differentiation. For instance, *her*-1 mutants lacking all activity are all hermaphrodites no matter what the X to autosome ratio is. The *her*-1 gene is said to be epistatic to the other six genes. Likewise, constitutive mutants that express the *tra*-1 gene in both XO and XX nematodes are always hermaphrodites even when the other genes are non-functional due to mutations.

Similar networks of interacting genes have been found to regulate sexual differentiation in yeast and fruit flies as is discussed in the following Specific Examples. As genetic analysis of other pathways in development is refined, many more cascades of regulatory gene products may be uncovered.

Summary

The primary germ cells can follow pathways of differentiation leading either to sperm or eggs. The choice is made by the sex of the gonads. Cells dissociated from male testes of mice reassemble into tubular configurations when gently rolled together. However, if the cells are first coated with antibodies reactive with H-Y surface glycoproteins, they reassemble into the configurtion seen in follicles of ovaries. The accessory organs that transport germ cells to the exterior of the individual in mammals are responsive to the hormones X and testosterone. X is produced in the testes and induces degeneration of the Müllerian duct. In females, no X is produced and the Müllerian ducts form the oviduct. The Wolffian duct degenerates spontaneously unless it is instructed by the presence of testosterone to form the vas deferens. Testosterone is produced in testes but not in ovaries and it recognized by the product of the *Tfm* gene. The unequal dosage of X-linked genes in females (XX) and males (XY) is equalized by the inactivation of all but a single X chromosome. Methylation of the DNA at sites along the X chromosome appears to play a role in inactivation. Sexual differentiation appears to be controlled by a hierarchical matrix of interacting gene products.

SPECIFIC EXAMPLES

1. Sex Determination in *Drosophila*

Genes responsible for some of the components involved in sex determination have been recognized in the fruit fly, *Drosophila melanogaster*. Mutations that result in altered sex-specific differentiation and in intersex flies have directed attention to a set of genes that play various roles in determining whether the individual fly will differentiate somatic body tissue as male or as female. Hormones do not seem to be involved in this case as the mutations all act in a cell-autonomous manner. That is, small clones of mutant cells in an otherwise wild-type individual display the mutant phenotype.

Adult flies exhibit several differences in their form depending on their sex (Fig. 12.6). They are said to be sexually dimorphic. The number of abdominal segments differ and males have darker posterior ones; the genitalia and analia differ and males have a row of heavy bristles parallel to the axis of the forelegs, referred to as sex combs. The courtship behavior also differs between males and females even when they are raised in isolation. At the molecular level, a few genes have been identified that are expressed in just one sex. Among these are the genes coding for yolk proteins, *Yp1*, *Yp2*, and *Yp3*, in the fat bodies of females but not of males. Finally, there is the phenomenon of dosage compensation in which almost every gene on the X (sex) chromosome in males (genotype XY) is transcribed at double the rate it is in females (genotype XX). Dosage compensation overcomes the fact that males

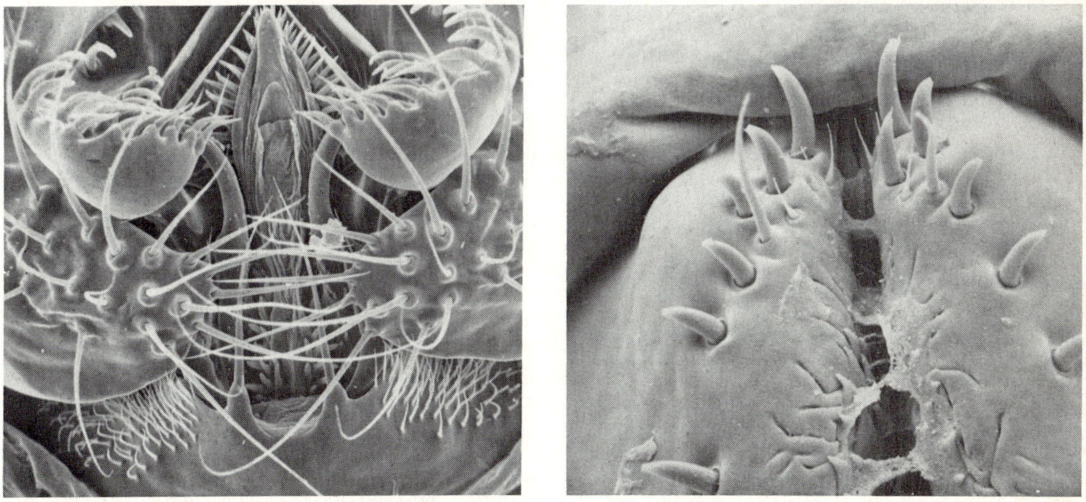

(A) (B)

Figure 12.6 External genitalia of *Drosophila*. *(A)* The male penis is flanked by claspers. *(B)* The female vulva is flanked by vaginal plates. The sexually dimorphic bristles can easily be seen in this SEM. The penis is about 150 μm long.

have only half the number of each X-linked gene. In mammals the disparity in gene dosage of X-linked genes between males (XY) and females (XX) is equalized by the repression of one complete X chromosome in females. In female flies, both X chromosomes are active but are transcribed at half the rate found in males.

The primary determinant for sex differentiation is the X-to-autosomal chromosome ratio (X:A). Both males and females have the same number of autosomes (two sets of nonsex chromosomes since they are diploids) whereas males have a single X and females have two X chromosomes. Thus, the ratio in males is X:2A (ratio 0.5) and in females is 2X:2A (ratio 1.0). The number of autosomal sets as well as the number of X chromosomes can be varied experimentally. It is found that when the ratio is 1.0 (e.g., 1X:1A, 3X:3A, or 4X:4A) the individual differentiates as a female. When the ratio is 0.5 the individual differentiates as a male. Individuals with an intermediate ratio (e.g., 2X:3A) are intersexual with some male and some female differentiation patterns. There appears to be a threshold level between ratios of 0.5 and 1.0 that determines the differentiation pattern.

No single locus on the X chromosome seems to be monitored for sex determination. The ratio is directive early in embryogenesis and the information used in a cell-autonomous manner. Each cell differentiates in the way it perceives its own X:A ratio and is unperturbed by the X:A ratio in surrounding cells.

Several genes are involved in measuring the X:A ratio. The *da* gene product is made during oogenesis under control of the maternal genes. It functions during early embryogenesis to regulate other genes depending on the X:A ratio. Mutations in the *da* gene *(daughterless)* result in females *(da/da)* having only sons. Female progeny all die during embryogenesis but the males show no abnormalities or disadvantages. No daughters are born even if the paternal genome donated a wild-type da^+ gene to the embryo. It has been reported that daughters can be rescued by injecting cytoplasm from wild-type da^+ eggs into the eggs of *da/da* mutant females. Thus, the da^+ gene product is a real substance, but egg cytoplasm represents too little material for its characterization by present techniques. Perhaps expression of a cloned *da* gene in bacteria will permit isolation of sufficient amounts of the product for analysis.

There is another gene, *Sxl* (sex lethal), in which mutations also result in all offspring being male. This gene is expressed in female (2X:2A) individuals as long as the embryo receives sufficient amount of the da^+ gene product to recognize that the X:A ratio is 1.0. It does not appear to be expressed in male (1X:2A) individuals, since mutations in the *Sxl* gene have no consequences in males. Individuals with the female complement of chromosomes (2X:2A) carrying an *Sxl* mutation die as embryos, perhaps as the result of a lack of proper dosage compensation. These results are summarized in Table 12.1. It takes at least two genes to keep a fly from differentiating as a male.

The next step in sex determiniation involves a complex genetic locus called double-sex *(dsx)*. Null mutations in *dsx* that result in the total loss of

Table 12.1 Sex Determination in *Drosophila*

Chromosomes	Ratio	Expression of *Sxl*	Phenotype
X: A ratio measured by *da* gene product	If 1.0	Activates *Sxl* gene	1. Female differentiation 2. Female expression of X-linked genes
	If 0.5	*Sxl* gene inactive	1. Male differentiation 2. Male expression of X-linked genes (dosage compensation)

activity in individuals carrying either 1X:2A or 2X:2A result in the development of morphologically indistinguishable intersexes. Differentiation of the germ cells into either sperm or egg is not affected in these mutant flies and is determined only by the X:A ratio. However, the cuticular structures are a curious mixture of both male-specific and female-specific structures in each cell.

A unique mutation, dsx^{136}, has been analyzed that affects males but not females. A detailed analysis of the transcription pattern has revealed that there are male-specific transcripts derived from *dsx* that are not present in wild-type female flies. There is also a *cis*-dominant mutation, dsx^{Mas}, that transforms chromosomally female individuals into intersexes but has no effect on males. It appears that this mutation results in the constitutive expression of the male-specific dsx^{+} function in XX individuals. Thus, irrespective of the X:A ratio and the genes that measure it, a fly heterozygous for dsx^{Mas} will form male structures. These are perfectly normal in an XY individual, but in an XX individual both male and female structures will be made as the result of expression of the male-specific *dsx* functions from the chromosome carrying dsx^{Mas} and the female *dsx* functions from the wild-type homolog (Fig. 12.7).

The choice of expression of the male or female functions of *dsx* appears to depend on the product of two other genes, *tra* and *tra*-2 (transformer and transformer-2). The *tra* gene is transcribed into a 1.1 kb mRNA that is most abundant in pupae. It is present in female flies but absent in male flies. Mutations in these genes transform females into males. Normally these genes are activated by the product of *Sxl* and function to modulate the pattern of expression of the *dsx* region to give the female-specific functions. In the absence of functional products from both *tra* and *tra*-2, as occurs in mutants as well as XY individuals, the male-specific functions of *dsx* are exclusively expressed.

The basis for this reasoning is the behavior of various double mutant individuals carrying 2X chromosomes (Fig. 12.7). Wild-type flies with 2X chromosomes are female, of course. Those with a mutation in *tra*-2 are male because the *dsx* female functions are not activated whereas the male-specific functions are expressed. Double mutants with mutations *tra*-2 and dsx^{136}

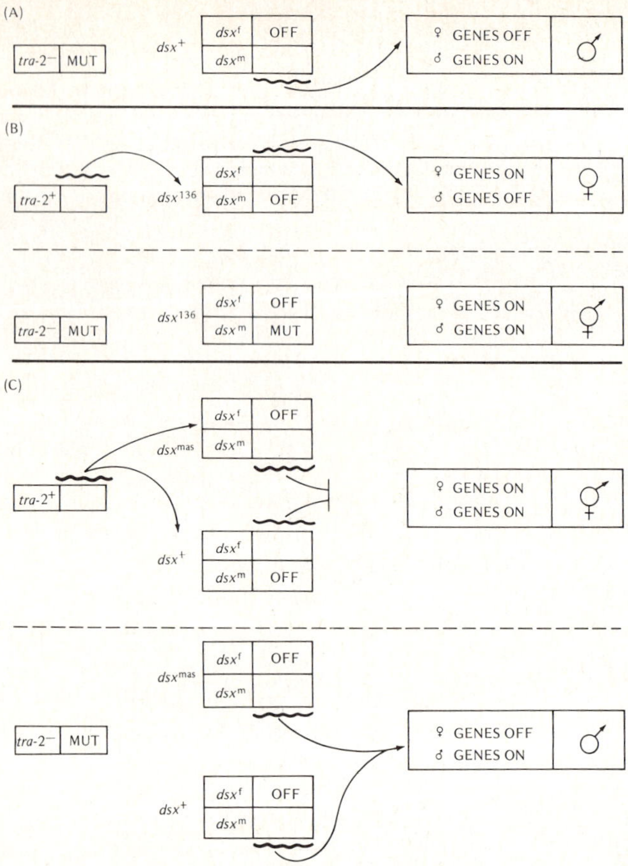

Figure 12.7 Mutations affecting XX individuals in *Drosophila*. Genes referred to as *tra*-2 and *dsx* in *Drosophila* are unrelated to genes affecting sexual differentiation in *C. elegans* but determine the sexual type of differentiation in somatic fly tissue. Differentiation of the primary sex cells is unaffected by mutations in these genes but the genitalia, abdomen, sex combs, and various other sexually dimorphic bodies are sensitive to these mutations. The phenotype of XX (genetically female) flies carrying mutations in *tra*-2 or either of two different mutant alleles of *dsx* are interpreted to result from interactions shown in this model. As yet there is no direct biochemical proof of these interactions, but these and other genetic analyses generate this hierarchical sequence. In wild-type females it is thought that the *da* gene product activates the *Sx1* gene, whose product in turn activates *tra*-2. The *tra*-2^+ function acts to cause the bifunctional dsx^+ locus to express the female-mode dsx^f function that represses male differentiation. Where an individual is homozygous for an allele only one of the two homologs is given in the figure *(A, B)*. Where an individual is heterozygous for alleles of *dsx*, both homologs are given in the figure *(C)*.

(A) In *tra*-2^- dsx^+ mutants the *tra*-2 mutant gene product cannot direct the female-mode dsx^f function and so the male mode dsx^m function, which represses female differentiation, is expressed. XX *tra*-2^- dsx^+ have male somatic differentiation (e.g., sex combs).

(B) The male-specific mutation, dsx^{136}, has no effect in tra^+ XX individuals and female somatic differentiations occur. But in *tra*-2^- XX individuals the dsx^{136} male-specific mutation results in intersex differentiation with both male and female somatic parts. The dsx^{136} mutation results in the lack of repression of male differentiations.

(C) XX *tra*-2^+ individuals carrying an allele of *dsx* (dsx^{Mas}) that constitutively expresses the male-mode of dsx^m function, as well as a wild-type allele of dsx^+, come out as intersexes with both male and female somatic differentiations. The simulataneous expression of dsx^m (from the dsx^{Mas} allele) and dsx^f (from the dsx^+ allele) prevents both from functioning. In *tra*-2^- XX individuals heterozygous for dsx^{Mas} and dsx^+, both alleles express dsx^m function and male development occurs.

are intersexes because the mutant male-specific functions of *dsx* are expressed. Double mutants with mutations *tra*-2^- and dsx^{Mas}/dsx^+ are male because the male-specific functions on the dsx^{Mas} chromosome are constitutive while the other chromosome also expresses the male-specific functions because of lack of functional *tra*-2 product. In the presence of a wild-type *tra*-2 gene, dsx^{Mas} results in intersexes.

It is likely that the function of the *dsx* gene product acts continuously throughout embryogenesis and into early adult life. A late loss of function results in changes in courtship behavior. Adult *Drosophila* males exhibit a stereotyped behavior before copulation that is genetically determined, not learned. By shifting individuals homozygous for a temperature-sensitive mutation in *tra*-2 to the nonpermissive temperature, it was found that the behavior changed within a day. A portion of the brain called the mushroom body is sexually dimorphic in *Drosophila* and continues to differentiate during adult life. The function of this organ may determine courtship behavior.

Transcription of the genes for yolk proteins Y_p1, Y_p2, and Y_p3 also requires the continuous expression of the *tra*-2 gene. If the temperature-sensitive mutation in *tra*-2 is inactivated in females, transcripts of the Y_p genes disappear within a few days. Upon return to permissive temperatures, the transcripts reappear within a day. These results indicate that fully differentiated fat body cells can switch genes for oogenesis on or off rapidly under the direction of the *tra*-2 gene.

Sexual differentiation has probably evolved separately several times. The genetic mechanisms to ensure that an individual is a functioning male or female and not intersex or infertile have been built up from less accurate processes. As the complexity of sexual dimorphism increased, various attributes had to be coordinated. Here in *Drosophila* we see how a sequence of genes functions within each cell to direct male and female differentiation. While the details most likely differ in other organisms, this example of hierarchical controls gives us some idea of the complexity of regulatory processes and an idea of how experimentally difficult it is to unravel them.

2. Mating Types in Yeast

Genetic exchange in yeast comes about by fusion of two morphologically identical haploid cells. The resulting diploids can grow as well as haploids but can also be induced to undergo meiosis and sporulation (Fig. 12.8). Nevertheless, yeast cells do not mate indiscriminately. Alpha (α) cells mate only with **a** cells, and **a** cells mate only with α cells. The resulting diploids (**a**/α) mate with neither cell type. Mating type is controlled by a single genetic locus called MAT on chromosome III. In α cells this locus contains a segment of 747 bp, referred to as Yα. In **a** cells this sequence is replaced by a segment of 642 bp, referred to as Ya. In both cell types the Y segment is flanked on the left by 707 bp, referred to as X, and on the right by 239 bp, referred to as Z1. The X and Z sequences at the MAT locus are identical in both **a** and α cells (Fig. 12.9). The 2.5 kb nucleotide sequence at MAT has been determined in wild-type and mutant strains.

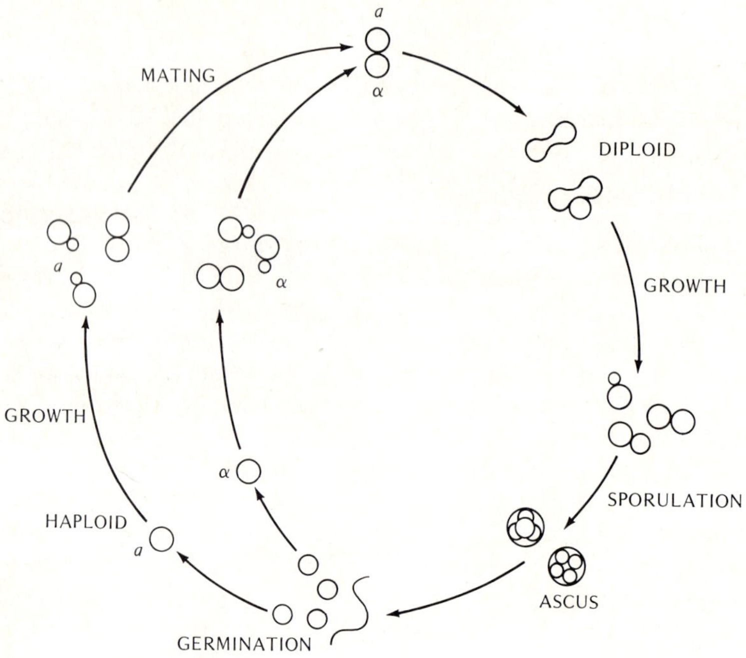

Figure 12.8 Life cycle of the yeast *Saccharomyces cerevisiae*. Heterothallic strains contain cells of two mating types, **a** and α. Haploid cells grow with a doubling time of about an hour in rich medium and multiply by budding. Haploid cells of opposite mating type conjugate and fuse to form diploid cells. These can grow exponentially as well. When starved, diploids can sporulate. Meoisis occurs during sporulation, producing four haploid spores within an ascus. Genetic analysis of these tetrads is used to determine linkage of genes and the mechanism of recombination.

An interesting thing about mating types in yeast is that the progeny of a single cell can have different mating types. Within as little as two rounds of cell division, an **a** cell can give rise to both **a** and α offspring. Likewise, an α cell can give rise to both **a** and α cells by the second cell generation. Thus, both cell types have the genetic information for **a** and α (i.e., Y**a** and Yα segments) but only one is expressed at a given time. There are unexpressed copies of XYZ sequences at both ends of chromosome III. Near the left telomere of chromosome III the $XY_{\alpha}Z$ is flanked by a W segment of 723 bp on the left and a Z2 segment of 88 bp on the right. The XYZ segments at the MAT locus are also flanked by W and Z2, but these segments are not found near the $XY_{\mathbf{a}}Z$ sequence at the right telomere. Thus, the left and right loci can be distinguished by molecular as well as by genetic techniques and are called HML_{α} and $HMR_{\mathbf{a}}$.

As frequently as once a generation the mating type changes. Both the mother cell and bud will change from α to **a** or vice versa. This suggests that the Y segment of the MAT locus converts from one mating type to the other before replication of the MAT locus.

There is a gene in yeast referred to as HO that codes for a site-specific nuclease. The HO nuclease cleaves both strands of DNA at the junction of the Y and Z segments of the MAT locus. The Y segment from either HML or HMR then displaces the Y segment of MAT and directs gene conversion within the MAT locus. The Y sequence that had previously occupied the MAT locus is degraded and replaced by repair DNA synthesis using the Y segment of either HML or HMR as template. Mutations in the HO gene result in failure of mating-type conversion to occur.

There are two genes specified by the MAT α locus and one by the MAT **a** locus. In **a** cells, the **Ya** segment promotes leftward transcription of the X sequence as well as the rightward transcription of the **Ya** sequence (Fig. 12.9). In α cells there is a leftward transcript that starts in Yα and proceeds through X and a rightward transcript of its own sequence as well as that of Z. These sequences at HML and HMR are silent. They are kept silent by the action of the products of at least four genes, SIR1, 2, 3, and 4. Mutations in any one of these SIR genes result in the expression of the HML and HMR loci. Both of the loci carry sequences of about 200 bp on their left that are essential for SIR-directed repression. These E(essential) sequences are high in A and T (approximately 85 percent) but are not homologous. The E_L of

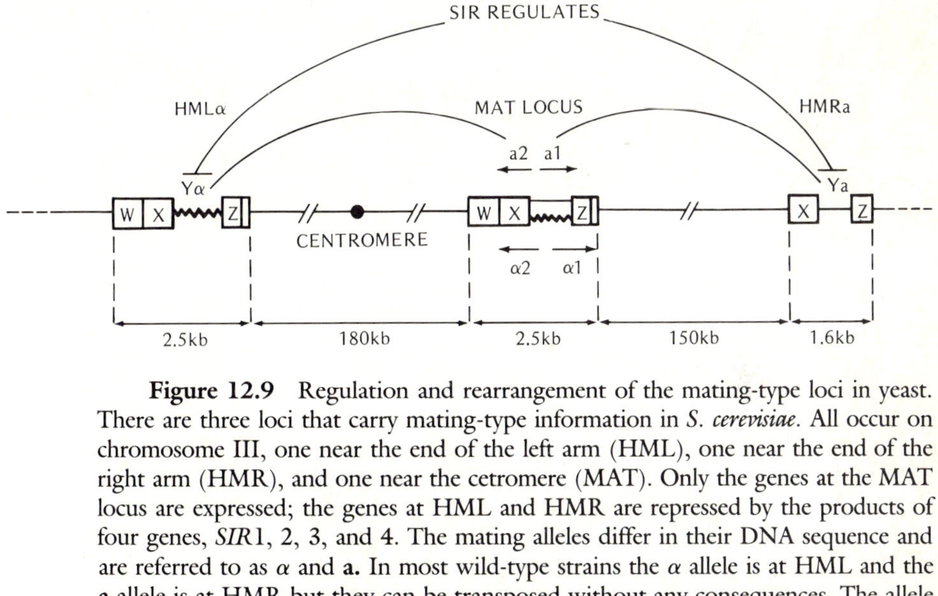

Figure 12.9 Regulation and rearrangement of the mating-type loci in yeast. There are three loci that carry mating-type information in *S. cerevisiae*. All occur on chromosome III, one near the end of the left arm (HML), one near the end of the right arm (HMR), and one near the cetromere (MAT). Only the genes at the MAT locus are expressed; the genes at HML and HMR are repressed by the products of four genes, *SIR*1, 2, 3, and 4. The mating alleles differ in their DNA sequence and are referred to as α and **a.** In most wild-type strains the α allele is at HML and the **a** allele is at HMR but they can be transposed without any consequences. The allele at MAT is either the α allele or the **a** allele and can change from one generation to the next in homothallic strains. These strains carry a gene, *HO,* that codes for a site-specific nuclease. The nuclease cuts out the resident allele at MAT and allows replacement of it from the allele of the opposite mating type. Strains carrying a mutation in *HO* (referred to as *ho* strains) are heterothallic. These strains cannot mate among cousins but only with independent strains of the opposite mating type. Both the **a** and α alleles are transcribed into two mRNAs (a1, a2 or α1, α2). The sequences are identical between the alleles in the X and Z regions but differ in the central Y region.

α-factor: TrpHisTrpLeuGlnLeuLysProGlyGlnProMetTyr-COOH.

a-factor: TyrIleIleLysGlyValPheTrpAlaAsxPro-COOH.
TyrIleIleLysGlyLeuPheTrpAlaAsxPro-COOH.

Figure 12.10 Mating factors of *Saccharomyces cerevisiae.* The α-factor produced by α cells is 13 amino acids long and is cleaved from a tandomly repeated protein precursor. There are two **a** factors produced by **a** cells from a pair of genes that have diverged at a single codon since duplication from a common precursor. The α factor arrests **a** cells in G_0 of the cell cycle; the **a** factor arrests α cells at the same stage.

HML does not hybridize with the E_R of HMR. Their orientation relative to the HM loci they control is crucial, indicating that there is a polarity of repression. SIR directs the E sequences to act *cis* to control about 3 kb of DNA no matter what that DNA codes for. The action of the SIR proteins on the HM loci not only represses transcription but also prevents endonucleolytic attack by the product of HO and alters the chromatin so that DNase I hypersensitive sites disappear. It seems to affect the structure of contiguous DNA by altering the degree of supercoiling. This is an interesting case of regulation that extends over a considerable distance to affect the availability of genetic information. Similar mechanisms have been implicated in the regulation of transcription in metazoan organisms.

The MAT locus determines three distinct cell types in yeast: (1) α cells, (2) **a** cells, and (3) **a**/α diploids. The mating behavior of these three cell types are controlled by regulation of **a** or α mating functions as well as mating functions common to both **a** and α cells. One of these mating functions is the production of a mating factor. α cells secrete a tridecapeptide (α factor) that blocks **a** cells in the G1 phase of the cell cycle and renders them adhesive to α cells (Fig. 12.10). The **a** cells also change their cell wall composition and take on the asymmetrical appearance of a schmoo, so called because they reminded certain scientists of Al Capp's cartoon creatures that resemble duck-pins or figure eights with the top smaller. **a** cells likewise secrete peptides (**a** factor) that have the same effects on α cells. There are two genes for **a** factor that produce precursor molecules that are cleaved to undecapeptides differing in a single amino acid (Fig. 12.10). α factor is cleaved from precursor proteins containing multiple copies of the factor. These secreted factors synchronize cells of the opposite mating type so that both cells are in G1 when they fuse and they can replicate the diploid complement in unison.

These mating factors are produced in a mutually exclusive manner. One cell never produces both; **a**/α diploids produce neither. The diploids also fail to produce many of the nonspecific mating functions.

How are the genes for **a** factor, α factor, and other mating functions regulated by the MAT locus? A large number of genetic analyses have indicated that the transcripts that start in the Y segment activate or repress transcription of the genes for mating-type function (Fig. 12.11). Recently it has been directly shown that the product of α2 is a DNA binding protein

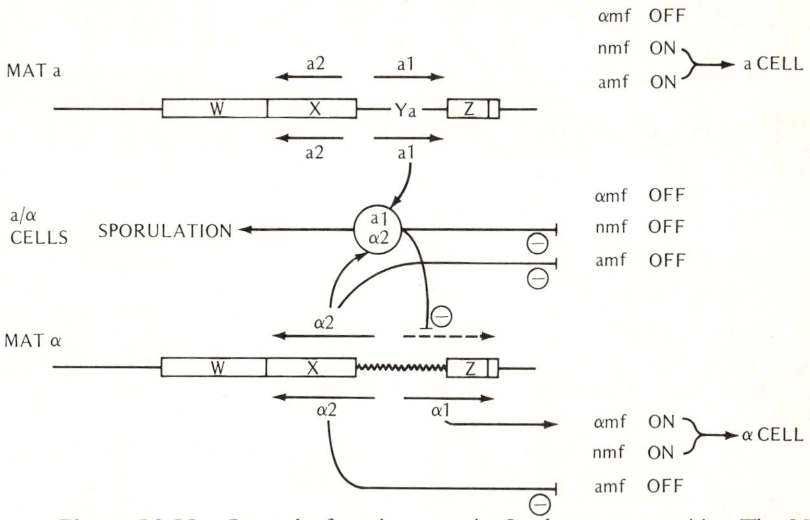

Figure 12.11 Control of mating type in *Saccharomyces cerevisiae*. The MAT locus has been subdivided into four regions covering the 2.5 kb of DNA at the locus. The W region is 739 bp, the X region is 707 bp, the Y region is 747 bp in MATα and 642 bp in MATa, the Z1 region is 239 bp. W, X, and Z are identical in MATa and MATα but Ya and Yα are nonhomologous.

In order for a cell to mate, it must express nonspecific mating functions (nmf) found in both **a** and α cells but not in diploids, as well as either **a**-specific mating functions (amf) or α-specific mating functions (αmf), but not both. The **a** mating functions are constitutive but repressible by the product of an α gene, α2. An example of such a mating function is the expression of a gene referred to as BAR1. It must be expressed for **a** cells to mate efficiently. The α mating functions, such as expression of a gene referred to as *STE* 3, require induction by the product of an α gene, α1. Therefore, of the two genes starting in Yα at the MAT locus in α cells, one (α2) represses **a** mating functions while the other (α1) induces α mating functions. Cells carrying Yα at MAT are α-mating haploids. The two genes starting in Ya at the MAT locus in **a** cells play no essential role in the mating-type function of haploids. Both the nonspecific mating functions, such as expression of a gene referred to as *STE* 5, and **a** mating functions are constitutive, and the α mating functions are not expressed because they require induction by the product of α1, which is not expressed in **a** cells being out at HML. Cells carrying Ya at MAT are **a** mating haploids.

A product of Ya is required to repress the mating functions of diploids and allow sporulation. The a1 gene product associates with the α2 gene product to make a complex that represses the nonspecific mating functions, such as *STE* 5 gene product. The α2/a1 complex binds to a 20 base sequence upstream of STE5 and represses transcription. Since diploids carry both MATa and MATα, both a1 and α2 are expressed and they become nonmating cells. The complex of α1/α2 also represses transcription of α1, and so the α mating functions (amf), such as expression of *STE* 3, fail to be induced. The α2 gene is expressed and besides contributing to the a1/α2 complex, represses the a mating functions (amf), such as expression of BAR1. The complex of a1/α2 is required for expression of several genes essential for sporulation. Thus, diploid cells made by fusion of MATa and MATα haploid cells are nonmating and able to sporulate under starvation conditions.

The a2 and α2 transcripts have identical 3′ halves, both being transcribed from X regions but differ in their 5′ ends since α2 starts in the Yα region. The a1 and α1 transcripts are nonhomologous. The MAT α2 protein is partially homologous to the homeo box sequence that is found in several regulatory proteins.

that recognizes a conserved sequence of 30 bases adjacent to **a** specific genes and represses their transcription. Therefore, a cell expressing $\alpha 2$ cannot be an **a** cell but can be an α cell or an **a**/α diploid. The α specific genes that are essential for a cell to be a mating type α strain are induced by the MAT$\alpha 1$ gene product. Since **a** cells do not express the $\alpha 1$ gene, they cannot be α cells but can be **a** cells.

The expression of the genes for α factor and other α-specific mating functions are activated by the product of the $\alpha 1$ transcript that starts in Yα and proceeds through Z. Of course, this transcript is not made in **a** cells that have Y**a** at the MAT locus and so **a** cells do not make α factor. The other transcript, $\alpha 2$, codes for a protein that represses the gene for **a** factor and therefore α cells do not make **a** factor. In the absence of these Y-promoted transcripts, as in **a** cells, the **a** factor gene is not repressed while the α factor gene is not activated. Therefore, **a** cells produce only **a** factor.

Diploids express neither **a** or α genes nor many of the nonspecific mating-type functions. They do not mate. In diploids the product of the $\alpha 2$ gene of MAT locus represses the **a** factor gene and, together with the product of the **a**1 transcript of the MAT-**a**, forms a complex (**a**1/$\alpha 2$) that represses the genes expressed in both haploid cell types essential for mating. This **a**1/$\alpha 2$ complex also represses transcription of the $\alpha 1$ gene that starts in Y and proceeds through Z. The product of this $\alpha 1$ transcript is necessary for expression of the α factor genes. Since it is absent in diploids, these cells do not make α factor. The **a**1/$\alpha 2$ complex represses HO and so stops switching. By these interactions the MAT locus controls a whole array of genes in a coordinated manner. It stands at the top of a cascade of regulatory processes controlling the flow of genetic information.

Rapid mating-type interconversion results in rapid diploidization of meiotic products to the apparent advantage of the yeast. The system of inserting cassettes into the MAT locus from distant silent copies achieves this goal. This system in yeast may have degenerated from one in which one of many mating-type cassettes could occupy the MAT locus. Many lower plants, fungi, and protozoa have multiple mating types to ensure outbreeding. Some yeast species have found an advantage in being inbred.

Mating-type interconversion occurs frequently (up to 80 percent of the time) in mother cells but never in daughter buds. The small buds do not have bud scars and can be distinguished from the larger, scarred mother cells. Daughter cells divide to produce a mother cell and a daughter cell always of the same mating type as before the division. Thus, the mating type of the original cell is propagated in a stem cell line. Experienced cells switch back and forth in mating type as frequently as once each generation. To a large extent this segregation of interconversion ability is the result of expression of the HO gene that initiates gene conversion at the Y-Z junction by a double-strand cut. The HO gene is only expressed for a brief period during the G1 phase of the cell cycle. The activity rapidly decays after its appearance. Daughter cells do not transcribe the HO gene until they have given birth to a bud. A factor necessary for expression of HO may segregate with maternal nuclear membrane and be unavailable to daughter cells until after they cross the G1 period. Such a segregation of potential has many parallels during determination in multicellular organisms.

Related Readings

Baker, B., and Belote, J. (1983). Sex determination and dosage compensation in *Drosophila melanogaster*. Annu. Rev. Gen. *17,* 345–393.

Brake, A., Julius, D., and Thorner, J. (1983). A functional prepro-α-factor gene in *Saccharomyces* yeasts can contain three, four or five repeats of the mature pheromone sequence. Mol. Cell. Biol. *3,* 1440–1450.

Brand, A., Breedan, L., Abraham, J., Sternglanz, R., and Nasmyth, K. (1985). Characterization of a "silencer" in yeast: a DNA sequence with properties opposite to those of a transcriptional enhancer. Cell *41,* 41–48.

Brunner, M., Moreira-Filho, C., Wachtel, G., and Wachtel, S. (1984). On the secretion of H-Y antigen. Cell *37,* 615–619.

Chandra, S. (1985). Sex determination: A hypothesis based on noncoding DNA. Proc. Natl. Acad. Sci. *82,* 1165–1169.

Delbos, M., Gipouloux, J., and Saidi, N. (1984). The role of the glycoconjugates in the migration of anuran amphibian germ cells. J. Embryol. Exp. Morph. *82,* 119–129.

Doniach, T., and Hodgkin, J. (1984). A sex-determining gene, *fem*-1, required for both male and hermaphrodite development in *Caenorhabditis elegans*. Devel. Biol. *106,* 223–235.

Feldman, J., Hicks, J., and Broach, J. (1984). Identification of sites required for repression of a silent mating type locus in yeast. J. Mol. Biol. *178,* 815–834.

Garabedian, M., Hung, M., and Wensink, P. (1985). Independent control elements that determine yolk protein gene expression in alternative *Drosophila* tissues. Proc. Natl. Acad. Sci. *82,* 1396–1400.

Klar, A., Strathern, J., and Hicks, J. (1984). Developmental pathways in yeast. In *Microbial Development,* R. Losick and L. Shapiro (eds.). Cold Spring Harbor, NY.

Magre, S., and Jost, A. (1984). Dissociation between testicular organogenesis and endocrine cytodifferentiation of Sertoli cells. Proc. Natl. Acad. Sci. *81,* 7831–7834.

McLaren, A., Simpson, E., Tomonari, K., Chandler, P., and Hogg, H. (1984). Male sexual differentiation in mice lacking H-Y antigen. Nature *312,* 552–555.

Miller, A., MacKay, V., and Nasmyth, K. (1985). Identification and comparison of two sequence elements that confer cell-type specific transcription in yeast. Nature *314,* 598–603.

Nakamura, D., and Wachtel, S. (1985). Vertebrate sex determination: an immunologic perspective. In *Biology of fertilization,* Vol. 1, ed. C. Metz and A. Monroy. Acad. Press, San Diego.

Nasmyth, K. (1982). Molecular genetics of yeast mating type. Annu. Rev. Gen. *16,* 439–500.

Page, D., de la Chapelle, A., and Weissenbach, J. (1985). Chromosome Y-specific DNA in human XX males. Nature *315,* 224–226.

Sairam, M., and Bhargavi, G. (1985). A role for glycosylation of the α subunit in transduction of biological signal in glycoprotein hormones. Science *229,* 65–67.

Sprague, G., Blair, L., and Thorner, J. (1983). Cell interactions and regulation of cell type in the yeast *Saccharomyces cerevisiae*. Annu. Rev. Microbiol. *37,* 623–660.

CHAPTER 13

Growth and Death

The final form of an organism is determined not only by the differentiation of diverse tissues but also by the relative rate of growth of cells within those tissues. The length of the right arm and the left arm, the right foot and the left foot are within 10 percent of each other in most mammals, yet they developed and grew quite independently of each other. The ratio of arm length to torso length differs considerably among even such closely related species as primates (apes have longer arms than humans). This clearly indicates that growth is genetically regulated.

Specific cell death also plays a central role in molding the final body plan. At various stages in development, certain cells die and are resorbed by the embryo. For example, as discussed in the previous chapter, Wolffian duct tissue spontaneously degenerates in female mammals. In males the increased level of testosterone prevents this cell death and leads to the differentiation of vas deferens. The system requires not only testosterone but also a testosterone-binding protein present in cells of the Wolffian ducts. In females the Müllerian duct differentiates into the oviduct. But in males the testes produce hormone X, which directs programmed cell death in the Müllerian duct tissue. Here we have dramatic cases of hormonal regulation of cell death.

Cells in Isolation

Cells from many tissues can be excised from embryos and plated in nutrient media as dissociated cells. Since cells were first cultured in glass Petri dishes they are referred to as *in vitro* (in glass) cultures. The requirements for growth *in vitro* differ from one cell type to the next and appear to give some insight into the factors that control growth in the embryo *(in situ)*. The standard media include all the required amino acids, vitamins, and salts in a pH-buffered solution containing glucose as the major source of both energy and carbon for biosynthesis. Most cell lines will not grow in this medium unless serum proteins are added. Serum (blood from which the red blood cells and other cells have been removed) contains a wide variety of growth factors and hormones that are only now being recognized.

Cells of some tissues can be grown in serum-free media if certain hormones are added. Insulin appears to be required for growth of many cell lines. Other growth hormones required by specific differentiated cell lines include somatomedin, testosterone, progesterone, lutenizing hormone, platelet-derived growth factor (PDGF), epidermal growth factor (EGF), and nerve growth factor (NGF).

The last two growth hormones, EGF and NGF, are of considerable interest because they elicit defined effects on whole embryos as well as stimulating growth *in vitro*. Both are small proteins of about 120 amino acids that function as dimers. Specific protein receptors exist that bind one or the other on the cell surface.

Administration of EGF results in precocious eruption of tooth rudiments in mice. NGF plays a role in the ramification of sympathetic nerve cells during embryogenesis. Mice injected with antibodies specific to NGF fail to develop their sympathetic nervous system. It is thought that the antibody binds to NGF in the embryo and sequesters the hormone. In the absence of NGF the sympathetic nerves do not extend processes normally. Nerve cells cultured *in vitro* respond to the addition of NGF by sending out long processes similar to those found in the nervous system of embryos. The

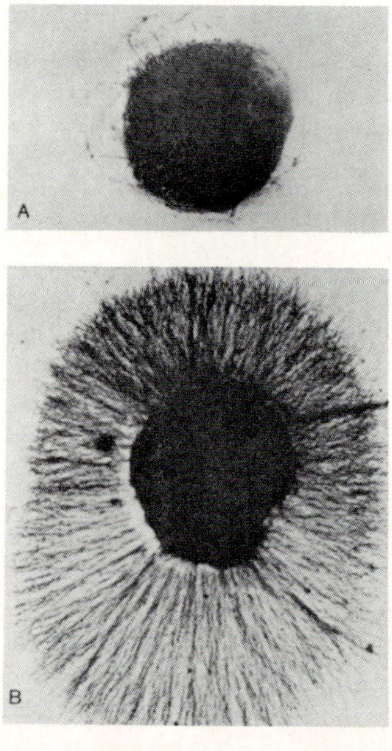

Figure 13.1 Effect of Nerve Growth Factor (NGF) on dendrite extension from a chick embryo sympathetic ganglion. *(A)* One day after explanting a ganglion into culture without NGF the ganglionic cells have not spread; *(B)* after one day in medium containing 10^{-6} *M* NGF, long processes have grown out.

response is specific to nerve cells and is elicited only by the specific NGF peptide (Fig. 13.1).

Differentiated cells proliferate far better when they are deposited on a layer of extracellular matrix than when put on naked glass or plastic. The surface of the cell appears to react to the underlying matrix in a manner conducive to growth. To a certain extent this may be the simple result of increased surface-to-volume ratio of cells on a matrix layer. Cells spread and flatten much better on such surfaces as a consequence of being able to apply adhesion mechanisms. However, the nature of the extracellular matrix determines the relative growth rate of different cell lines. For instance myoblasts grow well on a surface coated with collagen. Since muscle cells proliferate in the embryo in association with connective tissue producing collagen, the stimulatory effects of collagen *in vitro* may truly be mimicking a physiologically significant growth-controlling condition.

Although cells from many tissues will grow for several generations in appropriate media, most of them die after a few months. Either they were programmed to die or the *in vitro* conditions are still not ideal and an essential component slowly becomes limiting until death results. A few cells (1 in $\sim 10^6$) will often continue to grow, and these established cell lines turn out to have stable hereditary changes. These may be mutations in genes that normally regulate growth or they could be stable physiological states compatible with growth. A cell line that continues to grow is often referred to as transformed. In fact, many such cell lines will give rise to tumors when injected back into an individual. However, other established cell lines do not proliferate in an uncontrolled manner when placed back in a host organism. There appear to be several ways a cell population can adapt to the artificial conditions *in vitro* and become immortal.

The observation that cells taken from an embryo divide a limited number of times and then die has led some to believe that this process is similar to organismic senescence. The differences between cells in culture and an aging individual are more striking than the similarities. In some cases, cells in culture stop dividing as a consequence of massive terminal differentiation. It is not surprising that a population of myoblasts stops growing after the cells have all fused to form large syncytial myotubes. Likewise, keratinocytes stop dividing when they have filled up with keratin. Adipocytes no longer divide after they have swollen with lipids. In the whole organism similar differentiations also result in the loss of proliferative activity. Red blood cell precursors divide three or four times after the globin genes have been activated but by the time they have accumulated visible amounts of hemoglobin, cell division stops. After functioning in the circulatory system for many days, red blood cells lyse. This is a specialized form of programmed cell death.

Cells in Embryos

The period of most rapid cell division in an embryo is immediately following fertilization. The early cleavages in many eggs come every hour. In

Xenopus embryos there is an abrupt increase in the time between cell divisions at the midblastula stage. As discussed in Chapter 8, this midblastula transition (MBT) is characterized by the initiation of transcription of the embryos' genes and the loss in synchrony among divisions of different blastomeres. There is evidence that the ever-increasing number of nuclei titrate out a component initially present in the unfertilized egg. There is additional evidence that another consequence of using up this egg factor is that cell division slows. This indicates that specific components can regulate the rate of cell division. It is a plausible hypothesis that the availability of nutrients from the metabolism of reserves stored in yolk or provided from the placenta might determine the rate and extent of growth of different tissues.

Shortly after closure of the neural tube in the chick embryo, bulges can be seen on both sides of the body wall. These are the limb buds. Initially the bulges are symmetric but in the position of the future armpit of the wing the cells begin to die. This region is referred to as the posterior necrotic zone (PNZ). At about 4.5 days of incubation of a chick embryo, 2500 cells at the posterior of the wing bud lyse and are engulfed by macrophages. This ultimately leads to the asymmetry between the shoulder and the flanking sides of the wing (Fig. 13.2).

If small pieces of tissue are taken out of the future site of the PNZ at two days of incubation and put in culture, they die *in vitro* at the same time they would have if left in place, that is, two days later. These PNZ cells can be rescued only if put deep within mesodermal tissue. It appears that the PNZ cells are instructed before two days of embryogenesis to die two days later but that these instructions can be countermanded by conditions present in mesodermal tissues. Small pieces of tissue taken from other positions of the limb bud, such as the anterior portion (future shoulder), do not die soon after being put in culture. The death signal is clearly localized to the posterior margin of the limb bud.

Other appendages are also sculpted by programmed cell death. A fa-

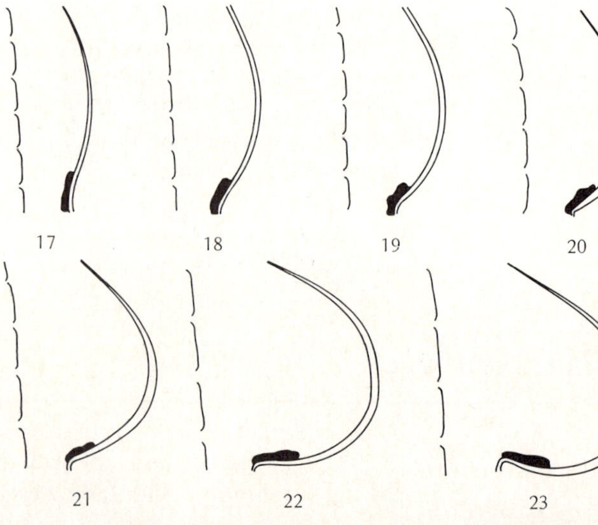

Figure 13.2 PNZ of the chick wing bud. At the stages indicated, the region of subsequent cell death is marked by shading. If these cells are transplanted elsewhere at stage 17 or 18, they will nonetheless lyse and their contents be resorbed two days later.

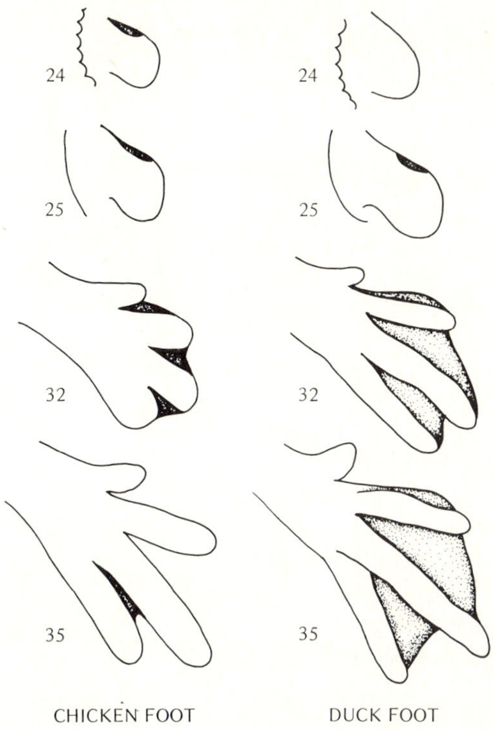

Figure 13.3 Regions of localized cell death in chick and duck feet. At the stages indicated, the regions of subsequent cell death are marked by shading. Epithelial and mesenchymal tissues between each of the digits die in chicken feet but only that between digits 1 and 2 die in duck feet.

miliar and dramatic case concerns the feet of chickens and ducks. The tissue between the toes of a chick die before hatching whereas that between the toes of a duck does not, thereby providing the webbing (Fig. 13.3). One can do transplants of limb buds between chick and duck embryos. A chick hind limb transplanted to a duck embryo develops into a chicken leg with no webbing between the toes. The genes necessary for the programmed cell death of the webbing cells clearly act within each affected cell autonomously of the rest of the embryo. Conversely a duck limb bud transplanted onto a chick embryo develops into a nicely webbed foot.

Metamorphosis

Some organisms have found a selective advantage of feeding in one body form and then radically changing that form before reproducing. A well-known case is the metamorphosis of tadpoles into frogs. Besides the dramatic tail resorption and leg extension, there are many internal changes as the way of life changes from an aquatic one to a terrestrial one. However, the best analyzed changes involve resorption of the tadpole tail. The tail tissue dies and the components enter the circulatory system of the froglet.

Metamorphosis in frogs is triggered by the thyroid hormone thyroxine. Simple injection of thyroxine into tadpoles initiates all the changes seen in

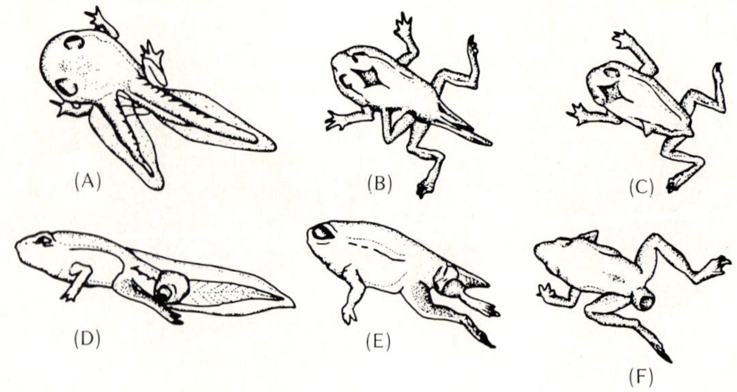

Figure 13.4 Fate of transplanted tail and eye during metamorphosis in the frog. Tail tissue was grafted to the flank of a tadpole *(A)*. It was resorbed at the same rate as the tail of the host *(B, C)*. An eye was grafted to the tail of a tadpole *(D)*. The transplanted eye was unaffected by the regressing tail tissue and came to lie on the frog's posterior *(E,F)*.

natural metamorphosis: cells in the tail die and cells in the limbs grow. Clearly cells of the different tissues were programmed to respond oppositely to the same trigger.

The eyes of a tadpole change very little during metamorphosis. An eye transplanted into the tail of a tadpole does not degenerate along with surrounding tissue during metamorphosis but is brought to the posterior of the froglet by the process of tail resorption (Fig. 13.4). The responses to thyroxine are clearly cell autonomous and not dependent on surrounding tissue.

Genetic Control of Proliferation

Almost all the cells in an adult have sufficient nutrients available for rapid growth. Yet only selected tissues such as the erythropoietic stem cells grow at an appreciable rate. The great majority of cells are kept in check by complex processes. Now and then one of the billions of cells sustains a mutation or other genetic change that releases it from growth control. If it proliferates rapidly, the mutant cell will give rise to a tumor. This pathological state of cancer may tell us a lot concerning the normal processes that regulate relative rates of growth during embryogenesis.

Many agents have been found that can transform a normal cell into a cancerous cell. The most informative have been the retroviruses that carry their genetic information as RNA. Upon infection of a cell, the RNA is reverse transcribed into DNA, which may then integrate into the host chromosomes. The retroviruses that cause neoplastic transformation often carry portions of cellular genes as well as their own viral genes. At some point in the history of these transforming viruses a recombinational event exchanged a portion of one of their own genes for a sequence from the host cell's genome. This hybrid genome can cause cells to be transformed. More than a dozen normal cellular genes are now known that cause cancer when carried by retroviruses. They are referred to as oncogenes. Several of the better understood oncogenes are described in some detail in the Specific Examples of this chapter.

Viral oncogenes such as v-*sis* or v-*ras* are not completely identical to their cellular counterparts c-*sis* and c-*ras* but have small rearrangements, deletions, or base changes. They are transcribed from the viral genome under the influence of viral control elements and so may be inappropriately expressed or overproduced. It is not yet clear which of these changes in expression of the oncogenes are responsible for causing cancer.

In several cases infection with a single virus carrying an oncogene does not transform a cell. Two different oncogenes must be added to the cell to break growth control. This finding was not completely unexpected since clinical and cell culture experiments had indicated that spontaneous cancers do not arise as the consequence of a single event. Kinetic analyses had indicated that two or more genetic alterations are necessary for a cell to lose sensitivity to natural growth regulatory mechanisms. There appears to be built-in redundancy in the system that keeps cell growth in check.

The physiological processes determined by the products of oncogenes are only now being glimpsed. The most striking property recognized so far for some oncogene products is that they catalyze the phosphorylation of specific proteins. The phosphorylation is somewhat unusual in that it modifies tyrosine residues rather than serine or threonine, the more common phosphorylated amino acids found in a variety of proteins. Phosphorylation of a protein can profoundly alter its function, stimulating some enzymes and inhibiting others. Structural proteins may change their associations when phosphorylated. It is not surprising, then, that introduction of an altered protein kinase or inappropriate expression of a protein kinase gene can have far-reaching effects on cell physiology since the protein kinase can modify a whole spectrum of other proteins. The consequences of these modifications may lead to uncontrolled growth.

Perhaps the normal cellular genes that are homologous to the viral oncogenes play important roles in allowing growth of some cells under some conditions (such as wound healing) and in keeping other cells quiescent. As cells grow, they pass through the stages of the cell cycle: G1, S, G2, and mitosis. The product of one of the cellular homologs of a viral oncogene, c-*myc,* a 65 kd protein, is found in the nucleus and binds DNA. It may be one of the components that determines the frequency with which a cell will progress from G1 to S and on around the cell cycle. Cancer directed research may soon provide some answers to basic developmental questions.

Summary

Embryos are shaped and reshaped to gain their final form. Often it is advantageous for certain tissues to degenerate and have the contents of their cells resorbed. Hormone X triggers degeneration of Müllerian ducts. Thyroxine stimulates the massive breakdown of connective tissue in the tails of tadpoles during metamorphosis. The posterior margin of wing buds (PNZ) is programmed for subsequent necrosis early in limb formation. The PNZ as well as the tadpole tail respond autonomously to their death knells even while other tissues are proliferating. Hormones also stimulate growth of specific cell types. Epidermal growth factor (EGF) acts preferentially on epidermal cells but also affects other cell types in culture. Nerve growth factor (NGF) preferentially stimulates growth of sympathetic nerves. The cellular receptors for these peptide hormones are associated with protein kinase activities that may transduce the signal by phosphorylation of internal cellular proteins. A complex set of genes regulates passage through the cell cycle to division. Some of these have been picked up and misused by viruses that cause cancers.

SPECIFIC EXAMPLES

1. Cancer Genes

The unregulated growth of cancer cells can result from a wide variety of genetic injuries, only some of which have been recognized. In the few cases in which abnormal expression or mutational alteration have indicated the gene responsible for the cancer, it has turned out that the normal gene plays a role in growth control as well.

Retroviruses use RNA as their genome in the infectious state. Because uninfected cells seldom replicate RNA, they do not contain the enzymes necessary for RNA-directed replication. The particles of retroviruses contain within themselves an enzyme, reverse transcriptase, that uses RNA as a template to direct DNA synthesis. The enzyme uses the deoxynucleotide triphosphates available in the cell to polymerize a DNA copy of the viral genome. The DNA copy, in turn, is used as a template for RNA synthesis by the host's transcriptional machinery. The RNA transcripts are used to direct synthesis of viral encoded proteins as well as serving as the genomes of progeny viruses. Sometimes the DNA copy is integrated into the host genome and is carried along as a provirus that can reappear as a viral infection at any time. When proviruses are reactivated they sometimes pick up a piece of the host genome, which is then replicated along with the viral genome.

Some retroviruses cause cancer. The range of species and cell types in which neoplastic transformation can occur is limited for each type of virus. One such virus, termed SSV (simian sarcoma virus), gives rise to sarcomas in monkeys. A single gene, v-*sis,* carried by SSV codes for a protein of 16 kd that is responsible for the transforming activity. It turns out that this protein is one of the two subunits of platelet-derived growth factor (PDGF). This growth factor is involved in stimulating fibroblast growth at the sites of surface wounds. It is carried by blood platelets to the site of the wound where it is liberated to stimulate wound healing. Thus, PDGF plays a necessary role in the ability of large animals to survive the pricks and thorns of life. But, when carried in a virus, the gene for one of the PDGF subunits is potentially lethal to the organism. The virus proliferates in the cell and generates an overabundance of the *sis* gene product, leading to unrestrained growth, especially in fibroblasts. The fact that v-*sis* of SSV is one of the PDGF genes has focused attention on the normal physiological role played by PDGF as well as clarified how the virus causes cancer.

Another virus has offered up a different component in the multistep process of growth control. Avian erythroblastosis virus causes cancer of the hematopoietic cells in birds due to a gene termed *erbB*. It turns out that this gene codes for a protein of 604 amino acids that is very similar to a portion of the surface receptor protein for EGF. The EGF receptor is a protein tyrosine kinase that phosphorylates cellular proteins when EGF is bound to its extracellular portion. The EGF receptor is a protein of 1250 amino acids that spans the cellular membrane. When the outside portion is stimulated by EGF, the inside portion catalyzes phosphorylation of proteins. The product of v-*erbB* looks like a protein kinase that has lost its outside portion but can still phosphorylate proteins. A plausible conjecture is that it does so without any need for stimulation by EGF. The cellular proteins regulating growth may then become phosphorylated all the time and stimulate the growth of a tumor.

The receptor for PDGF is also a protein tyrosine kinase, but so far no viruses carrying portions of the gene for the PDGF receptor have been found. However, seven different viruses have been found to carry cancer genes that have evolved from a common ancestor that coded for a protein kinase that also gave rise to the EGF receptor and *erbB*. Each of the viral oncogenes makes a protein kinase, as do two other related genes, *cadK* and *cdc*28 (Fig. 13.5). The *cadK* gene codes for a cellular protein kinase that is dependent on cAMP for activation. The *cdc*28 gene plays an essential role in regulating the cell cycle in yeast. The fact that *cdc*28 is related to viral oncogenes suggests that derepression of these protein kinases somehow allows the cell cycle to keep going in conditions under which it is normally restrained.

Evolutionary conservation of another viral oncogene, v-*ras,* has also been found in genes of yeast, *Dictyostelium,* and mammals. The normal cellular *ras* gene product of *Dictyostelium* is 60 percent homologous to v-*ras*. Both the viral and the yeast *ras* genes make a guanine nucleotide-binding protein that associates with the cytoplasmic face of the cell membrane. There is evidence that the human cellular *ras* gene codes for a protein related to a family of normal cellular proteins referred to as G. These proteins are involved in coupling surface receptors, such as the EGF receptor, to the production of the intracellular second message, cAMP. This cyclic nucleotide may then modulate many physiological functions by activating cAMP-dependent protein kinase. It is known that a mutation in the twelfth codon of c-*ras* can cause bladder cancer in humans, but how it does this has not yet been elucidated. The carboxy terminal portion of v-*ras* differs from the cellular gene, c-*ras,* at amino acid 59 (ala → thr) as well as at a few other places. The viral *ras* protein autophosphorylates amino acid 59 using GTP as the phosphate donor. Thus, the viral *ras* product is a protein threonine kinase.

There are two *ras* genes in yeast. By using sophisticated genetic techniques, it has been shown that growth of yeast requires transcription of at least one of the pair or an inserted mammalian *ras* gene. Thus, *ras* is a vital gene that is essential for growth that has conserved at least some of its vital functions for several hundred million years.

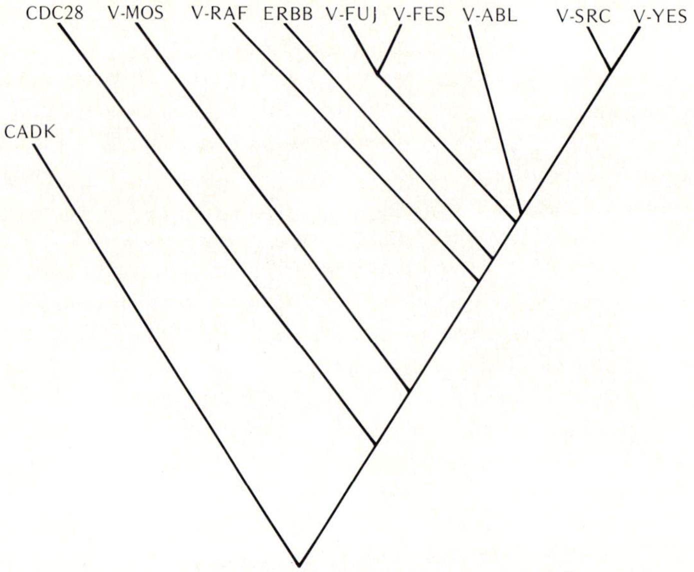

Figure 13.5 Evolutionary relatedness of protein kinase and cancer genes. Computer analysis of the primary amino acid sequence of various proteins has shown that they evolved from duplications of a common progenitor sequence. The major cellular cAMP-dependent protein kinase (*cad*K) is a distantly related member of this family. It phosphorylates serine and threonine residues in certain proteins when its regulatory subunit is dissociated by cAMP. The *cdc*28 gene was first found as a gene required for progress through the cell cycle in yeast and then subsequently shown to be a protein kinase that phosphorylates tyrosine in certain proteins. The viral oncogenes v-*mos,* v-*raf, erbB,* v-*fuj,* v-*fes,* v-*abl,* v-*src,* and v-*yes* are also protein tyrosine kinases. Some pairs have very recently evolved from a common ancestor such as v-*src* and v-*yes,* judging by the degree of homology of amino acid sequence between them. Other cancer genes such as v-*mos* and v-*raf* have considerable homology but have diverged further from each other and the rest of these cancer genes. This analysis was carried out by Russell Doolittle.

Transformation of some lines of cultured cells requires viral carried copies of both *ras* and *myc*. The *myc* gene was found in MC29 viruses. The normal cellular homolog of v-*myc,* c-*myc,* is expressed into RNA at a high rate in cultured human cancerous leukemia cells, and a related gene, N-*myc,* is amplified in cancerous neuroblastoma cells. Expression of c-*myc* is stimulated by PDGF. The *myc* gene product is concentrated in the nucleus and may play a role in triggering the onset of DNA synthesis as the cells progress around the cell cycle from G1 to S. Nongrowing cells are arrested after mitosis (M) in G1, the growth stage before replication of the genome (S). Once the cells enter S and synthesize DNA, they invariably complete replication, pass through the second growth phase (G2) of the cell cycle, and undergo mitosis (M). Growth-arrested cells are often said to be in G0 to emphasize that they are not progressing toward S. There is evidence that return from G0 to the cycling mode is mediated by successive factors. The first phase is known as priming for growth competence. The cellular *myc* gene product is implicated at this stage. Several aberrations result in excess *myc* expression, including transposition of the gene into an active immunoglobin region in cells of Burkitt's lymphomas as well as infection with viruses carrying the *myc* gene. In these cases it is thought that the unregulated growth results in part from excess production of the normal *myc* gene product.

These studies of retroviruses were initially directed toward the question of what genes cause cancer. Analyses of the evolution of the oncogenes have pointed at growth factors and their receptors that appear to be involved in normal growth control. The similarity of protein tyrosine kinases to a protein known to be required to regulate the growth cycle in yeast takes this approach a step forward. It is expected that many more of the viral oncogenes will turn out to be homologs of developmental genes.

2. Hydra Stem Cells

As far as we know, an individual hydra can live forever. It does not age because all of its tissues are replaced every month or so as long as it is able to feed. The cells die but the individual remains. Stem cells continue to divide and replenish the various tissues with epithelial cells, muscle cells, gland cells, nerves, and various specialized cell types. Moreover, the rate of replacement adapts to the rate of cell death.

Hydra is a small freshwater polyp that can be up to 20 mm in length (Fig. 13.6). Hydra are seldom at rest and continuously sway about, contract and stretch out, all the time waving their tentacles in search of small creatures they can capture and eat. Their dimensions change but their structures are constant.

At the top, the head consists of a mouth surrounded by a whorl of about six large tentacles. The gastric region makes up most of the body column. The lower region of the column gives rise to buds that develop into new individual hydra. A short stalk or peduncle connects this region with the foot that holds the individual in place. The body wall consists of two concentric epithelial layers surrounding the central gastric cavity. The ectoderm is separated from the endoderm by an extracellular matrix referred to as the mesoglea. Small living organisms, such as shrimp larvae, are stung and paralyzed by nematocytes on the tentacles and efficiently brought to the mouth (Fig. 13.7). Food is ingested and digested within the gastric cavity. Undigested material is finally regurgitated.

Hydra multiply both asexually and sexually. Well-fed individuals produce a bud every few days. The new young hydra can feed while still attached to the parental column and drops off when about one-quarter grown. Asexual growth by budding can result in stable, exponential growth with a doubling time of as little as 1.7 days in *Hydra littoralis*. All that is needed for growth of hydra is millimolar calcium in distilled water brought to pH 7.6 with 10^{-4} *M* sodium bicarbonate and a plentiful supply of brine shrimp larvae once a day. Under these conditions stem cells are producing the various cell types at maximal rates. In starved individuals the net cell number does not increase, although stem cells continue to divide. There is considerable turnover of cells in starving hydra. This might seem inefficient, but it

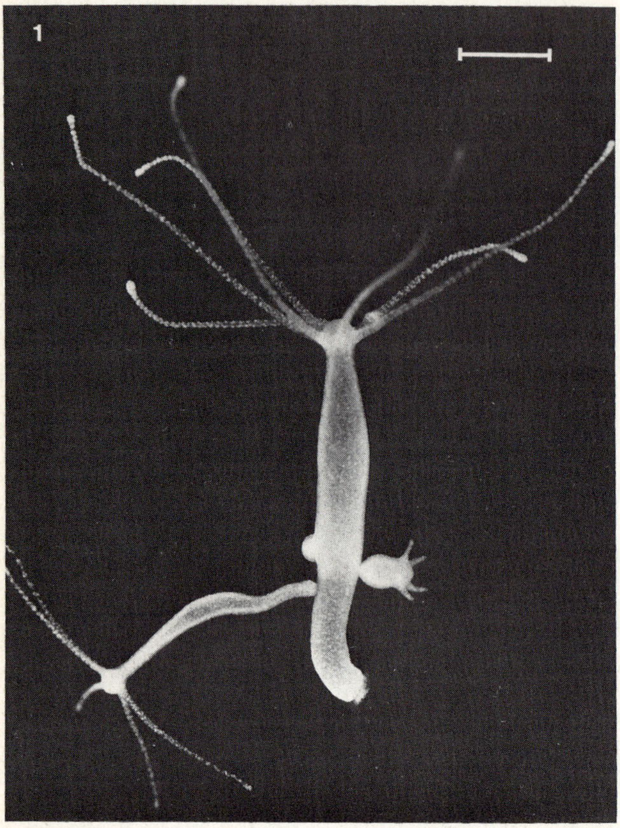

(A)

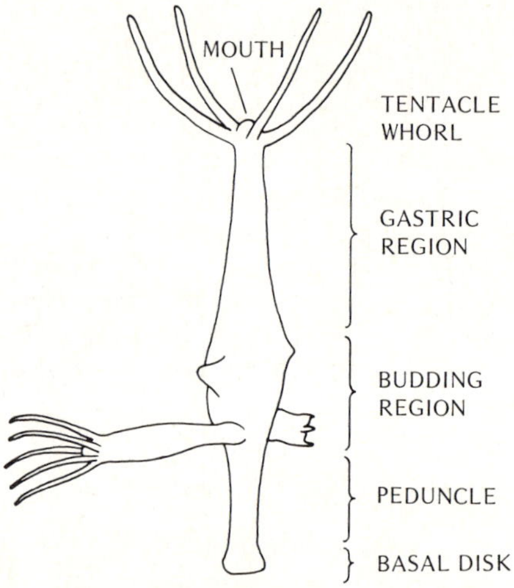

(B)

Figure 13.6 *Hydra attenuata.* The whorl of tentacles around the mouth contains batteries of stinging nematocysts. The mouth can open wide to engulf relatively large prey that is then digested in the gastric region. A hydra multiplies asexually by forming buds below the gastric region. Buds can capture and swallow food while still attached to the parent and thus give it a hydra-headed appearance. Hydra taper down to the peduncle, which ends in a basal disk used to attach the animal to solid supports. Hydra is a ctenophore, as are tubularians and jelly fish. Jelly fish have emphasized a different form of the general ctenophore life cycle, the radially symmetric medusa stage (scale bar: 1 mm).

keeps the motors of stem cell division running so that rapid growth can resume when food is once again available.

Starvation as well as certain microenvironmental conditions including increased partial pressure of carbon dioxide (pCO_2) induces sexual differentiation in hydra. Some species are hermaphroditic, producing both sperm and eggs, whereas others contain individuals that differentiate solely as males or females. The sperm are shed into the water and fertilize the eggs. Within about a month small polyps hatch and begin to feed. Unfortunately, the hatching rate is generally low and newly hatched animals take special care to survive. For these reasons little is known of the embryogenesis of hydra.

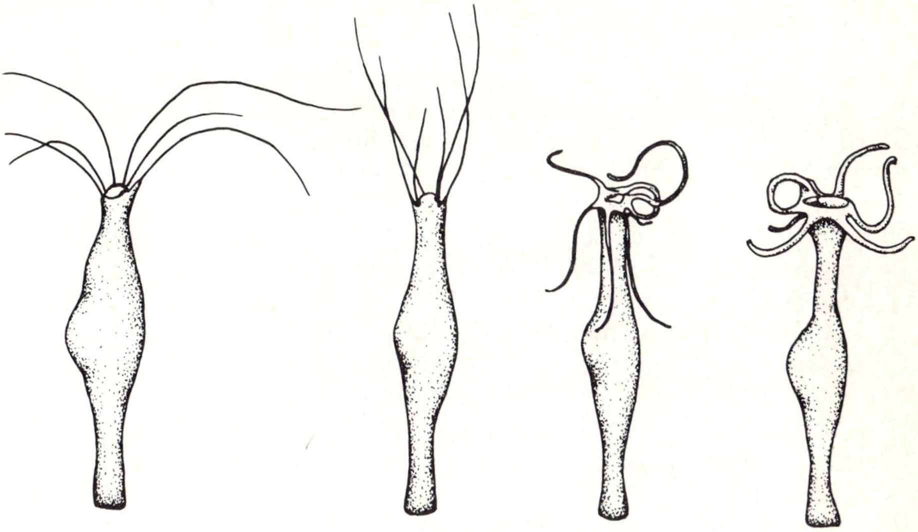

Figure 13.7 Feeding response in *Hydra*. Hydra eat only living organisms. They respond to reduced glutathione (GSH), a tripeptide found in all living cells. When an organism dies, its glutathione is rapidly oxidized. Thus, reduced glutathione signals living prey. Within a minute of adding GSH to the water, the tentacles sweep straight up, contract, and bend toward the mouth, which is opening in response to the signal. Frequently, GSH-stimulated hydra are seen with the tips of their tentacles in their mouths.

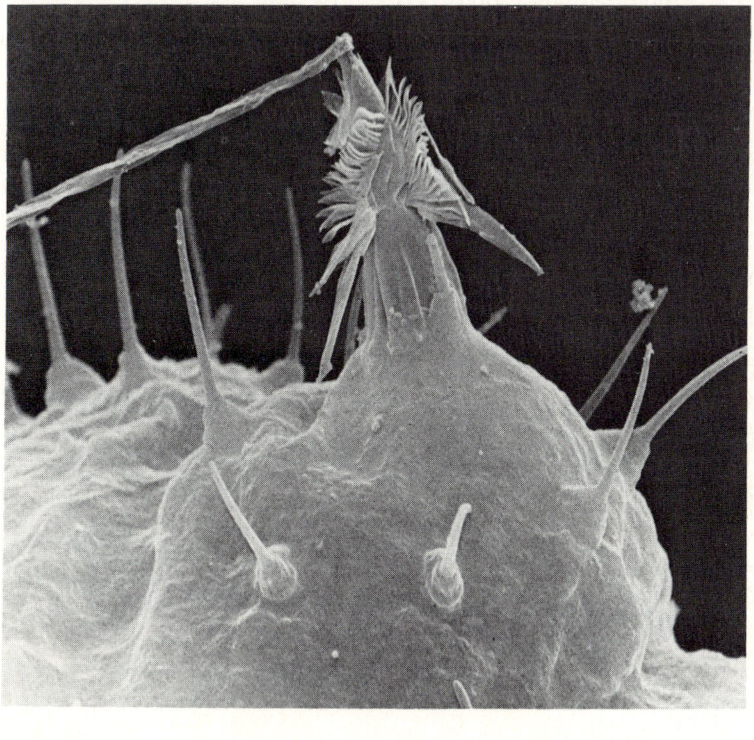

Figure 13.8 Nematocyst barb. A stenotele nematocyst has discharged in an attempt to spear prey (magnified 5000-fold).

Both ectodermal and endodermal cells are continuously formed in an individual hydra. New cells are born in the body column and displaced to the tentacles and the foot, where they are sloughed off. There are also directed migrations of certain specialized cells along the body axis.

Some of the most beautifully articulated cellular structures are found in the stinging cells on the tentacles of hydra (Fig. 13.8). These barbs, lassos, and stings are held coiled in four types of nematocytes. When chemically or mechanically stimulated, they discharge and uncoil at the amazing speed of 2 m/second, corresponding to an acceleration of 40,000 × g. The stem cells that give rise to nematocytes are the interstitial cells found in the body column ectoderm. Interstitial cells divide to produce nematocyte precursor cells, among other things. These precursor cells divide about four times while still in the body column to generate a nest of small nematocytes. Fully differentiated nematocytes then migrate rapidly to the tentacles where they have a brief residence before being used in capturing prey. They must be replaced after every feeding by newly generated nematocytes.

Interstitial cells also generate nerve cells as well as maintain their own population. When interstitial cells divide, the progeny often becomes committed to nerve cell differentiation. It may divide once again but will ultimately put out slender processes and become integrated into the primitive nerve net of hydra just above the mat of ectodermal muscle processes. Before terminal differentiation occurs, the nerve cell precursors migrate toward the head region so that many nerve cells are formed there. It appears that sorting out of committed cells rather than regional signals results in localized differentiation patterns: (1) nematocyte differentiation in the body column and (2) nerve cell differentiation in the head region. The stem cell population of interstitial cells that gives rise to both types of terminally differentiated cells is found in the body column but only one type of precursor, that for nerve cells, goes on extended migration before becoming the definitive cell type. Sorting out of this type may be involved in spatial patterning in many regenerating tissues.

The ectodermal and endoderm cells belong to two other self-renewing lineages. Interstitial cells can be eliminated from hydra by treatment with hydroxyurea, colchicine, or nitrogen mustard. Such individuals can be grown and will multiply if hand-fed and washed. They completely lack nerve cells, nematocytes, interstitial cells, and gland and mucous cells. Therefore, they cannot capture prey or coordinate feeding behavior and have to be force fed with crushed shrimp larvae and then be purged by hand, one at a time. Nevertheless, with this sort of tender care, hydra made of only the two epithelial cell layers grow and bud off young hydra (Fig. 13.9).

Hydra treated so as to preferentially destroy all interstitial cells cannot generate the cell types derived from this stem cell lineage (nerves, nematocytes, etc.) but continue to generate ectodermal and endodermal cells. Because these are completely separated by the extracellular mesoglea, they are thought to belong to separate lineages. Epithelial hydra have long tentacles and a translucent body column. Consistent with the lack of nerve cells, they fail to respond to stimuli and do not move.

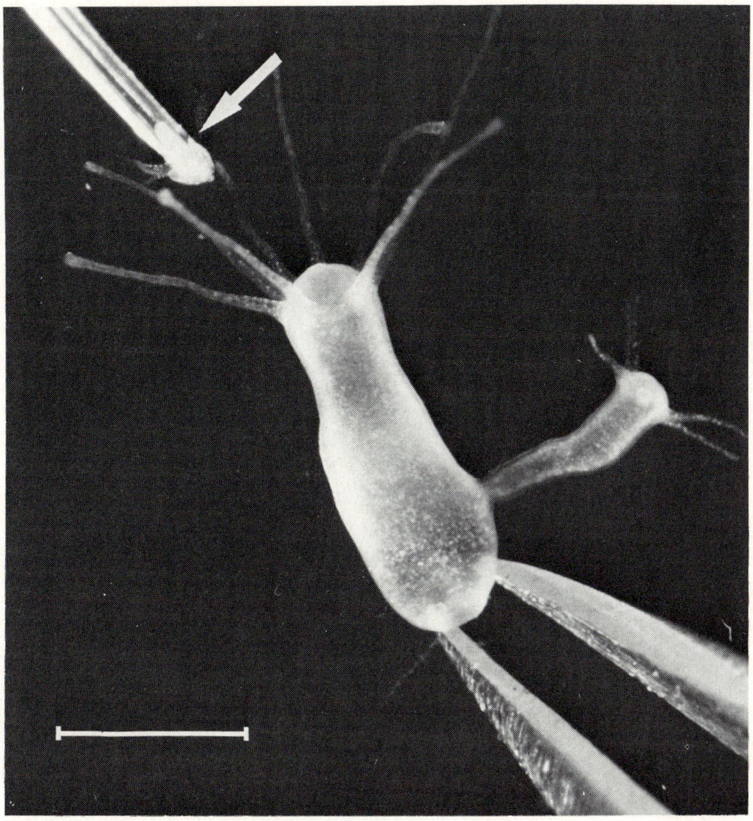

Figure 13.9 Hand feeding a nerveless hydra. This 2-mm long hydra has been treated with hydroxyurea and colchicine to preferentially block growth of interstitial cells. Only the epithelial cell layers are left, and the individual is paralyzed since it has no nerve cells. It also lacks nematocysts with which to capture food. It must be given minced shrimp on the end of a small rod while being gently held at its base with forceps (scale bar: 1 mm).

A normal hydra has about 90,000 cells in the body column of which 4 percent are interstitial cells. These 3600 stem cells generate 500 nerve cells and 1800 nematocyte clusters every day. They also give rise to about 3400 new interstitial cells per day. Thus, 60 percent of the daughter cells remain stem cells while 40 percent initiate differentiation. The decision to proliferate or differentiate occurs very early in clone development and appears to be regulated by short-range feedback signals. If the population of interstitial cells is experimentally depleted, a homeostatic mechanism triggers an increased growth rate in the stem cell population, resulting in the eventual recovery of the normal stem-cell concentration in the tissue. The probability of self-renewal is increased in animals depleted of stem cells and decreased in animals with an overabundance of stem cells. The change in probability occurs within a few days in small clones, indicating that rapid short-range controls are operating.

The head and the foot of hydra provide morphological landmarks that can be used to monitor polarity and cell differentiation. There appear to be gradients of both induction and inhibitory properties extending from both the head and the foot. When a hydra is decapitated, the tissue at the stump rapidly reorganizes to form a new head with tentacles. This regeneration involves respecification of differentiated cells but the details are still far from clear. Usually only the number of tentacles is recorded. Using this assay, a small peptide (Glu.Pro.Gly.Gly.Ser.Lys.Val.Ile.Phe) has been found that acts as a head activator. Addition of the peptide increases fractionally (about 10 percent) the number of tentacles that regenerate. It also stimulates the rate of bud formation in large, well-fed hydra. The effect is not great, but then the method of presentation to the cells—just adding it to the water—is not very specific. Head-activating peptide is found in highest concentration in nerve cells. As mentioned previously, the concentration of nerve cells is greatest in the head, so it follows that head activator is also greatest in the head. However, hydra from which interstitial cells have been removed contain no nerve cells yet still contain head-activating peptide. New heads regenerate on such hydra when they are decapitated. Thus, head activation appears to be a complex process.

A similar assay has led to the partial purification of a foot activator. This protein results in an acceleration in the number of amputated hydra that regenerate a foot able to stick them to a dish. Inhibitory substances that retard regeneration of head or foot have been found in extracts of hydra, but it is not yet clear how specific they are.

Results from various grafting experiments in which the frequency of head and foot formation are measured have been interpreted in the format

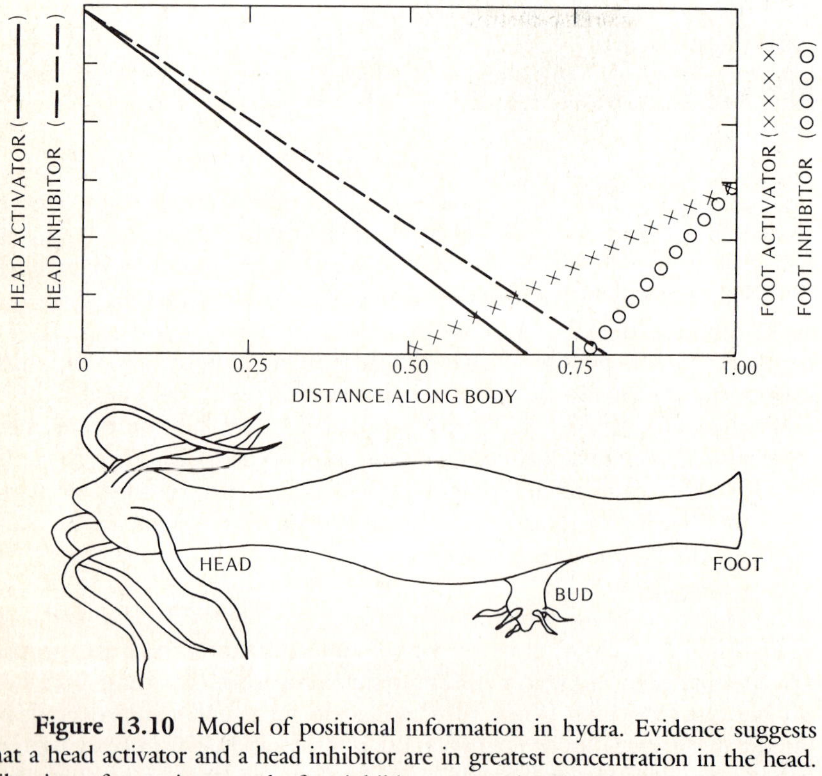

Figure 13.10 Model of positional information in hydra. Evidence suggests that a head activator and a head inhibitor are in greatest concentration in the head. Likewise, a foot activator and a foot inhibitor are produced in the foot and removed in the body.

of short-range activators and longer-range inhibitors for both head and foot (Fig. 13.10). This model has been successful in analyzing a series of aberrant inbred strains of *Hydra magnipapillata* as well. This model is adaptable to pattern formation in a variety of systems.

Head inhibition is a very labile property that decays with a half-life of only a few hours following decapitation. It appears to emanate from the head and be removed throughout the rest of the tissue. Assuming head inhibition is due to a small molecule, it can be calculated that it must diffuse down a *Hydra* at the rate of 10^{-6} cm^2/second, a value consistent with the measured diffusion rate of molecules less than 1000 daltons in aqueous solution at 20°C. It is interesting that both head inhibitor and foot inhibitor reach a minimum at the budding region, where new young hydra are formed. One of the aberrant strains studied, L4, buds at a very low rate and has a higher level of head inhibition that extends all the way to the foot. Another aberrant strain, mh-1, has a much reduced level of head inhibition and forms multiple buds close to the head.

Mathematical models suggest that activator production is an autocatalytic reaction that is antagonized by the inhibitor. Differences in stability and diffusion rates of these substances have been set to fit the measured properties of *Hydra*. The inhibitor acts as a size sensor and regulates the number of cells incorporated in the head.

The head itself is a complex tissue with six or seven regularly spaced tentacles radiating out around the mouth. During regulation of small fragments, the average number of tentacles increases from three to seven as the total number of cells increases exponentially by a factor of 20. The number of tentacles appears to depend on the circumference of the tentacle zone whereas the amount of tissue allocated to tentacles is proportional to the amount of body tissue. Spacing of tentacles is apparently patterned by a number-regulating system with about six cells separating the tentacles from one another. However, the details of these mechanisms are not yet well understood.

Hydra are very resilient organisms that survive well with a very simple body plan. They can be minced into small pieces and will then reassociate, sort out, and reform a functioning hydra. If the head is cut off, a new head is made. If the foot is cut off, a new foot is made. But even without these drastic injuries, a hydra must integrate the rate of replacement of nematocysts, nerves, ectodermal, and endodermal cells. These tissues are derived from three separate lineages in which stem cell proliferation and differentiation must be kept in balance by feedback signals.

Related Readings

Acherman, J., and Sugiyama, T. (1985). Genetic analysis of developmental mechanisms in *Hydra* X. Morphogenetic potentials of a regeneration-deficient strain (reg-16). Devel. Biol. *107,* 13–27.

Baldwin, G. (1985). Epidermal growth factor precursor is related to the translation product of the Moloney sarcoma virus oncogene *mos*. Proc. Nat'l Acad. Sci. *82,* 1921–1925.

Bosch, T., and David, C. (1984). Growth regulation in *Hydra.* Devel. Biol. *103,* 161–171.

Chiu, I., Reddy, P., Givol, D., Robbins, K., Tronick, S., and Aaronson, S. (1984). Nucleotide sequence analysis indentifies the human *c-sis* protooncogene as a structural gene for platelet-derived growth factor. Cell *37,* 123–129.

Doolittle, R., Hunkapiller, M., Hood, L., Devare, S., Robbins, K., Aaronson, S., and Antoniades, H. (1983). Simian sarcoma *onc* gene, *v-sis,* is derived from the gene (or genes) encoding a platelet-derived growth factor. Science *221,* 275–277.

Heidemann, S., Joshi, H., Schecter, A., Fletcher, J., and Bothwell, M. (1985). Synergistic effects of cAMP and nerve growth factor on neurite outgrowth and microtubule stability of PC12 cells. J. Biol. Chem. *100,* 916–927.

Heimfeld, S., and Bode, H. (1984). Interstitial cell migration in *Hydra attenuata.* Devel. Biol. *105,* 1–17.

Hurle, J., Colvee, E., and Fernandez-Terna, M. (1985). Vascular regression during the formation of free digits in the avian limb bud: a comparative study in chick and duck embryos. J. Embryol. Exp. Morph. *85,* 239–250.

Kataoka, T., Powers, S., McGill, C., Fasano, O., Strathern, J., Broach, J., and Wigler, M. (1984). Genetic analysis of yeast RAS1 and RAS2 genes. Cell *37,* 437–445.

Kris, R., Lax, I., Gullick, W., Waterfield, M., Ullrich, A., Fridkin, M., and Schessinger, J. (1985). Antibodies against a synthetic peptide as a probe for the kinase activity of the avian EGF receptor and v-erbB protein. Cell *40,* 619–625.

Lenhoff, H., and Loomis, W. F. (1961). The biology of hydra. University of Miami, Miami.

McGrath, J., Capon, D., Goeddel, D., and Levinson, A. (1984). Comparative biochemical properties of normal and activated human *ras* p21 protein. Nature *310,* 644–649.

Propst, F., and Van de Woude, G. (1985). Expression of c-mos proto-oncogene transcripts in mouse tissues. Nature *315,* 516–518.

Reed, S., Hadwiger, J., and Lorincz, A. (1985). Protein kinase activity associated with the product of the yeast cell division cycle gene CDC28. Proc. Natl. Acad. Sci. *82,* 4055–4059.

Schwab, M., Ellison, J., Busch, M., Rosenow, W., Varmus, H., and Bishop, J. M. (1984). Enhanced expression of the human gene N-*myc* consequent to amplification of DNA may contribute to malignant progression of neuroblastoma. Proc. Natl. Acad. Sci. *81,* 4940–4944.

Slamon, D., and Cline, M. (1984). Expression of cellular oncogenes during embryonic and fetal development of the mouse. Proc. Natl. Acad. Sci. *81,* 7141–7145.

Slamon, D. de Kernion, J., Verma, I., and Cline, M. (1984). Expression of cellular oncogenes in human malignancies. Science *224,* 256–262.

Tatchell, K., Chaleff, D., DeFeo-Jones, D., and Scolnick, E. (1984). Requirement of either of a pair of *ras*-related genes of *Saccharomyces cerevisiae* for spore viability. Nature *309,* 523–527.

Toda, T., Uno, I., Ishikawa, T., Powers, S., Kataoka, T., Brock, D., Cameron, S., Broach, J., Matsumoto, K., and Wigler, M. (1985). In yeast, RAS proteins are controlling elements of adenylate cyclase. Cell *40,* 27–36.

CHAPTER 14

Limb Development and Pattern

The basic plan of vertebrate limbs is fairly simple; it consists of one bone in the end closest to the body (proximal), followed by two bones, and ending with many bones at the distal end. In the vertebrate arm these bones are the humerus connecting the shoulder to the elbow, followed by the radius and ulna, and ending with the carpals and metacarpals of the wrist and hand. The same plan in the leg consists of the femur connecting the hip to the knee, followed by the fibula and tibia, and ending with tarsals and metatarsals. In all vertebrates the bones are surrounded by connective tissue and muscles (Fig. 14.1). Besides this inside to outside axis, limbs are asymmetric in the three spatial planes: proximal to distal, anterior to posterior (the thumb is anterior), and dorsal to ventral (the palm is ventral). The processes that lead to this pattern have been studied extensively and are the subject of imaginative theoretical proposals.

Limb buds form on both flanks of early vertebrate embryos. Transplantation of lateral mesoderm from the position at which a limb bud will form to another position on a second embryo will induce the formation of an extra limb bud. The mesoderm can be taken for transplantation even before the visible appearance of a bulge. Mesodermal but not overlying ectodermal cells at this position have already been instructed to give rise to a limb bud at this very early stage using signals that specify both anterior/posterior position and dorsal/ventral position relative to the axes of the whole embryo. Thereafter the limb bud differentiates autonomously as a secondary field. That is, a limb bud transplanted to an unusual position differentiates normally to

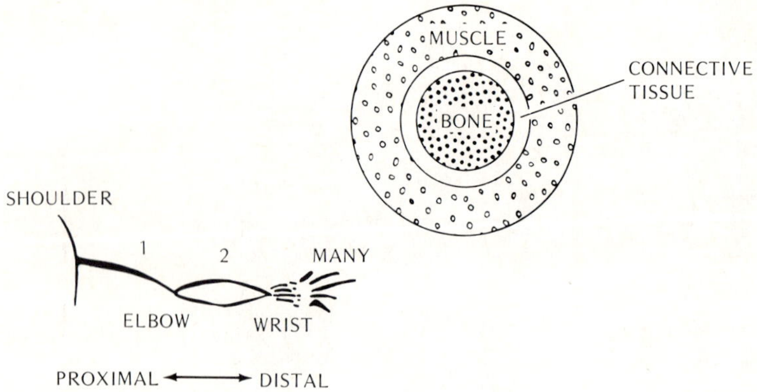

Figure 14.1 Morphological plan of a limb. All vertebrate limbs have first one bone, then two bones, and then many bones at the distal end. In cross section, calcified bone is found in the center surrounded by connective tissue that holds the muscles to the bone. An epithelial skin layer forms a protective covering.

form a bone pattern of one, two, many bones in the proximal to distal axis, while the dorsal/ventral and anterior/posterior axes are specified as soon as the bulge of a limb bud is visible. A limb bud rotated 180° will develop as a reversed limb, that is, the anterior bones will develop at the posterior in a rotated wing bud.

The AER

Limb bud mesoderm induces outgrowth of overlying ectodermal tissue as the bulge is formed. The bulge ends in a clearly observable ridge of cells referred to as the apical ectodermal ridge (AER) (Fig. 14.2). The AER consists of a thickened layer of epithelial cells with a somewhat different form of extracellular material separating it from the underlying mesoderm (Fig 14.3). The AER plays a controlling role in subsequent development of the limb. The pattern of bone formation is established from proximal to distal

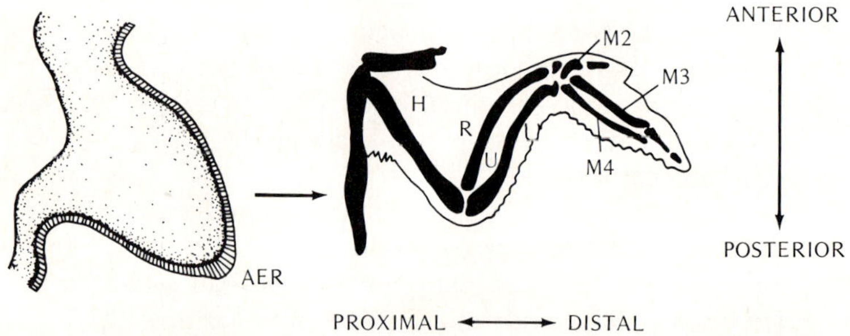

Figure 14.2 The apical ectodermal ridge (AER) of chick wing buds. At the distal end of the limb bud is a thickened ridge referred to as the AER. It directs cells that divide just beneath it to differentiation in progressively more distal patterns. The chick limb has a humerus (H) connected to the radius (R) and ulna (U). Only three metacarpals (M2, M3, M4) can be easily recognized at the distal end of the wing.

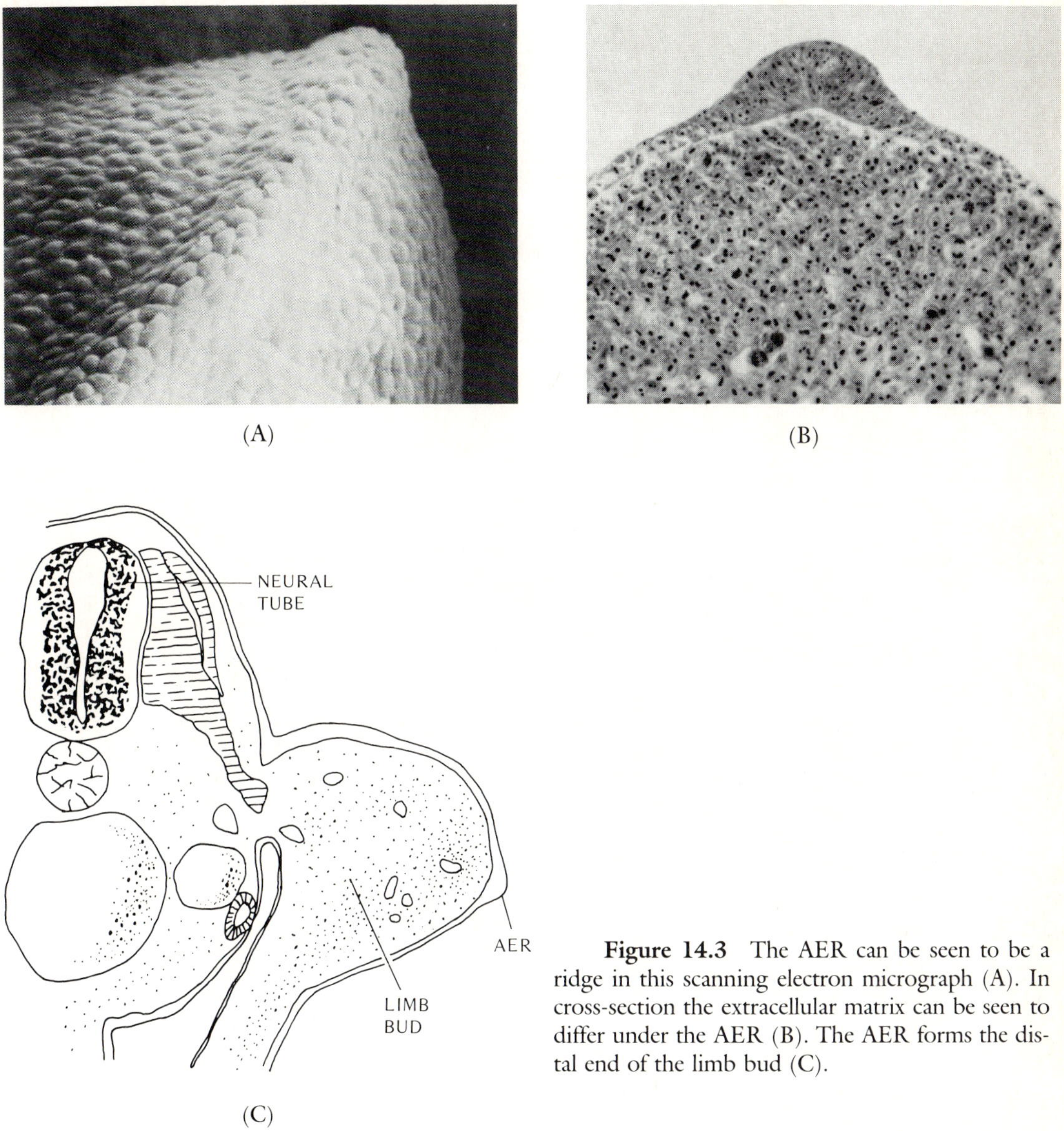

Figure 14.3 The AER can be seen to be a ridge in this scanning electron micrograph (A). In cross-section the extracellular matrix can be seen to differ under the AER (B). The AER forms the distal end of the limb bud (C).

in a temporal sequence. The first bones to be specified are the single long bones, the humerus in the arm and the femur in the leg. The digits of the hand are last. Microsurgical removal of the AER results in a truncated limb with only the proximal bones specified and distal structures missing. Removal of equal-sized pieces of epithelial cells from other places on the limb bud does not perturb normal limb development (Fig. 14.4).

The AER is clearly an essential structure for normal limb development. It is also an autonomous organizing force. When an AER is grafted onto a limb bud at an unusual position it instructs underlying mesodermal tissue to differentiate into a secondary set of bones. The duplication of structures includes those distal ones not yet specified in the host limb bud (Fig. 14.5).

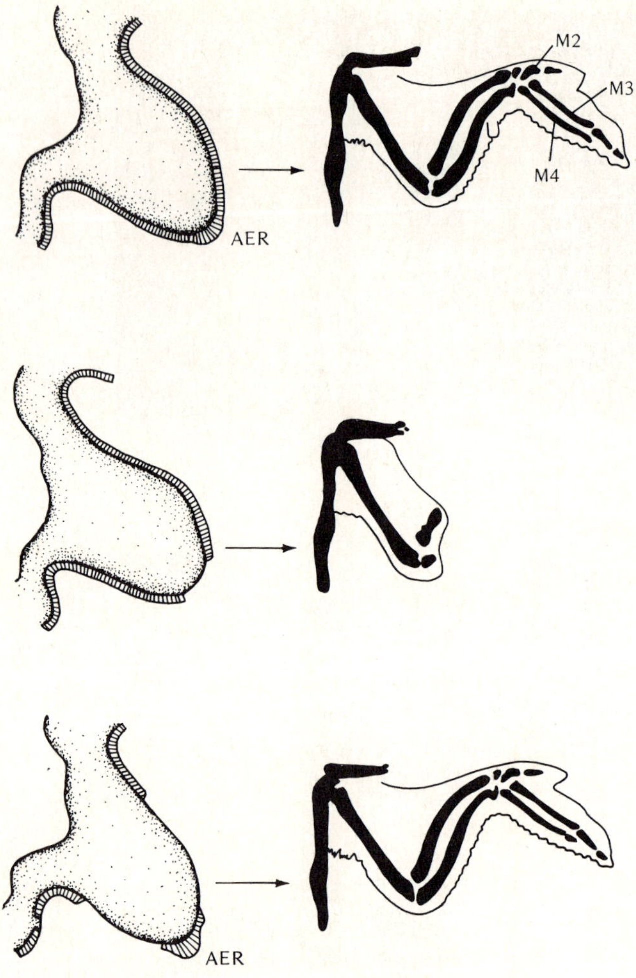

Figure 14.4 Removal of the AER results in formation of proximal bones but no distal structures. Removal of a similar-sized piece of epithelial tissue from a region other than the AER has no consequences; it heals.

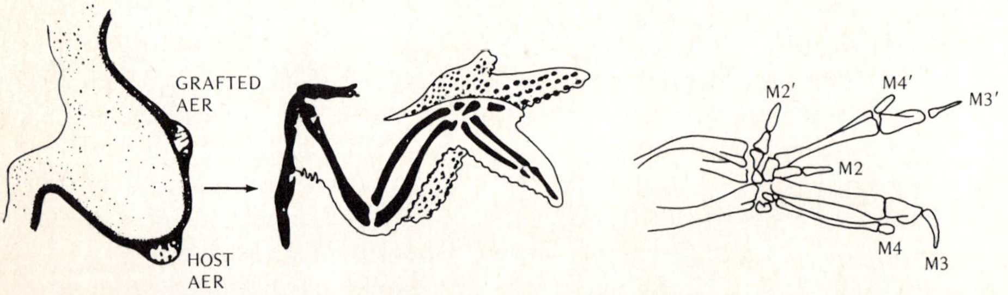

Figure 14.5 Grafting on a second AER results in duplication of distal structures. Note that the polarity of the duplicated structures in the anterior-to-posterior axis is normal; metacarpal 2′ is toward the head.

The AER is thus much more important than it appears. It guides and directs the type of differentiation in mesodermal tissue just behind it and lays out the whole architecture of the limb. As the limb grows out, the AER is maintained as a unique reference point and continues to specify the number and positioning of the bones that will form distally.

The ZPA

There is a second pattern-specifying region in limb buds that acts to control the pattern in the anterior/posterior axis. At the posterior margin of both wing and leg buds in chicks, there is a group of mesodermal cells with different properties than other mesodermal cells. It is referred to as the ZPA (zone of polarizing activity). Nothing visible distinguishes these cells but, when transplanted to the anterior margin of a host limb, bud cells of the ZPA will induce a mirror-image duplication of limb bones (Fig. 14.6). In chick wings the position of the ZPA is the same as the posterior necrotic zone (PNZ) in which the cells are programmed to die. However, in leg buds no PNZ is formed; the cells at the posterior margin of the leg do not die. Yet cells from this region of leg buds will induce supernumerary bones when transplanted to the anterior of a limb bud.

No matter where the ZPA is positioned it induces surrounding tissues to differentiate posterior structures. It gives a signal specifying the ante-

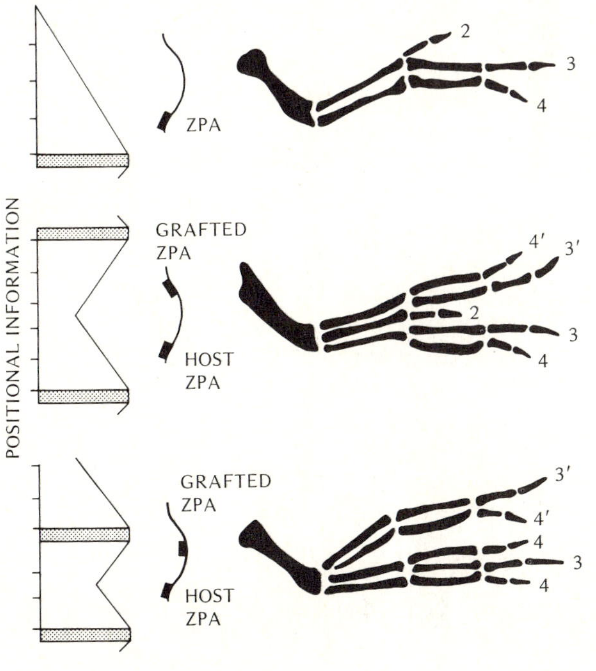

Figure 14.6 The ZPA affects the anterior/posterior pattern. Cells along the posterior margin of the limb bud direct nearby tissue to differentiate posterior structures. When grafted to the anterior margin, a mirror-image duplication occurs; digit 4′ is toward the head (4′, 3′, 2, 3, 4). When the ZPA is grafted to a medial position, duplication of digit 4 occurs on both sides of the graft and digit 2 fails to differentiate. An interpretation of these results in terms of graded positional information emanating from the ZPA is given. Treating the anterior margin with retinoic acid also results in mirror image duplication (4′3′2′234).

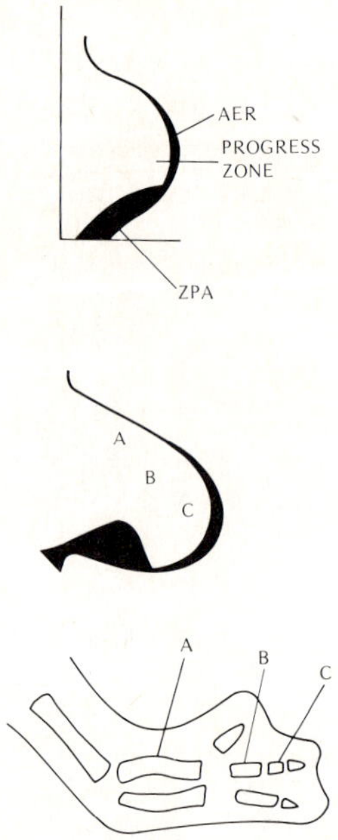

Figure 14.7 Sources of positional information. It is thought that the AER signals cells in a progress zone as to their position in the field. Proximal/distal positional information is specified by the time spent in the progress zone. Anterior/posterior information is specified by the distance from the ZPA. As the limb bud grows out, cells at *A* are already determined to proximal differentiation; cells at *B* are being specified; cells at *C* have not yet been specified.

rior/posterior axis. The signaling mechanism appears to be common among vertebrates since transplantation of tissue from the ZPA of a mouse limb into a host chick wing bud will give rise to supernumerary wing structures with the closest ones differentiating as posterior bones. The signals may be either mechanical or chemical, but they appear to have been conserved ever since mice and chickens shared a common ancestor more than 200 million years ago.

Recently it has been shown that small plastic beads impregnated with all-*trans* retinoic acid can mimic the effects of a ZPA in chick wing buds. When such beads are implanted in the anterior of a wing bud early in development, supernumerary bones are induced. The pattern of digits in limbs exposed to 10^{-9} M retinoic acid shows mirror-image duplication (432:234) just as in limbs that develop following insertion of a ZPA in the anterior margin of the limb bud. Related compounds such as 13-*cis* retinoic acid are inactive, suggesting that all-*trans* retinoic is acting specifically. It may be inducing surrounding cells to function as a ZPA or it may be mimicking the actual morphogen released from the ZPA. Retinoic acid also has dramatic effects on limb regeneration in amphibians and induces terminal differentiation in embryonal carcinomas of mammals. It is a good candidate for a general morphogen. In the limb bud it clearly affects the anterior/posterior pattern.

The general idea to come from these observations is that the AER plays a major role in specifying the pattern of limb development in the proximal/distal axis while the ZPA plays an equally important role in specifying the pattern in the posterior/anterior axis (Fig. 14.7). It is not yet known how the dorsal/ventral axis is specified.

Inside/Outside

Bones are centrally located in vertebrate limbs. The environment with respect to metabolic gases such as oxygen and carbon dioxide as well as nutrients most likely differs for cells deep within a developing limb bud relative to those near the surface. The early vascular system produces an array of capillaries on the periphery of developing limb buds but leaves the central region avascular. Cartilage and bone cells differentiate in this core. It is of interest that mesodermal cells cultured *in vitro* on surfaces of bone differentiate into connective tissue and secrete collagen. When the same cells are cultured *in vitro* on collagen surfaces, myotubes differentiate. This implies that, once bone is laid down, surrounding tissue will be instructed toward connective tissue, which, in turn, will instruct cells toward muscle differentiation.

Feathers and scales are formed by the ectodermal layer of cells on the surface of chick limbs. Thigh tissue produces characteristic feathers whereas foot tissue produces scales and claws. However, it is not the ectodermal tissue that is specified in this proximal-to-distal manner. It is the underlying mesoderm. This was shown by transplanting dermis from the thigh or foot of a chick hindlimb underneath the wing ectoderm. Thigh dermis induced thigh feather production in the wing while dermis from the foot induced scales and claws. Both structures are quite different from wing feathers (Fig. 14.8).

The wing ectodermal layer is clearly able to form either wing feathers, thigh feathers, or scales depending on the type of dermis beneath it. While positional information may come from diffusion of signaling molecules released from the dermis, it is also possible that the shape and structure of the dermal cells controls the shape of the ectodermal cells they are in contact with. Thus, information may be passed from one tissue to the next not by

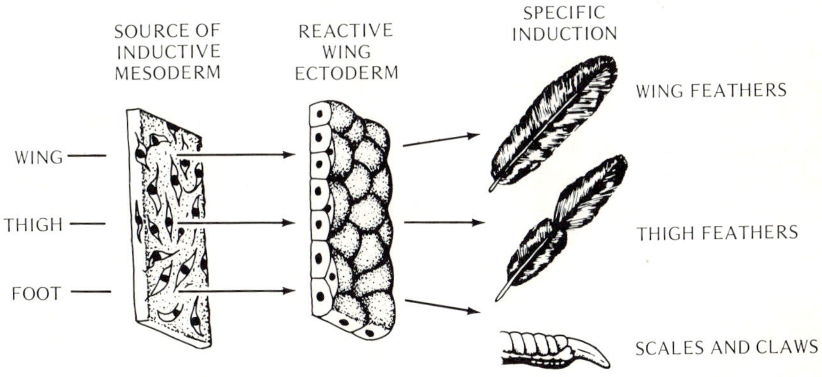

Figure 14.8 Mesoderm specifies ectodermal differentiation in the chick limb. Wing mesoderm placed under ectoderm induces wing feather formation; thigh mesoderm induces thigh feathers; foot mesoderm induces scales and claws.

diffusible molecules but by direct physical interaction. A line of dominos can be triggered to fall in one direction or the other by the fall of a single domino at one end. The signal to fall is passed down the length of the line of dominos, not by a diffusible signal but by physical interaction. Similar processes could lead one tissue to determine the type of differentiated structure an adjacent tissue will form.

Regeneration

Many vertebrates have the ability to regenerate portions of limbs following amputation. This ability is particularly good in Mexican axolotls, a type of amphibian that lives in murky ponds where neighbors inadvertantly snap off each other's limbs thinking they are food. A small but perfectly proportioned limb grows out of the stump over a period of several months and then subsequently grows to the size of the original limb.

During regeneration the proximal bone left in the stump elongates and progressively more distal bones are added. A normal limb is formed with the ratios of the bones being as accurately specified as in a newly developed embryonic limb.

Differentiated muscle and connective tissue present in the stump at amputation loses all signs of specialization and appears to dedifferentiate to the level of embryonic mesodermal cells. These dedifferentiated cells form a mass under the epithelial layer of cells that has grown over the wound (Fig. 14.9). This mass of mesodermal cells is referred to as a blastema. The cells stop producing type II collagen and, instead, secrete type I collagen into the extracellular matrix. In this regard they are similar to cells of developing limb buds. Upon cartilage redifferentiation they once again switch to production of type II collagen. New muscle and bones then differentiate from the blastema.

Chick foot epidermis normally makes scales but can be redirected to the formation of feather primordia by injecting embryos with retinoic acid. For a period of about 24 hours after injecting a 10-day embryo, feathers are induced. Thereafter, as the retinoic acid is eliminated from the tissue, scale development resumes. Most frequently the feet of treated embryos have one or two feathers on each scale at the distal tips seven days later. Recombining treated and untreated dermis and epidermis clearly showed that retinoic acid acts directly on epidermis to cause the formation of feathers on the feet. Thus, the epidermal tissue is responsive to at least one diffusible molecule.

The anterior/posterior axis is specified while the blastema is forming. If the end of the stump is rotated 180° before blastema formation, the tissues regulate and develop into a normal limb. However, rotation after blastema formation results in the development of a limb inverted in the anterior/posterior axis. So information concerning position is available during blastema formation and is then irreversibly interpreted long before any overt sign of anterior/posterior specialization can be observed.

The initial steps in regeneration of a limb occur by dedifferentiation

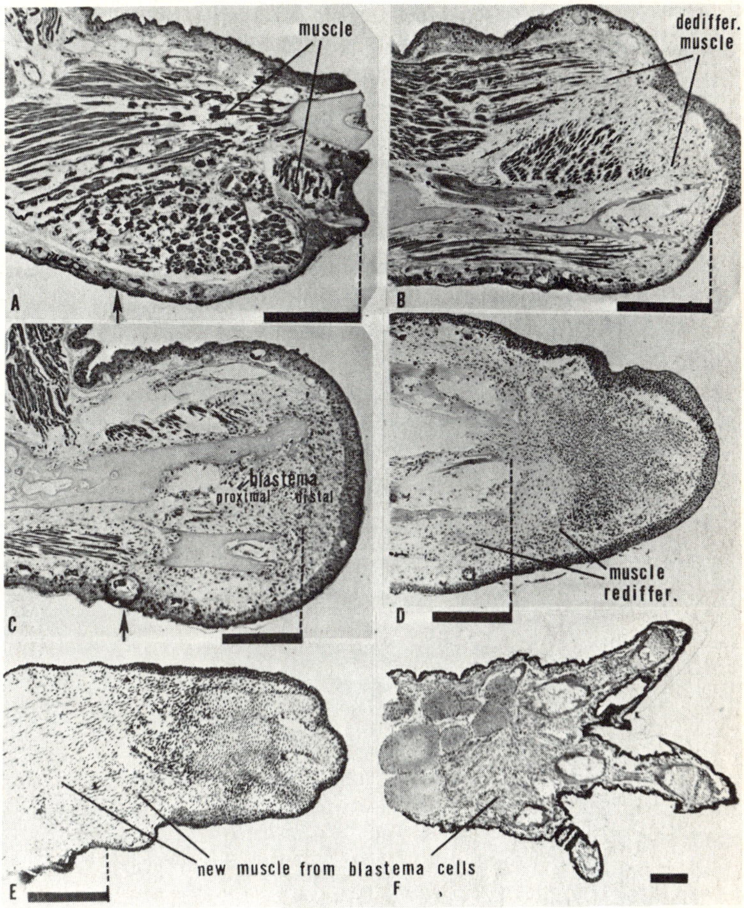

Figure 14.9 Regeneration of a newt limb. *(A)* Tissue at the stump has differentiated muscle tissue. *(B)* Eighteen days later the muscle has dedifferentiated. *(C)* twenty-one days after amputation the blastema tissue is growing out. The dotted line shows the position of the original wound. *(D)* As the blastema continues to grow, muscle redifferentiates. *(E)* Cartilage is deposited where bones will appear. *(F)* The limb is regenerated with the correct anterior/posterior and proximal/distal patterns of differentiation.

and redifferentiation without significant growth of the tissues in the affected field. This sort of regulation is referred to as morpholaxis as a small but normal differentiation field is reconstructed from the remaining cells. Sea urchin or frog blastulae bissected at the four-cell stage also show morpholaxis as they regulate to the new smaller size of the embryo and go on to make small but normally proportioned embryos. Likewise, isolated fragments from the anterior of *Dictyostelium* regulate in the absence of any cell growth, that is, by morpholaxis. In other cases of regeneration, such as that of the retina of developing frogs or imaginal disks in *Drosophila* embryos, the regeneration is by intercalation of new cells to fill the positions of the missing tissues. This type of regeneration is referred to as epimorphosis. It is not certain that the positional information mechanisms are the same between epimorphosis and morpholaxis. In morpholactic regeneration, random re-differentiation may be followed by sorting out of cells of different tissue types to re-form the pattern of the field.

Positional Signals

Cells within a field must be informed of their relative position to be able to differentiate appropriately. Embryonic fields in which position is specified are up to 1 mm in length and may be a hundred or so cells across. Diffusion of molecules across such a field is sufficiently rapid to account for the observed regulation of tissues. Cells secrete a large number of molecules the concentration of which could be used for positional information.

Various ions, such as Ca^{++}, have profound effects on specific cells. If previous differentiation had resulted in the ability of some cells to concentrate a given ion and sequester it and cells with a different developmental history secreted the same ion, then cells between the two would be exposed to a gradient in the concentration of the ion. Likewise, the presence of a large amount of yolk could be a source from which ions might be continuously drawn. Small ions diffuse rapidly in the interstitial fluid, and a stable gradient would be established across a field 100 cells wide in several hours. Cells in this gradient might differentiate differently if exposed to relatively high ion levels rather than relatively low levels.

Other rapidly diffusing molecules that have been considered as possible candidates for positional signals include cAMP, retinoids, and any of the dozen or so molecules recognized as neurotransmitters. Some cells might contain relatively high levels of adenyl cyclase and secrete cAMP while other cells might synthesize cAMP phosphodiesterase and actively remove extracellular cAMP. In between, cells would be exposed to a gradient in cAMP concentration. The same might be true for acetylcholine: the source cells might produce this molecule while sink cells might degrade it with acetylcholinesterase. If cells in the morphogenetic field had inserted the acetylcholine receptor into their membranes, they could respond to high or low acetylcholine levels by undergoing specific differentiations.

The reason that neurotransmitters are plausible candidates for signaling molecules is that there is no doubt that they carry out these functions in neural networks and at neuromuscular junctions. These highly differentiated cells must have evolved from less specialized cells that nevertheless had a signaling system. Perhaps nerves just amplified a system already in use to give positional information.

Many hormones are small peptides that diffuse from the circulatory system to the cells. There are specialized proteins on the surface of some cells that bind these peptides and pass the signal into the nucleus of the cells. This hormonal system may have evolved from earlier signaling mechanisms first selected to be able to integrate a morphogenetic field. One of the few molecules recognized so far that appears to carry positional information is a small peptide. The concentration of this peptide controls the number of tentacles formed in the head of a regenerating hydra, as was described in the previous chapter. Its action is very specific and it works at very low concentration (10^{-10} *M*) to stimulate tentacle formation; however, it has not been possible to find a significant gradient of this peptide in intact hydra. Clearly the search for positional signals must go on in diverse developmental fields.

Pattern Formation

The pattern in the left and right limbs, or any other paired system, can be compared to get an idea of the accuracy with which positional information is established and interpreted. In all cases analyzed in this way, the error is seen not to exceed ± 5 percent. It is a highly accurate system that results in the final body form. The mechanisms can function well over a considerable size range. Embryos derived from a single blastomere of a four-cell embryo develop normally although only a fourth the normal size. Likewise, cell proportioning in a *Dictyostelium* slug containing only 10^3 cells is as accurate as in slugs containing 10^4 or more cells. The mechanism is size invariant within these ranges.

There is good reason to consider that the number of signaling systems that have evolved to give positional information is small and that they may have been conserved in many diverse species. The extreme reductionist view is that a single system would be sufficient to account for positional information in all morphogenetic fields in all organisms. The differences seen between species would be the consequences of different genetically determined responses to the same signal. The differences between different morphogenetic fields—say, the forelimb and the hindlimb—would be due to different developmental histories. Cells of the forelimb bud obviously have the same genes as cells of the hindlimb bud of the same embryo, but the forelimb bud develops from cells further anterior along the whole embryo axis than the hindlimb bud. It is quite plausible that the cells that will form the forelimb were instructed to respond to a signaling molecule differently from hindlimb cells because of their differences in developmental history. Since so little is known of how cells respond to signaling molecules, this reductionist view cannot yet be tested. However, it certainly simplifies theoretical considerations.

Theoretical Considerations

From an abstract analysis of pattern formation, the signaling system can be separated into three independent processes: (1) Positional information. This consists of a gradient in the concentration of specific molecules or any other inhomogeneity. (2) Interpretation of positional information. This is a cellular process that determines whether or not a cell responds to positional information and how it responds. This process depends on the genetic makeup of the organism and may be regulated by previous developmental history. (3) Expression of the differentiated state specified by positional information. Once the information is received by a cell, differentiation may immediately start down a specific pathway or not. Once again, the specific differentiation chosen is determined by the genetic makeup and the previous developmental history of the cell.

This analysis points out that while positional information may be available to cells in a morphogenetic field it is up to them whether or not they recognize it and respond. Likewise, cells in two different fields such as forelimb bud and hindlimb bud may interpret the same positional information identically but they may express quite different differentiations due to previous events in their developmental history. The chick forelimb bud makes a wing while the chick hindlimb makes a leg.

Even the basic differences in the various spatial planes could all be consequences of different responses to the same positional information. The differences in expression could result from previous differences in axial position in the primary field of the developing embryo. Thus, sequential specification can lead from a simple separation into two cell types along a given axis to the final form in which hundreds of structures are accurately positioned in three spatial planes. However, it is equally plausible that separate systems of positional information regulate differentiation in the different axes.

Models

The ways in which positional information might be interpreted are most clearly visualized along a linear array of cells. If an inhomogeneity occurs from one end to the other, cells can respond when a threshold is reached (Fig. 14.10).

The concentration at the high end may be maintained by a source from which the hypothetical signal molecule, often referred to as a morphogen, diffuses toward the low end. The concentration of morphogen is kept low at the other end either by the presence of a large passive sink or by the ac-

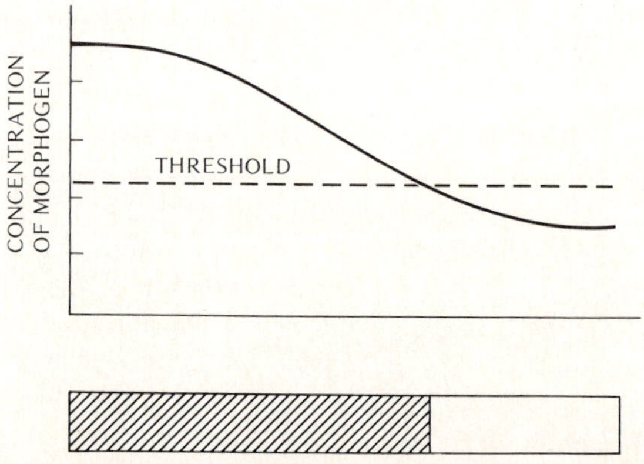

Figure 14.10 Source/sink model for positional information. If a diffusible substance is made at the left boundary and removed at the right boundary, the concentration of the substance will form a gradient across the field. Cells in the field may respond to a threshold concentration and differentiate from those that are in positions where the signal is below threshold.

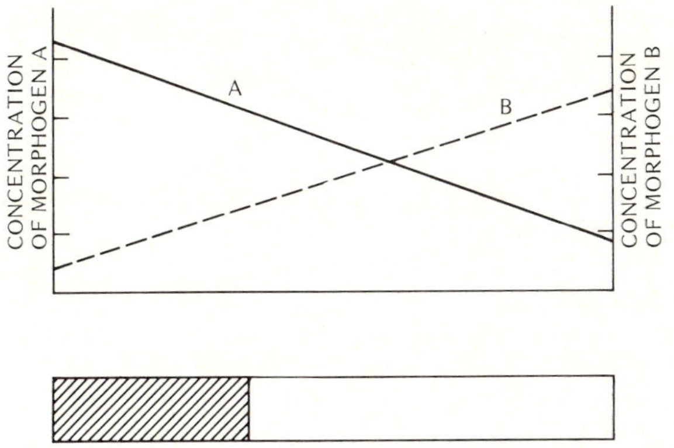

Figure 14.11 Double gradient. Some morphogenetic fields appear to have gradients of two morphogens. Cells can respond to the ratio of *A* to *B*, which will vary continuously across the field.

tive destruction of the morphogen by cells at that boundary. This analysis suggests that all the cells in this field will respond by the same differentiation events if they find themselves in an environment at which the concentration of the morphogen is above threshold. If the concentration of morphogen is below this level, they will pursue other differentiations. Variations on this theme include production of the morphogen by all the cells in the field and establishment of a gradient by a localized sink. The diffusion equations show that this situation would also result in a monotonic gradient within which some cells would be in an environment above threshold and others would be exposed to morphogen below the threshold necessary for response.

There are experimental results in some systems that are more easily interpreted by suggesting that cells respond to not just one morphogen but to the relative levels of two morphogens. Some analyses invoke a single source for both morphogens and a different rate of diffusion or stability of the two morphogens. Other analyses suggest that one morphogen is produced at one boundary while the other morphogen is produced at the other boundary (Fig. 14.11). In both of these schemes the ratio of the two morphogens will be continuously changing across the field.

There is no particular reason to assume that cells respond at only a single threshold. There could be a hundred thresholds for a hundred different responses. With two thresholds, a field could form three different regions as

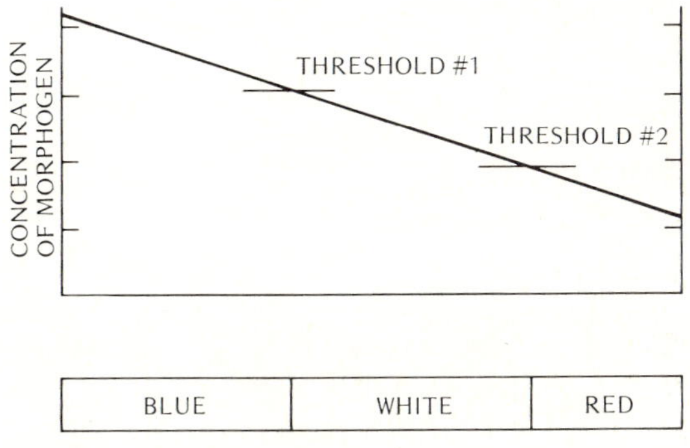

Figure 14.12 Multiple thresholds give multiple states in a single field. Cells may respond to morphogen above threshold 2 in one way and to morphogen above threshold 1 in another way. There might be several different receptors with different affinities to the morphogen, or a single receptor might have multiple binding sites.

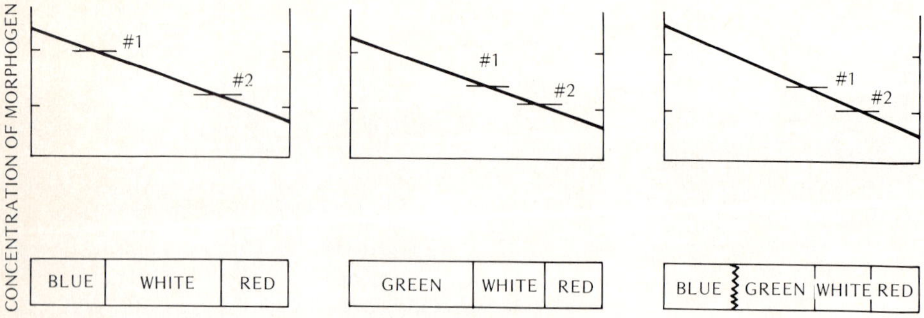

Figure 14.13 French flag metaphor. A French flag pattern (blue:white:red) is generated by two thresholds. Cells of another species or cells of another tissue with a different developmental history represented as an Italian flag (green:white:red) might respond to the same morphogen at different thresholds and express different responses. Grafts of these tissues would result in an organized chimeric pattern. Grafts of mouse limb bud tissue into chick limb buds suggest that the same positional information is recognized but interpreted differently by different cells.

in a French flag. Cells above threshold 1 would differentiate blue, cells above threshold 2 would differentiate white, and cells below both thresholds would differentiate red (Fig. 14.12). Lewis Wolpert has used this French flag metaphor to emphasize his reductionist views that a single system of positional information might be universal. Tissue from an organism that normally makes an Italian flag pattern (green, white, red) when inserted into a French flag field would respond to the positional information but because of its different genetic makeup would differentiate green above threshold 1 (Fig. 14.13).

Although there is presently little or no evidence to suggest how positional information is encoded or received, some indirect evidence in *Drosophila,* nematodes, and other organisms presented in the next two chapters indicates that most decisions of differentiation pathways are binary. At each decision point, the pathways bifurcate rather than splitting into three or more ways. Of course, two bifurcations that have not yet been separated will look like a four-way split.

There are also several analyses that suggest that positional information is expressed in a polar manner (Fig. 14.14). Unique points in two dimensions are specified by the portion of a circle separating them and their distance from the center. Most of these analyses have come from the study of circular patterns as seen in the cross section of cockroach legs or imaginal disks of *Drosophila*. Although they certainly account for many of the observations, they are just special cases of more general vector analyses. It has been difficult to conceive of inhomogeneities distributed in a polar fashion that can adapt to give rise to regulation and size invariance.

The initial appearance of inhomogeneities and their maintenance across a field have been the subject of sophisticated theoretical investigations based on the pioneering mathematics of nonlinear processes proposed by Alan M. Turing. The basic process is to consider an autocatalytic reaction in which

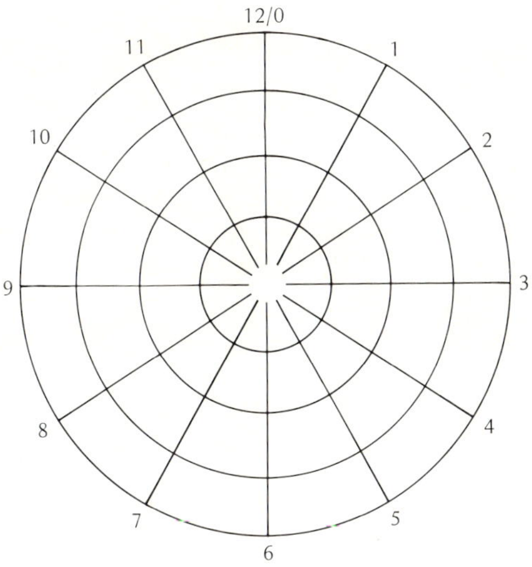

Figure 14.14 Clock face model. This system of polar coordinates gives positional information in two dimensions: location on the circle and distance from the center. Position 12/0 is not a discontinuity.

the product stimulates its own production either directly or indirectly. The rate of production will increase until it is infinite or other factors such as availability of substrate become limiting. Such a nonlinear process can start with random variations in the concentration of a morphogen across a field and amplify the differences until a monotonic gradient is established. Feedback regulation of related reactions is often invoked to keep the system stable enough to be a plausible system of positional information in biological terms. One particularly successful use of such equations invokes two morphogens, a short-range autocatalytic activator and a long-range inhibitor. Alfred Gierer and Hans Meinhardt have shown that, with appropriate parameters, equations defining these interactions can account for the development and regulation in hydra, *Dictyostelium,* and a variety of other simple systems. Of course, these theories can be confirmed or denied only when actual morphogens are recognized and quantitated with respect to the spatial coordinants of the systems.

Summary

Development of well-proportioned limbs requires communication of positional information concerning the length and breadth of the limb bud. Bones are laid down in a pattern of one, two, many in the proximal/distal axis. Cells beneath the apical ectodermal ridge (AER) are specified as to their position along this axis. As the limb grows out cells are left behind the progress zone under the AER, first in the proximal region and then at successively distal regions. The information they received while in the progress zone is later expressed in their pattern of differentiation. A zone of polarizing activity (ZPA) occurs along the posterior margin of the bud and appears to be the source of a diffusible morphogen. Cells exposed to high concentrations of morphogen differentiate into posterior structures while those exposed to lower concentrations differentiate into anterior structures. Bone matrix stimulates cells to differentiate into chondrocytes that secrete collagen while collagen matrices stimulate cells to differentiate into muscle cells. Regeneration of amputated limbs seems to result from dedifferentiation of cells at the stump followed by their respecification as to position. Positional information can be considered as monotonic concentration gradients of diffusible molecules produced at a localized source and removed at a localized sink. Cells may interpret the information and respond in ways specified by their genetic and developmental histories.

SPECIFIC EXAMPLE

Cockroach Leg Regeneration

Cockroaches, like other insects, have hard exoskeletons. This cuticular covering is deposited by an epidermal sheet one cell thick that determines the detailed sculpturing of the cuticle. As young cockroaches grow they periodically shed their old cuticle and lay down a new cuticle. Changes in the positional values assumed by individual cells in the epidermal layer can be scored in the pattern of bristles, ridges, and coloring along the length of a leg and around its circumference. The leg of a cockroach is surprisingly detailed when closely inspected (Fig. 14.15). Most experiments involve surgical operations on the femur or the tibia. The resulting changes in cuticle pattern are scored after the next molt, when new cuticle is laid down.

When a portion of the tibia is removed and the cut ends are stuck together, the wound heals rapidly, and after the next molt the new tibial cuticle is found to be of the normal length and to have the complete complement of bristles (Operation No. 1) (Fig. 14.16). The juxtaposition of proximal epidermal cells with distal epidermal cells from a position further along the leg than usual induces intercalary growth at the wound that fills in the missing part. Growth occurs until the normal number of cells separate the old epidermal cells and then stops. This result by itself suggests that there is a

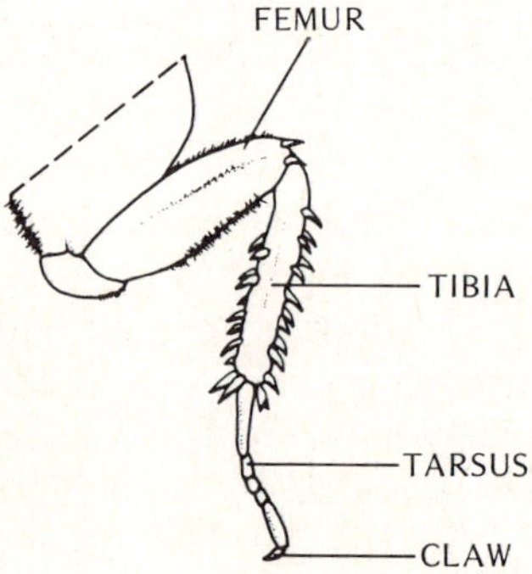

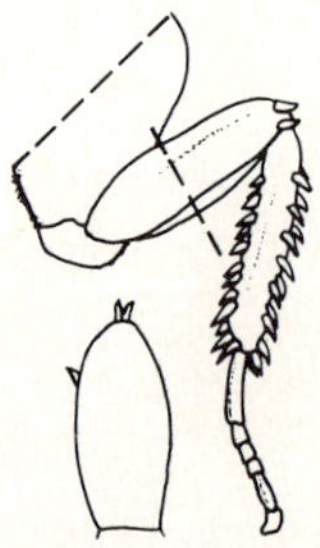

Figure 14.15 Structure of a cockroach leg. The femur, tibia, tarsus, and claws are each readily distinguished by shape and bristles. In cross section it can be seen that the femur is asymmetric around its circumference.

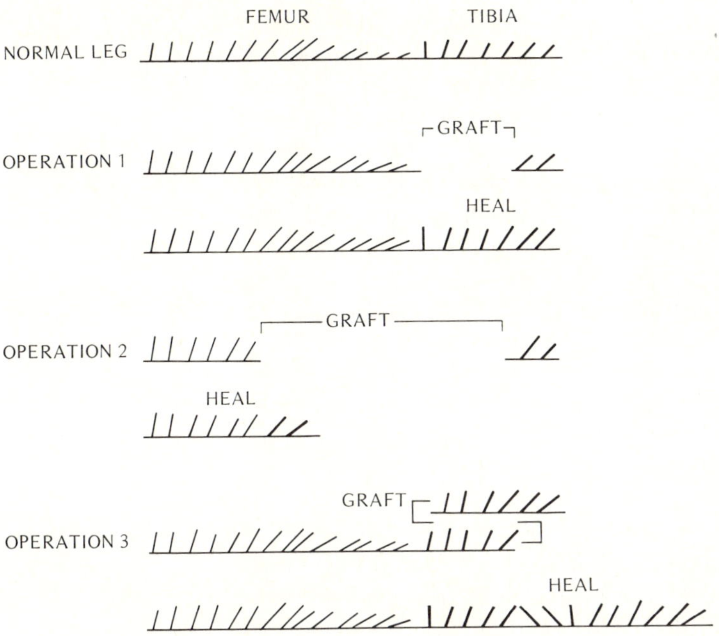

Figure 14.16 Bristle pattern in regenerating cockroach leg. Normally the bristles of the tibia point down the leg. When amputated, the tibia regenerates with bristles in the proper polarity (operation 1). When the femur–tibia joint is removed and the stumps grafted, there is no change in the bristle polarity (operation 2). When a long piece of tibia is grafted onto the end of a tibia, tissue is intercalated that has reversed polarity of bristles.

graded variable that continuously changes along the leg which regulates epidermal growth. Juxtaposition of cells that are normally separated triggers intercalation until the progression of positional values is once again smooth. The variable must be retained by the individual cells at least for the duration of a molt.

The variable that regulates growth appears to be repeated in different segments of the cockroach leg. Removal of the distal half of the femur and the proximal half of the tibia followed by grafting of the cut ends, gives a greatly shortened leg with the middle segment composed of both femur and tibia. Following the next molt, the leg that was operated on remains altered (Operation No. 2) (Fig. 14.16). No intercalary growth is stimulated because the positional values halfway down the femur match the positional values halfway up the tibia. This result suggests that the same graded variable regulates growth in both the femur and the tibia and that it is set at the same levels at the proximal ends of both the femur and the tibia and changes in the same manner along the length of the segment.

The variable itself might be a diffusible molecule. If so, there must be a barrier to its diffusion at the junction of the femur and the tibia. No barrier to diffusion has been seen but extracellular sheaths might separate the epidermal tissues of the femur and the tibia and have escaped detection so far. Alternatively, the epidermal cells of each segment might be coupled by intercellular junctions allowing diffusion of small molecules to be restricted to that segment. If this were so, this junctional coupling must be rapidly established between epidermal cells of juxtaposed femur and tibia in the manipulated leg.

The variable that regulates intercalary growth might also be a cellular parameter such as size, shape, or metabolic activity that is repeated in a proximal/distal distribution within each segment (Fig. 14.16). If so, direct interaction of adjacent cells with different positional values might trigger intercalary regeneration in the absence of exchange of any diffusible molecules. A variety of anatomical structures might account for changes along a segment that are expressed in cellular functions. The availability of nutrients in the hemolymph as well as gaseous exchange to the trachea may vary along the proximal/distal axis of each segment. Such differences could conceivably affect the physiology of the cells in a graded manner.

When the tibia of one cockroach is amputated near the distal end and the tibia of another cockroach is cut off near the proximal end and grafted to the stump, the resulting leg is unusually long. Rather than regulate the length of the tibia at the next molt, intercalary growth occurs at the joint. The regulated leg is even longer. Moreover, the bristle pattern in the intercalated portion is inverted. Normally the bristles point down the leg along the length of the tibia. After molting, the experimentally manipulated leg has bristles pointing down on the original proximal portion of the tibia, bristles pointing up on the intercalated tibial portion, and bristles pointing down on the grafted distal portion (Operation No. 3) (Fig. 14.16). Cells at the joint of host and graft tissue react to the discontinuity in positional values by intercalary growth. However, the juxtaposition of distal cells with more proximal cells inverts the normal pattern of bristle formation. The intercalated cells seem to have responded preferentially to the presence of proximal cells from the graft to generate a pattern extending backwards up the leg. This pattern is stable during subsequent molts.

The legs of cockroaches are not perfectly round but have bristles at the top and are flat underneath. This pattern is also able to regulate by intercalary growth following surgery. When one of the ridges of the flat bottom of a femur is cut off and the wound closed, the juxtaposed cells grow so that at the next molt the leg is as good as new with a flat bottom. However, if a flat bottom section is grafted into a cut in the side of a leg, intercalary growth occurs on both sides of the graft, making an extrathick femur (Fig. 14.17). The intercalated tissue sprouts bristles between the two flattened sections. These results have been interpreted to indicate that there is a graded change around the circumference of the epidermal cell layer at all positions down the leg. It is thought that the juxtaposition of cells with a discontinuity in this variable triggers intercalary growth until the change is smoothed

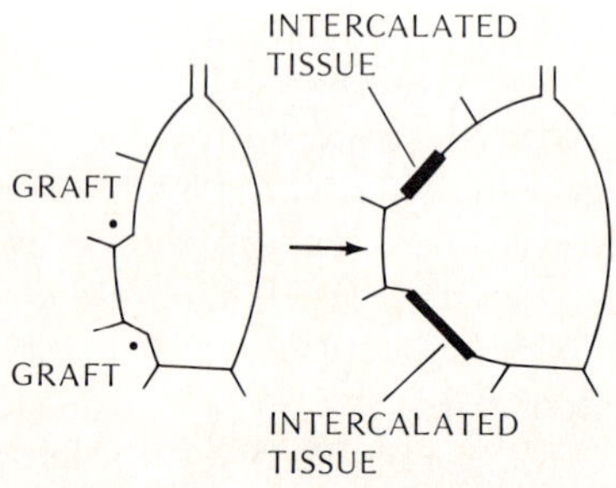

Figure 14.17 Regulation in a grafted femur. When the flat bottom portion of a femur is grafted to the side of another femur, the leg responds by intercalary growth of tissue that normally intervenes. The femur is wider at the next molt.

out. Although it is very difficult to equate a bristle with a precise positional value, the results have been interpreted in terms of polar coordinates to suggest that intercalation is stimulated by discontinuities in a circular parameter and functions to restore continuity by the shortest possible route. This analysis implies that positional information is received from both graft and host tissues and is continuously monitored during regeneration. Growth stops when the discontinuities are removed.

Cockroaches often seem almost indestructible. To some extent this is due to their ability to regenerate damaged parts at the next molt. A cockroach that has lost most of its leg will grow a new one. If the cut occurred in the femur, not only will the distal femur be regenerated but also the tibia, tarsus, and claw. This regeneration requires more than smoothing out discontinuities in the proximal/distal axis and around the circumference; it requires that regenerating cells change their segmental properties at the appropriate boundaries. Perhaps the epidermal cells that initially grow at the wound take on the positional values appropriate for all distal structures and then direct the fate of subsequently intercalated cells. Further regulation might then smooth out discontinuities and result in a normal functional leg. There is considerable mystery and beauty in seeing form and structure arise from undifferentiated cells, but the mechanisms that regulate this process are not yet clearly understood.

Related Readings

Carrino, D., and Caplan, A. (1985). Isolation and characterization of proteoglycans synthesized *in ovo* by embryonic chick cartilage and new bone. J. Biol. Chem. *260,* 122–127.

French, V. (1982). Leg regeneration in insects; cell interactions and lineage. Am. Zool. *22,* 79–90.

French, V., Bryant, P., and Bryant, S. (1976). Pattern regulation in epimorphic fields. Science *193,* 969–973.

Holder, N., and Weekes, C. (1984). Regeneration of surgically created mixed-handed axolotl forelimbs: pattern formation in the dorsal-ventral axis. J. Embryol. Exp. Morph. *82,* 217–239.

Lewis, J. (1981). Simpler rules for epimorphic regeneration: the polar-coordinate model without polar coordinates. J. Theor. Biol. *88,* 371–392.

Maden, M. (1982). Vitamin A and pattern formation in the regenerating limb. Nature *295,* 672–675.

Malacinski, G. (ed.) (1984). Pattern formation. Macmillan, New York.

Meinhardt, H. (1982). Models of biological pattern formation. Academic Press, New York.

Russell, M. (1985). Posititional information in insect segments. Devel. Biol. *108,* 269–283.

Sampath, T. Nathanson, M., and Reddi, A. (1984). In vitro transformation of mesenchymal cells derived from embryonic muscle into cartilage in response to extracellular matrix components of bone. Proc. Natl. Acad. Sci. *81,* 3419–3423.

Solursh, M. (1984). Ectoderm as a determinant of early tissue pattern in the limb bud. Cell. Diff. *15,* 17–24.

Tickle, C., Lee, J., and Eichele, G. (1985). A quantitative analysis of the effect of all-*trans*-retinoic acid on the pattern of chick wing development. Devel. Biol. *109,* 82–95.

Thoms, S., and Stocum, D. (1984). Retinoic acid-induced pattern duplication in regenerating urodele limbs. Devel. Biol. *103,* 319–328.

Todt, W., and Fallon, J. (1984). Development of the apical ectodermal ridge in the chick wing bud. J. Embryol. Exp. Morph. *80,* 21–41.

Turing, A. M. (1952). The chemical basis of morphogenesis. Phil. Trans. Roy. Soc. London (B) *237,* 37–72.

Wilson, D., and Hinchliffe, J. (1985). Experimental analysis of the role of the ZPA in the development of wing buds of wingless (ws) mutant embryos. J. Embryol. Exp. Morph. *85,* 271–283.

Wolpert, L. (1971). Positional information and pattern formation. Curr. Topics Devel. Biol. *6,* 183–224.

Wolpert, L., and Hornbruch, A. (1981). Positional signalling along the anteroposterior axis of the chick wing. The effects of multiple polarizing region grafts. J. Embryol. Exp. Morph. *63,* 145–159.

CHAPTER 15

Development of *Drosophila*

Thomas Hunt Morgan was a developmental biologist who recognized that a better understanding of genes was essential to the analysis of differentiation and pattern formation. He chose to study the genetics of the fruit fly, *Drosophila melanogaster,* because the individual flies are small and reproduce rapidly. Moreover, each fly presents a myriad of structures that can be phenotypically analyzed. There are antennae, wings, legs, and sex organs that are all subject to mutational alteration. On many of these structures individual cells form characteristically shaped hairs or bristles. For example, cells at the front of the wing make different structures from those at the back. Fifty years of intense genetic study by students of Morgan defined the *Drosophila* genome in considerable detail and proved that genes are arranged in a linear array on chromosomes. Further high-resolution genetics indicated that the *Drosophila* genome carries about 5000 genes. Recent molecular studies have described some of these genes at the nucleotide level and shown that mutations involve not only changes in single nucleotides but often insertions, deletions, inversions, or translocations. The consequences of loss of function of some of these genes on developmental processes are now being studied directly.

Embryogenesis

A generation in flies takes about two weeks (Fig. 15.1). During the first two days a fertilized egg develops into a feeding larva. Under optimal feeding conditions the larvae grow and molt once a day. They pass through

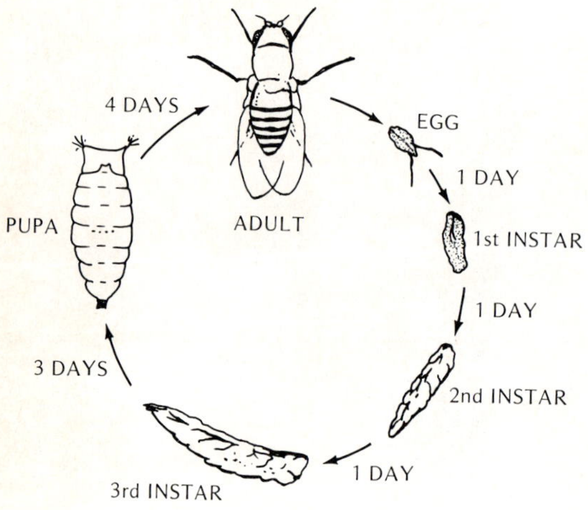

(A)

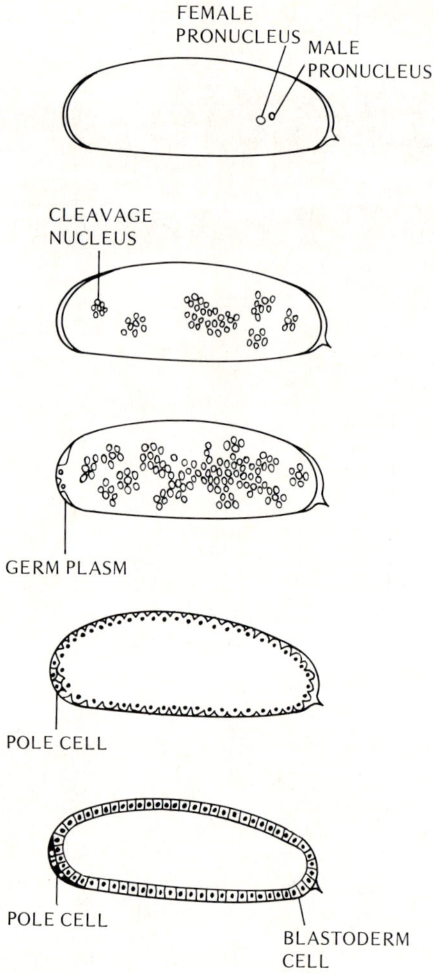

(B)

Figure 15.1. Life cycle of *Drosophila.* Within a day of emerging from the pupa, females are fertile. Eggs are laid and give rise to feeding grubs in a day. The first instar larva grows and molts to become a second instar larva within a day. A day later a third instar larva is made. This maggot feeds on yeast for about three days before pupating. Within the pupal case metamorphosis takes about four days. Eclosion of the newly formed fly from the pupa starts a new generation. During early development in *Drosophila* the nuclei divide without any cell division. When there are about 1000 nuclei, most of them migrate to the surface and divide twice again before cells are made just below the surface. At the posterior pole, cells are formed slightly before the rest of the cells.

three instars before pupating. Metamorphosis occurs in the pupa from which an adult fly emerges in about four days.

Following fertilization, the DNA replicates and nuclei divide every 10 minutes. After ten nuclear divisions, the blastoderm contains about 1000 nuclei. However, cell division has not yet occurred and the whole embryo is one large syncytial cell. At this point most of the nuclei migrate to the outer surface of the blastoderm and after four additional divisions produce 6000 nuclei that are surrounded by membranes that descend from the surface (Fig. 15.2). When the nuclei are fully enclosed in cells, the embryo is referred to as a cellular blastoderm. Cells at the posterior can be recognized as pole cells. Polar granules are concentrated in pole cells and appear to be essential for the subsequent differentiation of pole cells into germ cells. A gene, *grandchildless,* is essential for localization of polar granules and germ cell differentiation. Mutations in the *grandchildless* gene result in eggs that develop into sterile flies, which produce no children.

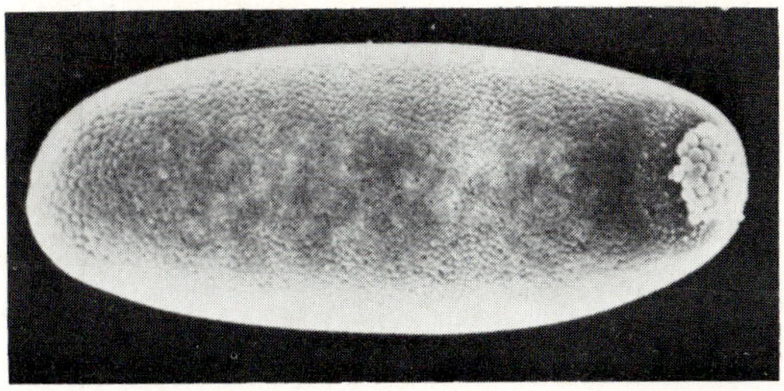

(A)

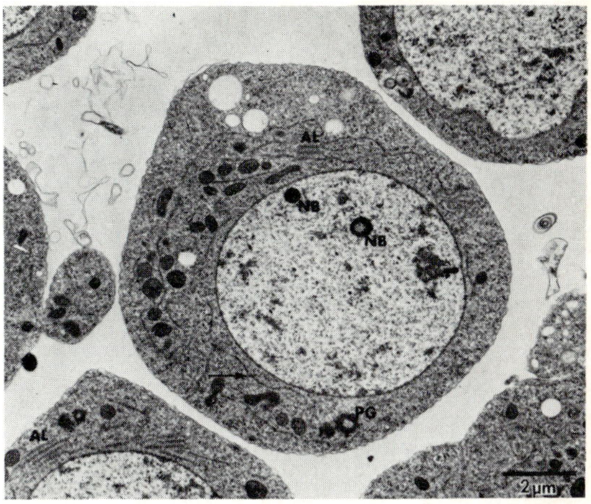

(B)

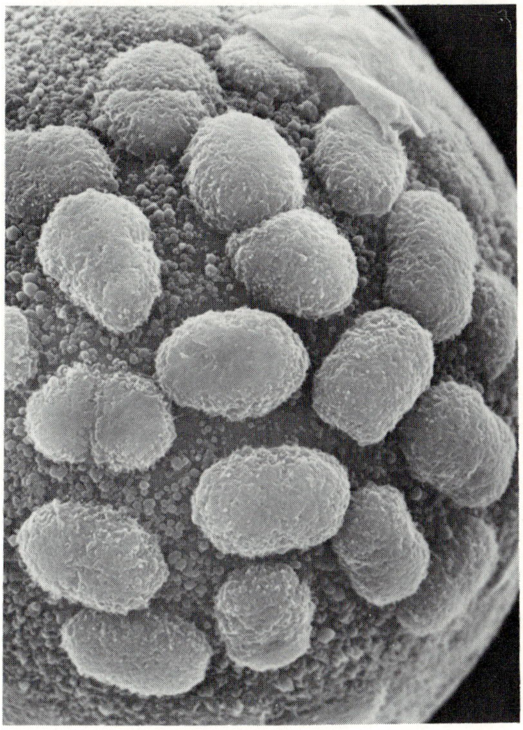

(C)

Figure 15.2. Pole cells stand out at the posterior pole. *(A)* They are the precursors of either sperm or eggs, depending on the sex of the individual. *(B)* Pole cells contain dense polar granules (PG) not seen in abundance in other cells. *(C)* Individual pole cells can be seen protruding.

Evidence that cytoplasm at the posterior pole actually directs germ cell differentiation utilized genetically marked strains of *Drosophila* (Fig. 15.3). Eggs of one strain that was wild-type for genes *y, w,* and sn^3 were injected at the anterior end with cytoplasm from the posterior end of another egg. Cells that normally form at the anterior never differentiate to give rise to germ cells. Cells that developed at the anterior of the injected eggs were transferred to the posterior of host blastoderms genetically marked at *y, w,* and sn^3. The host embryos gave rise to a few flies that carried germ cells with the wild-type genes at *y, w,* and sn^3 that must have come from the anterior of the injected eggs. Thus, the injection of polar plasm from the posterior directed anterior cells to form germ cells.

The pole cells initially divide less frequently than other cells of the blastoderm and protrude slightly from the posterior of the embryo (Fig. 15.2).

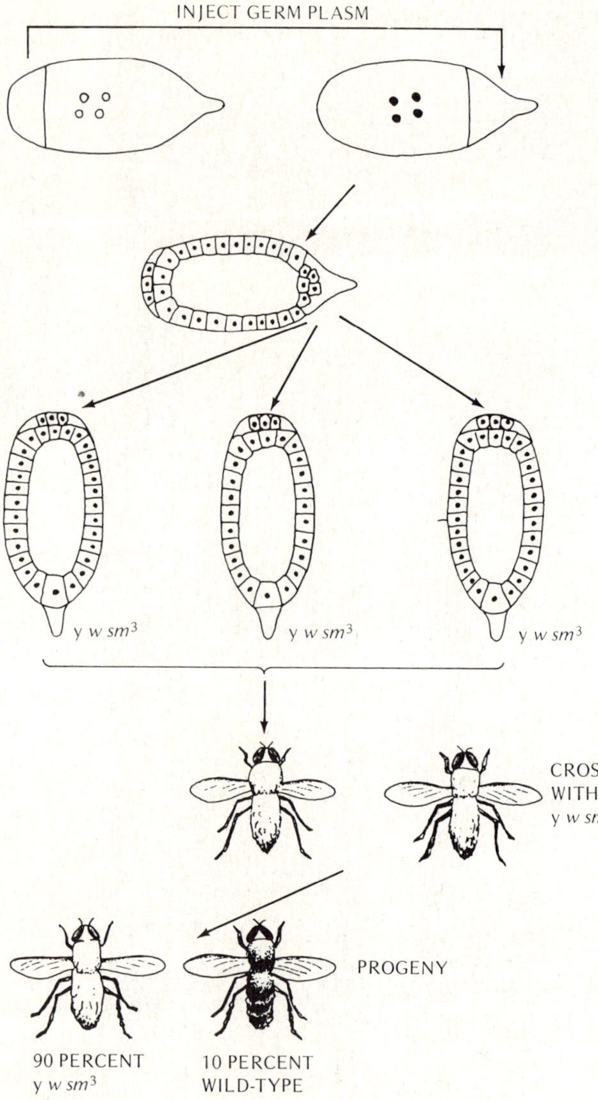

Figure 15.3. Experimental induction of germ cells in the anterior of *Drosophila* embryos. Germ plasma was taken from the posterior of early embryos and injected into the anterior near the micropyle of other early embryos. The recipient carried mutations *mwh* and *e* (multiple wing hair and ebony) but was wild-type for yellow, white, and singed. After the formation of the cellular blastoderm in recipient embryos, anterior cells that might have incorporated the injected germ plasma were placed in the posterior of cellular blastoderm embryos genetically marked by mutations in yellow, white and singed ($y\ w\ sn^3$ mutations). These flies were all yellow, white, and singed because only a few cells had been put in the posterior. They were mated to other $y\ w\ sn^3$ flies. About 10 percent of the progeny were wild type and so must have received the wild-type genes y^+, w^+, and *sn* from the anterior cells of the injected embryos. Anterior cells from uninjected embryos never formed germ cells.

They are fated to production of sperm or eggs at this stage. Moreover, the fate of many other cells with respect to adult segments is determined at the cellular blastoderm stage. Approximately 3.5 hours after fertilization, a ventral furrow forms. Mesodermal cells from the ventral furrow invaginate and will give rise to many of the larval structures. Shortly after the ventral furrow closes, a posterior midgut pocket forms on the dorsal side, which carries the pole cells to the interior (Fig. 15.4). By 6.5 hours the labial, maxillary, and mandibular rudiments are distinct, and evidence of segmentation can be seen along the ventral germ band. Late in gastrulation the parts that will make up the head of the adult fly are pulled into the developing larva only to emerge at metamorphosis. Within 24 hours after fertilization the first instar larva (L1) is ready to start eating.

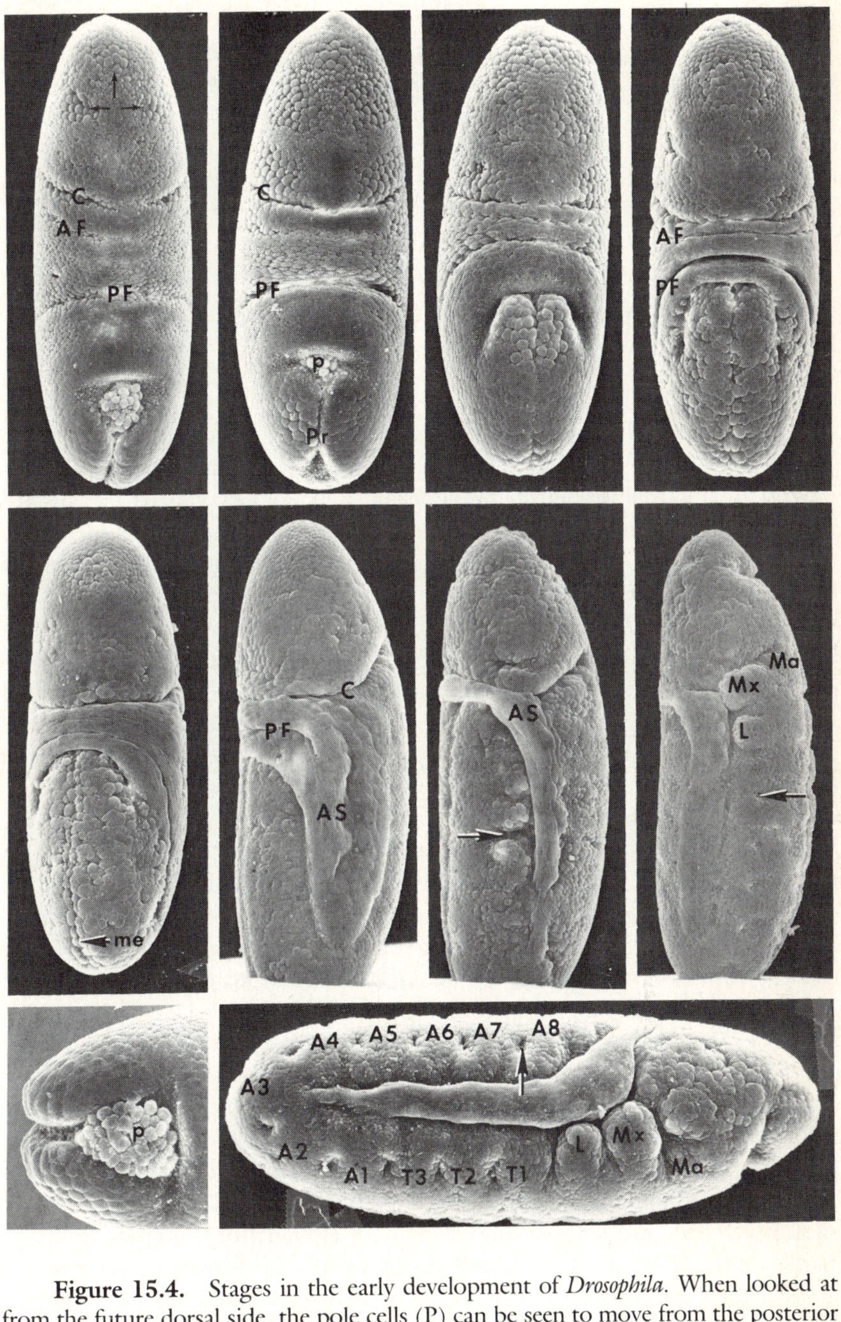

Figure 15.4. Stages in the early development of *Drosophila*. When looked at from the future dorsal side, the pole cells (P) can be seen to move from the posterior pole toward the posterior fold (PF) before being internalized. The anterior fold (AF) is lost during this period. The tissues that will give rise to the labia (L) and maxillary parts (Mx) and mandibules (Ma) can be discerned. Tracheal pits are indicated by arrows. The cephalic cleft (C) surrounds the head of the embryo until head involution begins. A lateral view of an embryo at the completion of gastrulation shows the head parts beginning to undergo involution, three thoracic segments, and eight abdominal segments. The embryo has curled up on itself.

Drosophila is a protostomal organism and appears to have evolved a mechanism of gastrulation quite different from that used by deuterostomes such as amphibians and mammals. Within 15 minutes after the appearance of the ventral furrow, 1000 cells invaginate to form the mesoderm. During the next 30 minutes three separate invaginations produce the major rudiments of the endoderm. An hour later another invagination occurs and joins the anterior midgut rudiment. The whole process is extremely rapid. Cells at the ventral furrow take up bottle shapes reminiscent of those seen at the blastopore of amphibian embryos. The cellular mechanisms of invagination may be quite similar, but the subsequent strategy for tissue formation appears completely different. Evolution of segmented protostomes may have occurred from ancestral species already diverged from the line that has given rise to deuterostomes.

An adult fly is composed of eight abdominal segments, three thoracic segments, and head structures derived from fusion of multiple segments. These same segments can be seen in larvae and are marked on the ventral surface by unique bands of bristlelike chaetae. The number of segments formed is determined in part by the action of a gene, *ftz*. Mutations in this gene result in embryos with only half the normal number of segments.

The *ftz (fushi tarazu)* gene maps at salivary chromosome position 84A within the Antennapedia complex, ANT-C, which includes homeotic genes that regulate the head and thoracic structures. The *ftz* gene is transcribed into a 1.9 kb mRNA at the time the nuclei migrate to the surface of the blastoderm. The *ftz* mRNA is initially present in all cells that will give rise to the embryo, but within an hour it is localized to seven stripes, each corresponding to the width of a perspective segment. Shortly thereafter the first indications of segmentation of the cellular blastoderm occur and the *ftz* mRNA disappears. The transient expression of this gene occurs just at the right time to have a causal role in the establishment of the periodic pattern. It is interesting that the pattern of *ftz* expression is initially continuous and then restricted to alternating segments. This localization may result from changes in the periodicity of resonating signals.

The *ftz* gene contains two exons. Within the second exon is a region that codes for a sequence of 60 amino acids that also occurs in at least ten other loci of the *Drosophila* genome. Many of these homologous regions occur within the Bithorax complex (BX-C) and Antennapedia complex (ANT-C) and appear to be portions of homeotic genes that affect morphogenetic decisions for different regions of the embryo. The conserved sequence has, therefore, been called the homeo box. Very similar but not quite identical homeo boxes have been found in the genomes of a beetle, an earthworm, an amphibian, chicks, mice, and humans. Perhaps this sequence arose early during evolution and has been retained in a wide diversity of organisms to establish positional specification.

One of the genes that contains the homeo box in the amphibian *Xenopus* is first transcribed late in gastrulation into two mRNA species, one of 1.2 kb and one of 1.6 kb. The 1.2 kb mRNA disappears at the tail-bud stage whereas the 1.6 kb mRNA is present throughout development. It is not yet known if expression of this gene is restricted to specific cells of the gastrula.

Metamorphosis

Larvae are kept from undergoing metamorphosis by a hormone produced at these stages. Juvenile hormone, an acyclic sesquiterpene (Fig. 15.5), is present during the first two molts and ensures that the larva transforms into the next instar larva. At the third molt, juvenile hormone is low and the larva pupates. Each molt is triggered by another hormone, 20-hydroxyecdysone (Fig. 15.5). This ecdysteroid has significant effects on the transcription pattern when present in the absence of juvenile hormone.

Within the pupa the whole body form is changed as the wormlike larva undergoes metamorphosis to form an adult fly. Most of the larval cells degenerate, and their components are assimilated by adult tissues. The structures of the fly are produced by groups of cells already organized in the first instar larva into imaginal discs (from imago). The imaginal discs are formed in the embryo by local invaginations. At each subsequent instar these tissues grow but do not express any structures of terminal differentiation. There are ten pairs of imaginal discs that make specific cuticular structures from the head to the genital parts (Fig. 15.6).

When juvenile hormone is low, ecdysteroids induce metamorphosis of these cells into the adult structures. In most cases the adult structures are formed by evagination of the imaginal discs. Legs pop out in a telescoping manner (Fig. 15.7).

The imaginal discs are collapsed bags that swell and extend to make legs, halteres, wings, antennae, and other appendages. The relative position of the cells in an imaginal disc do not change, and so a fate map of the adult structures can be drawn onto the surface of the disc (Fig. 15.8).

If small portions of a wing imaginal disc are removed just prior to metamorphosis, the structures that would have come from the missing piece are missing in the adult. This result indicates that each of the adult structures

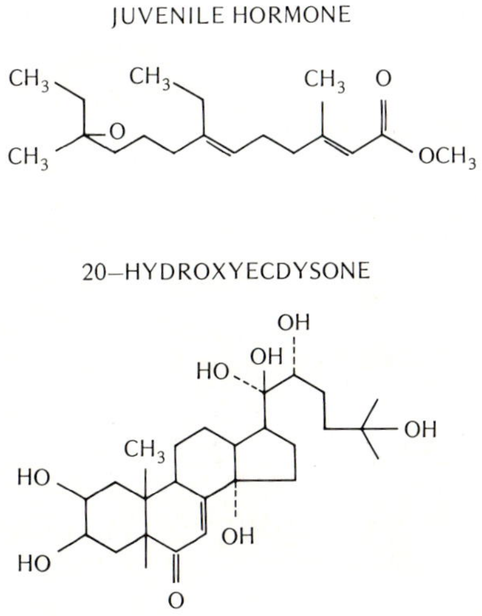

Figure 15.5. Structure of juvenile hormone and 20-hydroxyecdysone.

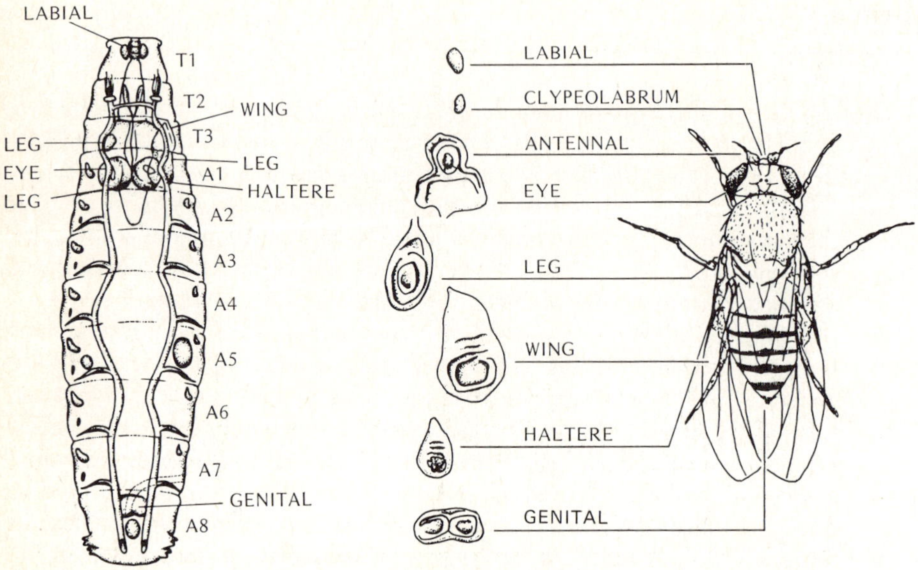

Figure 15.6. Imaginal discs of *Drosophila*. There are ten imaginal discs in larvae that give rise to adult structures. At metamorphosis the imaginal disc cells expand and take on their final shapes.

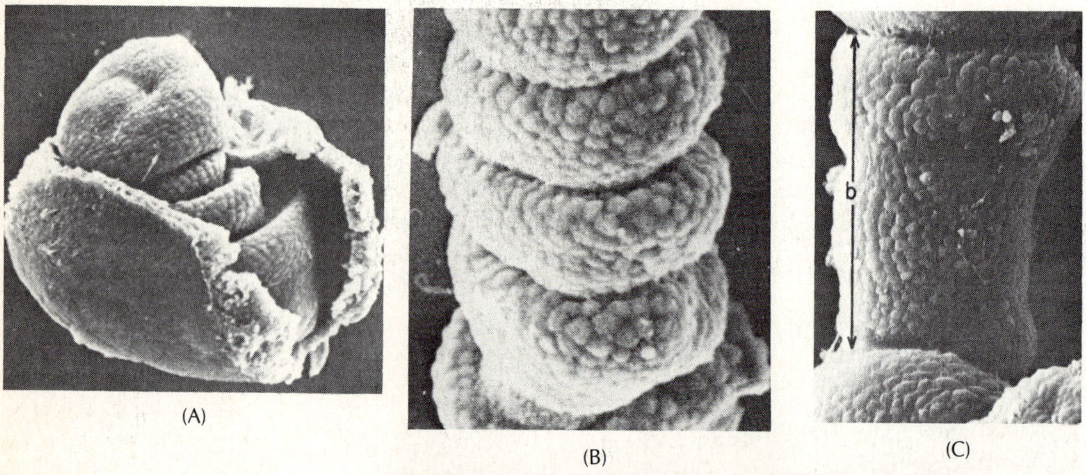

Figure 15.7. Telescoping out a leg from its imaginal disc. Small changes in the orientation and shape of the individual cells results in eversion of the disc tissue to form the leg. *(A)* Leg disc before eclosion; *(B)* partially expanded leg; *(C)* the distance between grooves has increased; *(D)* leg after metamorphosis.

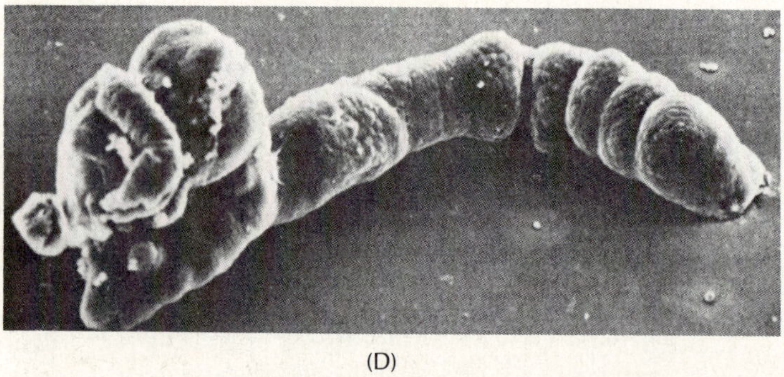

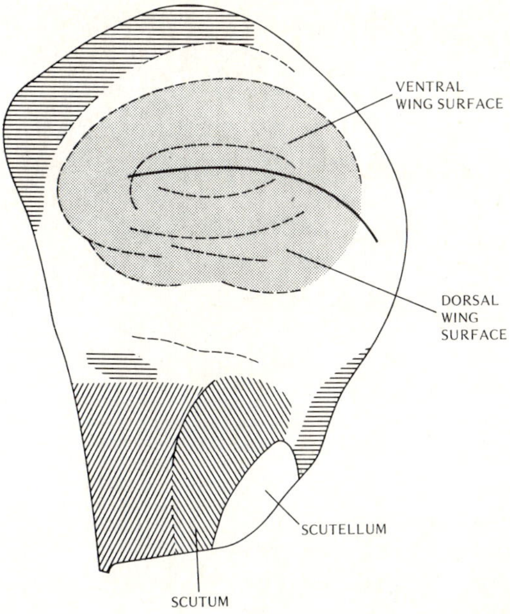

Figure 15.8. Fate map of the wing imaginal disc. Fragments cut from third-instar wing disc were induced to undergo metamorphosis by being put in the abdomen of larvae. Following metamorphosis, the structures formed were analyzed. Surprisingly detailed fate maps could be drawn from the data, suggesting that even small groups of cells are determined by the third instar.

down to individual hairs and bristles is determined in the cells of the pupal imaginal wing disc.

Transdetermination

Adult flies no longer molt and the ecdysteroid levels are very low. Imaginal discs are compact tissues that can be removed intact from larva and cultured in the abdomen of adult flies. They continue to grow there but do not undergo metamorphosis until they are replaced in a larva. In fact, imaginal discs can be serially passaged from one adult abdomen to the next for many generations and then finally analyzed by being inserted into a larva. When the host larva undergoes metamorphosis, cells of the much transplanted imaginal disc also differentiate into adult structures.

Leg imaginal disc cells cultured in an adult abdomen for a short period before being replaced in a larva give rise to only leg structures following metamorphosis. The leg disc is determined to form a leg although it has grown in an unusual environment and been put out of place in the larva. However, if the leg disc is serially cultured in adult abdomens so that many cell generations of growth have occurred, the cells will occasionally form wing structures. The larger the number of cell generations that the imaginal disc cells have undergone, the greater the frequency of formation of wing or antennal structures. This change in developmental pathways of the cells is referred to as transdetermination.

Transdetermination has been observed in most of the imaginal discs. However, the frequencies of transformations vary significantly. Antennal discs

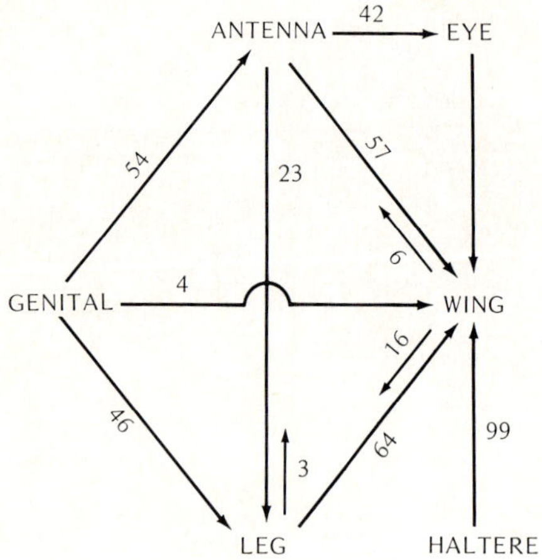

Figure 15.9. Pattern of transdetermination. Imaginal discs were cultured in the abdomens of adults and then put in the abdomens of larvae, where their hosts induced them to undergo metamorphosis along with them. The frequencies at which different discs transdetermined into other structures are given. Notice that genital discs formed other structures but other discs never made genital tissue. Leg disc transdetermined to antenna rarely but often formed wing. They never formed eye structures although antenna discs often did.

often form legs but leg discs seldom form antennae. Instead, leg discs most frequently transdetermine to form wing structures (Fig. 15.9).

The pattern of transdetermination frequencies among different imaginal discs can most easily be interpreted as the consequence of changes in a binary code. If only a single bit changes at each transdetermination and each disc is specified by a five-bit code, the pattern can be fairly well accounted for.

Transdetermination of imaginal discs is rare and determination is the rule. Although the cells in the discs do not express any known differentiations that distinguish one from another, a wing disc will form wing structures when metamorphosis is induced even if the disc is excised and placed in an abdomen. This experiment indicates that no local signals are necessary once discs are formed. Instructions are given during early embryogenesis that are stably retained for many cell divisions as the embryo develops. Stable transcriptional states may result from cooperative interactions among regulatory proteins binding to contiguous sites on chromatin. When the DNA replicates, each daughter strand will usually retain one of the interacting regulators, which will then facilitate binding of the other components. At each cell division there will be a small but significant chance that a given region of chromatin in one daughter cell will carry none of the regulatory proteins and so will not be able to attract the other components. This will result in a change in the state of determination of this cell and all of the cells it gives rise to in subsequent divisions. Direct evidence supporting this proposed mechanism for determination and transdetermination is scanty, but it is being actively explored.

Compartments

The fate and position of individual cells and clones derived by growth from single cells can be seen by genetically marking a few of the cells in a developing embryo. Mosaics of this sort can be constructed by irradiating with x-rays to induce mitotic recombination in a few cells of an embryo heterozygous for certain recessive genes. Cells of an individual genetically heterozygous for the recessive mutation *yellow* (+ /y) will be darkly pigmented due to the wild-type allele. If genetic crossing-over occurs during a mitotic division of an embryonic cell, one daughter cell will be brown (+ / +) and the other cell will be yellow (y/y). The y/y cell and all its progeny will be distinguishable by their lighter pigmentation. When the mitotic recombination is induced early during embryogenesis (about 3 hours post-fertilization), the patches of yellow cells on the cuticle of the adult are large, often covering a quarter of the wing. The mitotic recombination occurred in a cell that divided many times to give a clone in the wing disc. When mitotic crossing-over is induced at later times, the patches are smaller, indicating that few cell divisions occurred after the recombinational event and before formation of the adult structures. The patches are always coherent, indicating that the cells of a clone stay in close proximity with each other and do not wander off. When induced at late stages the patches are small but the outlines of hundreds of small patches on different flies can be overlaid to see if a clone can ever give rise to two different structures such as a wing and a leg. Alternatively, the heterozygous fly can be genetically manipulated so that a cell rendered homozygous for yellow (y/y) by mitotic recombination will also be genetically capable of growing faster than its neighbors. If the cross-over affects linked yellow and minute genes, a y + / + *min* cell will give rise to y + /y + and + *min*/ + *min* daughter cells. The minute *(min)* mutation slows growth even in the heterozygote. Therefore, the y + /y + yellow cell has a growth advantage in the mostly y + / + *min* embryo and can give rise to large patches of yellow tissue even when induced late in the life cycle.

These analyses have shown that clones grow only to give structures in the anterior or posterior of a single segment but never in both. This curious phenomenon was first seen in wings in which the anterior and posterior compartments are separated by a straight line running the length of the wing (Fig. 15.10). The line of demarcation in the wing follows no anatomical structure such as a vein or any discontinuity of the cells. But the progeny of a cell generated at the cellular blastoderm stage never cross that line. They

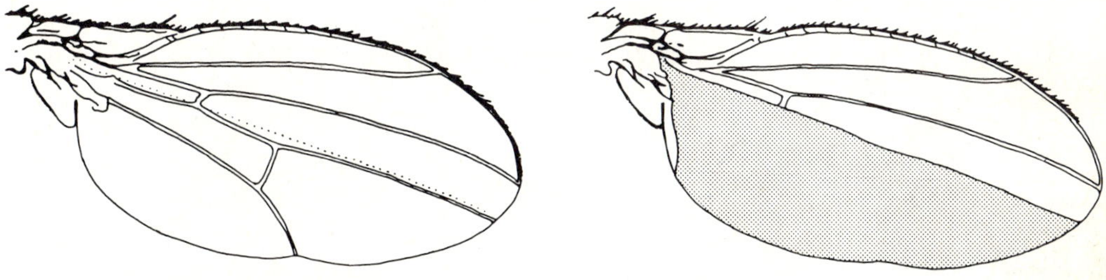

Figure 15.10. Anterior/posterior wing compartments. No anatomical structure in the adult wing separates the wing in half, yet clones derived from single cells of the cellular blastoderm early in the development of wing imaginal discs are never found to cross the boundary.

are either anterior or posterior. The basis must then reside at an earlier stage of embryogenesis. By injecting dye into imaginal disc cells and watching its passage to adjacent cells, some indication of compartments in imaginal wing discs has been found. The dye passes through junctions connecting the cells that will make up the structures of the anterior compartment but will not pass to cells that will make up the structures of the posterior compartment. Likewise, cells of the posterior wing disc compartment are connected to each other by ports but are not connected to the anterior compartment. It may be this restriction of intercellular junctions that establishes compartments in discs.

All the major adult appendages, such as wings, halteres, legs, or antennae, are divided into anterior and posterior compartments although it has often taken sophisticated genetics to demonstrate this. Marked clones generated in the first instar larva not only respect the anterior/posterior compartment boundary but also a dorsal/ventral boundary. Clones generated before this time will often mark both sides of the wing, but clones genetically marked after this time are seen to be present only on the dorsal or ventral surfaces, never on both. Thus, lineages leading to specific anterior or posterior compartments of segments are determined shortly after nuclei migrate to the surface of the blastoderm and appear to be established as cells are formed around each nucleus. The dorsal/ventral lineages separate somewhat later apparently by a different mechanism. At the cellular blastoderm stage the number of cells that are committed to form the wings is quite small but these divide and grow during embryonic and larval development. After several cell divisions have occurred a zone of more slowly dividing cells appears at the margin separating the precursors of the dorsal and ventral surfaces. This zone of nonproliferating cells then restricts clones to one side or the other. In this way the control of cell growth functions to shape the wing imaginal disc, which in turn determines the form of the wings.

Genes Affecting Compartments

A number of mutant flies have been isolated in which structures of one or more compartments are replaced by structures appropriate for other compartments. These flies are said to carry homeotic mutations (Table 15.1). Two clusters of such mutations have been studied in detail—the Bithorax Complex, BX-C, and the Antennapedia Complex, ANT-C. Both are large and complicated. They span several hundred kilobases and are both on the right arm of chromosome 3, BX-C at band 89E and ANT-C at band 84AB. Mutations in BX-C affect differentiations in the abdomen and thorax while those in ANT-C mainly affect differentiation in the head and thorax. Within each complex, mutations that fail to complement each other and give similar phenotypes extend over regions of up to 100 kb. These are much larger regions than are found for other genes and indicate that the genetic information in these complexes may not be involved only in coding for enzymes

Table 15.1 Some Homeotic Mutations of *Drosophila*

Name	Gene	Phenotype
BX-C *(band 89E)*		
1. bithorax	bx	anterior haltere→anterior wing
2. post bithorax	pbx	posterior haltere→posterior wing
3. Contrabithorax	Cbx	posterior wing→posterior haltere
4. bithoraxoid	bxd	1st abdominal segment→3rd thoratic segment
5. Ultrabithorax	Ubx	haltere→wing
6. infraabdominal-2	iab-2	2nd abdominal segment→1st abdominal segment
ANT-C *(band 84AB)*		
1. Antennapedia	Antp	antenna→2nd leg
2. Extra sex combs	$\text{Antp}^{\text{Extra sex combs}}$	2nd, 3rd leg→1st leg
3. proboscipedia	pb	labia→leg
Other		
1. engrailed	en	posterior compartments→anterior compartments
2. opthalmoptera	opt	eye→wing

or regulatory proteins that function in a simple linear manner to determine the pattern of differentiation. Within several of these loci (defined by lack of complementation of independent mutations) there are other mutations giving related but quite different phenotypes. The unusual and intriguing nature of these complexes has prompted their isolation by molecular cloning and initiated intensive work to recognize their products (Fig. 15.11).

BX-C

Calvin Bridges isolated a mutant fly more than 60 years ago that almost looked as if it had four wings rather than the two normally found on *Drosophila* (Fig. 15.12). On closer inspection it turned out that flies homozygous for this recessive mutation, *bithorax (bx/bx),* had made wing structures from the haltere. The haltere is a small appendage from the third thoracic segment (T3) that acts as a balancer during flight of *Drosophila.* The normal wings extend from the second thoracic segment (T2). Other mutations were found in another locus, *post bithorax (pbx),* that gave a similar but distinct phenotype. Genetic analysis showed that *bx* and *pbx* are closely linked but can be separated by recombination in about 3 percent of the progeny of appropriate crosses. Both *bx* and *pbx* mutations are recessive and complement each other. That is, a heterozygous diploid, *bx+/+pbx* has a wild-type phenotype. Molecular analyses have shown that *bx* mutations affect a 25 kb segment of DNA that is 50 kb away from the *pbx* mutations that turn out to be large deletions covering over 20 kb.

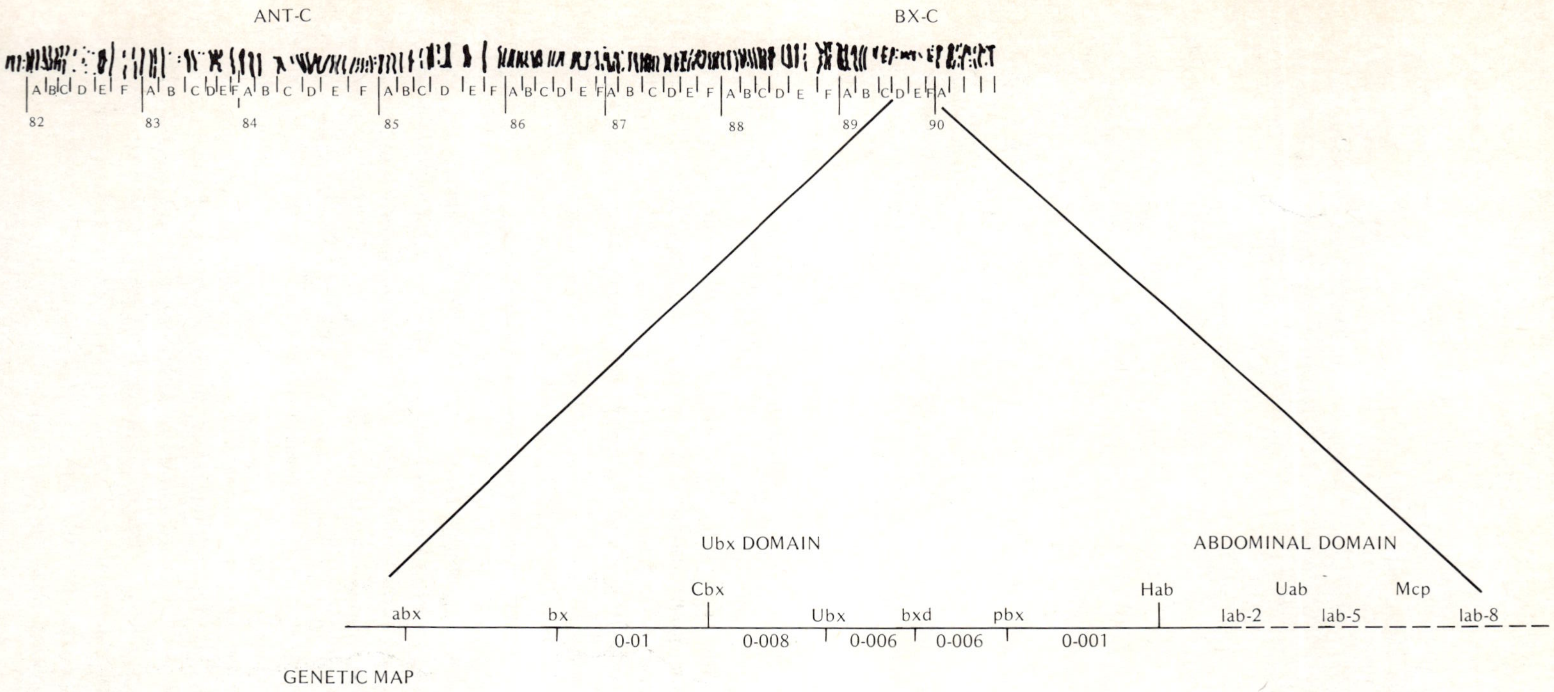

Figure 15.11. Genetic map of homeotic genes. Two regions covering several hundred kilobases of DNA each control segment differentiation in the head, thorax, and abdomen of *Drosophila*. They can be localized to chromosomal positions 84AB (ANT-C) and 89E (BX-C). Mutations in the bithorax complex (BX-C) have been mapped and their DNA alterations analyzed. The mutations are given in Table 15.1.

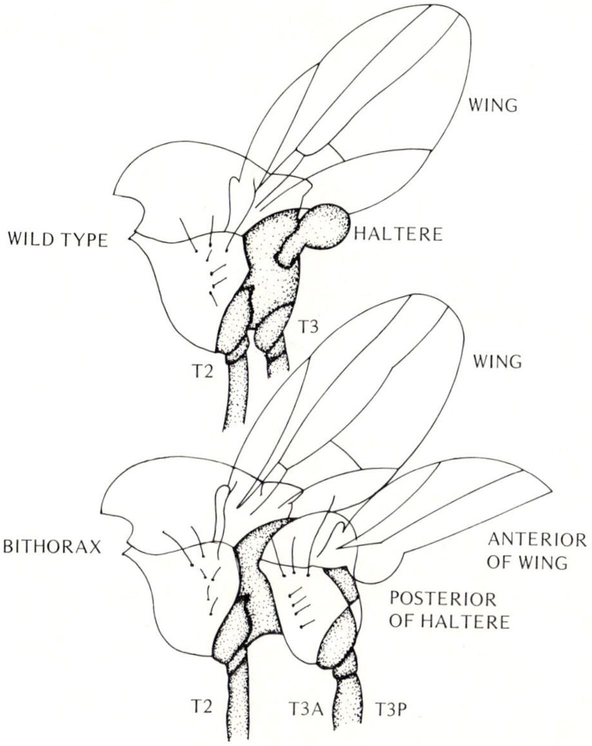

Figure 15.12. Homeotic transformation caused by the bx^3 mutation. Wild-type and mutant bx^3 *Drosophila* wing and haltere are compared. The mutation has transformed the anterior compartment of the haltere into the anterior compartment of the wing (T3A → T2A). The posterior compartment is unaffected. The anterior of the third leg is also transformed but is not shown in the drawing.

Mutations in both *bx* and *pbx* affect the third thoracic segment (the metathorax) by transforming halteres into wing structures. However, close inspection shows that *bx* mutations affect only the anterior compartment of the segment (T3A) (Fig. 15.12) whereas *pbx* mutations affect only the posterior compartment of the segment (T3P). Double mutants carrying both *bx* and *pbx (bx pbx/bx pbx)* transform both the anterior and posterior compartments of the third thoracic segment into a second thoracic segment: the haltere becomes a normal wing (Fig. 15.13). These four-winged dipterans remind one of four-winged dragonflies. It is likely that dipterans (meaning two-winged insects) evolved from an ancient four-winged insect by suppressing wing development from the third thoracic segment. A haltere was made instead and turned out to be useful as a balancer. This evolutionary change came about by the acquisition of the wild-type *bx* and *pbx* gene activities. Perhaps one evolved first and suppressed development of half of the backwings and further selection favored flies that had both genes and suppressed both halves of the back wings. Dipterans have more agile flying modes than four-winged insects such as butterflies and dragonflies.

Mutations in *bx* and *pbx* show that the action of these genes is confined to a single compartment. Wild-type *bx* genes function in the anterior compartment of segment T3; wild-type *pbx* genes function in the posterior compartment of segment T3. The action of these genes seems straightforward enough, but their products are not yet known. And other mutations indicate that the situation is much more complicated.

Figure 15.13. A four-winged fly. This *Drosophila* carries mutations bx^3 *pbx abi,* which transform the anterior and posterior compartments of the third thoracic segment and the first abdominal segment into the second thoracic segment. This fly was genetically constructed by Ed Lewis.

A transposition of the *pbx* sequence 30 kb toward the centromere results in a completely different phenotype. The wings that normally develop from the second thoracic segment (T2) are transformed into halteres. The transposition of wild-type *pbx* sequence is dominant (referred to as *Cbx*). The transposed *pbx* appears to suppress wing development in both compartments of T2 rather than affecting T3P as it had before transposition. Flies carrying *Cbx* have no wings and two sets of halteres.

Mutations in and around *pbx* not only transform T3P but also affect the first abdominal segment, A1, transforming it into T3. These recessive mutants, bithoraxoid *(bxd),* might just be strong alleles of *pbx;* however, *bxd* mutations are complemented in heterozygotes carrying both *bxd* and *pbx* *(bxd+/+pbx)*. Some of the *bxd* mutations are 20 kb away from the *pbx* deletions, leaving room for four or five genes in between. However, it is possible that this region of the *Drosophila* genome gives rise to exceptionally long transcripts.

There are dominant mutations in and around the *bx* region that result in extra wings. By looking at the legs of flies carrying these *Ultrabithorax* mutations *(Ubx),* it is clear that the third thoracic and first abdominal segments have been transformed into the first and second thoracic segments (T3, A1→T1, T2). The mutations are dominant and spread over 73 kb of DNA bracketing *bx* and *Cbx* mutations. *Ubx* mutations are lethal when homozygous and fail to complement the homeotic characteristics of *bx, pbx,* and *bxd*. There appears to be a necessary interaction of the sequences spanned by *Ubx* and *pbx*. Perhaps this region codes for a variety of gene products that function in a network of reactions to determine the pattern of differentiation of thoracic segments. The region is transcribed in the thorax and abdomen but not in the anterior of the organism. The highest levels of transcripts are found in the third thoracic and first abdominal segments (T3 and A1). The primary transcript from the *Ubx* region appears to be ten times bigger than most transcripts (about 75 kb) and is processed to give various normal-sized RNAs. The proteins coded for by the *Ubx* transcripts accumulate in nuclei of the cells possibly due to nucleic acid binding by the amino acid domains coded for by the homeo boxes found in these genes.

Deletion of the whole BX-C as well as sequences toward the chromosomal telomere [Df(3R)P9] are lethal, but larva are formed in which it can be seen that most of the thoracic and abdominal segments are identical. They may be similar to the posterior of the first thoracic segment and the anterior of the second thoracic segment (TIP + T2A). This may be a basic ground state for thoracic and abdominal segments. The phenotype of flies carrying the deletion Df(3R)P9 suggests the presence of genes distal to *pbx* on chromosome 3 that regulate differentiation of abdominal segments. Several mutations that genetically map in this region support the notion that each abdominal segment is specified from more anterior ones by a unique gene. These mutations, infraabdominal *(iab),* are recessive. Mutations in the *iab*2 gene transform the second abdominal segment into the first (A2→A1); *iab*5 transforms the fifth abdominal segment into the fourth (A5→A4); the *iab*8 mutation transforms the eighth abdominal segment into the seventh (A8→A7). Linked dominant mutations (Hab, Uab, Mcp) lead to depression of the *iab* genes and result in transformation of abdominal segments to more posterior ones (A1→A2; A4→A5). At least that is the present interpretation.

Mutations in the Bithorax Complex (BX-C) reduce the complexity of the organism. In *Drosophila* each segment differentiates differently. Insects of this sort may have evolved from organisms constructed in a manner similar to modern-day centipedes and millipedes in which hundreds of segments, all identical, are strung in a line. Genes may have evolved to suppress or add developmental potential in adjacent segments. As each gene was added to the genetic repertoir of a newly evolved species, another segment was differentiated. A schematic model of such a process has been proposed by Ed Lewis (Fig. 15.14).

The model proposes that the genes of the BX-C function only in the posterior compartment of the second thoracic segment and more posterior ones. In fact, mutations in BX-C have little or no effect on head segments or T1. This is true even if structures normally made by T2 are formed in the head due to other homeotic mutations such as *ophthalmoptera* (a wing comes out of the eye) or *antennapedia* (legs replace antennae). In the thoracic compartments (T2P to T3P) and the abdominal compartments, an increasing number of BX-C genes must be active to ensure normal differentiation. Segment assignment is thus thought to be coded by a sequence of

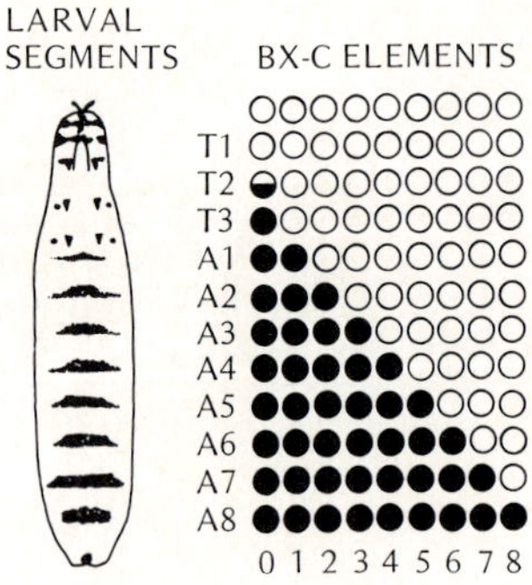

Figure 15.14. Model for BX-C function. The BX-C locus may have nine functions (0–8). Element 0 is marked by *bx* and *pbx* mutations; element 1 is marked by *bxd* mutations. Elements 2 to 8 are partially marked by *iab* mutations. Segment compartments in which an elememt is expressed are filled. Note that each segment differs from its neighbors only in a single element. This model has been proposed by Ed Lewis.

discrete genes. If the sequence is perturbed by mutation, nonsense code words are formed that are misinterpreted.

Direct support for this model has come from analysis of the RNA transcripts of *Ubx* and the protein coded by this gene. Both *Ubx* RNA transcripts and the proteins they code for are absent in the head and the T1 segments but present in T2, T3, and the abdominal segments. The major product of *Ubx* is a 44 kd protein found predominantly in nuclei where it may regulate the expression of segment-specific genes. The *Ubx* gene is expressed early in larval embryogenesis when it functions to specify compartments. Thus, several aspects of the model appear to be confirmed and there is considerable excitement as other predictions are tested.

ANT-C

A few thousand kilobases away from BX-C is another complex wherein mutations affect the head and thoracic segments. Mutations in a 100 kb region of ANT-C result in transformation of antennae into second legs (Fig. 15.15). Other mutations in this region result in transformation of head into thorax (Antp^{cephalothorax}) and transformation of second and third legs into first legs (Antp $^{\text{extra sex combs}}$). These mutations are dominant and are associated with rearrangements and insertions into the region. When homozygous, these mutations are lethal.

Within a few hundred kilobases there is a locus, *proboscipedia* (*pb*), necessary for normal labial development. Mutations in *pb* give rise to legs or antennae replacing the labial palps. Development of the labia are also affected by certain mutations at another locus, *Scr,* between *pb* and *ANTp*. All of these loci are surrounded by genes required during embryogenesis. It appears that the pathways of development for the head and anterior thoracic segments are determined in part by genes of ANT-C. How these genes may function has not yet been indicated.

The Antp region is transcribed over 100 kb and the product spliced down to give two prominent transcripts of 3.5 kb and 5 kb that are polyadenylated. The initial transcript is enormous and even the spliced mRNAs are exceptionally large. These mRNAs are present throughout embryogenesis and up to pupation. They are absent in adult flies. Mosaic flies con-

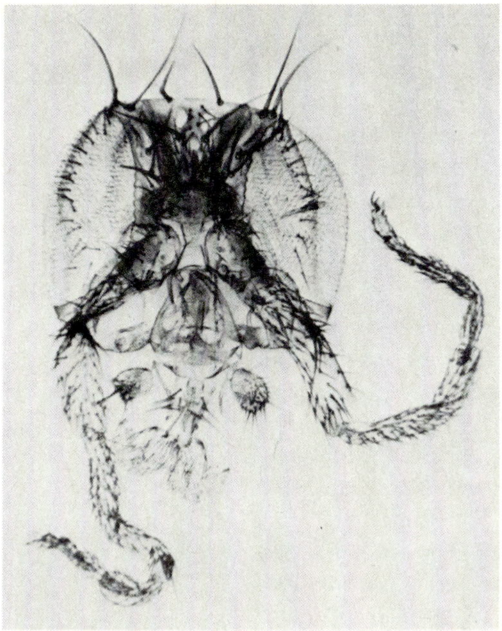

Figure 15.15. Fly with legs replacing antennae. This mutant individual carries the dominant antennapedia mutation *Antp*^{73b} and has transformed the antennal-aristal appendage into a leg. A fly head is 660-μm wide.

structed by mitotic crossing over in heterozygous flies have shown that patches of cells homozygous for mutations in either BX-C or ANT-C will express the mutant defect. The action of the genes in these complexes appears to be essential throughout embryogenesis to maintain the assignments of each compartment.

Engrailed

A mutant fly was found in which sex combs normally seen on the anterior of the first leg of males were partially duplicated on the posterior of the first leg. Subsequent inspection of flies carrying this mutation, *engrailed (en),* showed that not only was the anterior of the first leg duplicated but so were the hair patterns of the other two legs and the wing. In each of the thoracic segments, portions of the posterior compartment had differentiated into anterior structures rather than posterior ones. By studying double mutants homozygous for the recessive *engrailed* mutation together with a homeotic mutation such as *Antp,* asymmetric structures could be analyzed that were derived from the head. In all cases, homozygous *en*/*en* mutants had portions in all posterior compartments transformed into anterior ones. These studies have shown that *en* works in both the head and thoracic segments.

The wild-type *en* gene product must function in posterior compartments to maintain their specification and keep them from differentiating into structures appropriate for the anterior of their segment. The *en* mutations do not change segmental specification but affect the anterior/posterior decision. Therefore, the *en* gene product is thought to work in parallel with other homeotic genes such as those of BX-C and ANT-C to give the complete address of segment and compartment.

Clones of *en/en* cells generated late in embryogenesis by mitotic recombination of heterozygous flies express the engrailed phenotype. That is, a patch of *en/en* cells in the posterior compartment of an otherwise wild-type fly will differentiate only anterior structures. Like the other homeotic mutants studied to date, the *engrailed* gene acts in a cell-autonomous fashion. Mutant cells are unaffected by adjacent wild-type cells.

When a patch of *en/en* cells in the posterior compartment occurs near the compartment boundary in wings, the clone is seen to cross the boundary. Unlike wild-type cells in which a clone always stops abruptly at the line demarking the compartments, posterior *en/en* cells are unaware of the restriction. Whatever specifies compartments appears to depend on the function of the *engrailed* gene product. The *engrailed* gene is transcribed over a 70-kb region to give rise to mRNA transcripts of 1.3 kb and 2.6 kb that first accumulate at the start of gastrulation (3.5 hours). They are found predominantly in the posterior compartments of segments seen at 6 hours. Initially, the stripes are only two cells wide. At later stages the gene is expressed in the posterior but not in the anterior compartment of wing imaginal discs. Interestingly, there is a sequence of several hundred bases within the *en* gene related to the homeo box found in *ftz, Antp, Ubx,* and other homeotic genes. The engrailed gene is transcribed at the appropriate stage and in the appropriate cells to specify an anterior/posterior divergence of differentiation. Expression of *en* appears to be regulated by positional information indicating posterior compartments and, in turn, *en* products direct appropriate posterior differentiation.

Regeneration of Imaginal Discs

Surgical removal of fragments of imaginal discs results in loss of the structures normally derived from the missing fragment. However, if pieces of imaginal discs are cultured in the abdomens of adult flies for a few days before metamorphosis is induced, cells are intercalated into the wound. When the fragment removed is less than half of the wing imaginal disc, intercalated cells differentiate to give the missing structures. Put another way, the wounded disc regulates by epimorphosis to regenerate the normal wing pattern. If more than half of the wing disc is removed, the intercalated cells differentiate to give a mirror-image duplication of the remaining piece. The position of the small piece removed from the wing disc makes no difference. Taken from any place a small piece will differentiate into the structures of its fate and direct newly intercalated cells to do likewise. The fact that the duplicated structures are arranged in a mirror-image fashion has significant

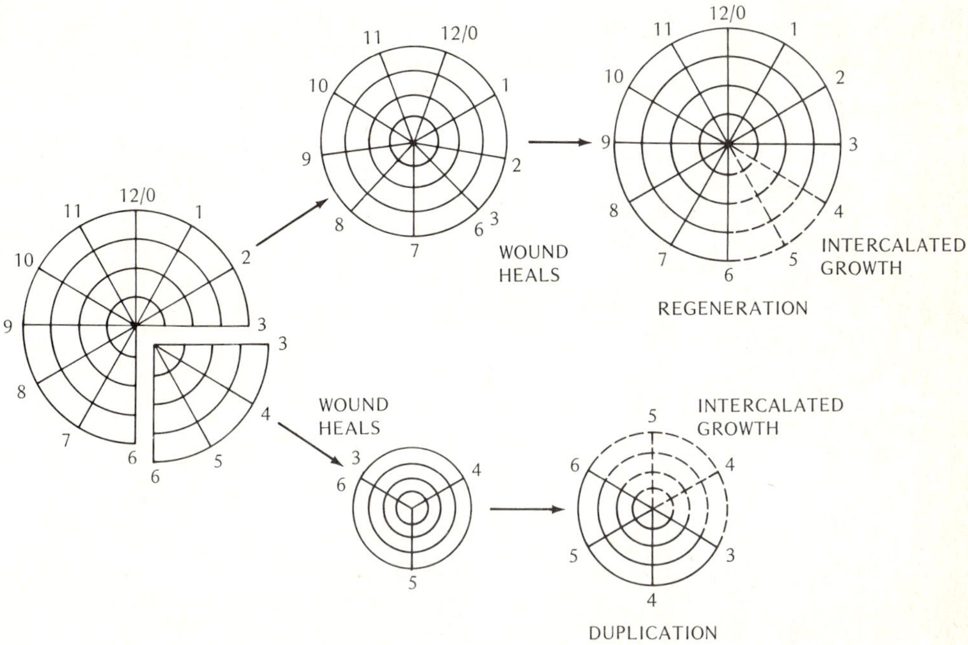

Figure 15.16. Analysis of regeneration and duplication of fragments of wing disc of *Drosophila* in polar coordinates. Shortest route intercalary growth gives regeneration of the large fragment and duplication of the small fragment no matter how the cut is made.

bearing on the mechanisms that can transmit positional information from the original piece to the new cells that fill the wound.

These results have been interpreted assuming a polar coordinate system for pattern specification (Fig. 15.16). When analyzed in this way, the results can be seen to follow two rules:

1. Intercalated cells fill intermediate positions between the cut edges by the shortest route—for example (a) removal of a small fragment of positions 3 to 6 leaving edges of 3 and 6 will fill in by differentiating intercalated cells as 4 and 5 (regenerated), and (b) removal of a large fragment of positions 6 to 3 leaving edges of 6 and 3 will result in the small remaining piece filling in by differentiating intercalated cells as 4 and 5 (duplication).
2. A complete circle is reformed in either case.

With these two rules, a large number of results with manipulated imaginal discs as well as newt limbs and cockroach legs are accounted for. However, this does not mean that positional information actually goes in circles or that this is the only way to describe the results.

Summary

Embryogenesis in insects follows quite different paths from those seen in vertebrates. Initially, only the nuclei divide in a fertilized egg. When there are about a thousand nuclei, they migrate to the surface of the cell, where certain aspects of their subsequent differentiation are specified. Those cells that incorporate cytoplasm of the posterior pole have the potential to form germ cells whereas others do not. Groups of cells invaginate to form the larval organs whereas others are committed to the formation of adult structures. These imaginal discs are kept from proceeding to their terminal differentiation by juvenile hormone. They can be passaged in larvae producing juvenile hormone for many generations and will still differentiate into their normal determined structures when ecdysone triggers metamorphosis. In a few cases, however, they will transdetermine so as to differentiate into other structures. Wing and leg imaginal discs have been shown to be divided into anterior and posterior compartments. *Drosophila* homozygous for a mutation in the *engrailed* gene do not correctly delineate posterior compartments so that in mutant flies they take on anterior characteristics. Mutations in two complex genetic regions, ANT-C and BX-C, affect the segment specification of certain anterior and posterior compartments. Regeneration of pieces of imaginal discs is characterized by mirror-image duplication of structures in small pieces and by replacement of all normal structures in large pieces. The results have been interpreted in terms of a clock-face model. It is thought that opposition of mismatched cells induces intercalary growth of new cells that fill in with intermediate positional values in the shortest sequence that re-forms a complete circle without discontinuities.

SPECIFIC EXAMPLE

Expression of Transformed Genes

There are some genes that are especially amenable to study because they are dispensable for growth or development under laboratory conditions. The alcohol dehydrogenase gene of *Drosophila (Adh)* is one of these. The enzyme is necessary for survival in an environment of 6 percent alcohol such as is found in fermenting fruit, but it is not required when the fruit flies are fed yeast cells in the laboratory. One can select for flies with active alcohol dehydrogenase by spiking the vials they are kept in with 6 percent alcohol. It is also possible to select for *Adh* mutations by presenting the flies with a suicide compound (1-pentyne-3-ol) that kills them when metabolized by alcohol dehydrogenase. Only flies with inactive alcohol dehydrogenase survive this treatment. These selection techniques have generated a large number of interesting mutations.

Alcohol dehydrogenase is a homodimer composed of two identical 255-amino acid polypeptides. It is coded for by a single gene *(Adh)* that is found at salivary chromosome band 35B2 on the left arm of the second chromosome. The gene has been isolated, cloned in bacterial plasmid, and sequenced. The transcribed portion is about 1.1 kb and contains two short intervening sequences 65 and 70 bp long within the coding region that are spliced out in the processing steps to form functional ADH mRNA. Point mutations and deletions within the coding region result in flies with no active ADH enzyme that are fine unless given alcohol.

The *Adh* gene is expressed at two stages in the life cycle of *Drosophila* that are separated by a period during which the gene is repressed. The specific activity of ADH rises throughout larval development to a peak just before pupation. During metamorphosis of the pupa the specific activity falls as the tissues are rearranged. It begins to rise again just before the emergence of adult flies and continues to increase for several days. Clearly a mechanism functions to direct transcription of *Adh* in larval tissues, represses the gene during pupal development, and reactivates it in adult tissues.

Adh is not expressed in all cells in either larvae or adults. The enzyme cannot be detected in larval hypoderm, imaginal discs, salivary glands, foregut, or hindgut. It is present in midgut of both larvae and adults. It is present in larval somatic muscles but not in adult ones. The highest activity is found in the fat body that forms in larvae and persists until two days after eclosion. The *Adh* gene is therefore subject not only to developmental regulation but also to tissue-specific regulation.

It turns out that expression of *Adh* during larval development initiates transcription at a different position at the 5′ end of the gene than occurs in adult flies. A comparison of the *Adh* mRNA sequences with the genomic sequence pointed out that transcription in adults starts at a site 707 bases further upstream than larval transcription (Fig. 15.17).

Most of the adult transcript that precedes the larval transcript is spliced out in the processing of the mRNA (Fig. 15.18). The first 87 bases are retained as nontranslated 5′ sequences, whereas an intron of 654 bases is spliced out leaving 36 bases before the AUG codon at which translation of ADH starts. The larval transcript starts 70 bases upstream of the initial AUG codon, and thus is 46 bases shorter than the adult transcript. The adult transcriptional initiation point is preceded by a TATA box about 30 bases upstream, consisting of TATTTAA, while the larval transcriptional initiation point is preceded by TATAAATA positioned between bases 32 and 25 upstream. These TATA boxes most likely direct RNA polymerase II to the transcriptional initiation sites. In an *in vitro* assay using human transcriptional enzymes, the adult promoter functions well to initiate transcription at the same position as is found in flies.

What would happen if the upstream adult promoter were deleted? Would the larval promoter function in both larval development and adults? Would the proper tissue specificity be maintained in adults? It is necessary to isolate the gene to perform such exact surgery on a specific nucleic acid sequence and then return it to the organism so that it is reincorporated. The first step, isolation of the gene, had been done by cloning it in a bacterial plasmid,

```
 -260                 -240                  -220                  -200                  -180
GACTCTTTTTGATTTTGGAATATTTTCGTTCGTTTTATGTTTTTACGTTTTCGCATATTTGTTTCACAGTGCACTTTCTGGTGTTCCATTTTCTATTGG
 -160                 -140                  -120                  -100                  -80
GCTCTTAACCCCGCATTTGTTTGCAGATCACTTGCTTGCGCATTTTTATTGCATTTTACATATTACACATTATTTGAACGCCGCTGCTGCTGCATCCGTC
                                                                       ADULT RNA
                      -40                   -20                        →                +20
GACGTCGACTGCACTCGCCCCCACGAGAGAACAG[TATTTAA]GGAGCTGCGAAGGTCCAAGTCACCGATTATTGTCTCAGTGCAGTTGTCAGTTGCAGTTC
 +40                  +60                   +80                   +100                  +120
AGCAGACGGGCTAACGAGTACTTGCATCTCTTCAAATTTACTTAATTGATCAA[GTAAGTAGCAAAGGGCACCCAATTAAAGGAAATTCTTGTTTAATTG
 +140                 +160                  +180                  +200                  +220
ATTTATTATGCAAGTGCGGAAATAAAATGACAGTATTAATTAGTAAATATTTTGTAAAATCATATATAATCAAATTTATTCAATCAGAACTAATTCAAG
 +240                 +260                  +280                  +300                  +320
CTGTCACAAGTAGTGCGAACTCAATTAATTGGCATCGAATTAAAATTTGGAGGCCTGTGCCGCATATTCGTCTTGGAAAATCACCTGTTAGTTAACTTCT
 +340                 +360                  +380                  +400                  +420
AAAAATAGGAATTTTAACATAACTCGTCCCTGTTAATCGGCGCCGTGCCTTCGTTAGCTATCTCAAAAGCGAGCGCGTGCAGACGAGCAGTAATTTTCCA
 +440                 +460                  +480                  +500                  +520
AGCATCAGGCATAGTTGGGCATAAATTATAAACATACAAACCGAATACTAATATAGAAAAAGCTTTGCCGGTACAAAATCCCAAACAAAAACAAACCGTG
 +540                 +560                  +580                  +600                  +620
TGTGCCGAAAAATAAAAATAAACCATAAACTAGGCAGCGCTGCCGTCGCCGGCTGAGCAGCCTGCGTACATAGCCGAGATCGCGTAACGGTAGATAATGA
                                                                                 LARVAL RNA
 +640                 +660                                        +700           →
AAAGCTCTACGTAACCGAAGCTTCTGCTGTACGGATCTTCC[TATAAATA]CGGGGCCGACACGAACTGGAAACCAACAACTAACGGCGCCCTCTTCCAATT
 +740                 +760                  +780                  +800                  +820
GAAACAG]ATCGAAAGAGCCTGCTAAAGCAAAAAAGAAGTCACCATGTCGTTTACTTTGACCAACAAGAACGTGATTTTCGTTGCCAGTCTGGGAGGCATT
                                                  Int Ser Phe Thr Leu Thr Asn Lys Asn Val Ile Phe Val Ala Gly Leu Gly Gly Ile
```

Figure 15.17. DNA sequence flanking the *Adh* gene. The start of the adult ADH mRNA is designated nucleotide +1. The start of the larval ADH mRNA occurs at nucleotide +708. Both are preceded by TATA boxes about 30 bases upstream. The adult mRNA intron that is excised is marked at nucleotide +87 and +741. Translation starts at +778 on both the adult and larval mRNA.

but the final step requires understanding of techniques only recently developed, that is, P-element mediated transformation.

It has been known for some time that certain races of *Drosophila melanogaster* (P strains) give rise to small numbers of offspring when mated to other races (M strains). Even the few progeny that result from such crosses show an abnormally high frequency of mutations. This hybrid dysgenesis results only when the male in the cross is a P strain and the female is an M strain. In reciprocal crosses (M strain male by P strain female) the brood size is normal.

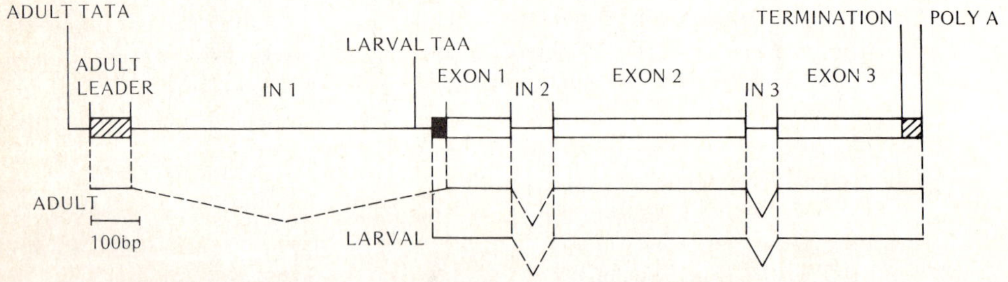

Figure 15.18. Adult and larval transcripts of *Drosophila Adh* gene. Intron 1 (in1) is only transcribed in the adult. It is excised from the transcript to make the adult ADH mRNA. Introns 2 and 3 are excised from both adult and larval transcripts. Translation starts at the same position in both adult and larval mRNA.

Hybrid dysgenesis results from a sequence carried in multiple copies in P strains referred to as the P element. When sperm carrying P elements enter eggs of M strains, the element is mobilized and inserts into a wide range of sites in the genome of the zygote. Many of these sites are in the middle of functional genes and so the inserted P element causes mutations by disrupting the genes. Some of these mutations are lethal and thus reduce the brood size. Others are recessive or innocuous and so give rise to mutant but viable offspring. Within a generation the P element stabilizes and no further mutants arise. By comparison with transposons in bacteria, it has been suggested that P elements code for a specific recombinational enzyme (transposase) that inserts them at sites in the genome. They are also thought to code for a system that represses these recombinational enzymes. When a P sperm first enters an M egg, the repressor is absent because M strains do not carry the P elements. P elements are inserted here and there until the repressor builds up and stabilizes the situation. When eggs of such an individual are fertilized by P sperm, the repressor keeps the P elements in check. A similar process results in the movement of unstable elements such as *Ac/Ds* in maize. These mobile elements cause mutations but are not known to be directly involved in development. However, they are exceptionally useful for analysis of genes in development.

The *Adh* gene, along with 6 kb of genomic DNA on the 5′ side, was placed in the middle of a P element carried on a bacterial plasmid. This construct was microinjected along with an intact cloned P element into very early embryos, which were genetically *Adh* negative. The injected DNA was incorporated into the pole cells, which then went on to differentiate into germ cells. Adults from injected embryos were mated to null *Adh* flies individually and the progeny were selected for ethanol resistance. Several independent lines were found in which the transformed *Adh* gene was stably integrated into the genome.

Three transformants carried the transformed gene on chromosome 2 but all had integrated it at positions other than the chromosomal location of the resident null *Adh* gene on the left arm of chromosome 2. Two lines integrated *Adh* in the X chromosome and one line integrated it in the third chromosome. The *Adh*-specific activity of five of these lines was identical to that of untransformed Adh^+ flies in both the larval and adult stages, indicating that the chromosomal location has little or no effect on expression and that all essential *cis*-acting elements required for transcription are contained in the 6 kb 5′ flanking sequences. The tissue specificity of ADH activity was also retained in the transformed lines (see Table 15.2). Thus, *cis*-acting controls for tissue specificity are also present on the cloned fragments and are not affected by the surrounding chromosomal environment.

The *Adh* transcripts were isolated from both larvae and adults of one of the transformed lines. The larval transcript was shown to initiate 70 bases upstream of the translated sequence while the adult transcript was shown to initiate 777 bases upstream of the translated sequence just as occurs in transcription of the wild-type *Adh* gene. Therefore, it can be concluded that the chromosomal position does not affect the developmental switch in promoters and that the *cis*-acting signals are carried on the cloned fragment.

Table 15.2 Histochemical Staining of Transformed Larvae and Adults

Strains	Larval ADH Distribution										
	Salivary Glands	Foregut	Brain	Imaginal Discs	Fat Body	Anterior Midgut	Somatic Muscles	Posterior Midgut	Malpighian Tubules	Hindgut	Hypoderm
Adh^{fn23}	−	−	−	−	−	−	−	−	−	−	−
Adh^{F}	−	−	−	−	+	+	+	+	+	−	−
tAP-1	−	−	−	−	+	+	+	+	+	−	−
tAP-3	−	−	−	−	+	+	+	+	±	−	−
tAP-5	−	−	−	−	+	+	+	+	+	−	−
tAP-6	−	−	−	−	+	+	+	+	+	−	−

	Adult ADH Distribution							
	Crop	Midgut	Fat Body	Somatic Muscles	Malpighian Tubules	Rectum	Testes and Paragonia	Seminal Vesicles and Vas Deferens
Adh^{fn23}	−	−	−	−	−	−	−	−
Adh^{F}	+	+	+	−	+	+	−	+
tAP-1	+	+	+	−	+	+	−	+
tAP-3	+	+	+	−	+	+	−	+
tAP-5	+	+	+	−	+	+	−	+
tAP-6	+	+	+	−	+	+	−	+

**Adh*^{fn23} and *Adh*^{F} are strains of *Drosophila melanogaster* with marked *Adh* genes. The tAP strains carry transformed *Adh* genes at various positions in the genome. Histochemical staining of tissues was positive (+) in those organs with ADH activity and negative (−) in those in which the *Adh* gene was not expressed.

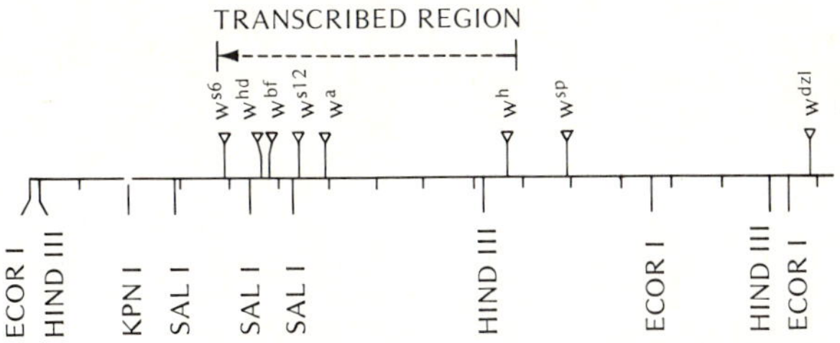

Figure 15.19. Mutations affecting eye pigmentation in *Drosophila*. The *white* locus is defined by a series of mutations that affect more than 12 kb of DNA, 6 kb of which is transcribed. The sequence can be recognized by the arrangement of restriction sites. The restriction enzymes that cut at various sites are shown. Mutation W^{dzl} is an insertion of a transposon that results in the lack of transcription of *white* in tissues of the head.

Paring down of the cloned *Adh* will indicate the recognition sequences involved. The search for *trans*-acting regulatory components will lead into tissue specificity as well as temporal alteration.

The first mutation analyzed in *Drosophila* affected eye color and fell in the *white (w)* locus. Wild-type flies have deep red eyes. However, white-eyed flies are just as healthy in the laboratory. Thus, the *white* gene is amenable to study. A cloned sequence of 11.7 kb was inserted into a P element and transformed into host embryos from a stock carrying a deletion of the *white* locus (Fig. 15.19).

Adults from the transformed embryos were mated to w^- flies and the red-eyed progeny collected. The *white* locus is normally found on the X chromosome, where it is subject to dosage compensation so that females (XX) with two copies make about as much pigment as males (XY) with only one copy. The transformed *white* gene landed on the second and third chromosomes as well as on the X and yet was dosage compensated even on the autosomes. The *cis*-acting sequences required for dosage compensation of this normally sex-linked gene appear to have been carried within the 3 kb segment on the 5′ end of the gene. Although this is the only X-linked gene studied so far in this detail, it would appear that dosage compensation is regulated on a gene-by-gene basis.

Dopa Decarboxylase Gene

Drosophila homozygous for mutations in the *Ddc* locus of the second chromosome have pupae of a distinctive yellow-green that can be readily distinguished from the reddish brown pupae normally seen. The wild-type *Ddc* locus codes for an enzyme, dopa decarboxylase, that catalyzes a reaction leading to compounds that pigment and cross-link the cuticle. The gene is expressed almost exclusively in cells of the hypoderm and thus is subject to tissue-specific regulation. The enzyme first appears when the embryos hatch. The specific activity rises again to a peak during pupariation when the pupal case is laid down, then falls until it rises to a second peak at eclosion just before the adult flies emerge (Fig. 15.20).

Transcription of the *Ddc* gene covers a 4 kb sequence and the resulting transcript is processed to a 2.1 kb and a 2.3 kb mRNA by splicing out two

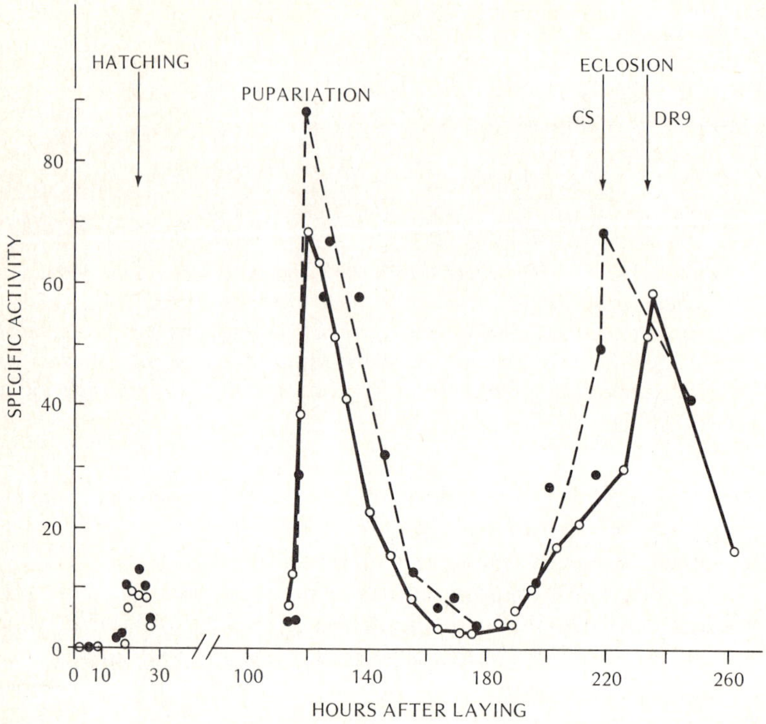

Figure 15.20. Stage-specific expression of dopa decarboxylase. At various times during synchronous development of wild-type Canton S (●—●) and transformant DR9 (o—o) samples were taken for determination of DDC specific activity. Between 10 and 20 animals were collected at each point. Eclosion of DR9 flies occurred slightly later than that of Canton S (CS) flies, as indicated.

introns near the 5′ end. The end point of the first intron that is spliced out is 200 bp further upstream for the 2.3 kb mRNA than for the 2.1 kb mRNA. The synthesis of dopa decarboxylase at pupariation and again at eclosion is the result of accumulation of the 2.1 kb and 2.3 kb mRNA from *Ddc* respectively at these developmental stages. Thus, this gene is subject to dramatic temporal regulation of transcription as well as cell-type regulation. It is of interest to see what is required for these control processes.

A fragment of the wild-type genome cut with the restriction enzyme, Pst, was cloned into the P element transposon and shown to carry 2.5 kb of DNA upstream of the 5′ end of the *Ddc* gene as well as the gene itself and 1 kb of 3′ flanking DNA. A mutant strain of *Drosophila* with altered dopa decarboxylase (Ddc^{ts2}) was transformed with this P element and progeny with reddish-brown pupae isolated. The transformed gene was found to have integrated at various positions on the chromosomes of different transformants although only a single *Ddc* gene was integrated in each line. One of the lines (DR9) had integrated the gene on the right arm of the second chromosome on the other side of the centromere from the mutated endogenous *Ddc* gene. Nevertheless, the temporal pattern of expression of the transformed *Ddc* gene was essentially identical to that found in wild-type flies (Fig. 15.19). Moreover, dopa decaboxylase was found almost exclusively in cells of the hypoderm. These results show that all *cis*-acting regulatory components lie within 2.5 kb of the *Ddc* gene. Further pairing down of the gene showed that sequences between −200 bp and +400 bp of the transcriptional start site were sufficient for both temporal and cell-type-specific regulation.

Related Readings

Beachy, P., Helfand, S., and Hogness, D. (1985). Segmental distribution of bithorax complex proteins during *Drosophila* development. Nature *313,* 545–551.

Bender, W., Akam, W., Karch, F., Beachy, P., Peifer, M., Spierer, P., Lewis, E., and Hogness, D. (1983). Molecular genetics of the bithorax complex in *Drosophila melanogaster.* Science *221,* 23–29.

Benyajati, C., Spoerel, N., Haymerle, H., and Ashburner, M. (1983). The messenger RNA for alcohol dehydrogenase in *Drosophila melanogaster* differs in its 5′ end in different developmental stages. Cell *33,* 125–133.

Colberg-Poley, A., Voss, S, Chowdhury, K., and Gross, P. (1985). Structural analysis of murine genes containing homeo box sequences and their expression in embryonal carcinoma cells. Nature *314,* 713–718.

Doctor, J., Fristom, D., and Fristom, J. (1985). The pupal cuticle of *Drosophila:* biphasic synthesis of pupal cuticle proteins *in vivo* and *in vitro* in response to 20-hydroxyecdysone. J. Cell Biol. *101,* 189–200.

Goldberg, D., Posakony, J., and Maniatis, T. (1983). Correct developmental expression of a cloned alcohol dehydrogenase gene transduced into the *Drosophila* cell line. Cell *34,* 59–73.

Karess, R., and Rubin, G. (1984). Analysis of P transposable element function in *Drosophila.* Cell *38,* 135–146.

Kaufman, S. (1971). Gene regulation networks: a theory. Curr. Topics Devel. Biol. *6,* 145–182.

Kaufman, S., Shymke, R., and Trabert, K. (1978). Control of sequential compartment formation in *Drosophila.* Science *199,* 259–270.

Kornberg, T., Siden, I., O'Farrell, P., and Simon, M. (1985). The *engrailed* locus of *Drosophila:* In situ localization of transcripts reveal compartment-specific expression. Cell *40,* 45–53.

Lawrence, P., and Morata, G. (1983). The elements of the bithorax complex. Cell *35,* 595–601.

Lewis, E. (1978). A gene complex coordinating segmentation in *Drosophila.* Nature *276,* 565–570.

McGinnis, W., Hart, C., Gehring, W., and Ruddle, F. (1984). Molecular cloning and chromosomal mapping of a mouse DNA sequence homologous to homeotic genes of *Drosophila.* Cell *38,* 675–680.

Nusslein-Volhard, C. (1979). Maternal effect mutations that alter the spatial coordinates of the embryo of *Drosophila melanogaster.* Symp. Soc. Devel. Biol. *37,* 185–211.

O'Brochta, D., and Bryant, P. (1985). A zone of non-proliferating cells at a lineage restriction boundary in *Drosophila.* Nature *313,* 138–141.

Scholnick, S., Morgan, B., and Hirsch, J. (1983). The cloned dopa decarboxylase gene is developmentally regulated when reintegrated into the *Drosophila* genome. Cell *34,* 37–45.

Scott, M., Weiner, A., Hazelrigg, T., Polisky, B., Pirrotta, V., Scalenghe, F., and Kaufman. T. (1983). The molecular organization of the *antennapedia* locus of *Drosophila.* Cell *35,* 763–776.

Struhl, G., and Brower, D. (1982). Early role of the esc^{+} gene product in determination of segments in *Drosophila.* Cell *31,* 285–292.

Turner, F. R., and Mahowald, A. P. (1977). Scanning electron microscopy of *Drosophila melanogaster* embryogenesis. Devel. Biol. *57,* 403–416.

Weir, M., and Lo, C. (1982). Gap junctional communication compartments in the *Drosophila* wing disc. Proc. Natl. Acad. Sci. *79,* 3232–3235.

Weir, M., and Lo, C. (1985). An anterior/posterior communication compartment border in *engrailed* wing discs: possible implications for *Drosophila* pattern formation. Devel. Biol. *110,* 84–93.

White, R., and Wilcox, M. (1984). Protein products of the bithorax complex in *Drosophila.* Cell *39,* 163–171.

CHAPTER 16

Development of Nematodes

The ultimate fate map follows each cell from the zygote to the individual differentiated tissues of the adult and gives the time of division and movements of the intermediate cells along the way. This had recently been done for the free-living nematode, *Caenorhabditis elegans*. The adult worm contains only 808 somatic cells and is sufficiently transparent that interference-phase microscopy (Nomarski optics) can be used to follow each nucleus from its birth during embryogenesis to its place in the adult. Up to nine cell generations lead from the zygote to the hatched larva (L1), and the ancestors of each differentiated cell have been traced and named. Interestingly, many cells die during embryogenesis and give rise to no progeny cells.

The life of a *C. elegans* nematode is short and fast. Fourteen hours after fertilization of an egg, the embryo hatches as a feeding larva. The larva grows and passes through four molts in the next day and a half. The worm that emerges from the fourth molt is fully active sexually and starts laying fertilized eggs within an hour. A single nematode will produce about 250 progeny before the cessation of egg laying four days later (Fig. 16.1).

C. elegans is a self-fertilizing hermaphrodite; individuals produce both eggs and sperm. The sperm are stored in a specialized structure, the spermatheca, through which the eggs must pass. Eggs are fertilized in this structure and immediately begin development. Initially the cells divide every 20 to 30 minutes; gastrulation occurs at between 2 and 3 hours of develop-

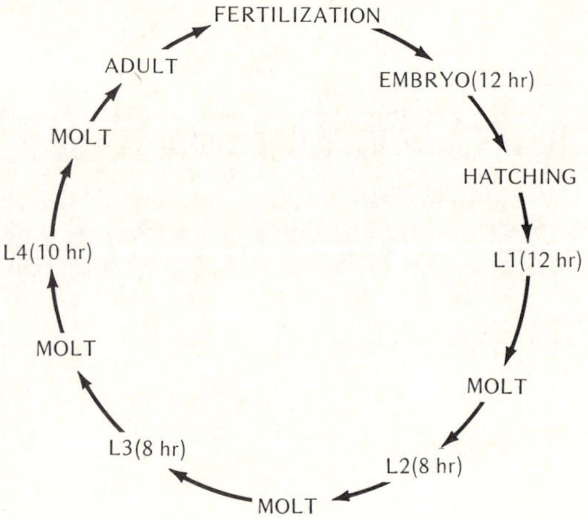

Figure 16.1. Life cycle of *Caenorhabditis elegans*. Embryogenesis takes only about 12 hours. The larva hatches and molts 12 hours later to give the L2 form. L2 molts to give an L3. The L3 gives rise to the L4 larva, which molts to finally emerge as an adult. Fertilization in hermaphrodites occurs when the eggs pass through the spermatheca.

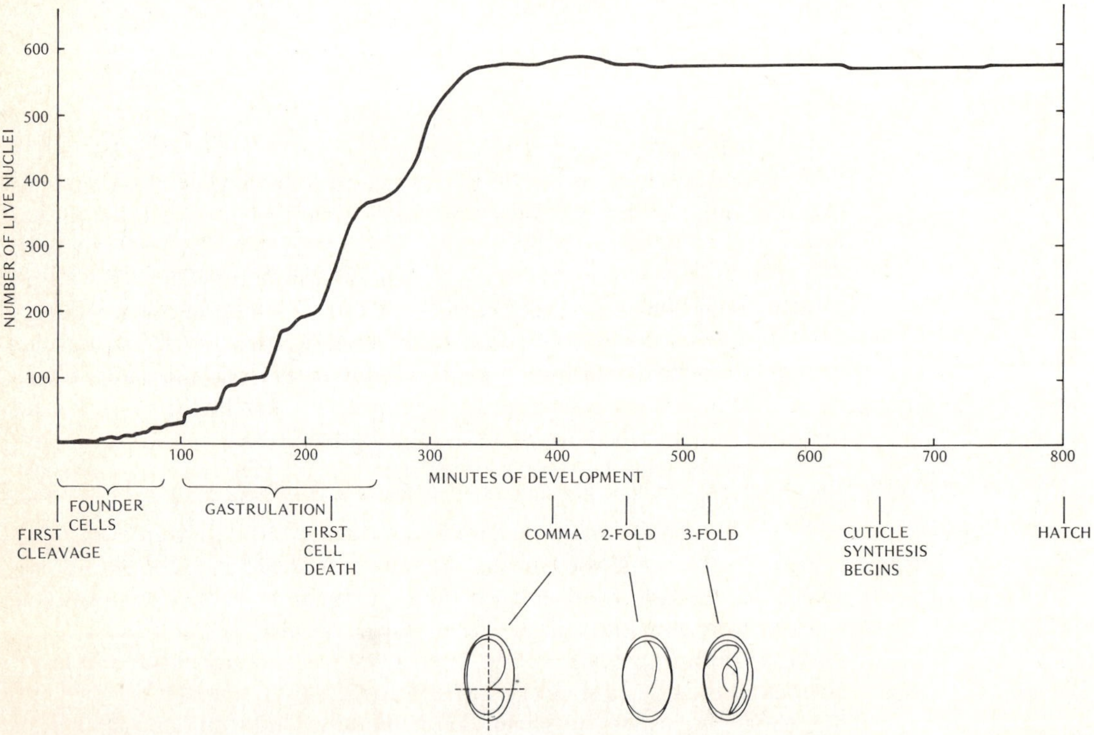

Figure 16.2. Number of living nuclei during embryogenesis of *C. elegans*. The first cleavage occurs 50 minutes after fertilization. Gastrulation occurs when there are about 100 cells. Most cell division has occurred by the comma stage at 400 minutes after the first cleavage. The embryo hatches at 800 minutes.

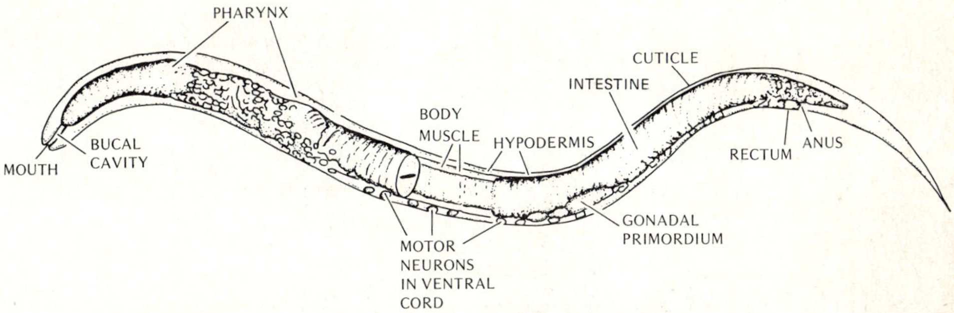

Figure 16.3. Anatomy of the L1 larva. The worm at this stage is not much more than a tube from the mouth to the anus. However, every cell is specifically positioned in exactly the same place in every wild-type individual. Muscle, intestine, and nerve cells have differentiated. The adult gonad will result from dramatic growth of the gonadal primordium. In hermaphrodites a vulva and associated muscles will develop.

ment. Exactly 671 cells are formed during embryogenesis of which 113 cells die, leaving exactly 558 cells in the L1 larva that hatches (Fig. 16.2).

Some of the larval cells continue to divide during growth through the L2, L3, and L4 larval forms before the final adult forms. These 55 cells are referred to as blast cells, and most give rise to the hypodermis that covers the worm.

At a frequency of about 1 in 700, nondisjunction of X chromosomes in germ cells reduces the normal hermaphrodite diploid to an XO male. These worms produce only sperm and no eggs. They also differentiate specialized structures near the tail to deposit sperm in the spermatheca of hermaphrodites. The L1 larval form of males is almost identical to that of hermaphrodites (Fig. 16.3) but two fewer cells die during embryogenesis leaving 560 intact cells. Adult males have 72 more somatic cells than hermaphrodites.

Adults also contain germ cells that produce the eggs and the sperm. These are derived from two germ cell blasts present in the L1 larva.

Cleavage

The first few divisions of the zygote are asymmetric, producing a larger cell at the anterior than at the posterior. Subsequent division planes are not orthogonal as in sea urchin embryos and so the arrangement of the cells is more complex. However, the cleavage pattern is just as regular as in other embryos occurring in an invariant manner in every zygote. There are 22 known maternal effect genes that, when mutated, affect the timing or positioning of the early cleavage planes. These genes must function in the hermaphordite during oogenesis to give normal cleavage patterns. In only one case could sperm from a heterozygous male supply the missing function. This is somewhat surprising since in *C. elegans* the sperm are large cells and sup-

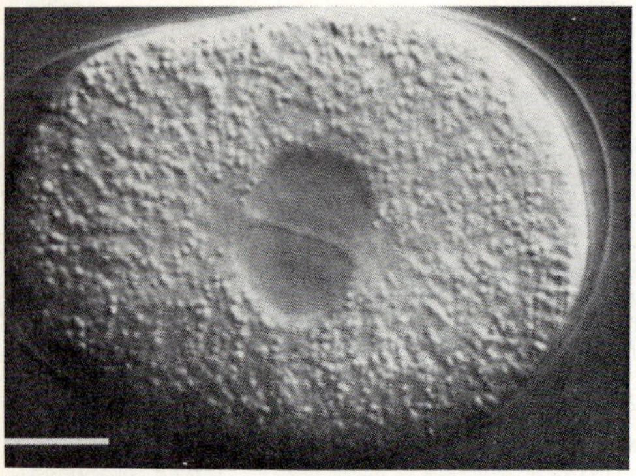

(A)

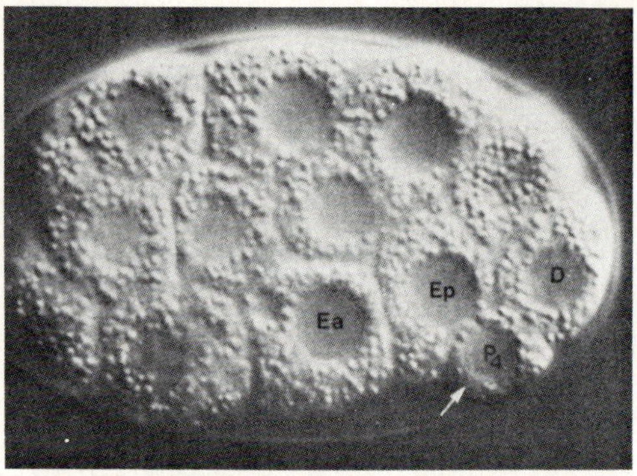

(B)

Figure 16.4. Asymmetric cleavage in *C. elegans*. *(A)* Even the first division produces unequal-sized cells: the larger AB cell and the smaller P1 cell. *(B)* By the 16-cell stage the germ line progeny of P1 cell, P4, is smaller than the rest of the embryonic cells. However, cleavage stops after one more division of this small cell. The progeny of P4, Z2 and Z3, do not divide again until after hatching when they generate all of the gametes. The progenitor of Z2 and Z3, P4, is marked by an arrow.

ply a considerable amount of cytoplasm to the zygote. Nevertheless, it appears that the pattern of early cleavage is set up in the egg before fertilization.

After four divisions one of the cells is considerably smaller than the other cells (Fig. 16.4). This cell, referred to as P4, will divide again in about an hour to give the two germ cell blasts, Z2 and Z3, that will not divide again during embryogenesis. During postembryonic development in the larva they will divide repeatedly to give rise to eggs and sperm.

P-granules

A specific antigen has been recognized by monoclonal antibody that is localized exclusively to the germ line cells. It is found in Z2 and Z3 and later, in adults, in sperm and eggs (Fig. 16.5). The antigen is found in structures referred to as P-granules. They are dispersed in unfertilized eggs but move to the posterior pole before the first division (Fig. 16.5). Their localization is inhibited by treatment of zygotes with cytochalasins B and D, indicating that microfilaments play an essential role in their movement. At the two-cell stage P-granules are seen only in cell P1, where they again concentrate in the posterior before the second cleavage. At each subsequent division P-granules are found exclusively in cells P2, P3, and P4 and finally reside in the two daughter cells of P4, Z2 and Z3. They uniquely mark the ancestors of the germ cells. It is not yet known what P-granules might do. They appear to surround the nucleus in a cloud and are reminiscent of the polar granules associated with the portion of the *Drosophila* egg that in-

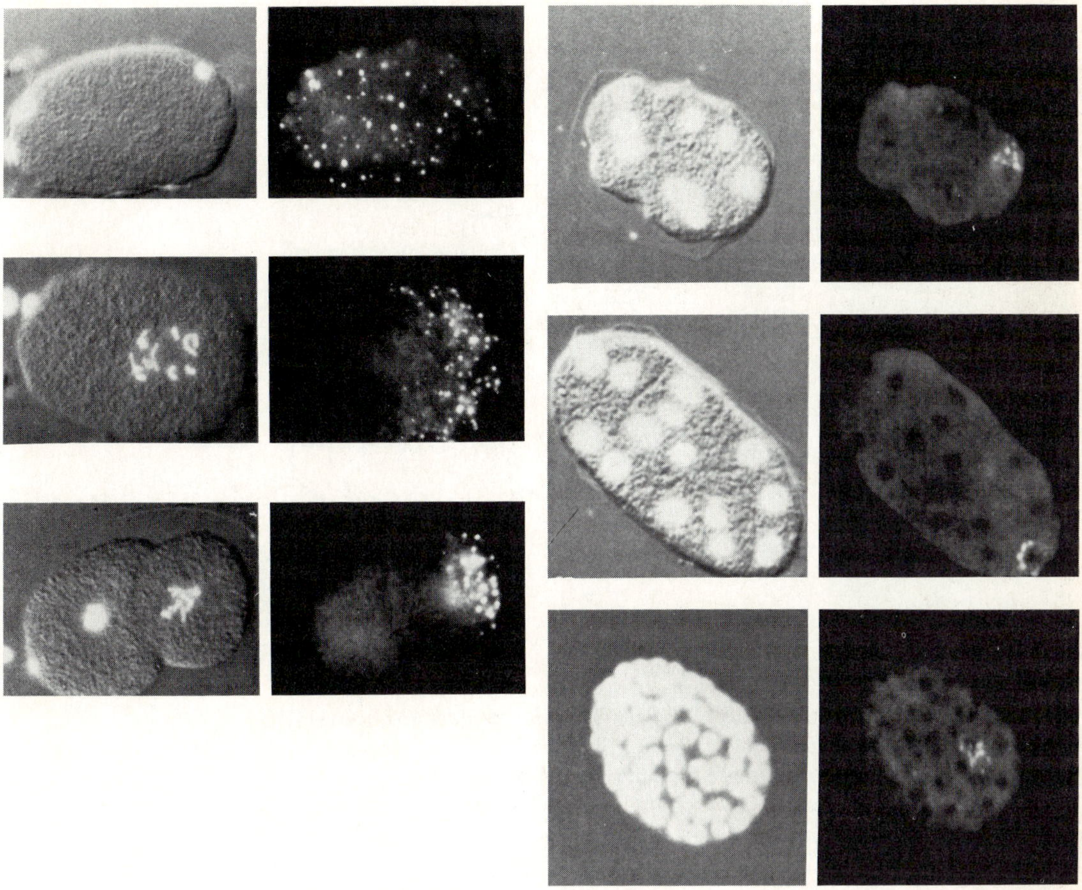

Figure 16.5. Localization of P-granules to the germ line. P-granules can be recognized by fluorescent-coupled antibodies specific to these nuage structures. Even before the first division, P-granules are segregated to the side that will form cell P1. By the 6-cell stage the P2 cell is in early prophase and P-granules are in the ventral region of this cell. By the 15-cell stage P-granules surround the nucleus of cell P3. In 100-cell embryos only cells Z2 and Z3 have accumulated P-granules. Nuclear staining on the left; P-granules staining on the right.

structs nuclei toward the developmental pathway to germ cell differentiation. Discovering whether or not P-granules are instructive as well as indicative will require considerable more analysis.

Founder Cells

The other cells produced by the first four divisions have been referred to as founder cells because they give rise predominantly to specific cell types (Fig. 16.6).

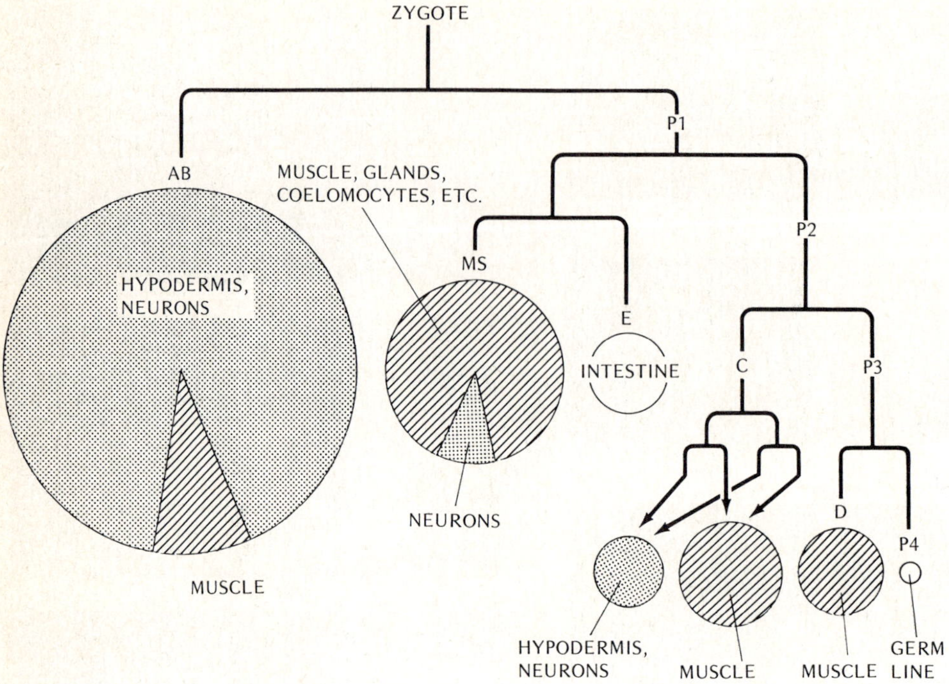

Figure 16.6. Founder cells of *C. elegans*. The cell types derived from six early embryonic cells are presented proportionally. The hypodermis is formed from descendants of cell AB generated at the first division. This cell also gives rise to neurons and a small proportion of muscle cells. Cell MS generated at the third division gives rise to muscle cells and neurons. Cell E makes cells of the intestine. Cell P4 gives rise to the germ line. Typically mesodermal tissue is striped.

The anterior cell produced at the first cleavage, referred to as AB, differentiates predominantly into surface hypodermal cells and neurons, although some muscle cells are derived from this cell. Some neurons and hypodermal cells can also trace their lineage to cells MS and C formed at the third and fourth division. This points out that a given cell type does not always descend from the same progenitor. Moreover, some neurons are sister cells to muscle cells.

Besides the unique heritage of germ cells from P4, there are two cells produced early in embryogenesis that give rise to a single tissue: cell E, produced at the third division, gives rise to all of the intestinal cells and nothing else, whereas cell D, produced as a sister to P4 at the fourth division, gives rise only to muscle cells.

The E cell lineage gives rise to 20 cells during embryogenesis (Fig. 16.7). During postembryonic development most of the nuclei of these cells divide to form the large syncytial cells of the larval and adult intestine. The progeny of E are arranged in a row running from anterior to posterior along the embryonic axis connecting the pharynx to the anus (Fig. 16.8). The germ cell precursors Z2 and Z3 extend lobes between these intestinal cells and may be nourished by them during embryogenesis. At hatching Z2 and Z3

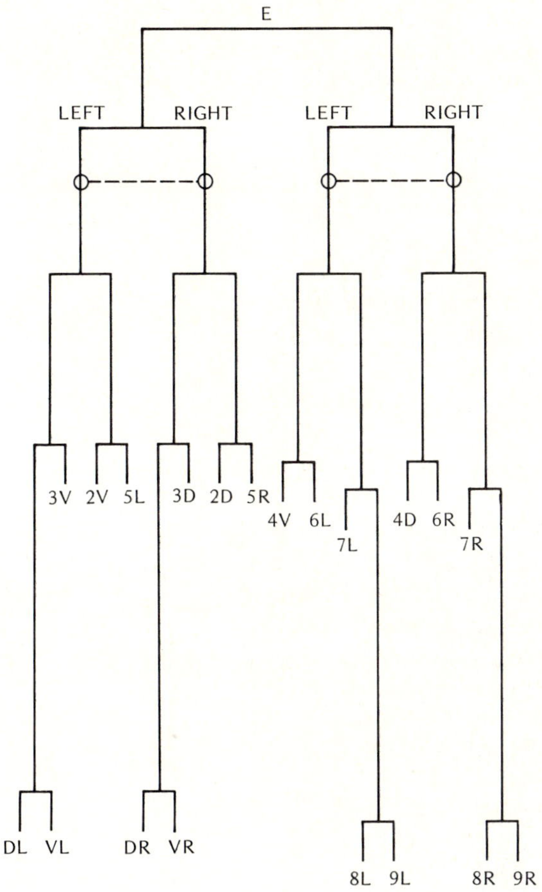

Figure 16.7. E cell lineage. Each cell can be unequivocally recognized and has been named. These cells all line the intestine. Notice the symmetry of the lineages between the cells connected by dashed lines and the mirror image of the division pattern between the two lines derived from the division of cell E.

retract their lobes and are nurtured by two somatic cells, Z1 and Z4, descendents of founder cell MS.

The time of cell division of cell E and its progeny during embryogenesis is invariant. It can be predicted within a few minutes in any embryo. Mutations are known that affect the timing, indicating strict genetic control. The orientation of each cleavage plane is also invariant. Most divisions separate daughter cells into an anterior one and a posterior one, but the di-

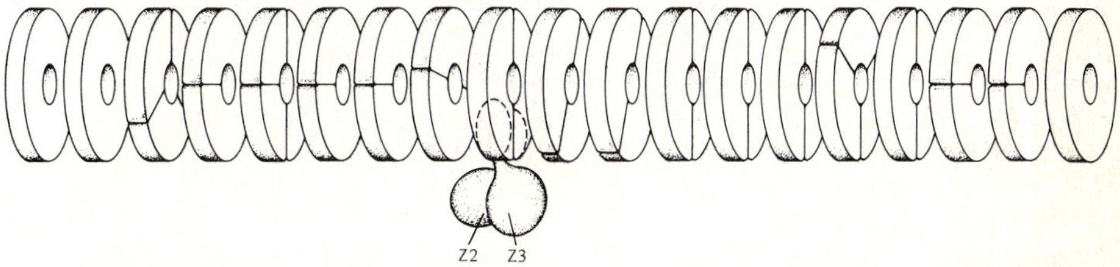

Figure 16.8. Alimentary tract of *C. elegans.* This drawing is of an embryo at 430 minutes but the arrangement is similar in L1 larva except that the lobes of Z2 and Z3 are retracted as cells Z1 and Z4 come up.

visions of the two daughter cells of E are in the left/right axis. The last embryonic division of the most anterior descendent of E is dorsal/ventral. That of a homologous second-cousin cell is also dorsal/ventral. All other cells divide anterior/posterior.

The differentiation of each cell is also invariant. The most anterior cells make up the dorsal and ventral portions of the alimentary tract (int 1D and V). The next cell will become the ventral portion of the third unit (int 3V). The arrangement of the larval cells roughly follows the pattern of anterior/posterior divisions but cell movements occur during embryogenesis to establish the final positions of many cells. No simple logic can be seen in the pattern, suggesting that it has been tinkered with during evolution.

Rhabditin

After cell E has divided three times to give rise to eight gut cells (190 minutes after the first division) catabolism of tryptophan in these cells gives rise to fluorescent granules referred to as rhabditin. These granules make convenient markers of the E cell descendents since they can be easily seen in a fluorescence microscope. These granules appear in no other cell. When cleavage is arrested by adding cytochalasin B and colchicine at the eight-cell stage (70 minutes after the first division) only cells ABal, ABar, ABpl, ABpr, MS, E, C, and P3 are present. No further division takes place. About two hours later cell E accumulates rhabditin granules. Clearly, division of cell E is not necessary to trigger rhabditin formation. The E cell is already determined to produce rhabditin and this property is restricted to the E cell.

When cleavage is blocked at the four-cell stage only cells ABa, ABp, EMS, and P2 are made. Rhabditin accumulates two and a half hours later in cell EMS, the progenitor of cells E and MS (Fig. 16.9).

Two-cell embryos blocked from further cleavage accumulate rhabditin only in the posterior cell, P1, that would have given rise to the E cell lin-

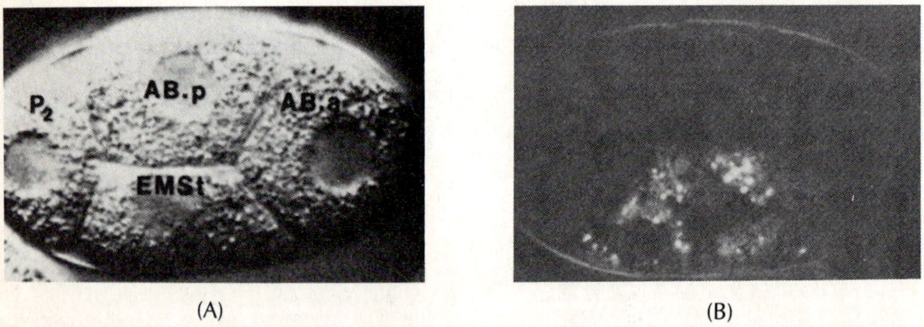

Figure 16.9. Four-cell embryo of *C. elegans.* *(A)* If cleavage is blocked at this stage, rhabditin accumulates in cell EMS two and a half hours later. Normally rhabditin accumulates only in the progeny of cell E. *(B)* Rhabditin can be seen by fluorescence.

eage. Thus, right at the first cleavage, processes are set in train that will result in this specialized metabolism of tryptophan. These instructions are localized sequentially at each division to occur uniquely in P1, EMS, E, and all its descendents. Both these instructions and the P-granules are localized in P1 but at the next division the rhabditin instructions go to daughter cell EMS while the P-granule instructions go to daughter cell P2.

The ability of cell E and its descendents to accumulate rhabditin does not seem to depend on interaction with other cells in the embryo. Many of the other cells can be killed by laser ablation or broken by applying pressure, and the E cell will nevertheless accumulate rhabditin. The determinants necessary for this differentiation are localized and tightly associated with the E cell.

D Cell Lineage

When P3 divides at 90 minutes of development it gives rise to cells D and P4. The D cell divides to form 20 cells of the body muscle (Fig. 16.10).

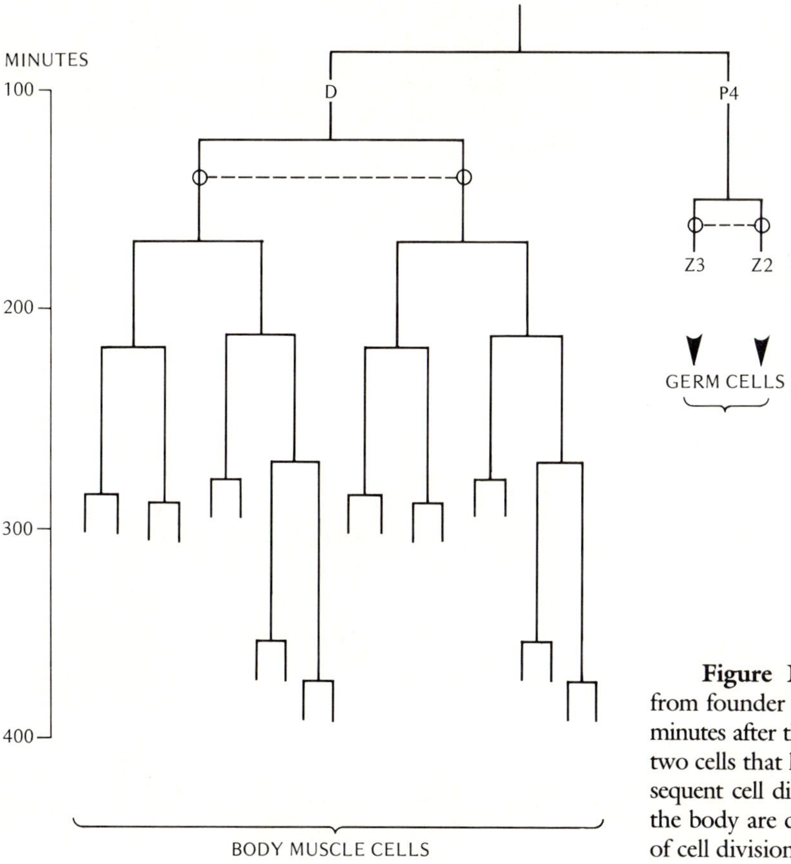

Figure 16.10. Muscle cells derived from founder cell D. The D cell is born 90 minutes after the first division. It divides into two cells that have identical patterns of subsequent cell division. Twenty muscle cells of the body are derived from cell D. The time of cell division is the same in every embryo.

As in the E cell lineage the time and orientation of each of the divisions is invariant. The patterns of cleavage of the two daughter cells derived from cell D are exactly the same with respect to both time and place. It is as if the instructions given to each daughter cell were identical. These mesodermal cells all differentiate into the muscles of the larva. If either daughter cell is killed by laser ablation, the muscles derived from that cell are missing in the larva. The embryo does not regulate to supply the missing cells from other lineages. For this reason it is said that development is strictly mosaic.

Gastrulation

Individual cells enter the interior to form the endoderm and mesoderm. The first cells to do so are the progenitors of the intestine, Ea and Ep. They sink inwards from the ventral side, near the posterior end of the embryo about 100 minutes into embryogenesis (Fig. 16.5). No clear blastopore is seen nor do the cells move as a sheet. Since the fate of each embryonic cell is set, there is no need for tissue interaction or polarity to be established.

The entry zone lengthens as cells P4, the progeny of MS, and myoblasts derived from C and D migrate toward the center of the embryo, which now contains about 200 cells. During the next hour cells that will form the pharynx (derived from cell AB) enter through the ventral cleft. The cleft then closes over. All but two of the myoblasts enter during gastrulation. Surprisingly, two myoblasts (ABprppppa and ABplppppa) do not enter until their terminal division at 5 hours of development, almost as it they were afterthoughts. The movements of these cells into the embryo and thereafter are also invariant from one embryo to the next.

Mutations that affect gastrulation in *C. elegans* have been found. They alter the timing of the initial cell division by only 10 to 20 percent but have dramatic consequences at gastrulation. It appears that minor early perturbations are amplified at subsequent steps to throw embryogenesis off course.

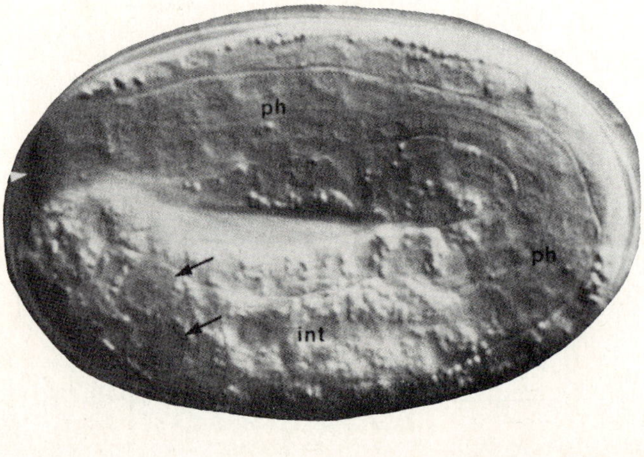

Figure 16.11. Twofold embryo of *C. elegans*. Five hundred minutes after fertilization the embryo is doubled over. The pharynx (ph) and intestine (int) can be seen. Cells Z2 and Z3 are marked by arrows. These germ line cells ceased dividing 5 hours earlier and so are relatively large.

By 7 hours of development almost all embryonic cell divisions have occurred. Programmed cell death winnows out unneeded cells, which are then rapidly engulfed, usually by their sister cell. Cells destined to be phagocytosed are born with very little cytoplasm and die without differentiating. It can be thought of as a way to eliminate the previous cell division. Perhaps during evolution a mechanism arose to eliminate a cell before a mechanism evolved that could block the division.

At this stage the embryo has elongated to such an extent that it cannot lie straight in its shell; it curls its tail to take on a characteristic shape referred to as comma. Subsequent elongation of the embryo doubles it up on itself. The germ cells Z2 and Z3 can be seen as large prominent cells at this time (Fig. 16.11).

Cuticle

At 11 hours of development the hypodermal cells derived from founder cells AB and C cover the surface of the embryo and start to secrete the tough outer cuticle that protects the larva from dessication and provides an exoskeleton. The cuticle is constructed from proteins homologous to collagen and elastin. So far 36 genes have been recognized that, when mutated, result in abnormalities in cuticle anatomy. Mutations in these genes give phenotypes named "dumpy," "roller," and "blister."

A few hours later the finished embryo hatches as an L1 larva and proceeds to feed voraciously on available bacteria. Through the next three molts, successively larger cuticles are laid down to cover the growing larvae. These larval cuticles are all very similar, but during the final molt the adult cuticle that is laid down is quite distinct. Mutants that carry recessive mutations (*lin*-14(o)) resulting in premature appearance of adult cuticles in the larval stage have been found. Other mutations (*lin*-14(d)) have been found that result in a replay of the larval program of gene expression at the fourth molt so that sexually mature worms have a larval cuticle. Interestingly, these juvenilizing mutations are dominant. Genes that determine the temporal pattern of gene expression have been referred to as heterochronic, in analogy to the homeotic genes that determine the spatial pattern of gene expression. It has been speculated that evolution of new species often results from mutations in such genes that, on rare occasions, give rise to organisms better adapted to a new strategy for survival.

Regulation

As has been mentioned, an embryo missing a specific cell develops into an embryo missing the tissue derived from that cell. Killing cell P4 results in sterile worms lacking germ cells. Killing either of the daughter cells of D results in larvae missing the body muscle cells that normally arise from that

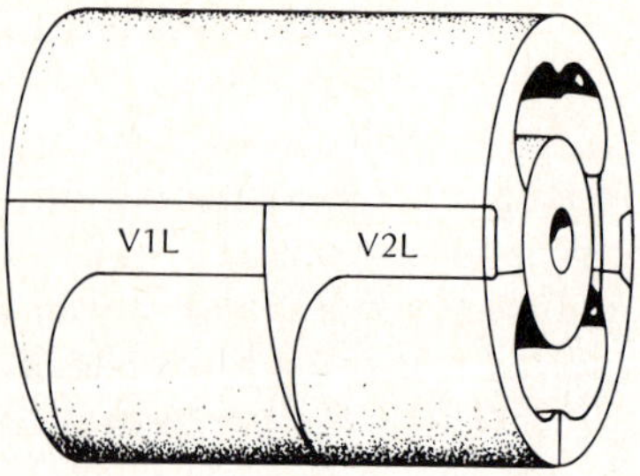

Figure 16.12. Blast cells V1L and V2L. These cells give rise to the lateral cuticular ridges by cell division during postembryonic growth. They are referred to as blast cells because they continue to divide after embryogenesis. If killed they are not replaced by other blast cells or their progeny.

cell. Killing a cell (ABarppa) that arises at the sixth division results in the loss of five hypodermal seam cells as well as several neuronal cells. The seam cells in wild-type larvae remain separate from the rest of the hypodermis and are responsible for making the lateral cuticular ridges (Fig. 16.12)

Hypodermal seam cells such as V1L and V2L normally divide during postembryonic growth. They are therefore referred to as blast cells. Although V1L, V2L, V4L, and V6L, all descendents of the ablated cell (ABarppa), were missing in laser ablated embryos, surviving blast cells behaved normally during postembryonic development and did not replace the missing seam cells. Each blast cell appears to function autonomously.

Only two cases in which the descendents of a killed cell are replaced have been found. In both cases the cell that is replaced normally meets the cell that can replace it at the ventral midline. Although derived from different lineages, these pairs appear to compete for a primary fate. If the first does not get there, the other will take over. These exceptions only emphasize the point that almost every step in embryogenesis of *C. elegans* is hardwired by its genes. The timing and orientation of each division are set, as are the fates of every cell. Even the time of death of certain cells in invariant. It is efficiency taken to an extreme.

Nematodes may have evolved from an organism with embryonic processes less rigidly detailed. The regulation seen in embryos of sea urchins, frogs, chicks, and mammals allows adaptive processes to counter mistakes or accidents. Their development can fine tune a process to redirect random fluctuations back onto the track leading to formation of a normal embryo. But these processes take time. Perhaps the selective pressures that gave rise to the extraordinarily rapid development of a nematode embryo sacrificed adaptability for speed. Genetically programming each division and subsequent differentiation allows each lineage to give rise to its final products autonomously, independent of what is going on around it.

The pattern of cell deaths, cell movements, and in some cases final differentiated states does not give the appearance of a plan designed with the final larval structure in mind. It looks as if selective pressures have resulted in a more competitive form by making minor adjustments on preexisting plans. The changes have come about by tinkering with complex processes that had previously been selected to make the best organism for that time or place. Rather than start over with a newly evolved zygote, the plan was adapted to make minor variations giving selective advantage. The same is probably true of embryonic processes seen in more regulative embryos but cannot yet be so clearly seen.

Summary

Unlike vertebrates, nematodes have highly determined embryos where the fate of almost every cell is irreversibly set as soon as it is born. The timing and orientation of the next division is specified as well as its differentiation or death. The lineage of each cell has been traced and shown to be invariant from individual to individual. The hermaphrodite larva has exactly 558 cells whereas the male larva has exactly 560 cells. They pass through four molts before becoming sexually active adults. Sperm and eggs are all descendents of two sister cells, Z2 and Z3, formed early in embryogenesis. The lineage leading to these cells (P1→P2→P3→P4→Z2, Z3) is marked by localization at each cell division of P-granules. Other cells do not receive these granules and go on to other fates. The E cell produced at the third cell division gives rise to all the cells of the intestine. This lineage is marked by accumulation of rhabditin, a fluorescent product of tryptophan catabolism. The D cell produced at the fourth division produces two lines whose progeny all differentiate into muscle cells. The pattern of timing and orientation of the cell generations in both lines is identical, indicating that each daughter of the D cell received identical instructions. Laser ablation of almost any cell results in larvae missing that cell and all its descendents. There is no filling in as in regulative embryos. Only two cases have been found in which a cell takes over the function of a missing cell.

SPECIFIC EXAMPLE

Postembryonic Development

The hatched larva of *C. elegans* contains 558 cells including two germ cell precursors (Z2 and Z3). During larval development in hermaphrodites 55 blast cells of various lineages divide until there are 959 cells in the adult. The germ cell precursors divide to make several thousand eggs and sperm. The most dramatic change is the growth of the gonads and accessory organs such as the spermatheca and vulva in hermaphrodites (Fig. 16.13).

The blast cells that grow and differentiate during larval development to form the somatic tissue of the gonads are chiefly descendents of founder cell MS (Fig. 16.14).

The embryonic lineages giving rise to blast cells M and *mu int R* are parallel from the left and right daughter cells of founder cell MS as are those of blast cells Z4 and Z1. The patterns of division foreshadow the bilateral symmetry of the gonads in hermaphrodites. However, before larval development starts these four cells undergo invariant migrations to arrange

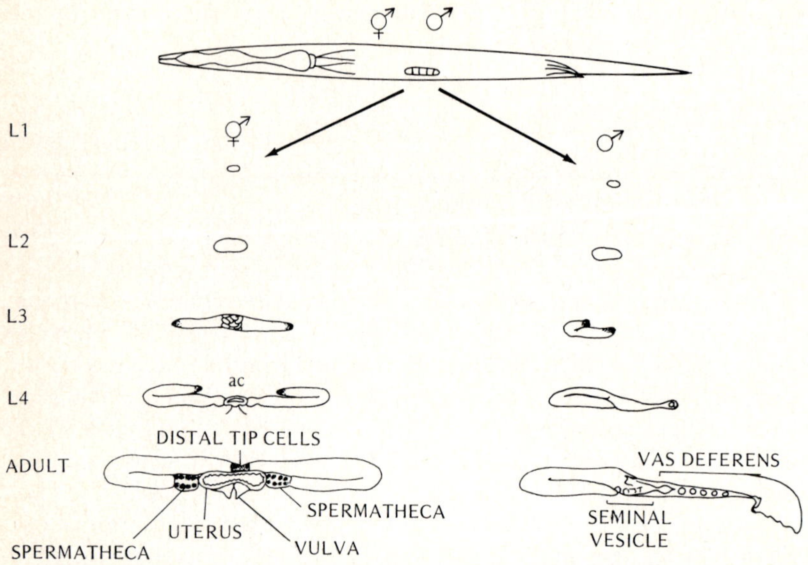

Figure 16.13. Development of the gonads in hermaphrodites and males of *C. elegans*. In the newly hatched worm the gonad is made of only four cells, Z1, Z2, Z3, and Z4. In hermaphrodites (⚥) the gonad elongates both anterioraly and posteriorly during the larval stages. The tip of the gonad reflexes at the L3–L4 molt. In males (♂) the gonad grows only toward the anterior and reflexes at the L2–L3 molt. Oogenesis is dependent on the distal tip cells. The anchor cell (ac) induces the formation of a vulva. Eggs pass through the spermatheca and begin developing in the uterus. In males the vas deferens carries the sperm to the copulatory bursa.

Figure 16.14. Lineage of the gonadal blast cells. Somatic gonadal tissue is derived from cells M, Z4, and Z1, all derived from founder cell MS. Note the similarities in the trees giving rise to these cells. Cell mu int R cannot replace a missing M cell: it forms a muscle of the intestine. The M cell gives rise to muscles of the uterus and vulva. The somatic structures of the gonads are derived from cells Z1 and Z4.

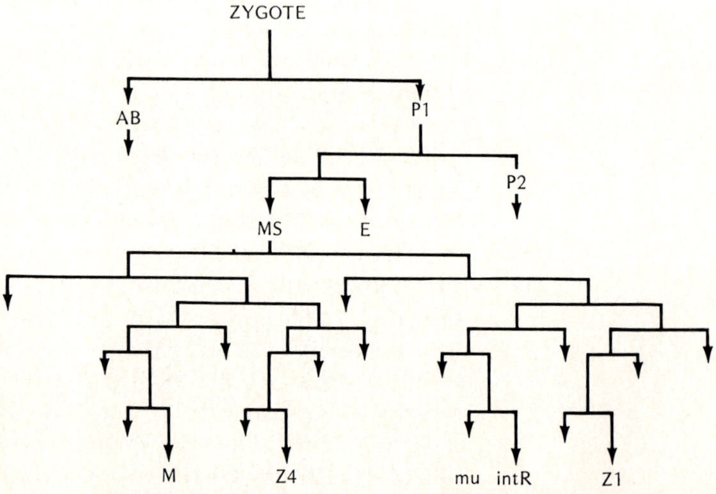

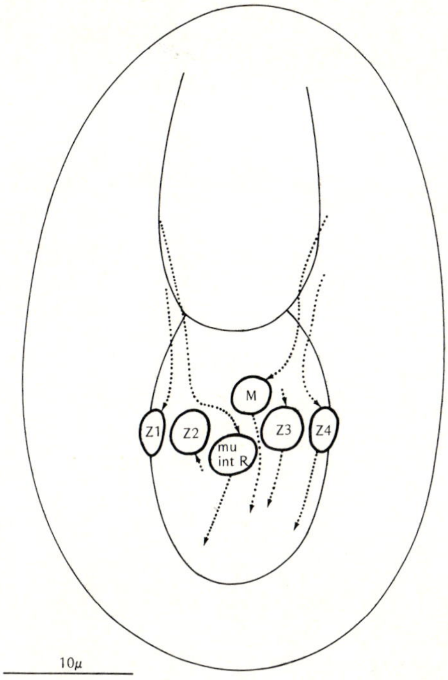

Figure 16.15. Migration of gonadal blast cells. The positions of the nuclei are drawn 320 minutes after the first division. The paths started near the pharynx at 250 minutes and end at 400 minutes. The pathways are invariant. This is a ventral view.

themselves around the germ cell precursors, Z2 and Z3, that are descendents of founder cell P4 (Fig. 16.15). Blast cell M is born on the left, next to the pharynx. It migrates posteriorly along the midline and then shifts to the right-hand side of the intestine. The contralateral homolog of M *(mu int R)* migrates in a similar way but then differentiates into a muscle cell of the intestine. If either cell M or *mu int R* is killed by laser ablation, the remaining homolog does not replace the missing cells in the adult. Their fates are set.

The large M cell gives rise to all the postembryonically derived muscle cells of the body wall as well as muscles of the uterus and vulva. Two descendents differentiate into coelomocytes (Fig. 16.16). The M cell divides in the anterior/posterior plane and each daughter divides left to right to give four cells with bilateral symmetry. Each of these cells divide dorso-laterally twice to generate a total of 16 progeny of the M blast cell. Most of these differentiate into body muscle cells, but two migrate anteriorly and divide further to give rise to the vulval and uterine muscles (Fig. 16.16). However, growth and differentiation of these cells require interactive processes with the germ cell line since laser ablation of the gonadal primordium results in the lack of vulval divisions. In fact, a specific cell of the gonadal line, the anchor cell, is both necessary and sufficient for vulval induction.

The somatic structures of the gonads themselves are derived from the progeny of the blast cells Z1 and Z4. They undergo an invariant series of up to eight divisions to generate two symmetrical structures consisting of 143 cells in hermaphrodites. The individual structures consist of descen-

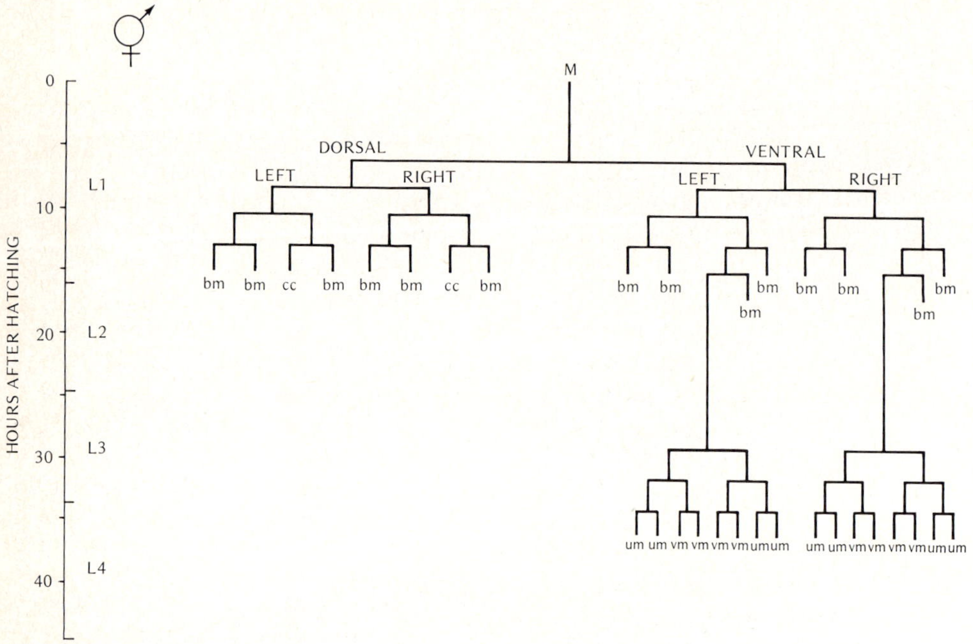

Figure 16.16. The M cell chandelier. The M cell divides repeatedly in the L1 larva to produce coelomocytes (cc) and body muscle (bm) cells. Two progeny cells divide again in the L3 larva to generate 16 muscle cells of the uterus (um) and vulva (vm). All divisions except the first two are anterior/posterior. Division of these cells requires interaction with progeny of the germ line cells Z2 and Z3.

dents of more than one lineage. The orientation of division planes and time of division are identical in the Z1 lineage and the Z4 lineage, indicating that these cells share the same developmental history and environmental clues. The positions taken up by all but two of the gonadal cells are invariant from worm to worm. The ventral uterus is derived from two cells (Z1ppp and Z4aaa) that take up either of two configurations in the L2 larva that are related by twofold rotational symmetry. Individuals with either configuration develop similar vulval structures, but the ancestry of the cells that occupy equivalent positions in the structure is different. Thus, the topological arrangement of these two cells affects the number of divisions that the four daughter cells will each take and the differentiation fate of the 37 progeny cells. This is one of the few cases of indeterminacy in postembryonic development in the nematode. The observation that the four daughter cells, as a group, follow one of two alternative lineage pathways suggests that neighboring cells can influence each other's fate in this case.

Shortly after hatching, the germ cell precursors Z2 and Z3 retract the lobes that protrude into two intestinal cells (Fig. 16.8) and attach to the somatic gonadal cells. If these somatic cells are absent due to laser ablation, the germ cells fail to grow. It is likely that the germ cells derive nutrients directly from the somatic gonadal cells. Z2 and Z3 cells divide continuously from L1 through adulthood. Their descendents populate both the anterior and posterior gonadal arms, differentiating into sperm in the spermatheca and into eggs in the uterus starting from the distal tip of the gonads.

The gonadal structures in males are quite different from those in hermaphrodites. There is a single, asymmetric gonad producing sperm as well as specialized body structures such as the hook for depositing the sperm in the uterus of hermaphrodites. Somatic gonadal tissue in males is also derived from cells Z1 and Z4. They give rise to 56 cells in males by a different pattern of divisions from that seen in hermaphrodites. However, the two programs have several features in common, suggesting that they are controlled in part by similar genetic mechanisms.

A variety of mutations have been recovered that affect postembryonic development. One of the most dramatic, *lin*-5, causes failure of all postembryonic nuclear and cell divisions but not replication of the chromosomes. This phenotype indicates that postembryonic division is subject to different genetic control from embryonic division. The larval blast cells continue to replicate their DNA the set number of times seen in the normal lineage, but the chromosomes form abnormal metaphase plates that decondense until the next attempted round of division. The resulting cells end up with large polyploid nuclei. Clearly, the mechanism that counts the number of rounds of chromosomal replication is independent of concomitant cell and nuclear division. In spite of the lack of division, cells express several aspects of the differentiation expected of their descendents.

More specific effects occur from mutations in the *lin*-12 gene that codes for a protein related to epidermal growth factor (EGF). A series of semidominant mutations isolated in this locus result in the formation of a series of small, nonfunctional vulvas along the ventral midline. Worms heterozygous for these mutations fail to lay eggs. Therefore, revertants to egg laying could be fairly easily selected. Thirty-two revertants were found to carry secondary mutations in the *lin*-12 locus, two of which were shown to be translational stop signals (amber mutations). Thus, the original *lin*-12(d) multivulva mutation was reverted by inactivating the *lin*-12(d) gene in heterozygous individuals. When homozygous, *lin*-12 null mutations (*lin*-12[o]) were found to result in scrawny, sterile worms that tend to explode as young adults. Hermaphrodite survivors have a large protrusion at the normal position of the vulva.

The *lin*-12 mutations affect cells of many different lineages; neuroblasts, ectoblasts, neurons, sex mesoblasts, and body muscles are affected. In the somatic gonadal cells of wild-type worms a single precursor cell (Z1ppp or Z4aaa) gives rise to the anchor cell. Which of these cells follows the pathway leading to formation of an anchor cell is determined by the orientation they take up in the L2 larva. In hermaphrodites homozygous for *lin*-12[o], both of these cells gives rise to anchor cells. In hermaphrodites carrying the semidominant mutations *lin*-12(d), both give rise to ventral uterine precursor cells and no anchor cells are formed. From this and observations on other cell lineages in both *lin*-12[o] and *lin*-12(d) mutants, it has been suggested that the *lin*-12 product specifies cell fates.

The *lin*-12 gene may be active only in certain cells. The protein it specifies then directs a change in pattern of differentiation in those cells. When the *lin*-12 protein is inactivated as in *lin*-12[o] mutants, no change occurs in those cells. In *lin*-12(d) mutants it is thought that the protein is produced constitutively in all cells and changes the pattern of differentiation in cells susceptible to its action. Thus, this homeotic gene may regulate the pattern of differentiation by binary switches.

Although Z1 and Z4 have different lineage patterns in males than in hermaphrodites, the fate of their progeny is susceptible to mutations in *lin*-12. The *lin*-12[d] mutations change linker cells into vas deferens precursor cells whereas *lin*-12[o] do the opposite. Likewise, ventral hypodermal cells in males are constitutively switched to a lineage pattern that forms male hooks in *lin*-12(d) mutants whereas cells that normally form the male hook do not do so in *lin*-12(o) mutants.

By shifting a temperature-sensitive *lin*-12(o) mutant to nonpermissive temperatures, it was found that the *lin*-12 protein must act at the time at which cell fates are determined during the second and third larval stages. It will be fascinating to see the mechanism by which cell fates are determined in this system.

The Number of Genes to Make a Worm

Adults of *C. elegans* have hundreds of different kinds of cells, each positioned exactly to make functional tissues. Specific nerves integrate their complex behavior by connecting peripheral parts to carefully constructed ganglia where the information is processed and responses are determined. In these organisms it is particularly clear that each step, from fertilization through the larval instars to the final form, is genetically determined. The time of division, the plane of division, the migration of cells, and the terminal differentiation of almost every one of the 959 cells are controlled by specific genes. Yet there is direct evidence that the genome of *C. elegans* carries fewer than 5000 genes. Since at least half of these are housekeeping genes responsible for simple cellular functions, that leaves only a few thousand genes to direct all the details of development. A significant proportion of these have already been discovered by analysis of mutations affecting development.

A few thousand genes seem a small number for development of a complex organism. However, when they are combined in different sequences a large number of unique directions can be given. This text uses only a few thousand different words but contains many varied sentences. The number of developmental genes in *Drosophila* and *Dictyostelium* has also been estimated from mutational frequencies to be only a few thousand. It is quite conceivable that development of a mouse or human could be carried out with only about twice this number of developmental genes.

We may not know exactly how most of these genes determine the fate of embryonic cells, but we know they are responsible. The isolation and characterization of these genes and their products will shed light on what they are doing. When their control is understood, it may become apparent how they are regulated into the whole process of development. The complexity is not so great that this is beyond reasonable expectations. It may be that the whole picture becomes clear only when the last piece of the puzzle is fit into place, but the excitement of adding each piece to a meaningful representation will surely keep the players at it.

Related Readings

Brenner, S. (1974). The genetics of *Caenorhabditis elegans*. Genetics *77,* 71–94.

Cassada, R., Isnenghi, E., Culotti, M., and von Ehrenstein, G. (1981). Genetic analysis of temperature-sensitive embryogenesis mutants in *Caenorhabditis elegans*. Devel. Biol. *84,* 193–205.

Cox, G., and Hirsh, D. (1985). Stage-specific patterns of collagen gene expression during development of *Caenorhabditis elegans*. Mol. Cell Biol. *5,* 363–372.

Eide, D., and Anderson, P. (1985). Transposition of Tc1 in the nematode *Caenorhabditis elegans*. Proc. Natl. Acad. Sci. *82,* 1756–1760.

Ferguson, E., and Horvitz, R. (1985). Identification and characterization of 22 genes that affect the vulval cell lineages of the nematode *Caenorhabditis elegans*. Genetics *110,* 17–72.

Hedgecock, E., and White, J. (1985). Polyploid tissues in the nematode *Caenorhabditis elegans*. Devel. Biol. *107,* 128–133.

Herman, R., and Karl, C. (1985). Muscle-specific expression of a gene affecting acetylcholinesterase in the nematode *Caenorhabditis elegans*. Cell *40,* 509–514.

Hodgkin, J. (1984). Switch genes and sex determination in the nematode *C. elegans*. J. Embryol. Exp. Morph. *83,* suppl. 103–107.

Kimble, J., Edgar, L., and Hirsh, D. (1984). Specification of male development in *Caenorhabditis elegans:* the *fem* genes. Devel. Biol. *105,* 234–239.

Kimble, J., and White, J. (1981). On the control of germ cell development in *Caenorhabditis elegans*. Devel. Biol. *81,* 208–219.

Schierenberg, E., and Wood, W. (1985). Control of cell-cycle timing in early embryos of *Caenorhabditis elegans* Devel. Biol. *107,* 337–354.

Sigurdson, C., Spanier, G. and Herman, R. (1984). *Caenorhabditis elegans* deficiency mapping. Genetics *108,* 331–345.

Strome, S., and Wood, W. (1983). Generation of asymmetry and segregation of germ-line granules in early *C. elegans* embryos. Cell *35,* 15–25.

Sulston, J., and Horvitz, H. (1981). Abnormal cell lineages in mutants of the nematode *Caenorhabditis elegans*. Devel. Biol. *82,* 41–55.

Sulston, J., Schierenberg, E., White, J., and Thomson, J. (1982). The embryonic cell lineage of the nematode *Caenorhabditis elegans*. Devel. Biol. *100,* 64–119.

Wood, W., Hecht, R., Carr, S., Vanderslice, R., and Hirsh, D. (1980). Parental effects and phenotypic characterization of mutations that affect early development in *Caenorhabditis elegans*. Devel. Biol. *74,* 446–469.

Study Questions Part III

1. a. Give three molecular and two functional differences between anterior and posterior cells in slugs of *Dictyostelium*.
   b. What mechanisms could generate this heterogeneity?
2. Cells in the anterior 20 percent of *Dictyostelium* slugs normally differentiate into stalk cells but not spores. When the posterior 80 percent is removed, the anterior cells form only stalk cells when terminal differentiation occurs within 8 hours. However, if terminal differentiation is not induced for 24 hours, many of the anterior cells form spores. Give plausible reasons to account for these two observations.
3. What is the evidence that a dependent sequence regulates biochemical differentiation in *Dictyostelium?*
4. What are the molecular differentiations necessary for *Dictyostelium* cells to acquire the ability to be chemotactic?
5. What mechanism may account for cell–cell adhesion in *Dictyostelium?*
6. a. What phenotype would you expect in (a) XX and (b) XY rabbits that are homozygous for a mutation in a gene which is essential for production of the hormone X?
   b. What phenotype would you expect in (a) XX and (b) XY rabbits homozygous not only for the above mutation but also for the *Tfm* mutation?
7. Give three lines of evidence that show that testosterone stimulates the differentiation of Wolffian ducts into vas deferens.
8. How is the differentiation of the sexual apparatus genetically controlled in mammals? Please give all known details for both males and females.
9. What is the evidence that the H-Y antigen determines the type of differentiation a germ cell will follow?
10. What is the phenotype of mice that carry mutations inactivating the testosterone receptor *(Tfm)* that are genetically (a) XX; (b) XY?
11. Does an eye transplanted to the tail of a tadpole become resorbed at metamorphosis? Explain why.
12. Give three cases of regulated cell death in development.
13. Describe the consequences to limb formation of removal of the AER from limb buds.
14. Are the feet formed from a duck leg bud transplanted to a chick embryo webbed or not?
15. What is the experimental evidence that the ZPA determines the anterior/posterior axis of limbs?
16. a. Differentiation of *Drosophila* germ line cells has been correlated with the segregation of cytoplasmic granules into this cell lineage. What is the evidence that the posterior region of the embryos which contain these granules directs germ cell differentiation?

b. What can you conclude about the role of the cytoplasmic granules in directing germ cell determination?

17. Give a plausible mechanism by which a zebra could get its stripes.
18. What phenotypic results would you expect of a fly homozygous for both *engrailed* and *ophthalmoptera?*
19. What is the experimental evidence that indicates the existence of anterior and posterior compartments in wings?
20. What phenotype would you expect for a *Drosophila* that was homozygous for a mutation which results in the lack of production of polar granules?
21. What is the phenotype of flies carrying the mutation *bithorax (bx/bx)?*
22. The enzyme alcohol dehydrogenase accumulates in different tissues in *Drosophila* larvae and in *Drosophila* adults. What is the molecular mechanism that accounts for this change in cell-type expression of the *Adh* gene?
23. What is the evidence that accumulation of rhabditin in E cell progeny of nematodes is independent of cell division or interaction with other cell types?
24. How can one measure dedifferentiation in *Dictyostelium?*
25. What is the primary determinant for sex differentiation in *Drosophila?* Give two genes necessary for responding to this determinant.
26. Why is it thought that the *tra*-2 gene must function even in adult *Drosophila* to regulate sex-specific functions?
27. What are the molecular events that occur when a yeast cell of α mating type gives rise to **a** mating-type daughter cells?
28. What is the consequence of treating a yeast cell of α mating type with **a** factor? With α factor?
29. What genes function to repress transcription of HMR and HML?
30. Why do mutations in *HO* genes of yeast (*ho* mutants) block sporulation in clonal populations?
31. Discuss how some retroviruses may cause cancer.
32. What mechanism leads to nerve cells being formed predominantly in the head region of *Hydra?*
33. Why are the healed legs of cockroaches longer at the molt following removal of a small piece at the joint of the tibia and the femur and grafting of the stumps?
34. What is the evidence that the number of rounds of DNA replication of chromosomes is determined independently of the number of cell divisions in postembryonic development of nematodes?
35. About how many genes do you think are selectively advantageous in:
    a. yeast
    b. *Dictyostelium*
    c. *C. elegans*
    d. *Drosophila*
    e. humans

    Please give your reasoning in each case.
36. Why is there reason to think that homeo boxes may indicate the genes that determine pattern in segmented organisms?

37. What is the phenotype of nematodes *(C. elegans)* homozygous for the following mutations:

Chromosomes	*Mutations*
a. XO	*tra*-1^-
b. XX	*tra*-1^-
c. XO	*tra*-1^-, *tra*-2^-
d. XO	*tra*-2^-, *lin*-12[d]

Guide to Subjects

This presentation has covered development in a fairly large set of organisms and tissues. Wherever specific genes or molecular mechanisms have been implicated in the processes of differentiation, they have been pointed out. A variety of recent techniques and concepts have also been discussed. Since many of these subjects have bearing on more than just the specific example with which they were raised, they are compiled so as to bring them all together (Tables A-1 to A-6). These brief tables may be useful in seeing the interrelations of the individual components.

The diversity of form in living organisms is a source of continuous delight and amazement to naturalists. It is also somewhat daunting to those who would like to understand how it is generated. The causal pathways to specialized structures have followed several routes. For instance, the wing has been reinvented in insects, birds, and bats. This indicates that there is more than one way that leads to an extended surface. Development in each case is not simple but uses preexisting forms to adapt for optimal survival in a newly entered ecological niche. Nevertheless, the mechanisms for such adaptations use variations on a fairly small number of themes: differential cell growth, differential cell death, reexpression of a previously used developmental program (reiteration), alteration in relative timing of events, modification of cellular adhesivity, modulation of the extracellular matrix, and differential gene expression.

Living organisms are autocatalytic in that successful forms breed more of themselves. Once a small change from preexisting forms has proven advantageous, more of that new form is generated. Thus, evolution can rapidly capitalize on a new process that results in an altered developmental sequence. In complex multicellular organisms, efficiency of production of form is secondary to the effectiveness of the form itself. Thus, we see several strategies that lead to the same end.

For any new developmental strategy to be inherited for generations, the changes must be laid down in the primary sequence of DNA in the germ line. It is the genes that ultimately direct morphogenesis and not the mother's tongue, as was thought a thousand years ago, as mentioned in the Introduction. We have made progress since the days of frustration when the workings of an embryo were so astounding that they were considered outside the realm of the laws of physics and chemistry and had to be directed by "vital forces." An appreciation of the many possible interacting steps in an interconnected circuit of a fairly small number of genes makes attainable the goal of understanding the rules of development and how they are applied in specific organisms.

Table A–1 Featured Species

Organism	Special Interest	Chapters
Vertebrates		
*Homo sapiens**	Mammalian development	1, 4, 5, 8, 9, 10, 12, 13
*Xenopus laevis**	Amphibian development	1, 2, 8, 9, 10, 14
*Gallus domesticus**	Avian development	1, 8, 9, 10, 14
Invertebrates		
*Drosophila melanogaster**	Genetic analysis	4, 12, 15
*Strongylocentrotus purpuratus**	Gene expression	6, 7, 8, 9
*Caenorhabditis elegans**	Cell lineage	12, 16
*Hydra attenuata**	Stem cell regulation	13, 14
Protists		
Tetrahymena pyriformis	Cortical inheritance	5
*Dictyostelium discoideum**	Biochemical differentiation	6, 11
Molds		
Neurospora crassa	Genetic regulation	3, 6
Saccharomyces cerevisiae	Mating type control	3, 12, 13
Plants		
Nicotiana tobacum	Totipotency	1, 7
Volvox carteri	Photosynthetic alga	1
Ulva stenophylla	Alteration of haploid and diploid	6
Zea maize	Gene expression in seeds	10, 15

*Fate maps presented.

Table A–2 Differentiated Cell Types

Tissue	Specialized Function	Major Interest	Chapters
Heart	Circulation	Rearrangement	1, 10
Lymphocyte	Immune response	Immunoglobins	2, 5
Oocyte	Fertilization	rRNA, yolk	3, 6
Red blood cell	Oxygen carrier	Globins	3, 5
Pancreas	Secretion	Amylase, chymotrypsinogen	4
Muscle	Contraction	Actin, myosin	4, 9, 10, 14, 16
Lens	Light focusing	Crystallins	4, 10, 11
Nerve	Signal transfer	Neuroreceptors	5, 10, 13, 16
Epithelium	Surface	Adhesion	5, 16
Follicle cells	Egg shell	Chorion proteins	6
Sperm	Fertilization	Tubulin	6, 7, 12, 16
Spores	Dissemination	Spore coat	6, 11
Ovule	Fertilization	Multiple nuclei	7
Pollen	Fertilization	Meiosis	7
Neural plate	Spinal cord	Tube formation	10
Neural crest cells	Many	Migration	10
Somites	Repeated structures	Compaction	10
Seed	Germination	Protein bodies	10
Aleurone	Endosperm metabolism	Amylase	10
Scutellum	Nutrient uptake	Catalase	10
Primary sex cells	Gametes	Tissue interaction	10, 12
Skin	Protection	Keratins	10, 14, 16
Gut	Absorption	Endoderm	10, 15, 16
Iris	Light control	Dedifferentiation	11
Testes	Sperm	H-Y antigen	12
Vas deferens	Sperm transport	Testosterone response	12
Fallopian tube	Egg transport	X response	12
Limbs	Locomotion	Pattern	14, 15

Table A–3 Defined Genes

Organism	Gene	Product	Role	Chapter
Human	globin	Globin	Oxygen carrier	3, 5
(*Homo sapiens*)	*hMT*-IIA	Metallothionein	Heavy metal binding	4
	c-*myc*	58 kd	DNA binding	13
	Tfm	Testosterone receptor	Male differentiation	12
Mouse	kappa	Immunoglobin light chain	Antibody	2
(*Mus musculus*)	TAT	Tyrosine amino transferase	Metabolism	3
Rat	chymo	Chymotrypsinogen	Digestion	4
(*Rat ratus*)	insulin	Insulin	Hormone	4
Chick (*Gallus domesticus*)	δ-crystallin	δ-crystallin	Lens	4
South African clawed toad (*Xenopus laevis*)	5*S* RNA	5*S* RNA	Ribosome structure	3
Fruit fly	actin	Actin	Muscle	4
(*Drosophila melano-*	myosin	Myosin	Muscle	4
gaster)	tubulin	Tubulin	Sperm	2
	S-36-1; *S*-38-1	Chorion proteins	Egg shells	6
	da, Sx1, dsx		Sex chromosome response	12
	tra, tra-2		Sex chromosome response	12
	grandchildless		Polar plasm	15
	BX-C, ANT-C		Segment specification	15
	engrailed		Posterior compartment	15
	Adh	Alcohol dehydrogenase	Alcohol metabolism	15
	Ddc	Dopa decarboxylase	Cuticle pigment	15
Snail (*Lymnaea perega*)	*D*		Dextral cleavage	8
Nematode	*her, tra*-1, *tra*-2		Sex chromosome response	12
(*Caenorhabditis elegans*)	*lin*-5		Postembryonic division	12
	lin-12		Lineage determination	12
Sea urchin	CyIIIa	Actin	Ectodermal	9
(*Strongylocentrotus* *purpuratus*)	Spec 1	15 kd	Ca^{++} binding	9
Slime mold	*modB*	Glycosyltransferase	Protein modification	11
(*Dictyostelium*	*SP*96	96 kd	Spore coat protein	11
discoideum)	*gp*80	80 kd	Adhesion	11
	Upp	Pyrophosphorylase	UDPG synthesis	11
Bread mold (*Neurospora crassa*)	*qa*	Catabolic enzymes	Quinate metabolism	3

Table A–3 Defined Genes (*Continued*)

Organism	Gene	Product	Role	Chapter
Corn (*Zea maize*)	*CAT*	Catalase	H_2O_2 removal	10
Yeast (*Saccharomyces cerevisiae*)	*gal*	Enzymes and activators	Galactose metabolism	3
	MAT	a1, a2, or α1, α2	Mating type	12
	SIR		Control of HML, HMR	12
Bacteria (*Escherichia coli*)	*lac*	β-Galactosidase	Lactose metabolism	3
Viruses	v-*sis*, v-*ras*	Tyrosine kinase	Cancer	13

Table A–4 Specific Mechanisms of Differentiation

Organism	Tissue Used as Example	Mechanism	Chapter
Vertebrates	Lymphocyte	DNA rearrangement	2
	Lymphocyte	mRNA splicing	2
	Mammary gland	Nucleosome alteration	3
	Oviduct	Steroid induction	3
	Oocyte	Polymerase factor	3
	Erythropoietic	Transcriptional initiation	3
	Lens	Protein stability	4
	Oocyte	rDNA amplification	6
	Nephric ducts	Hormonal control of death	12
	Fibroblast	Matrix control of growth	14
Invertebrates	Follicle cells	Chorion gene amplification	6
	Sperm	Lectin binding	7
Bacteria and molds	*E. coli*	Protein/nucleic acid interaction	3
	Hyphae	*cis*-acting DNA sequences	3
	Myxamoebae	Biochemical circuits	11

Table A–5 Analytical Techniques

Techniques	Chapters
Nucleic acids	
Southern analysis of DNA	2, 3
Restriction site mapping of DNA	2, 3, 4, 9, 15
Cloning of DNA	3, 4, 9, 10, 11, 13, 15, 16
Sequencing of DNA	3, 4, 11, 13, 15
Northern analysis of RNA	4, 9, 11, 13, 15
mRNA hybridization	4, 8
Dot blots	4, 10
in situ hybridization to chromosomes	4, 15
Electron microscopy of DNA	6
Reannealing of DNA	9
cDNA hybridization	9, 11, 12
Proteins	
Gel electrophoresis	4, 5, 6, 10, 11, 15, 16
Immunofluorescent localization	5, 7, 10, 11, 16
Western analysis of antigens	11
Genetic	
Cell cloning	1, 2, 3, 11, 13
Transformation of genes	13, 15
Mutant analysis	all
Organismic	
Cytology	4, 5, 7, 8, 9, 10, 12, 14, 15, 16
Microsurgery	5, 8, 9, 10, 11, 14, 15
Tissue grafting	10, 11, 13, 14

Table A–6 Developmental Concepts

Concepts	Chapter
Differential gene expression	2, 3, 4, 9, 10, 11, 15, 16
Specific gene amplification	2, 6
Unchanging complement of genetic information	2, 11
Coordinate induction of genes	3, 4, 11, 15, 16
cis-Acting DNA sequences	3, 4, 15
Hormonal control	3, 6, 12, 15
Tandemly repeated genes	3, 11
Temporal sequence of gene expression	3, 11, 15, 16
Evolutionarily related genes	3, 13
Coordinate accumulation of proteins	4, 5, 6, 10, 11
Tissue-specific gene expression	4, 5, 6, 9, 10, 11, 15
Cell fusion	4, 7
Determination	4, 8, 9, 10, 15
mRNA stability	4, 8, 10, 12
DNA methylation in relationship to transcription	4, 12
In vitro differentiation	4, 12, 13
cis-Acting enhancers	4, 13
Stem cells	5, 6, 10, 13, 14, 16
Ionic control of membrance function	5, 7, 11
Cortical inheritance	5, 8
Regulation of developmental potential	8, 9, 10, 11, 12, 14, 15, 16
Adhesion changes	8, 9, 10, 11, 13
Induction of developmental pathway	8, 14, 15
Cleavage planes	8, 16
Biochemical specialization	9, 10, 11, 15, 16
Dedifferentiation	11, 12, 13, 14
Causal sequences	11, 12, 15, 16
Tissue proportioning	11, 13, 14, 15
Pattern formation	11, 13, 14, 15
Number of genes required for development	11, 15, 16
Programmed cell death	12, 13, 16
Genetic circuits	12, 15
Metamorphosis	13, 15
Cell cycle control	13, 16
Germ line determinants	15, 16
Cell lineages	16

Illustration Acknowledgments

I thank the following sources for supplying original photographs and for permission to reprint illustrations. I am especially grateful to Peter Bryant, Ed Lewis, Anthony Mahowald, Jean-Paul Revel, Victor Vacquier, and William Wood for esthetic as well as informational illustrations.

Part I

1.4 Photo by Bridges and Sturtevant, from Stent, G., and Calendar, R. (1978). Molecular genetics. Freeman, San Francisco.

1.5 Beckwith, J., and Zipser, C. (1970). The lac operon. Cold Spring Harbor Laboratory, New York.

1.7 Photo by Lloyd Beidler from Life on earth (1978). Sinauer Associates, Sunderland, Massachusetts.

1.12 Nilsson, L. (1973). Behold man. Bonnier Fakta Bokförlag, Stockholm.

1.13 Bloom, R. and Fawcett, D. (1975). A textbook of histology, 10th ed. W. B. Saunders, Philadelphia.

1.17–1.20 Nilsson, L. (1973). Behold man. Bonnier Fakta Bokförlag, Stockholm.

1.27 Vasil, V., and Hildebrandt, A. (1965). Science *150,* 889–892.

2.1 Lefevre, G. (1976). In The genetics and biology of *Drosophila,* Ashburner, M., and Novitski, E. (eds.). Academic Press, San Diego.

3.10 Life on earth (1978). Sinauer Associates, Sunderland, Massachusetts.

3.17 Stent, G., and Calender, R. (1978). Molecular genetics, 2nd ed. Freeman, San Francisco.

4.1 Bloom, R. and Fawcett, D. (1975) A textbook of histology, 10th ed. W. B. Saunders, Philadelphia.

4.2 Konigsberg, I.

4.5 Bloom, R. and Fawcett, D. (1975). A textbook of histology, 10th ed. W. B. Saunders, Philadelphia.

4.8, 4.9 Fyrberg, E., Mahaffey, J., Bond, B., and Davidson, N. (1983). Cell *33,* 115–123.

4.10 Rozek, P. and Davidson, N. (1983). Cell *32,* 23–34.

4.12 Pictet, R., Clark, W., Williams, R., and Rutter, W. (1972). Devel. Biol. *29,* 436–467.

4.13 Bloom, R., and Fawcett, D. (1975). A textbook of histology, 10th ed. W. B. Saunders, Philadelphia.

4.14 Van Nest, G., Raman, R., and Rutter, W. (1983). Devel. Biol. *98,* 295–303.

5.1–5.3, 5.6 Bloom, R. and Fawcett, D. (1975). A textbook of histology, 10th ed. W. B. Saunders, Philadelphia.

5.7 Revel, J-P.

5.14 Alberts, B., et al. (1983). Molecular biology of the cell. Garland Publishers, New York.

5.16–5.21 Hay, E. (1981). Cell biology of extracellular matrix. Plenum Press, New York.

5.22 Revel, J-P.

Part II

6.2 Nilsson, L. (1973). Behold man. Bonnier Fakta Bokförlag, Stockholm.

6.3 Phillips, D. (1980). In Browder, L. (ed.), Developmental biology, W. B. Saunders, Philadelphia.

6.6 Bloom, R. and Fawcett, D. (1975). A textbook of histology, 10th ed. W. B. Saunders, Philadelphia.

6.8 Brandhorst, B.

6.10 Mahowald, A. (1968). J. Exp. Zool. *167,* 237–262.

6.12, 6.13 Osheim, R., and Miller, O. (1983) Cell *33,* 543–553.

7.1–7.2 Summers, R. G. (1975). Am. Zool. *15,* 523.

7.3 Epel, D. (1977). Sci. Amer. *237,* 128–136.

7.5 Revel, J. P.

7.7 Gilkey, J., Jaffe, L., Ridgeway, E., and Reynolds, G. (1978). J. Cell Biol. *76,* 448–466.

7.8 Weier, E., Stocking, R., and Barbour, M. (1970). Botany. John Wiley and Sons, New York.

7.9 Knox, R. (1973). J. Cell Sci *12,* 421–443. (1976). The developmental biology of plants and animals. W. B. Saunders, Philadelphia.

7.9 Revel, J. P.

7.12 Norstog, K., and Long, R. (1976). Plant biology. W. B. Saunders, Philadelphia.

8.2 Kessel, R., Beams, H., and Shin, C. (1974). Amer. J. Anat. *141,* 341–359.

8.3 Beams, H., and Kessel, R. (1976). Amer. Sci. *64,* 279–290.

8.7 Bloom, R., and Fawcett, D. (1975). A textbook of histology, 10th ed. W. B. Saunders, Philadelphia.

8.11 Schroeder, T. (1972). J. Cell Biol. *533,* 419–434.

8.12, 8.13 Longo, F., and Anderson, E. (1968). J. Cell Biol. *39,* 339–368.

9.1 Revel, J-P.

9.14 Balinsky, B. I. (1981). An introduction to embryology, 5th ed. W. B. Saunders, Philadelphia.

9.19 Chamberlin, M., Britten, R., and Davidson, E. (1975). J. Mol. Biol. *96,* 317–333.

9.20 Shott, R., Lee, J., Britten, R., and Davidson, E. (1984). Devel. Biol. *101,* 295–306.

9.21 Brushkin, A., Bedard, P., Tyner, A., Showman, R., Brandhorst, B., and Klein, W. (1982). Devel. Biol. *91,* 317–324.

10.1 Revel, J-P.

10.3 Revel, J-P.

10.11, 10.12 Sawyer, R., O'Guin, M., and Knapp, L. (1984). Devel. Biol. *101,* 8–18.

10.16 Scandalios, J., Tsaftaris, A., Chandlee, J., and Skaden, R. (1984). Devel. Gen. *4,* 281–293.

10.18, 10.19, 10.20 Vacquier, V.

Part III

11.9 Dowds, B., and Loomis, W. (1984). Mol. Cell Biol. *4,* 2273–2278.

11.19 Finney, R., Mitchell, L., Soll, D., Murray, B., and Loomis, W. (1983). Devel. Biol. *98,* 502–509.

12.3 Jones, H., and Scott, W. (1958). Hermaphrodite genital anomalies and related endocrine disorders. Williams & Wilkins Co., Baltimore.

12.4 Bartalos, M., and Baramki, T. (1967). Medical cytogenetics. Williams & Wilkins Co., Baltimore.

12.6 Bryant, P.

12.7 Baker, B., and Belote, J. (1983). Ann. Rev. Gen. *17,* 345–393.

13.1 Levi-Montalcini, R., and Amprino, R. (1947). Arch. Biol. *58,* 265–288.

13.6 Marcum, B.

13.8 Wood, R.

13.9 Marcum, B.

14.3 Saunders, J.

14.9 Hay, E.

15.2, 15.4 Mahowald, A.

15.7, 15.15 Yund, M., and Gemerand, S. (1980). In The molecular genetics of development, T. Leighton and W. F. Loomis (eds.). Academic Press, New York.

15.13 Lewis, E.

15.20 Scholnick, S., Morgan, B., and Hirsh, J. (1983) Cell *34,* 37–45.

16.4, 16.5, 16.9, 16.11 Wood, W.

Index